International Encyclopaedia of

NANO TECHNOLOGY, SCIENCE AND PHYSICS

International Encyclopaedia of NANO TECHNOLOGY, SCIENCE AND PHYSICS

Volume 1

M.C. Narayan

ANMOL PUBLICATIONS PVT. LTD.
NEW DELHI - 110 002 (INDIA)

ANMOL PUBLICATIONS PVT. LTD.
H.O.: 4374/4B, Ansari Road, Daryaganj,
New Delhi-110 002 (India)
Ph.: 23278000, 23261597
B.O.: No. 1015, Ist Main Road, BSK IIIrd Stage
IIIrd Phase, IIIrd Block,
Bangalore - 560 085 (India)
Visit us at: www.anmolpublications.com

International Encyclopaedia of Nano Technology, Science & Physics

First Edition, 2008
ISBN 978-81-261-3680-3 (Set)

PRINTED IN INDIA

Printed at Mehra Offset Press, Delhi.

Contents

Preface

In general terms, the physics of nano-scale and molecular systems studied with scanning probes is called Nano Physics. The nanoscale physics uses various experimental techniques to examine the physical properties of objects in the nanoscale size range, that is, a little bit larger than the size of atoms. Some interesting physical properties at this range include conductivity of small numbers of atoms and molecules, forces arising between objects on this scale, and the transition between the quantum nature of a few atoms and a large number of atoms. In Physics, the prefix "Nano" is used to describe a billionth of a unit, so a nanometer is a billionth of a meter. This is the equivalent of approximately five to ten atoms laid end-to-end. These dimensions are also reflected in the English translation of the Greek word "nanos"—"dwarf". A single human hair is around ten thousand times thicker than a nanometer, and another impressive comparison – a nanoparticle is to a soccer ball as a soccer ball is to the Earth. But nanotechnology is not just microtechnology on a smaller scale; it is much more than a mere improvement on previous forms of technology. Nanotechnology is a completely new production process in a molecular world, opening up possibilities far beyond what we have known before. Nature provides the inspiration for this new technology – from time immemorial, nanomachines have been at work in all organic cells, building our world of plants, animals and people from a store of chemical elements. The discovery in recent years of the mechanisms for accessing individual building blocks of matter, and with it the increasing understanding of how these building blocks organize themselves has sparked off the industrial conquest of the nanosphere throughout the world. For instance, Nanotube is real. Strictly speaking, any tube with nanoscale dimensions, but generally used to refer to carbon nanotubes (a commonly mentioned non-carbon variety is made of boron nitride), which are sheets of graphite rolled up to make a tube. The dimensions are variable (down to 0.4 nm in diameter) and you can also get nanotubes within nanotubes, leading to a distinction between multi-walled and single-walled nanotubes. Apart from remarkable tensile strength, nanotubes exhibit varying electrical properties (depending on the way the graphite structure spirals around the tube and other factors), and can be insulating, semiconducting or conducting (metallic). Carbon nanotubes are considered by some to be fullerenes (strictly speaking, this is probably correct, but they are more commonly treated as distinct) and are sometimes referred to as buckytubes. Microelectromechanical systems (MEMS) are generally used to refer to systems that can respond to a stimulus or create physical forces (sensors and actuators) and that have dimensions on the micrometer scale. They are almost exclusively made using the same lithographic techniques that are used to make silicon chips for computing. Miniature accelerometers are the most successful product in this field and are used to trigger air bags in cars. When such systems can be made with nanoscale dimensions they can be classified as NEMS. The success story of the giant magnetoresistance effect (GMR) shows in what a short space of time nanotechnology can make it from research laboratory to complete market penetration. Some of the contemporary research areas in global Nano Technology, Science & Physics research include:

- Low temperature measurements on structured graphene
- Low-Temperature Microscopy
- Semiconductor nanowires
- Mesoscopic transport through multi-terminal nanostructures
- Time-resolved single electron transport in mesoscopic structures
- Hole transport in p-type GaAs nanostructures
- Two-dimensional electron gas systems
- Quantum dots and quantum point contacts
- Two-dimensional electron gas systems
- Nanoparticles and clusters
- Nanotubes
- Spin injection and magnetic semiconductors
- Superconductors
- Molecular Beam Epitaxy
- Scanning Tunneling Microscopy
- Spintronics
- III-V Materials
- X-STM
- Optical Transitions

This book titled "International Encyclopaedia of Nano Technology, Science & Physics" deals briefly with the subject areas of nano-optics, nanotube, nanofibre, nano-sensor, nanowire and nanoribbon. Keeping in mind the applications and utilities of Nano Physics for meeting the contemporary global energy requirements, it seems that this subject will become the core area of future nanoscience and nanotechnology research. Nanostructured materials do not represent a new phenomenon. However, the ability to probe, manipulate, understand and engineer matter at atomic scales has only recently come within our grasp. Developments such as the invention of the

Scanning Tunneling Microscope have since made nanoscale science a reality. Today, the term nanotechnology is used to describe specific construction and manipulation performed on an atomic and molecular scale. Generally, systems with a maximum size of less than a hundred nanometers are classified as nanotechnology. This book tends to cover the aforementioned subject areas in their broadest possible terms. All useful supplementary research and reference tools have been included for readers' further investigations in the subject area of Nano Physics. Subject matters of Nano Physics are covered in following chapters in this Encyclopaedia: Basics of Nano Physics and Nanotechnology; Integrating Physics, Computers and Nanotechnology; Historical Evolution and Development of Nanotubes; Applications of Nanotubes and Buckytubes in Conductive Plastics, Adhesives, Connectors and Other Areas; Recent Advances in Other Major Fields of Nanoscience, Nano Physics and Nanotechnology; Use of Nano Physics in Nano Labs and Nanotechnology Products; Major Applications & Utilities of Nano-optics; and Towards Explaining the Differences among Nanoscale Scanning, Nanosensors, Nanowires and Nanoribbons. This Encyclopaedia is research friendly and future oriented.

—EDITOR

Glossary

Acetylcholine: Neurotransmitter that functions at cholinergic chemical synapses, found both in the brain and in the peripheral nervous system. It is the neurotransmitter at vertebrate neuromuscular junctions.

Acrosomal Process: Long, thin, actin-containing spike produced from the head of certain sperm when they make contact with the egg. Seen in sea urchins and other marine invertebrates whose eggs are surrounded by a thick gelatinous coat.

Active Component: A device that adds intelligence in some manner to the signal or data that passes through it. For example, in networking, an active hub regenerates fading input pulses into new, strong output pulses. In contrast, a passive hub is just a junction box that does not affect the passing data. Non-mechanical circuit component that has gained or switched current flow, such as a diode, transistor, etc.

Active Device: Device requiring a source of energy for its operation and has an output that is a function of present and past input signals. Examples include controlled power supplies, transistors, LEDs, amplifiers, and transmitters.

Adsorption: Formation of molecules of a gas similar to a thin film that binds on the surface of a solid. Unlike absorption, the binding to the surface in adsorption is usually weak and reversible. Adsorption is a method for separating mixtures and purifying fluids.

Affinity: A thermodynamic measurement of the strength of binding between molecules, say between an antibody and antigen. Each antibody/antigen pair has an association constant, Ka, expressed in L/mol.

Akebono: A Japanese scientific studies satellite that became operational in 1989. It contained only one non-Japanese instrument: Canada's Suprathermal Ion Mass Spectrometer, sent to investigate the Earth's outer atmosphere.

Algorithm: In mathematics, computing, linguistics, and related disciplines, an algorithm is a definite list of well-defined instructions for completing a task; that given an initial state, will proceed through a well-defined series of successive states, eventually terminating in an end-state. The transition from one state to the next is not necessarily deterministic; some algorithms, known as probabilistic algorithms, incorporate randomness.

Amide: A molecule containing an amine bonded to a carboxyl group; the resulting bond has substantial double-bond character. Also termed a peptide; amide bonds link amino acids in proteins.

Amino Acid: A molecule containing both an amine and a carboxylic acid group; in the 20 genetically encoded amino acids in biology, both groups are bound to the same C. Amino acids joined by amide bonds form peptides and proteins; these do not contain amino acids as such, and are instead said to contain *amino acid residues*. Organic molecules that are the building blocks of proteins. There are some two hundred known amino acids, of which twenty are used extensively in living organisms.

Amperometric Sensor: Amperometric sensors involve a heterogeneous electron transfer as a result of an oxidation/reduction of an electro-active species at a sensing electrode surface. A current is measured at a certain imposed voltage of the sensing electrode with respect to the reference electrode. Analytical information is obtained from the current-concentration relationship at that given applied potential.

Analyte: An analyte is a substance or chemical constituent that is determined in an analytical procedure, such as a titration. For instance, in an immunoassay, the analyte may be the ligand or the binder, while in blood glucose testing, the analyte is glucose. In medicine, analyte often refers to the type of test being run on a patient, as the test is usually determining a chemical substance in the human body.

Anechoic Chamber: Anechoic chambers duplicate the silence of space, where there is no medium (like air or water) through which the sound waves can travel. The walls of the anechoic chambers are made of carbon compound pyramids which absorb any microwave and radio signals that hit them.

Angle of Inclination: The angle at which a satellite's orbit is tilted in relation to the Earth's equator. A 90 degree angle of inclination is a polar orbit. A zero degree angle of inclination is an *equatorial orbit.*

Antenna: A piece of equipment that allows transmission and reception of radio signals. Satellites need antennas to communicate with Earth. A satellite may need to receive instructions and transmit the information it collects, or it may relay the information sent to it to another site on Earth. Since the information is transmitted using radio waves, which move at the speed of light, this method allows for very fast communications (only a very small time lag).

Aperture: In optics, an aperture is a hole or an opening through which light is admitted. More specifically, the aperture of an optical system is the opening that determines the cone angle of a bundle of rays that come to a focus in the image plane. An optical system typically has many openings, or structures that limit the ray bundles (ray bundles are also known as *pencils* of light).

Aromaticity: Aromaticity is a chemical property in which a conjugated ring of unsaturated bonds, lone pairs, or empty orbitals exhibit a stabilisation stronger than would be expected by the stabilisation of conjugation alone. It can also be considered a manifestation of cyclic delocalisation and of resonance. This is usually considered to be because electrons are free to cycle around circular arrangements of atoms, which are alternately single- and double-bonded to one another.

Artificial Intelligence (AI): A field of research that aims to understand and build intelligent machines; this term may also refer to an intelligent machine itself.

Atmosphere: The many layers of gases that surround a planet. The Earth's atmosphere is composed of several layers of gases that separate our planet from space. The major gases in the Earth's atmosphere are nitrogen and oxygen. The air we breathe is part of the atmosphere.

Atmospheric Studies Satellite: A type of scientific satellite that studies the Earth's atmosphere. They were some of the very first satellites launched into space, including Canada's first satellite Alouette.

Atom: An atom is the smallest particle characterizing a chemical element. An atom consists of a dense nucleus of positively charged protons and electrically neutral neutrons, surrounded by a much larger electron cloud consisting of negatively charged electrons. When the number of protons matches the number of electrons, an atom is electrically neutral; otherwise it is an ion and has a net charge. An atom is classified according to its protons and neutrons: the number of protons determines the chemical element and the number of neutrons determines the isotope of that element. The concept of the atom as an indivisible component of matter was first proposed by early Indian and Greek philosophers.

Atomistic Simultations: Atomic motion computer simulations of macromolecular systems are increasingly becoming an essential part of materials science and nanotechnology. Recent advances in supercomputer simulation techniques provide the necessary tools for performing computations on nanoscale objects containing as many as 300,000 atoms and on materials simulated with 1,000,000 atoms. This new capability will allow computer simulation of mechanical devices or molecular machines using nanometer size components.

Bacteriorhodopsin: Bacteriorhodopsin is a protein used by archaea, most notably halobacteria. It acts as a proton pump, i.e. it captures light energy and uses it to move protons across the membrane out of the cell. The resulting proton gradient is subsequently converted into chemical energy. Bacteriorhodopsin is an integral membrane protein usually found in two-dimensional crystalline patches known as "purple membrane", which can occupy up to nearly 50% of the surface area of the archaeal cell.

Ballistic Magnetoresistance (BMR): Ballistic Magnetoresistance is yet another way in which spin orientation, encoding information on a storage medium such as a hard drive, can modify electrical resistance in a nearby circuit, thereby accomplishing the sensing of that orientation.

Band Gap Energy (E_g): For semiconductors and insulators, the energies that lie between the valence and conduction bands; for intrinsic materials, electrons are forbidden to have energies within this range.

Banoelectronics: Nanoelectronics in Japan and Korea tends to focus on next generation semiconductor devices and *single electron devices (SED).*

Banopositioning: The means of controlling motion on the nanometre scale—is now a key enabling technology in high- tech fields such as semiconductor test and measurement, photonics alignment, scanning microscopy and microlithography.

Banoscale: At the nanoscale, physics, chemistry, biology, materials science, and engineering converge toward the same principles and tools. The nanoscale is not just another step toward *miniaturisation*, but a qualitatively new scale. The new behaviour is dominated by quantum mechanics, material confnement in small structures, large interfacial volume fraction, and other unique properties, phenomena and processes. Many current theories of matter at the microscale have critical lengths of nanometer dimensions.

Banoscience: Nanoscience is primarily the extension of existing sciences into the realms of the extremely small (*nanomaterials, nanochemistry, nanobio, nanophysics*, etc.) while *nanoengineering* represents the extension of the engineering fields into the nano- scale realm (*nanofabrication, nanodevices*, etc.). The exponential growth of nanoscience is largely due to the development of new instruments and related techniques that are used to "routinely" probe and manipulate material at the atomic and molecular level. *Scanning probe microscopies,* analytical electron- beam techniques, epitaxial growth facilities, and *synchrotron* radiation sources are all opening huge opportunities.

Banoshells: Procedures that target cancer cells while leaving normal cells untouched, patients controlling the release of medicine in their bodies with an infrared light, and medical test results produced in seconds rather than days – these are three new medical technologies currently being tested by nanotechnology researchers at Rice University.... The research focuses on nanoshells, a new type of nanoparticle invented by Naomi Halas, Rice professor of electrical and computer engineering and chemistry. Nanoshells are layered nanoparticles whose ability to manipulate light and color can be designed into the nanoparticle by varying the thickness of the nanoparticle's layers.

Baronatation: In medical nanorobotics, locomotion through a frozen fluid by applying mechanical pressure along the path travelled to induce melting ahead, followed by regelation (refreezing) behind.

Bionanotechnology: Bionanotechnology is the intersection of biology and nanotechnology. Bionanotechnology is a broad and somewhat vague term which is sometimes used interchangeably with nanobiotechnology, which usually refers more specifically to the use of nanotechnological devices for applications in biotechnology. Bionanotechnology may also refer to the use of biomolecules for applications in nanotechnology. A major example of this is DNA nanotechnology, which uses self-assembling nucleic acid structures to control matter at the nanoscale. In a wider sense, bionanotechnolgy refers to synthetic technology based on the principles and chemical pathways of living organisms.

Bearing: A mechanical device that permits the motion of a component (ideally, with minimal resistance) in one or more degrees of freedom while resisting motion (ideally, with a stiff restoring force) in all other degrees of freedom.

Bilayer Lipid Membrane (BLM): The structure found in most biological membranes, in which two layers of lipid molecules are so arranged that their hydrophobic parts interpenetrate, whereas their hydrophilic parts form the two surfaces of the bilayer. The reduction in the free energy of a system that occurs when a ligand binds to a receptor. Generally used to describe the total energy required to remove something, or to take a system apart into its constituent particles—for example, to separate two atoms from one another, or to separate an atom into electrons and nuclei.

Binding: The process by which a molecule (or ligand) becomes bound, that is, confined in position (and often orientation) with respect to a receptor. Confinement occurs because structural features of the receptor create a potential well for the ligand; van der Waals and electrostatic interactions commonly contribute.

Biocatalysis: The application, both actual and potential, of biological catalysts (including whole cells or isolated components thereof, natural and modified enzymes and catalytic antibodies) for the synthesis, interconversion or degradation of chemical species. In addition to papers describing the synthetic applications of biotransformations the journal particularly welcomes papers focusing on the mechanistic principles, kinetics and thermodynamics of biocatalytic processes, the chemical or genetic modification of biocatalysts and the activity and stability of biocatalysts in non- aqueous and multi- phasic environments, including the design of large scale biocatalytic processes. The scope of the journal also encompasses biomimetic systems and environmental applications of biocatalysis where the mechanistic principles are understood or novel biocatalytic activities are involved.

Biochauvinism: The prejudice that biological systems have an intrinsic superiority that will always give them a monopoly on self-reproduction and intelligence.

Biocompatible Coated Materials: Biocompatible materials usually used in dental and bone implants that enhance biologic fixation, thereby increasing the bond strength between the coated material and bone, and minimize possible biological effects that may result from the implant itself. MeSH, 1999

Biodefense: The need for improved national defenses against biological attacks are urgently needed. A real-time reporting system needs to be developed, whereby officials can be informed about an emerging threat before an agent has a chance to affect thousands of people. The development of integrated systems for detecting and monitoring biological agents is instrumental to this goal. The development of vaccines and novel detection platforms will enable preparedness, while also benefiting human health.

Biomaterials: The development of *biomaterials* is not a new area of science, having existed for around half a century. The study of biomaterials is called biomaterial science. It is an exciting field of science, having experienced steady and strong growth over its history with companies such as Smith and Nephew investing large amounts of money in new products. Biomaterials science encompasses elements of medicine, biology, chemistry and materials science. While a definition for biomaterials has been difficult to formulate, a widely accepted definition for biomaterials is that: "A biomaterial is any material, natural or man-made, that comprises whole or part of a living structure or biomedical device which performs, auguments, or replaces a natural function".

Biomechanics: Biomechanics is mechanics applied to biology (Fung). This includes research and analysis of the mechanics of living organisms and the application of engineering principles to and from biological systems. This research and analysis can be carried forth on multiple levels, from the molecular, wherein biomaterials such as collagen and elastin are considered, all the way up to the tissue and organ level. Some simple applications of Newtonian mechanics can supply correct approximations on each level, but precise details demand the use of continuum mechanics.

Biomedical Nanotechnology: The science and technology of diagnosing, treating, and preventing disease and traumatic injury, of relieving pain, and of preserving and improving human health, using molecular tools and molecular knowledge of the human body.

Biomimetics: The design of systems, materials, and their functionality to mimic nature. Current examples include layering of materials to achieve the hardness of an abalone shell or understanding why spider silk is stronger than steel.

Biomimetics: Imitating, copying, or learning from nature. Nanotechnology already exists in nature; thus, nanoscientists have a wide variety of components and tricks already available.

Biomolecular Materials: An emerging discipline, materials whose properties are abstracted from biology. They share many of the characteristics of biological materials but are not necessarily of biological origin. For example, they may be inorganic materials that are organised or processed in a biomimetic fashion. A key feature of biological and biomolecular materials is their ability to undergo *self- assembly*.

Biorobotics: Our research focuses on the role of sensing and mechanical design in motor control, in both robots and humans. This work draws upon diverse disciplines, including biomechanics, systems analysis, and neurophysiology. The main approach is experimental, although analysis and simulation play important parts. In conjunction with industrial partners, we are developing applications of this research in biomedical instrumentation, teleoperated robots, and intelligent sensors.

Biotin: A vitamin of the B complex. It is a co-enzyme for various enzymes that catalyse the incorporation of carbon dioxide into various compounds. It is essential for the metabolism of fats. Biotin is attached to pyruvate carboxylase by a long, flexible chain like that of lipoamide in the pyruvate dehydrogenase complex. Adequate amounts are normally produced by the intestinal bacteria in animals.

Biotinylated-DNA: A DNA molecule labelled with biotin by incorporation of biotinylated -dUTP into a DNA molecule. It is used as a non- radioactive probe in hybridisation experiments, such as Southern transfer. The detection of the labelled DNA is achieved by complexing it with streptavidin (an antibiotic with a high affinity for biotin) to which is attached a colour-generating agent such as horseradish peroxidase that gives a fluorescent green colour upon reaction with various organic reagents. Biotinylation: To label a probe with biotin.

Bond: Two atoms are said to be bonded when the energy required to separate them is substantially larger than the van der Waals attraction energy. Ionic bonds result from the electrostatic attraction between ions; covalent and metallic bonds result from the sharing of electrons among atoms; hydrogen bonds are weaker and result from dipole interactions and limited electron sharing. When used without modification, "bond" usually refers to a covalent bond.

Bone Substitutes: Synthetic or natural materials for the replacement of bones or bone tissue. They include hard tissue replacement polymers, natural coral, hydroxyapatite, beta- tricalcium phosphate, and various other biomaterials. The bone substitutes as inert materials can be incorporated into surrounding tissue or gradually replaced by original tissue.

Broadband Mode: In this mode of operation the error signal E representing bending of the cantilever not corrected by the feedback is scaled and added to the Z signal to provide a mathematically correct image (I=E+Z).

Brownian Assembly: Brownian motion in a fluid brings molecules together in various position and orientations. If molecules have suitable complementary surfaces, they can bind, assembling to form a specific structure. Brownian assembly is a less paradoxical name for self-assembly.

Brownian Motion: Motion of a particle in a fluid owing to thermal agitation. Motion of a particle in a fluid owing to thermal agitation, observed in 1827 by Robert Brown. Stochastic motion of a particle suspended in a surrounding gas or liquid of other particles, molecules or atoms, which is in thermodynamical equilibrium. The motion of the particle is due to the collisions with the particles from its surrounding and as such is 'random', i.e. the change of particle's speed in any of the encounters is random. Although the Brownian motion is stochastical, it has some statistical properties which are fixed.

Buckminsterfullerene: A fullerene with the chemical formula (C_{60}). A broad term covering the variety of buckyballs and carbon nanotubes that exist. Named after the architect Buckminster Fuller, who is famous for the geodesic dome, which buckyballs resemble.

Buckyball: The buckyball (fullerenes) are a recently-discovered family of carbon allotropes named after Buckminster Fuller. They are molecules composed entirely of carbon, in the form of a hollow sphere, ellipsoid, or tube.

Spherical fullerenes are sometimes called buckyballs, the C_{60} variant is often compared to the typical white and black soccer football, the Telstar (football) of 1970. Cylindrical fullerenes are called buckytubes. Fullerenes are similar in structure to graphite, which is composed of a sheet of linked hexagonal rings, but they contain pentagonal (or sometimes heptagonal) rings that prevent the sheet from being planar. In molecular beam experiments, discrete peaks were observed corresponding to molecules with the exact mass of sixty or seventy or more carbon atoms. In 1985, Harold Kroto (of the University of Sussex), James Heath, Sean O'Brien, Robert Curl and Richard Smalley, from Rice University, discovered C_{60}, and shortly after came to discover the fullerenes. Kroto, Curl, and Smalley were awarded the 1996 Nobel Prize in Chemistry for their roles in the discovery of this class of compounds. C_{60} and other fullerenes were later noticed occurring outside of a laboratory environment (e.g. in normal candle soot).

Canadarm: The nickname for Spar Aerospace's SRMS (Space shuttle Remote Manipulator System). The first Canadarm was a gift from Canada to NASA's space shuttle programme; NASA subsequently bought Canadarms to equip the rest of the shuttle fleet. The Canadarm is used during shuttle flights to release satellites into orbit, retrieve them if they malfunction, and aid in their repair.

Cantilever Vibration Mode: An atomic force microscope (AFM) combined with ultrasonic frequency vibration of a cantilever excited at its support. This method enables both topography and elasticity imaging of stiff samples such as metals and ceramics, without a need for bonding a transducer to the sample.

Capacitance: The charge-storing ability of a capacitor, defined as the magnitude of charge stored on either plate divided by the applied voltage. A 1-F capacitor charged to 1 V contains C of charge and 1 C is an amount of charge equal to that of about 6.24×10^{18} electrons.

Capillaries: Capillaries are the smallest of a body's blood vessels, measuring 5-10 ìm in diameter, which connect arterioles and venules, and enable the interchange of water, oxygen, carbon dioxide, and many other nutrient and waste chemical substances between blood and surrounding tissues.

Carbon Black: Carbon black is a powdered form of elemental carbon. The primary use of carbon black is in rubber products, mainly tyres and other automotive products, but also in many other rubber products such as hoses, gaskets and coated fabrics. Much smaller amounts of carbon black are used in inks and paints, plastics and in the manufacture of dry-cell batteries.

Carbon Nanofoam: Carbon nanofoam is the fifth known allotrope of carbon discovered in 1997 by Andrei V. Rode and co-workers at the Australian National University in Canberra. It consists of a low-density cluster-assembly of carbon atoms strung together in a loose three-dimensional web. The large-scale structure of carbon nanofoam is similar to that of an aerogel, but with 1% of the density of previously produced carbon aerogels — only a few times the density of air at sea level. Each cluster is about 6 nanometers wide and consists of about 4000 carbon atoms linked in graphite-like sheets that are given negative curvature by the inclusion of heptagons among the regular hexagonal pattern. This is the opposite of what happens in the case of buckminsterfullerenes, in which carbon sheets are given positive curvature by the inclusion of pentagons. Unlike carbon aerogels, carbon nanofoam is a poor electrical conductor. The nanofoam contains numerous unpaired electrons, which Rode and colleagues propose is due to carbon atoms with only three bonds that are found at topological and bonding defects.

Carbon Nanotubes: These tiny straw-like cylinders of pure carbon have useful electrical properties. They have already been used to make tiny transistors and one-dimensional copper wire. They were developed by using nanotechnology, a relatively new field that involves building electronic circuits and devices from single atoms and molecules. Nano means one thousand millionth of a unit. A nanometer is therefore one thousand millionth of a meter. The first nanofabrication experiments occurred in 1990 when individual xenon atoms were placed on a nickel substrate and used to spell out a company logo. One primary goal of nanotechnology is to build computer chips and other devices that are thousands of times smaller than they are now Carbon nanotubes have enormous theoretical possibilities but have not lived up to the hype surrounding their development. Researchers have continued to look for ways to use them, however, as successful applications have the potential to be highly lucrative. Scientists have recently succeeded in altering carbon nanotubes so that they supply electrons when exposed to light. This was done by having two flat rings of carbon molecules sandwich a ferrocene (iron) molecule. Ferrocene is known for its tendency to relinquish electrons. When exposed to visible light, the carbon atoms accepted the ferrocene molecule.

Carbonyl: A chemical moiety consisting of O with a double bond to C. If the C is bonded to N, the resulting structure is termed an amide; if it is bonded to O, it is termed a carboxylic acid or an ester linkage.

Carboxylic Acid: Carboxylic acids are organic acids characterized by the presence of a carboxyl group, which has the formula -C(=O)OH, usually written -

COOH or -CO_2H. Carboxylic acids are Bronsted/Lowry acids — they are proton donors. Salts and anions of carboxylic acids are called carboxylates. The simplest series of carboxylic acids are the alkanoic acids, R-COOH, where R is a hydrogen or an alkyl group. Compounds may also have two or more carboxylic acid groups per molecule. Carboxylic acids are polar, and form hydrogen bonds with each other.

Cathode: The electrode in an electrochemical cell or galvanic couple at which a reduction reaction occurs; in other words the electrode receiving electrons from an external circuit.

Cation: An ion consists of one or more atoms and carries a unit charge of electricity. Those that are positively electrified (hydrogen and the metals) are called cations (cf. anion).

Cell Repair Machine: Nanocomputers including system and molecular scale sensors and tools, programmed to repair damage to cells and tissues. Molecular and nanoscale machines with sensors, nanocomputers and tools, programmed to detect and repair damage to cells and tissues, which could even report back to and receive instructions from a human doctor if needed.

Charge Coupled Device: A highly sensitive camera, this counts individual photons of light to put together an image. It can take very detailed photographs of small areas.

Charged Particle: In physics, a charged particle is a particle with an electric charge. It may be either a subatomic particle or an ion. A collection of charged particles, or even a gas containing a proportion of charged particles, is called a plasma, which is called the *fourth state of matter* because its properties are quite different from solids, liquids and gases (plasma is the most common state of Matter in the universe). Particles either have a positive, negative or no charge (being neutral). Their effects go beyond the laboratory though: see aurora borealis

Chemotactic Nanosensor: In medical nanorobotics, a nanosensor used to determine the chemical characteristics of surfaces, possibly configured as a pad coated with an array of reversible, perhaps reconfigurable, artificial molecular receptors.

Chromatography: The physical method of separation in which the components to be separated are distributed between two phases, one of which is stationary while the other moves in a definite direction. Chromatography is a widely used for the separation, identification, and determination of the chemical components in complex mixtures.

Chronocyte: In medical nanorobotics, a theorized mobile, mass-storage (nanorobotic)

device, similar to a communicyte, that may be used as a mobile source of precisely synchronized universal time inside the human body.

Clays: A family of finely crystalline materials, composed primarily of silicates, with layered crystal structures. Clays may be refined from geological materials or may be synthetic.

Cobots: Collaborative robots designed to work alongside human operators. Prototype cobots are being used on automobile assembly lines to help guide heavy components like seats and dashboards into cars so they don't damage auto body parts as workers install them.

Cold Plasma Analyzer: An instrument on board the Swedish scientific satellite Freja. By sweeping its sensor through plasma, it obtains measurements of the charged particles in a layer of the atmosphere, the magnetosphere and upper ionosphere, and thus helps scientists understand Earth's atmosphere and how to protect it from pollutants.

Combinatorial Biology: Involves genetic manipulation of bacteria and fungi that produce complex natural products. This technology includes construction of large libraries of recombinant microbes capable of generating novel organic molecules and engineering secondary metabolite biosynthetic pathways to modify valuable biologically active microbial metabolites.

Communications Satellite: A type of satellite used for communications on Earth by allowing radio, television, and telephone transmissions to be sent live anywhere in the world. Before satellites, transmissions were difficult or impossible at long distances. The signals, which travel in straight lines, could not bend around the round Earth to reach a destination far away. Because communications satellites are in orbit, the signals can be sent instantaneously into space and then redirected to another satellite or directly to their destination.

Communicyte: In medical nanorobotics, a theorized mobile, mass-storage (nanorobotic) device that can be used for information transport throughout the human body.

Composites: Combinations of metals, ceramics, polymers, and biological materials that allow multi-functional behaviour. One common practice is reinforcing polymers or ceramics with ceramic fibres to increase strength while retaining light weight and avoiding the brittleness of the monolithic ceramic. Materials used in the body often combine biological and structural functions (e.g., the encapsulation of drugs).

Computational Nanotechnology: Permits the modeling and simulation of complex nanometer-scale structures. The predictive and analytical power of computation is critical to success in nanotechnology: nature required several hundred million years to evolve a functional "wet" nanotechnology; the insight provided by computation should allow us to reduce the development time of a working "dry" nanotechnology to a few decades, and it will have a major impact on the "wet" side as well.

Computronium: A highly (or optimally) efficient matrix for computation, such as dense lattices of nanocomputers or quantum dot cellular automata.

Confocal Microscopy: A light microscopic technique in which only a small spot is illuminated and observed at a time. An image is constructed through point- by- point scanning of the field in this manner. Light sources may be conventional or laser, and fluorescence or transmitted observations are possible.

Conjugation: In medical nanorobotics, the docking of two or more nanorobots for the purpose of exchanging information, energy or materials, or to establish a larger multirobotic structure; in biology, the union of two unicellular organisms accompanied by an interchange of nuclear material, as in Paramecium.

Coupling: A connection between more than one object or energy pathway so that together they function as a single unit.

Covalent Bond: A bond formed by sharing a pair of electrons between two atoms. A primary interatomic bond that is formed by the sharing of electrons between neighboring atoms. An attachment between two atoms resulting from the sharing of a pair of electrons

Covalent Radius: Given a set of N elements that can form covalent single bonds in molecules, with $N(N - 1)$ possible elemental pairings, it has proved possible to define a covalent radius for each element such that the actual bond length between any two elements that form a covalent single bond is roughly equal to the sums of their covalent radii.

Creep: The time-dependent permanent deformation that occurs under stress; for most materials it is important only at elevated temperatures.

Crosslinked Polymer: A polymer in which adjacent linear molecular chains are joined at various positions by covalent bonds.

Cryogenic: In physics or engineering, cryogenics is the study of the production of very low temperatures (below –150 °C, –238 °F or 123 K) and the behaviour of materials at those temperatures. (Rather than the familiar temperature scales of Fahrenheit and Celsius, cryogenicists use the Rankine and Kelvin scales.) The word *cryogenics* literally means "the production of icy cold"; however the term is used today as a synonym for the low-temperature state. It is not well-defined at what point on the temperature scale refrigeration ends and cryogenics begins. The workers at the National

Institute of Standards and Technology at Boulder, Colorado have chosen to consider the field of cryogenics as that involving temperatures below – 180 °C (93.15 K). This is a logical dividing line, since the normal boiling points of the so-called permanent gases (such as helium, hydrogen, neon, nitrogen, oxygen, and normal air) lie below -180 °C while the Freon refrigerants, hydrogen sulfide, and other common refrigerants have boiling points above -180 °C.

Crystal Grain Boundary: Crystalline materials are composed of atoms arranged in an orderly way, in precise relationships to one another. A grain boundary is a narrow region of transition for one ordered area to another of different crystallographic orientation—an interface between one crystal grain and another.

Crystal Structure: For crystalline materials, the manner in which atoms or ions are arrayed in space. It is defined in terms of the unit cell geometry and the atom positions within the unit cell.

Crystallescence: In medical nanorobotics, the crystallisation of solid solute that is offloaded by nanorobot sorting rotors at a concentration that exceeds the solvation capacity of the surrounding solvent.

Cyclotron: A type of particle accelerator in which an ion introduced at the center is accelerated in an expanding spiral path by use of alternating electrical fields in the presence of a magnetic field.

Cytocarriage: In medical nanorobotics, the commandeering of a natural motile cell, by a medical nanorobot, for the purposes of in vivo transport, or to perform a herding function, or for other purposes.

Cytopenetration: In medical nanorobotics, entry into cells by penetrating the plasma membrane.

Deoxyribonucleic Acid (DNA): Deoxyribonucleic acid are long chains consisting of four kinds of nucleotides; the order of these nucleotides encodes the information needed to construct protein molecules. These in turn make up much of the molecular machinery of the cell. DNA is the genetic material of cells.

Directed-Assembler: A specific type of assembler that makes use of directed-assembly, such that the assembly process requires external energy or information input.

Disassembler: In molecular nanotechnology, a nanomachine or system of nanomachines able to take an object apart while at each step recording the structure and composition of that object at the molecular level.

Disasterbation: Idly fantasizing about possible catastrophes (ecological collapse, full-blown totalitarianism) without considering their likelihood or considering their possible solutions and preventions.

DNA Probes: A DNA or nucleic acid probe is a short strand of DNA that locates and binds to its complementary sequence in samples containing single strands of DNA or RNA enabling identification of specific sequences. Nucleic acid probe assays exploit the fundamental hybridisation reaction that occurs spontaneously between two complementary DNA.

DNA Sequencing: There are two main classical methods for sequencing DNA: The first method, developed by Allan Maxam and Walter Gilbert, involves chemicals used to cleave the DNA at certain positions, generating a set of fragments that differ by one nucleotide. The second method, developed by Fred Sanger and Alan Coulson, involves enzymatic synthesis of DNA strands that terminate in a modified nucleotide. Analysis of fragments is similar for both methods and involves gel electrophoresis and autoradiography or fluorescence. The enzymatic method has largely replaced the chemical method as the technique of choice, although there are some situations where chemical sequencing can provide data more easily than the enzymatic method.

DNA: A molecule encoding genetic information, found in the cell's nucleus. Deoxyribose Nucleic Acid. DNA is a molecule used in biological systems to store genetic information. It has a long, double stranded structure with the two strands wrapped around each other to form a right handed helix. There are about 10 nucleotide pairs per helical turn. Each strand is composed of a sugar phosphate backbone and attached bases. It is connected to a complementary strand by hydrogen bonding (non-covalent) between paired bases, adenine (A) with thymine (T) and guanine (G) with cytosine (C). DNA is an abbreviation of DeoxyriboNucleic Acid. It is the genetic 'blueprint' of all life on this planet. DNA typing is often applied to Human Identification which has also been called DNA Fingerprinting. DNA Fingerprinting has been attributed to Alex Jeffries, who with his colleagues in 1985, discovered a sequence of nucleotides, subsequently termed short tandem repeats (STR) when comparing the DNA sequence of the human á-globin chromosomes isolated from different individuals.

Dnhancer: Regulatory DNA sequence to which gene regulatory proteins bind, influencing the rate of transcription of a structural gene that can be many thousands of base pairs away.

Domain: A region of a ferromagnetic or ferrimagnetic material in which all atomic or ionic magnetic moments are aligned in the same direction.

Dopeyballs: Superconducting Buckyballs (they) have the highest critical temperature of any known organic compound.

Doppler Effect: An apparent shift in the *frequency* of a wave. For example, when someone is listening to the sound of an ambulance siren, and that person is staying still but the ambulance is driving by, the person will hear a change in pitch of the ambulance siren. That change in pitch is caused by the doppler effect. The frequency of a sound wave determines the pitch, and the distance of the source of the sound from the sound's observer determines the amount that the frequency seems to have shifted, known as the doppler shift.

Drug Design: Includes not only ligand design, but also pharmacokinetics (Pharmacogenomics) toxicity, which are mostly beyond the possibilities of structure- and/ or computer- aided design. Nevertheless, appropriate chemometric (Chemoinformatics) tools, including experimental design and multivariate statistics, can be of value in the planning and evaluation of pharmacokinetic and toxicological experiments and results. Drug design is most often used instead of the correct term "ligand design".

Drug Discovery and Molecular Imaging: Molecular imaging's core role for the pharmaceutical industry is in drug discovery and development. Pure research is already profiting from cell- based molecular imaging, which will continue to be based on fluorescence, bioluminescence, and confocal microscopy. Applications in small animal imaging, lead characterisation, and lead optimisation are also discussed. The insights into basic cell biology that this research is yielding today will form the basis of drug development during the second half of the decade, as the results are absorbed by the pharmaceutical industry.

Drug Ontology: A three year project to develop a highly structured drug knowledge base. Unlike existing reference sources, it is intended solely for use by software applications.

Druggable: Able to be modulated by a small molecule to produce a desired phenotypic change in cell targets.

Dry Nanotechnology: Derives from surface science and physical chemistry, focuses on fabrication of structures in carbon silicon, and other inorganic materials. Unlike the 'wet' technology, 'dry' techniques admit use of metals and semiconductors. The active conduction electrons of these materials make them too reactive to operate in a 'wet' environment, but these same electrons provide the physical properties that make 'dry' nanostructures promising as electronic, magnetic, and optical devices. Another objective is to develop 'dry' structures that possess some of the same attributes of the self-assembly that the wet ones exhibit.

Ductility: A measure of a material's ability to undergo plastic deformation before fracture; expressed as percent elongation (%EL) or percent area reduction (%AR) from a tensile test.

Elongation Factor: Protein required for the addition of amino acids to growing polypeptide chains on ribosomes.

Emergence: A complex whole created by simple parts, as in the brain where billions of neurons work individually, but collectively make up our consciousness and give us the ability to think, rationalize, and create.

Encryption Device: An instrument often used in spy satellites. It encodes digital images or data before it is sent back to Earth.

Endocrine Cell: Specialized animal cell that secretes a hormone into the blood; usually part of a gland, such as the thyroid or pituitary gland.

Endocytosis: Uptake of material into a cell by an invagination of the plasma membrane and its internalisation in a membrane-bounded vesicle.

Engulf Formation: In medical nanorobotics, a configuration that may be adopted by a metamorphic nanorobot, in which the nanorobot reshapes itself to create an interior cavity capable of trapping a living cell, virion, or other biological particle.

Enhancer: Regulatory DNA sequence to which gene regulatory proteins bind, influencing the rate of transcription of a structural gene that can be many thousands of base pairs away.

Entanglement: From quantum mechanics, entanglement is a relationship between two systems in which they exist in more than one state simultaneously (by superposition), and the state of one system determines the state of the other.

Enternal Computer: The satellite's method of storing and analyzing data collected by the satellite, and a way of controlling its various systems. The technical term for this satellite subsystem is *TT&C* (Telemetry, Tracking, and Control).

Enzyme immunoassay/EIA: In an enzyme immunoassay (EIA), an enzyme-labeled antibody or antigen is used for the detection and quantification of the antigen-antibody reaction. In an electrochemical EIA, the enzyme-catalyzed reaction is monitored electrochemically (amperometric, potentiometric, voltametric or conductometric). In EIA, the antibody-antigen reaction furnishes the needed specificity. The enzyme label provides the sensitivity via chemical amplification. A large molecule, usually a protein, that catalyzes biological reactions. A protein that acts as a catalyst in a biochemical reaction. Protein that catalyzes a specific chemical reaction.

Equator: The equator is an imaginary circle drawn around a planet (or other astronomical object) at a distance halfway between the poles. The equator divides the planet into a Northern Hemisphere and a Southern Hemisphere. The latitude of the equator is, by definition, 0°. The length of Earth's equator is about 40,075.0 km, or 24,901.5 miles. The equator is one of the five main circles of latitude based on the relationship of the Earth's rotation and plane of orbit around the sun. Additionally, the equator is the only line of latitude which is also a great circle. The Sun, in its seasonal movement through the sky, passes directly over the equator twice each year on the March and September Equinoxes. At the equator, the rays of the sun are perpendicular to the surface of the earth on these dates. Places near the equator experience the quickest rates of sunrise and sunset in the world, taking minutes. Such places also have a relatively constant amount of day/night time on every day throughout the year compared with more northerly or southerly places.

Fatty Acid: Compound such as palmitic acid that has a carboxylic acid attached to a long hydrocarbon chain. Used as a major source of energy during metabolism and as a starting point for the synthesis of phospholipids.

Fault-tolerant: Describes a system that can suffer failure in a component or subsystem, yet continue to function correctly.

Femtoengineering: Will involve engineering using mechanisms within a quark.

Femtosecond: Femtosecond is one quadrillionth of a second, and is to a second what a second is to 32,700,000 years. At 186,000 miles per second, in one femtosecond light travels only far enough to traverse about 1,000 silicon atoms. When used to time a laser pulse, it allows for ultra-precise micromachining, with virtually no damage to surrounding material.

Femtotechnology: Femtotechnology the art of manipulating materials on the scale of elementary particles (leptons, hadrons, and quarks). The next step smaller after picotechnology, which is the next step smaller after nanotechnology.

Fermi Energy: The energy level in a solid at which the probability of finding an electron is 1/2. For a metal, the energy corresponding to the highest filled electron state in the valence band at 0 K.

Ferroelectric Material: A dielectric material such as Rochelle salt and barium titanate with a domain structure containing dipoles (asymmetric distributions of electrical charge) which spontaneously align. Their domain structure makes them analogous to ferromagnetic materials. They exhibit hysteresis and usually the piezoelectric effect.

Ferromagnetism: Permanent and large magnetisations found in some metals (e.g. Fe, Ni and Co), resulting from the parallel alignment of neighboring magnetic moments.

Fibrin Sealant: A biological tissue glue derived from human plasma. Sometimes referred to as fibrin glue, it contains the components necessary for the last step of coagulation: i) fibrinogen and other coagulation factors and ii) thrombin. It is used in surgical procedures to arrest bleeding and as an adjunct to wound healing.

Fluidic System: Device for synthesis or screening in which fluids such as reagents or assay buffers may be directed to specified locations by the opening and closing of valves in a stationary network of tubes and wells.

Fluorescence Microscopy: Microscopy of specimens stained with fluorescent dye (usually fluorescein isothiocyanate) or of naturally fluorescent materials, which emit light when exposed to ultraviolet or blue light. Immunofluorescence microscopy utilizes antibodies that are labeled with fluorescent dye.

Fluorescence Polarisation: Based on the observation that emission signals from small fluorescent molecules are relatively depolarized, while binding to a larger molecule reduces the tumbling rate of the fluorescer, resulting in a relatively polarized emission signal. First described in 1926 (Perrin) and has been a powerful tool in the study of molecular interactions. When fluorescent molecules are excited with plane polarized light, they emit light in the same polarized plane, provided that the molecule remains stationary throughout the excited state (4 nanoseconds in the case of fluorescein). However, if the excited molecule rotates or tumbles out of the plane of polarized light during the excited state, then light is emitted in a different plane from that of the initial excitation.

Fluorescent Dye: Molecule that absorbs light at one wavelength and responds by emitting light at another wavelength; the emitted light is of longer wavelength (and hence of lower energy) than the light absorbed.

Focal Length: The distance needed between a lens and an object to make the object visible and in focus. The human eye has different focal lengths depending on how much the muscles in the eye around the lens are contracted. That is how the human eye focuses.

Footprint: The area with which a satellite in *geostationary orbit* can communicate. A footprint can be as large as an entire country; for example, many Canadian satellites have footprints almost the entire size of Canada, from coast to coast. Click here for a picture of a satellite's footprint.

Free Energy: Energy that can be extracted from a system to drive reactions. Takes into account changes in both energy and entropy.

Frequency Bands: The grouping together of electromagnetic waves of a similar frequency. In spectroscopic studies, one type of wave from the electromagnetic spectrum will show up as a band of colour that can be interpreted as a frequency band. This allows scientists to figure out what kind of energy is coming from a particular source For example, X-rays would make up a frequency band. Some bands are wider than others.

Genetic Algorithm: Any algorithm which seeks to solve a problem by considering numerous possibilities at once, ranking them according to some standard of fitness, and then combining ("breeding") the fittest in some way. In other words, any algorithm which imitates natural selection.

Genomic Arrays: Allow toxicologists to look at cellular behaviour in a completely new light. In a sense, recording individual gene responses to powerful insults such as alkylating agents was akin to studying the effects of poverty by monitoring a person's bank account—the complete picture is much larger than what is actually being measured. But genomic arrays simultaneously report indicators of multiple dimensions of the cellular response to stimuli. Now, in addition to gaining insight into basic cellular mechanisms of repair, researchers looking at a variety of indicators and responses of toxicity may gain some predictive power regarding individual compounds—and individual humans. Both academic and private laboratories have already begun work on finding genes that induce protection or sensitivity to toxicants in individual cells and people.

Geostationary Orbit: An orbit in which a satellite appears to remain in the same spot in the sky all the time. When a satellite is in geostationary orbit, it travels at exactly the same speed as the Earth is rotating below it. A satellite in geostationary orbit is very high up, at 35 850 km above the Earth. Geostationary satellites are always located directly above the equator. The area with which a satellite in geostationary orbit can communicate is called its *footprint*.

Green Ceramic Body: A ceramic piece, formed as a particulate aggregate, that has been dried but not fired.

Green Goo: Nanomachines or bio-engineered organisms used for population control of humans, either by governments or eco-terrorist groups. Would most probably work by sterilizing people through otherwise harmless infections.

Hemidesmosome: Specialized cell junction between an epithelial cell and the underlying basal lamina.

Henry: Unit of inductance. One henry (H) is the inductance of a closed circuit in which an electromotive force of 1 volt is produced when the electric current in the circuit varies uniformly at the rate of 1 ampere per second.

Heterochromatin: Region of a chromosome that remains unusually condensed and transcriptionally inactive during interphase.

Histone: A protein present in the eukaryotic nucleus that is bound to the DNA at regularly spaced intervals.

Immuno-electron Microscopy: This technique allows the investigator to identify antibody/antigen complexes that localize to a particular subcellular organelle or compartment by using the Protein A gold technique. The tissue is first fixed in 4.0% para-formaldehyde and processed for embedding in a low temperature embedding resin (Lowicryl). Paraffin embedded tissue (same tissue) sections are cut and screened by light microscopy using a streptavidin-biotin technique to determine the appropriate antibody concentration to be used for the EM immunogold procedure. Ultrathin sections from the same block(s) are cut, incubated with primary antibody and then later incubated with Protein A gold particles (size range is 5 nm to 20 nm). The gold particles bind to the Fc portion of the antibody and are detected by EM. A variation of the large block "pop-off" technique for immunoelectron microscopy is also available.

Immunoglobulin: An antibody molecule. Higher vertebrates have five classes of immunoglobulin—IgA, IgD, IgE, IgG, and IgM—each with a different role in the immune response.: : immunoglobulin like (Ig-like) domain: : Characteristic protein domain of about 100 amino acids that is found in antibody molecules and in many other proteins that form the Ig superfamily.

Immunohistochemistry: Histochemical localisation of immunoreactive substances using labeled antibodies as reagents. Immunohistochemistry involves using antibodies (typically visualized via an enzyme- linked antibody assay) that specifically bind to proteins of interest. This method allows one not only to assess levels of a protein but also to localize the protein within cells in the tissue sample.

Inductance: The property of an electric circuit which tends to oppose change in current in the circuit. One henry (H) is the inductance of a closed circuit in which an electromotive force of 1 volt is produced when the electric current in the circuit varies uniformly at the rate of 1 ampere per second.

Inflammatory Response: When a bacterial infection is established in the body, it is the function of the immune system to control or eradicate it. The initial reaction of the immune system to an infection differs, depending on the site which has been invaded, and on the nature of the invader. There can be many "triggers", that can kick the immune system into action. Here are some of the ways in which the immune system can be activated.

Infrared Spectromicroscopy: Synchrotron Radiation is a powerful tool in research, broadly applied in the VUV and x-ray spectral regions. The three main advantages of synchrotron radiation are: broadband characteristics, high collimation and the possibility to calculate its properties with Schwinger's equation. Synchrotron radiation can also be superior in the infrared spectral region.

Inmessaging: In medical nanorobotics, conveyance of information from a source external to the human body, or external to working nanodevices, to a receiver located inside the human body.

Input Device: A digitizer, scanner, line art, clip art, keyboard, video camera or any other device that is used to generate and send a design or other instruction into a computer for eventual production on an output device.

Intrinsic: Characterizes pure undoped semiconductor; electrical conductivity depends only on temperature and the band gap energy.

Ion Exchange: A chemical reaction involving the exchange of hydrated ions in a solid for similarly charged but different element ions in solution.

Ion Microscopy: Use of the Secondary Ion Mass Spectrometry SIMS technique to obtain micrographs of the elemental (or isotopic) distribution at the surface of a sample with a spatial resolution of 2 mm or better.

Ion: An atom or a molecule that has lost or gained an electron. An ion is also known as a charged particle. Ions are usually more reactive than neutral, uncharged particles. They cause all sorts of things from static electricity and chemical reactions like smog, to body processes like cellular respiration and digestion.

Ionophore: A macro-organic molecule capable of specifically solubilizing an inorganic ion of suitable size in organic mediums.

Kinetic Energy: Kinetic energy (SI unit: the joule) is energy that a body possesses as a result of its motion. It is formally defined as *the work needed to accelerate a body from rest to its current velocity*. Having gained this energy during its acceleration, the body maintains this kinetic energy unless its speed changes. Negative work of the same magnitude would be required to return the body to a state of rest from that velocity.

Lectin: Protein that binds tightly to a specific sugar. Abundant lectins derived from plant seeds are often used as affinity reagents to purify glycoproteins or to detect them on the surface of cells.

Leucine Zipper: Structural motif seen in many DNA-binding proteins in which two a helices from separate proteins are joined together in a coiled-coil, forming a protein dimer.

LIF Laser Induced Fluorescence: The optical emission from molecules that have been excited to higher energy levels by absorption of electromagnetic radiation. The main advantage of fluorescence detection compared to absorption measurements is the greater sensitivity achievable because the fluorescence signal has a very low background. For molecules that can be resonant excitated, LIF provides selective excitation of the analyte to avoid interferences. LIF is useful to study the electronic structure of molecules and to make quantitative measurements of analyte concentrations. Analytical applications include monitoring gas-phase concentrations in the atmosphere, flames, and plasmas; and remote sensing using light detection and ranging (LIDAR).

Ligand: In protein chemistry, a small molecule that is (or can be) bound by a larger molecule is termed a ligand. In organometallic chemistry, a moiety bonded to a central metal atom is also termed a ligand; the latter definition is more common in general chemistry.

Ligand-directed Drug Discovery: The first goal is to select a small number of likely targets (which at least in some cases, such as targets in pathogens, are validated targets) rapidly, without the need for costly and time-consuming biologically based target validation and functional determination. The second goal is to obtain lead compounds for these targets, instead of starting with a biologically validated target and then attempting to screen libraries for lead compounds that modulate it.

Light Emitting Diodes: Light Emitting Diodes (LEDs) work on a completely different concept. Traditionally LEDs are created from two semiconductors. By running current in one direction across the semiconductor the LED emits light of a particular frequency (hence a particular color) depending on the physical characteristics of the semiconductor used. The semiconductor is covered with a piece of plastic that focuses the light and increases the brightness. These semiconductors are very durable, there is no filament, they donít require much power, theyíre brighter and they last a long time. By densely packing red, blue and green LEDs next to each other on a substrate one can create a display.

Limit of Detection: The smallest measurable input. This differs from resolution, which defines the smallest measurable change in input. For a temperature measurement, this would provide an indication of the lowest temperature a sensor could generate an output in response to.

Limited Assembler: An assembler with built-in limits that constrain its use (for example, to make hazardous uses difficult or impossible, or to build just one thing). Assembler capable of making only certain products; faster, more efficient, and less liable to abuse than a general-purpose assembler.

Linde Scenario: A scenario for indefinite survival of intelligent life. It assumes it is possible to either create basement universes connected to the original universe with a wormhol*e or t*he existence of other cosmological domains.

Intelligent life continually migrates to the new domains as the old grow too entropic to sustain life.

Linear: Aside from its geometric meaning, linear describes systems in which an output is directly proportional to an input. In particular, a linear elastic system is one in which the internal displacements are (at equilibrium) directly proportional to applied forces.

Liposome: Type of nanoparticle made of lipids, or fat molecules, surrounding a water core. Liposomes, several of which are widely used to treat infectious diseases and cancer, were the first type of nanoparticle to be used to create therapeutic agents with novel characteristics.

Lofstrom Loop: An beanstalk-like megaconstruction based on a stream of magnetically accelerated bars linked together. The stream is sent into space, where a station rides it using magnetic hooks, redirects it horizontally to another station, which sends it downwards to a receiving station on the ground. From this station the stream is then sent back to the launch station (a purely vertical version is called a space fountain). This structure would contain a large amount of kinetic energy but could be built gradually and would only require enough energy to compensate for losses when finished. Elevators could be run along the streams, and geostationary installations could be placed along the horizontal top.

Macroscale: Larger than nanoscale; often implies a design that humans can directly interact with; too large to be built by a single assembler (one cubic micron of diamond contains 176 billion atoms).

Macrosensing: In me*dical nanorobotics, the detecti*on of global somatic states and extrasomatic states (sensory data originating outside of the human body) by in vivo nanorobots.

Magnetic Resonance Force Microscopy MRFM: A new microscopic imaging technique that combines aspects of atomic force microscopy AFM with magnetic resonance imaging MRI. Interest in this technique is driven by the possibility of reaching the "Holy Grail" of microscopy: true three-dimensional, sub- surface imaging with atomic resolution and chemical specificity.

Magnetostrictive Material: A material that changes dimension in the presence of a magnetic field or generates a magnetic field when mechanically deformed.

Martensite: A metastable iron phase supersaturated in carbon that is the product of a diffusionless (athermal) transformation from austenite.

Mass Spectrometry: This technique can be used to both measure and analyze molecules under study. It involves introducing enough energy into a target molecule to cause its ionisation and disintegration. The resulting fragments are then analyzed, based on the mass/ charge ratio to produce a "molecular fingerprint."

Mass to Charge Ratio: A number defining how a particle will respond to an electric or magnetic field that can be calculated by dividing the mass of a particle by its charge. For instance, mass spectrometry provides information on the mass to charge ratio of a molecular species. In this case, the ratio is expressed by m/z where m is the molecular weight and z is the elementary charge.

Massometer: In medical nanorobotics, a nanosensor device for measuring the mass of individual molecules or small physical objects to single-proton resolution.

Materials Science: Science of ceramics, glass, metals, plastics, semiconductors.

Meridian Lines: The imaginary lines that were drawn on the globe for navigational purposes. Meridian lines mark longitude and run through both the north and the south poles. They separate the Earth into time zones and they are used to indicate a coordinate on the Earth. The Greenwich meridian is the central meridian line from which all time zones are set.

Metal Nanoshells: A new type of nanoparticle composed of a semiconductor or dielectric core coated with an ultrathin conductive layer.. By adjusting the relative core and shell thicknesses, metal nanoshells can be fabricated that will absorb or scatter light at any wavelength across the entire visible and infrared range of the electromagnetic spectrum.

Metastable: A classical system is metastable if it is above its minimum-energy state, but requires an energy input before it can reach a lower-energy state; accordingly, a metastable system can act like a stable system, provided that energy inputs (e.g., thermal fluctuations) remain below some threshold. Systems with strong metastability are commonly described as stable. Quantum mechanical effects can permit metastable states to reach lower energies by tunneling, without an energy input; an associated, broader definition of *metastable* embraces all systems that have a long lifetime (by some standard) in a state above the minimum-energy state.

Meteosat: A geostationary weather satellite launched by the European Space Agency (ESA) and now operated by the company Eumetsat. The most recent version of Meteosat was launched in June 1988. It provides weather imaging of the Earth at both visible light and infrared wavelengths.

Micellar Nanocontainers: Block copolymer micelles are water- soluble biocompatible nanocontainers with great potential for delivering hydrophobic drugs. An understanding of their cellular distribution is essential to achieving selective delivery of drugs at the subcellular level.

Micelles: Micelles are small, spherical structures composed of molecules that attract one another to reduce surface tension. The head of the molecule is hydrophilic, meaning it likes water, while the interior portion is hydrophobic, meaning it avoids water.

Microbubbles: Very small encapsulated gas bubbles (diameters of micrometers) that can be used in diagnostic and therapeutic applications. Upon exposure to sufficiently intense ultrasound, microbubbles will cavitate, rupture, disappear, release gas content, etc. Such characteristics of the microbubbles can be used to enhance diagnostic tests, dissolve blood clots, and deliver drugs or genes for therapy. MeSH 2004

Microinjection: The insertion of a substance into a cell through a microelectrode. Typical applications include the injection of drugs, histochemical markers (such as horseradish peroxidase or lucifer yellow) and RNA or DNA in molecular biological studies. To extrude the substances through the very fine electrode tips, either hydrostatic pressure (pressure injection) or electric currents (ionophoresis) is employed. A technique for introducing a solution of DNA, protein, or other soluble material into a cell using a fine microcapillary pipet.

Microparticles: Applications include calibration of flow cytometers, particle and hematology analyzers, confocal Laser scanning microscopes and zetapotential measuring instruments; flow measurements in gases and liquids like Laser Doppler Anemometry (LDA); Particle Dynamics Analysis (PDA), and Particle Image Velocimetry (PIV); medical diagnostics; separation phases for chromatography; support for immobilized enzymes; spacer in liquid crystal displays (LCD's); peptide synthesis; cell separation; tracers in environmental science; model systems in medicine, biochemistry, colloid chemistry, and aerosol research.

Microphone: A microphone, sometimes referred to as a mike or mic (pronounced "mike"), is an acoustic to electric transducer that converts sound into an electrical signal. Microphones are used in many applications such as telephones, tape recorders, hearing aids, motion picture production, live and recorded audio engineering, in radio and television broadcasting and in computers for recording voice, VoIP and numerous other computer applications.

Microplate Reader: Created from the tube spectrophotometer designs of the 1970s to save precious antibody samples. At first clumsy and inaccurate, absorbance microplate readers have evolved to pack unbelievable power and precision, replacing cuvette spectrophotometers for most multisample applications. Continuous improvement is enhancing the classic designs to embrace the world of high- throughput (HT) screening and to allow complete analytical automation. To handle the HT range (more than 10 microplates a day or 1,000 assays), many instruments now allow robotic handling of plates "stacked" in accessory plate handlers.

Microprocessor: A microprocessor incorporates the functions of a central processing unit (CPU) on a single integrated circuit (IC). The first microprocessors used a word size of only 4 bits, so that the transistors of its logic circuits would fit onto a single part. One or more microprocessors typically serve as the processing elements of a computer system, embedded system, or handheld device. Microprocessors made possible the advent of the microcomputer in the mid-1970s. Before this period, CPUs were typically made from small-scale integrated circuits containing the equivalent of only a few transistors. By integrating the processor onto one or a very few large-scale integrated circuit packages (containing the equivalent of thousands or millions of discrete transistors), the cost of processing capacity was greatly reduced.

Microscopy: Microscopy primer, microscope basics, special techniques, tutorials, virtual microscopy Molecular Expressions, National High Field Magnetic Laboratory, Florida State Univ. US, Olympus America, Inc. 2003 http://micro.magnet.fsu.edu/primer/ Narrower terms: atomic force microscopy AFM, Confocal Scanning Laser Scanning Microscopy CLSM, confocal microscopy, cryoelectron microscopy, electron microscopy, fluorescence microscopy, immunoelectron microscopy, ion microscopy, Laser Fluorescence Microscopy LFM, laser scanning microscopy, Multiphoton Laser Scanning Microscopy MLSM, Magnetic Resonance Force Microscopy MRFM, multiple-photon excitation fluorescence microscopy, Near- field Scanning Optical Microscopy NSOM, Scanning Electron Microscopy SEM, Scanning Transmission Electron Microscopy STEM, Scanning Tunneling Microscopy STM, scanning probe microscopy, Surface Plasmon Resonance microscopy, Total Internal Reflectance Fluorescence Microscopy TIR-FM, Transmission Electron Microscopy TEM, two- photon Laser Fluorescence Microscopy, virtual microscopy

Microspectrophotometry: Analytical technique for studying substances present at enzyme concentrations in single cells, in situ, by measuring light absorption. Light from a tungsten strip lamp or xenon arc dispersed by a grating monochromator illuminates the optical system of a microscope. The absorbance of light is measured (in nanometers) by comparing the difference between the image of the sample and a reference image. MeSH 1990

Microtubule: A cylindrical biological structure made of tubulin (which is a specific type of protein) dimers, and found in cells. Their role is to mechanically support cellular structure (cell rigidity) and the cell movement. Additionally, biological molecular motors, such as kinezin, use microtubules to move along them while transporting proteins in cells. Tubulin dimers in a microtubule are organised in protofilaments along which they bond relatively strongly. The bonds among protofilaments are weaker. Protofilaments can be thought of as one-dimensional chains of tubulin dimers positioned on a cylinder. In a 13-protofilament microtubule (which occurs in human cells), protofilaments are parallel to a microtubule cylindrical axis. In other types of microtubules, with different number of protofilaments, protofilaments spiral around the microtubule, i.e. they can be thought of as cylindrical geodesics. Microtubules are typical example of bilogical self assembly since tubulin dimers are spontaneously assembled in microtubules in cellular surrounding (cytoplasma). The diameter of a microtubule is about 25 nanometers.

Microwave: A form of electromagnetic radiation that is beyond the range of the *visible light* spectrum. Microwaves have very high frequencies and wavelengths of 1 mm to 50 cm.

Miller Indices: A set of 3 integers (4 for hexagonal) that designate crystallographic planes, as determined from reciprocals of fractional axial intercepts.

Miniaturisation: Desirable for many technologies for overall cost reduction (including reduction in the amount of reagents and analytes). Important to remember that building space is often the least available and most expensive component of an overall laboratory budget.

Molecular Assembler: A molecular assembler has been defined as a single molecule machine that would assemble individual atoms or molecules according to specific instructions to construct a desired product. Some biological molecules the such as a ribosomes fit this definition since working within cell's environment, it receives instructions from mRNA and then assembles specific sequences of amino acids to construct protein molecules. However, the term "molecular assembler" usually refers to theoretical man-made or synthetic devices.

Molecular Beam Epitaxy (MBE): Process used to make compound (multi-layer) semiconductors. Consists of depositing alternating layers of materials, layer by layer, one type after another (such as the semiconductors gallium arsenide and aluminum gallium arsenide).

Molecular Combing: A method ... which can straighten and align molecules of genomic DNA on a solid surface. The technology also includes a battery of novel statistical methods developed for analyzing the large amounts of data obtained from FISH analyses made on individual DNA molecules. Molecular combing relies on the action of a receding air/ water interface, or meniscus, to uniformly straighten and align DNA molecules on a solid surface.

Molecular Design: The application of all techniques leading to the discovery of new chemical entities with specific properties required for the intended application.

Molecular Electronics (ME): Molecular electronics offers the tantalizing prospect of eventually building circuits with critical dimensions of a few nanometers. Some basic devices utilizing molecules have been demonstrated, including tunnel junctions with negative differential resistance, rectifiers and electrically configurable switches that have been used in simple electronic memory and logic circuits. A major challenge that remains is to show that such devices can be fabricated economically using a process that will scale to circuits with large numbers of elements while maintaining their desired electronic properties.

Molecular Machine: A mechanical device that performs a useful function using components of nanometer scale and defined molecular structure; includes both artificial nanomachines and naturally occurring devices found in biological systems. Any machine with atomically precise parts of nanometer dimensions; can be used to describe molecular devices found in nature.

Molecular Manipulator: A device combining a proximal probe mechanism for atomically precise positioning with a molecule binding site on the tip; can serve as the basis for building complex structures by positional synthesis. A programmable device able to position molecular tools with high precision, for example, to direct a sequence of mechanosynthetic steps; a molecular assembler.

Molecular Manufacturing: Manufacturing using molecular machinery, giving molecule-by-molecule control of products and by-products via positional chemical synthesis. Exponential general-purpose molecular manufacturing — that's a mouthful, but what does it mean? Let's take the phrase apart to see why it is so important. The building of complex structures by mechanochemical processes. The production of complex structures via nonbiological mechanosynthesis.

Molecular Mechanics Models: Many of the properties of molecular systems are determined by the molecular potential energy function. Molecular mechanics models approximate this function as a sum of 2-atom, 3-atom, and 4-atom terms, each determined by the geometries and bonds of the component atoms. The 2-atom and 3-atom terms describing bonded interactions roughly correspond to linear springs.

Molecular Medicine: A variety of pharmaceutical techniques and gene therapies that address specific molecular diseases or molecular defects in biological systems. A variety of pharmaceutical techniques and therapies in use today. Studying molecules as they relate to health and disease, and manipulating those molecules to improve the diagnosis, prevention, and treatment of disease.

Molecular Mimicry: The process in which structural properties of an introduced molecule imitate or simulate molecules of the host. Direct mimicry of a molecule enables a viral protein to bind directly to a normal substrate as a substitute for the homologous normal ligand. Immunologic molecular mimicry generally refers to what can be described as antigenic mimicry and is defined by the properties of antibodies raised against various facets of epitopes on the viral protein.

Molecular Motor: Protein or protein complex that transforms chemical energy into mechanical work at a molecular (nanometer) scale. The chemical process that produces energy is hydrolysis of adenosine triphosphate (ATP) to adenosine diphosphate (ADP) and phosphate (P). Rotatory and translatory motors are known to exist. Motor proteins move in eukaryotic cells along filaments (actin filaments, microtubules)—periodic and relatively rigid protein structures with a periodicity of the order of 10 nanometers. Some of the known motors are myosin that moves along actin filaments, and kinesin and dynein that move along microtubules (tubulin filaments). They play a role in muscular contraction, cell division, cellular material transport, cell division, etc. The fact that such structures exist is a great inspiration for nanotechnological applications (including nanobots, of course).

Molecular Nanoscience: An emerging interdisciplinary field that combines the study of molecular/ biomolecular systems with the science and technology of nanoscale structures and systems. The potential applications for this research are very broad and include such possibilities as the use of biomolecules and cellular systems to self- assemble nanoelectronic circuitry and other nanoscale structures and the use of lamellar host frameworks containing nano pores that can be tailored to include guest molecules for separation of chemicals for pharmaceutical and other applications.

Molecular Nanotechnology (MNT): Molecular nanotechnology (MNT) is the engineering of functional systems at the molecular scale. An equivalent definition would be "machines at the molecular scale designed and built atom-by-atom". This is distinct from nanoscale materials. Based on Richard Feynman's vision of miniature factories using nanomachines to build complex products (including additional nanomachines), this advanced form of nanotechnology (or *molecular manufacturing*) will make use of positionally-controlled mechanosynthesis guided by molecular machine systems.

Molecular Recognition: The ability of biological macromolecules such as proteins and DNA to recognize selectively and to bind to other species to form larger supramolecular complexes is a key element in the extraordinarily diverse and controlled chemistry exhibited by nature. Chemists today are increasingly interested in mimicking these processes which involve 'molecular recognition' of one molecule by another via formation of specific nonequivalent bonds between them and spontaneous binding together of two to many thousands of molecules into well defined supramolecular systems with new chemical properties. Industrially important applications already include drug design, synthesis of stereo regular polymers, affinity chromatography, crystallisation, epitaxial growth and liquid crystal displays.

Molecular Sorting Rotor: A class of nanomechanical device capable of selectively binding (or releasing) molecules from (or to) solution, and of transporting these bound molecules against significant concentration gradients.

Molecular Systems Engineering: Design, analysis, and construction of systems of molecular parts working together to carry out a useful purpose.

Moledular Technology: Moledular technology promises to create a new class of imaging agents that offer distinct advantages and can be used for the early detection and diagnosis of disease. Diagnostic molecules have the potential to act as biomarkers in drug development and diagnostics, and can be used in the imaging of cancer in living subjects. This meeting will address the challenges in implementing nanotechnology for drug delivery systems and imaging agents, and promote dialogue between diagnostic and therapeutic development. The production and application of structures, devices and systems by controlling shape and size at nanometre scale. The development and use of techniques to study physical phenomena and construct structures in the nanoscale size range or smaller. The creation of functional materials, devices and systems through control of matter at the scale of 1 to 100 nanometers, and the exploitation of novel properties and phenomena at the same scale. Nanotechnology is emerging as a field critical for enabling essential breakthroughs that may have tremendous potential for affecting biomedicine. Moreover, nanotechnologies developed in the next several years may well form the foundation of significant commercial platforms. Emerging as a new field enabling the creation and application of materials, devices, and systems at atomic and molecular levels and the exploitation of novel properties that emerge at the nanometer scale. Many areas of biomedicine are expected to benefit from nanotechnology including sensors for use in the laboratory, the clinic, and within the human body; new formulations and routes for drug delivery; and biocompatible, high- performance materials for use in implants.

Moledulcar Manufacturing: The automated building of products from the bottom up, molecule by molecule, with atomic precision. This will make products that are extremely lightweight, flexible, durable, and potentially very 'smart'.

Monkeywrenching: In medical nanorobotics, the mechanical or chemical jamming of cellular equilibrium processes, with a cytocidal objective.

Monoclonal Antibodies: Produced by injecting animals to elicit a response from lymphocytes to produce antibodies. Lymphocytes which produce antibodies with strong binding capability can be isolated and used to produce only one kind of antibody (monoclonal) on a permanent basis once the lymphocytes are immortalized. This is accomplished by fusing them (combining them genetically) with cancer cells which have the distinction of living indefinitely in a culture. Monoclonal antibodies can be produced repeatedly and collected for use in immunodetection.

Nano Cubic Technology: An ultra-thin layer coating that results in higher resolution for recording digital data, ultra-low noise and high signal-to-noise ratios that are ideal for magneto-resistive (MR) heads. It is capable of catapulting data cartridge and digital videotape to one-terabyte native (uncompressed) capacities and floppy disk capacities to three gigabytes. To help visualize the potential, 1TB can store up to 200 two-hour movies.

Nanoarray: An ultra-sensitve, ultra-miniaturized array for biomolecular analysis. BioForce Nanosciences' Nanoarrays utilize approximately 1/10,000th of the surface area occupied by a conventional microarray, and over 1500 nanoarray spots can be placed in the area occupied by a single microarray domain.

Nanoassembler: The Holy Grail of nanotechnology; once a perfected nanoassembler is availble, building anything becomes possible, with physics and the imagination the only limitation (of course each item would have to be designed first, which is another small hurdle).

Nanobubbles: Nanobubbles are tiny gas bubbles that have been found on some surfaces in liquids. Typically composed of atmospheric gases, they occur on hydrophobic (water repellent) surfaces in water. The existence of nanobubbles has also been inferred in other solvents, (in this case the surfaces are called lyophobic). The nanobubbles are from 10-100nm in diameter, and, although they appear to be forbidden by thermodynamics, their lifetime is at least of the order of hours. They likely arise by nucleation of gas at the surface from a supersaturated solution. Nanobubbles appear to be responsible for the long-range attraction that occurs between hydrophobic colloid particles and surfaces. Nanobubbles bridge apposite surfaces and draw the particles into contact, leading to a capillary adhesion that makes aggregation of such particles practically irreversible.

Nanocantilever: The simplest micro-electro-mechanical system (MEMS) that can be easily machined and mass-produced via the same techniques used to make computer chips. The ability to detect extremely small displacements make nanocantilever beams an ideal device for detecting extremely small forces, stresses and masses. Nanocantilevers coated with antibodies, for example, will bend from the mass added when substrate binds to its antibody, providing a detector capable of sensing the presence of single molecules of clinical importance.

Nanocentrifuge: In medical nanorobotics, a proposed nanodevice that can spin materials at very high speed, imparting rotational accelerations of up to one trillion gravities (g's), thus permitting rapid sortation.

Nanochemistry: The goal of the Nanochemistry group is to develop chemistry beyond the realm of individual molecules by exploring how systems of several molecules or other electronically confined entities self-organise into distinct structures on the nanometer scale. The new insight is used to study and manipulate biological nanosystems and to create novel nanoarchitectures with unique structural, electronic, optical and magnetic properties. Specific research areas include bionanotechnology and self-assembling molecular electronics.

Nanochip: Nanochip is developing the next generation of removable, portable memory chips for consumer, computer, and server applications. Using MEMs technology along with arrays of atomic force probe tips, Nanochip is developing storage chips that will store tens of Gigabytes per chip. Our technology is not bound by lithography constraints so we will be able to offer storage chip products through our customers that will exceed 100 GB per chip set in the beginning and grow to Terabytes per chip set in the future. Nanochip is an integrated circuit (IC) that is so small, in physical terms, that individual particles of matter play major roles. Miniaturisation of electronic and computer components has always been a primary goal.

Nanoclusters: A nanocluster or *nanocrystal* is a fragment of solid comprising somewhere between a few atoms to a few tens of thousands of atoms. Nanoclusters are therefore a novel state of matter with properties that are neither those of a bulk crystal nor those of individual atoms and molecules. Over the past 10 years huge advances have been made both in the synthesis of size- tunable, monodisperse (i.e. size- selected) nanoclusters of various chemical compositions and in the development of techniques for their assembly into nanostructured solids (facilitating the synthesis of what have been termed "designer materials").

Nanocones: Nonplanar graphitic structures. Carbon-based structures with five-fold symmetry that form due to disclination defects in two-dimensional graphene sheets. They have been observed as nanotube caps and as freestanding structures.

Nanocrystals: A nanocrystal is a crystalline material with dimensions measured in nanometers; a nanoparticle with a structure that is mostly crystalline. These materials are of huge technological interest since many of their electrical and thermodynamic properties show strong size dependence, and can therefore be controlled through careful manufacturing processes. Nanocrystals are also of interest since they often provide single domain crystalline systems that can be studied to provide information that can help explain the behaviour of macroscopic samples of similar materials, without the complicating presence of grain boundaries and other defects. Semiconductor nanocrystals in the sub 10nm size range are often referred to as quantum dots. Nanocrystals made with zeolite are used as a filter to turn crude oil onto diesel fuel at an ExxonMobil oil refinery in Louisiana, a method cheaper than the conventional way. A layer of nanocrystals is applied to new type of solar panel named SolarPly made by Nanosolar. It is cheaper than other solar panels, more flexible, and claims 12% efficiency. (Conventional solar panels convert 9% of the sun's energy into electricity.) Crystal tetrapods 40 nanometers wide convert photons into electricity, but only have 3% efficiency.

Nanocrystalline Silicon: Nanocrystalline silicon (nc-Si) - an allotropic form of silicon - is similar to amorphous silicon (a-Si), in that it has an amorphous phase. Where they differ, however, is that nc-Si has small grains of crystalline silicon within the amorphous phase. This is in contrast to polycrystalline silicon (poly-Si) which consists solely of crystalline silicon grains, separated by grain boundaries. nc-Si is sometimes also known as microcrystalline silicon (μc-Si). The difference comes solely from the grain size of the crystalline grains. Most materials with grains in the micrometre range are actually fine-grained polysilicon, so nanocrystalline silicon is a better term. nc-Si has many useful advantages over a-Si, one being that if grown properly it can have a higher mobility, due to the presence of the silicon crystallites. It also shows increased absorption in the red and infrared wavelengths, which make it an important material for use in a-Si solar cells. One of the most important advantages of nanocrystalline silicon, however, is that it has increased stability over a-Si, one of the reasons being because of its lower hydrogen concentration. Although it currently cannot attain the mobility that poly-Si can, it has the advantage over poly-Si that it is easier to fabricate, as it can be deposited using conventional low temperature a-Si deposition techniques, such as PECVD, as opposed to laser annealing or high temperature CVD processes, in the case of poly-Si.

Nanodevices: Scientists believe nanoscale devices may lead to computer chips with billions of transistors, instead of millions—which is the typical range in today's semiconductor technology. The more transistors crammed on a chip, the more powerful it is. "This technology has the potential to replace existing manufacturing methods for integrated circuits, which may reach their practical limits within the next decade when Moore's Law eventually hits a brick wall," said physicist Bernard Yurke of Bell Labs. DNA, which provides the molecular blueprints for all living cells, is an ideal tool for making nanoscale devices. "We took advantage of how pieces of DNA—with its billions of possible variations—lock together in only one particular way, like pieces of a jigsaw puzzle," Yurke said.

Nanodots: Nanodots, also known as quantum dots, consist of 100s-1000s of atoms of inorganic semiconductor nanoparticles and are approximately one billionth of a meter in size. Developed in the mid-1980s for optoelectronic applications, they have interesting structural, electronic, and optical properties—they strongly absorb light in the near UV range and re-emit visible light that has its color determined by both the nanodot size and surface chemistry. And as the size of nanodots can be controlled during synthesis with nanoscale precision, so the optical properties can be manipulated. In addition, nanodots have a longer life than organic fluorophores, and have a broad excitation spectrum. These factors combined make the use of quantum dots as light-emitting phosphors a strong candidate for a major application of nanotechnology in the future.

Nanofluidics: Controlling nano-scale amounts of fluids Researchers at Cornell University are using nanotechnology to build microscopic silicon devices with features comparable in size to DNA, proteins and other biological molecules — to count molecules, analyze them, separate them, perhaps even work with them one at a time.

Nano-genomics: The main objective of the group is to develop experimental tools that will accelerate the finding of genomic factors (e.g. genes, polymorphisms, methylation patterns) that contribute to complex traits. In the long term this may lead to predictive medicine based on personalized genomic information. Currently available molecular tools are too slow, cumbersome and expensive to allow comprehensive genomic information to be easily obtained from a large number of individuals in a population (e.g. for case/

control studies). A radical re-invention of the way genes and DNA are analyzed is needed. We believe that the analysis of single DNA molecules can overcome many of the present limitations and in addition will reveal information which is inaccessible by conventional methods.

Nanoguitar: Smallest guitar, about the size of a human blood cell, carved out of crystalline silicon, illustrates new technology for nanosized electromechanical devices. The world's smallest guitar is 10 micrometers long — about the size of a single cell — with six strings each about 50 nanometers, or 100 atoms, wide. Just one of several structures that Cornell researchers believe are the world's smallest silicon mechanical devices. Researchers made these devices at the Cornell Nanofabrication Facility, bringing microelectro-mechanical devices, or MEMS, to a new, even smaller scale — the nano-sized world."

Nanoharvesting: Nanoharvesting agents are being designed to act as molecular mops for these biomarkers and can be directly queried by mass spectrometry. Dr. Emanuel F. Petricoin, Co-Director, NCI-FDA Clinical Proteomics Programme, Senior Principal Investigator, Center for Biologics Evaluation and Research, U. S. Food and Drug Administration.

Nanohorns: One of the single walled carbon nanotube types, with an irregular horn-like shape, which may be a critical component of a new generation of fuel cells. "The main characteristic of the carbon nanohorns is that when many of the nanohorns group together an aggregate (a secondary particle) of about 100 nanometers is created. The advantage being, that when used as an electrode for a fuel cell, not only is the surface area extremely large, but also, it is easy for the gas and liquid to permeate to the inside. In addition, compared with normal nanotubes, because the nanohorns are easily prepared with high purity it is expected to become a low-cost raw material."

Nanoindentation: Nanoindentation is similar to conventional hardness testing performed on a much smaller scale. The force required to press a sharp diamond indenter into a material is measured as a function of indentation depth. As depth resolution is on the scale of nanometers (hence the name of the instrument), it is possible to conduct indentation experiments even on thin films. Two quantities which can be readily extracted from nanoindentation experiments are the material's modulus, or stiffness, and its hardness, which can be correlated to yield strength. Investegators have also used nanoindentation to study creep, plastic flow, and fracture of materials.

Nanomachining: Like traditional machining, where portions of the structure are removed or modified, nanomachining involves changing the structure of nano-scale materials or molecules.

Nanomedicine: Nanomedicine is the medical application of nanotechnology and related research. It covers areas such as nanoparticle drug delivery and possible future applications of molecular nanotechnology (MNT) and nanovaccinology. Current problems for nanomedicine involve understanding the issues related to toxicity and environmental impact of nanoscale materials. Direct funding for nanomedicine projects has begun, and the US National Institute of Health received funding in 2005 to set up four nanomedicine centres. In April 2006, the journal Nature Materials estimated that 130 nanotech-based drugs and delivery systems were being developed worldwide.

Nanonewtons: Forces 1 billion times smaller than the force required to hold an apple against Earth's gravity. Nanonewton forces are estimated with atomic force microscopes and instruments that measure the properties of ultrathin coatings like those used on computer hard drives or turbine blades. Nanotechnology: Cracking the nanonewton force barrier NIST, US, 2003.

Nanopens & Nanopencils: (AKA: Atomic Pencil) *"Analogous to using a quill pen but on a billionth the scale"*, and may transform dip-pen nanolithography. Allows for drawing electronic circuits a thousand times smaller than current ones. The "pen" is an atomic force microscope (AFM).

Nanopipettes: "Cantilevered/Straight Nanopipettes can be used as *nanopens* for controlled chemical delivery or removal from regions as small as 100 nanometers. They can also be used as vessels for containing molecules whose optical properties change in response to their chemical environment." Other uses include "controlled chemical etching with the precision of atomic force microscopy; chemical imaging of surfaces; delivering femtosecond laser pulses; and performing NSOM/SNOM imaging using a UV excimer laser."

Nanoplotter: A multi-tip nanopen. "A device that can draw patterns of tiny lines just 30 molecules thick and a single molecule high. ... produces eight identical patterns at once and extends ... dip-pen nanolithography towards mass producing nanoscale devices and circuits by converting what was a serial process to a parallel one. May be use to "... miniaturize electronic circuits, pattern precise arrays of organic and biomolecules such as DNA and put thousands of different medical sensors on an area much tinier than the head of a pin."

Nanopore Technology: Nanopore technology is an elegant concept. A membrane with very small channels (a few nanometers in diameter), called nanopores, separates two solutions. When a voltage is applied across the membrane, charged biomolecules migrate through the pores in a controlled manner....

As each nucleotide passes through the nanopore, an electronic signature is produced that can be used to characterize it. The nanopore's size is such that it permits only a single nucleic acid strand to pass through it at any one time. Therefore the technology can be used to sequentially measure a biopolymer's properties along its length. Research in the field of *computational biology* will play a significant role in the development of nanopore technology. It will be used to help optimize experimental strategies and designs, and to create the mathematical methods needed to interpret the data generated. The potential capabilities of nanopore technology are very broad. Nanopore technology can distinguish between and count a variety of different molecules in a complex mixture. For example, nanopores could discriminate between hybridized or unhybridized unknown RNA and DNA molecules that differ by a single nucleotide only. Compared to existing techniques, nanopore technology is expected to provide direct characterisation of individual nucleic acid and protein molecules directly derived from biological samples, thereby making it applicable to a wide set of analyses.

Nanoplumbing: Illustrations of projects that would be considered relevant to this PA. o Nanoplumbing components such as valves, microfluidic channels, and motors (e.g., to be used as pumps).

Nanopores: Involves squeezing a DNA sequence between two oppositely charged fluid reservoirs, separated by an extremely small channel. Essentially itty bitty tiny holes. Nanoscopic pores found in purpose-built filters, sensors, or diffraction gratings to make them function better. As activated carbon, they may also be used as an alternative fuel storage medium, due to their massive internal surface area. "Scientists believe nanopores, tiny holes that allow DNA to pass through one strand at a time, will make DNA sequencing more efficient."

Nanorobotics: Nanorobotics is the technology of creating machines or robots at or close to the scale of a nanometres (10^{-9} metres). More specifically, nanorobotics refers to the still largely hypothetical nanotechnology engineering discipline of designing and building nanorobots. Nanorobots (nanobots, nanoids or nanites) would be typically devices ranging in size from 0.1-10 micrometers and constructed of nanoscale or molecular components. As no artificial non-biological nanorobots have so far been created, they remain a hypothetical concept at this time.

Nanoscience: The prefix "nano" originates from the Greek word for dwarf, and thus refers to something small. As a prefix for a unit of time or length, it means one billionth of that unit. Thus, a nanometer (nm) is 10-9 meter. The dot over this letter "i" is approximately one million nanometers in diameter.

The fundamental building blocks of nature, atoms and molecules, have dimensions in the nanometer domain (i.e., the nanoscale). Many water molecules can easily occupy a sphere 1 nm in diameter. The DNA double helix is approximately 2 nm wide. The way molecules, for example, assemble into larger, supramolecular entities on the nanoscale determines important material properties (e.g., electrical, optical, and mechanical properties). Of course, it has long been recognized that nature performs this assembly very well in the creation of the sophisticated molecular machinery that supports our life on earth. In short, by controlling structure on the scale of ~ 1 – 100 nm, one can, in principle, ultimately design new materials with specific properties. Scientists have long imagined the possibility of manipulating individual atoms and molecules. Nanoscience is primarily the extension of existing sciences into the realms of the extremely small (nanomaterials, nanochemistry, nanobio, nanophysics, etc.) while nanoengineering represents the extension of the engineering fields into the nanoscale realm (nanofabrication, nanodevices, etc.).

Nanoshell: A nanoparticle composed of a metallic shell surrounding a semiconductor. When nanoshells reach a target cancer cell, they can be irradiated with near-infrared light or excited with a magnetic field, either of which will cause the nanoshell to become hot, killing the cancer cell.

Nanospheres: Self-assembling nanospheres that fit inside each other like Russian dolls are one form of a broad range of submicroscopic spheres ... The durable silica spheres, which range in size from 2 to 50 nanometers, form in a few seconds, are small enough to be introduced into the body, and have uniform pores that could enable controlled release of drugs. The spheres can absorb organic and inorganic substances including small particles of iron, which means they can be controlled by magnets and the contents released as needed.

Nanosprings: A nanowire wrapped into a helix. Speculation is that they "may someday make highly sensitive magnetic field detectors, perhaps finding application in hard drive read heads. Alternatively, nanosprings could serve as positioners, or even as tiny conventional springs, for nanomachines of the future."

Nanostructures: Nanometer sized objects. Nanostructures may be considered as small, familiar, or large, depending on the view point of the disciplines concerned. To chemists, nanostructures are molecular assemblies of atoms numbering from 10^3 to 10^9 and of molecular weights of 10^4 to 10^{10} Daltons. Thus, they are chemically large supramolecules. To molecular biologists, nanostructures have the size of familiar objects from proteins to viruses and cellular organelles. But to material scientists and electrical engineers, nanostructures are the current limit of microfabrication and thus are rather small. Nanostructures are complex systems which evidently lie at the interface between solid- state physics, supramolecular chemistry, and molecular biology.

Nanosystem: A eutactic set of nanoscale components working together to serve a set of purposes; complex nanosystems can be of macroscopic size.

Nanotubes: Nanometer-sized tubes composed of various substances including carbon, boron nitride, or nickel vanadate. A one dimensional *fullerene* (a convex cage of atoms with only hexagonal and/ or pentagonal faces) with a cylindrical shape. Carbon nanotubes discovered in 1991 by Sumio Iijima resemble rolled up graphite, although they can not really be made that way. Depending on the direction that the tubes appear to have been rolled (quantified by the 'chiral vector'), they are known to act as conductors or semiconductors. Nanotubes are a proving to be useful as molecular components for nanotechnology.

Nanowire: A nanometer-scale wire made of metal atoms, silicon, or other materials that conduct electricity. Nanowires are built atom by atom on a solid surface, often as part of a microfluidic device. They can be coated with molecules such as antibodies that will bind to proteins and other substances of interest to researchers and clinicians. By the very nature of their nanoscale size, nanowires are incredibly sensitive to such binding events and respond by altering the electrical current flowing through them, and thus can form the basis of ultra sensitive molecular detectors.

National Nanotechnology Initiative: US federal government agencies participating include the National Science Foundation, the Department of Defense, the National Institute of Health, NASA, and NIST.

Navigation Satellite: A type of satellite that gives ships and aircraft their coordinate positions on the Earth. Navigation satellites were developed in the 1950's, and they rely on the *doppler effect* to calculate the position of vessels emitting a radio signal. Navigation satellites are also widely used by the military.

Navstar: A series of American *navigation satellites*. The first satellite in the Navstar system was launched in 1978. Currently, there are 24 orbiting Navstar satellites. The three-dimensional satellite navigational system of Navstar enables a traveller to find out his or her position anywhere on or above the planet.

Neophile: The trait of being excited and pleased by novelty. Common among most hackers, SF fans, and members of several other connected leading-edge subcultures, including the pro-technology 'Whole Earth' wing of the ecology movement, space activists, many members of Mensa, and the Discordian/neo-pagan underground. All these groups overlap heavily and (where evidence is available) seem to share characteristic hacker tropisms for science fiction, music, and oriental food. The opposite tendency is 'neophobia'.

Omega Point: Also called the Quantum Omega Point Theory. A possible future state when intelligence controls the Universe totally, and the amount of information processed and stored goes asymptotically towards infinity.

Organic Moledule: A molecule containing carbon; the complex molecules in living systems are all organic molecules in this sense.

Organotypic: Despite their wide use, the physiological relevance of organotypic slices remains controversial. Such cultures are prepared at 5 days postnatal. Although some local circuitry remains intact, they develop subsequently in isolation from the animal and hence without plasticity due to experience. Development of synaptic connectivity and morphology might be expected to proceed differently under these conditions than in a behaving animal.

Outmessaging: In medical nanorobotics, conveyance of information from a transmitter located inside the human body, especially from working nanodevices, to the patient or to a recipient external to the human body.

Overcoming Cytoxicity: Studies have shown that quantum dot toxicity is due to cadmium ions being released into cells. This is exacerbated by oxygen, UV exposure and the large exposed surfaces of the spherical quantum dots. To avoid these problems, surfaces are added to the quantum dots, traditionally consisting of Zinc-Sulfide compounds to create a more benign surface. Polymers are used as well, which have been shown to be stable, but the polymers themselves may kill cells.

Phagocytosis: Process by which particulate material is endocytosed ("eaten") by a cell. Prominent in carnivorous cells, such as *Amoeba proteus*, and in vertebrate macrophages and neutrophils. (From Greek *phagein*, to eat.)

Pharmacodelivery: Site-directed pharmacodelivery is a desirable but elusive goal. Endothelium and epithelium create formidable barriers to endogenous molecules as well as targeted therapies *in vivo*.

Pharmacophore Generation: A procedure to extract the most important common structural features relevant for a given biological activity from a series of molecules with a similar mechanism of action.

Phonon: A particle of sound. The energy E of a phonon is given by the Einstein relation, $E = hf$. Here f is the frequency of the sound and h is Planck's constant. The momentum p of a photon is given by the de Broglie relation, $p = h/ë$. Here ë is the wavelength of the sound. A quantum of acoustic energy, analogous to the quantum of electromagnetic radiation, the photon. Thermal excitations in a crystal or in an elastic continuum can be described as a population of phonons (analogous to blackbody electromagnetic radiation). In highly inhomogeneous solids, a description in terms of phonons breaks down and localized vibrational modes become important.

Photon: A particle of light. The energy E of a photon is given by the Einstein relation, $E = hf$. Here f is the frequency of the light and h is Planck's constant. The momentum p of a photon is given by the de Broglie relation, $p = h/ë$. Here ë is the wavelength of the light.

Photonic Crystal: Composite material made as periodical arrangement of pieces of material (e.g. spheres) embedded in a background matrix of another material that has very different (much lower) dielectric constant and/or magnetic permittivity (electromagnetic properties of material). Such structures do not allow for propagation of light of specific frequencies, due to a destructive interference of photons of such frequencies (this is a wave phenomenon also known as photonic band gap). Photonic crystals offer a possibility of manipulating the emission, propagation and splitting of electromagnetic waves which is expected to be useful for technologies of all-photon telecommunications. The production of such materials relies on nanotechnological methods (e.g. lithography, self assembly approaches...).

Piezoelectric Material: A ferroelectric material in which an electrical potential difference is created due to mechanical deformation, or conversely, in which the application of a voltage causes dimensional changes in the material.

Pinhole: The term pinhole embraces a wide variety of oxide defects and is used in a broad sense today. Listed in this category are cracks caused by thermal contraction after oxidation or by handling, and regions of oxide with low dielectric strength caused by dust particles, inadequate masking, contamination, or poor resist adhesion.

Polar Orbit: Usually has an angle of inclination of 90 degrees to the equator. On every pass around the Earth, it passes over both the north and south poles. Therefore, as the Earth rotates to the east underneath the satellite which is travelling north and south, it can cover the entire Earth's surface. A polar orbiting satellite covers the entire globe every 14 days.

Polar: Satellite launched on February 24, 1996 by NASA in the Global Geospace Science project. Polar is an atmospheric studies satellite in polar orbit. One purpose of Polar is to gather information that will help scientists protect future satellites from radiation and other atmospheric dangers.

Polymeric Materials: Materials composed of large molecules, generally based on carbon, that have been formed from the chemical bonding of smaller units (monomers).

Polymerisation: A chemical reaction resulting in the bonding together of small molecular units (monomers) to form a much larger molecule (polymer). There are many different routes, or reaction mechanisms, for polymerisation.

Polymers-biomedical: More and more therapeutic problems are relevant to the use of polymer- based therapeutic aids for a limited period of time, namely the healing time related to the outstanding capacity of living systems to self-repair, e.g. bone fracture fixation with screws and plates, of wound closure by sutures and also of drug delivery from implants or similar systems based on polymeric matrices, or on aqueous dispersions or solutions of polymers.

Posthuman: Person's of unprecedented physical, intellectual, and psychological capacity, self-programming, self-constituting, potentially immortal, unlimited individuals.

Quantum Cryptography: A system based on quantum- mechanical principles. Eavesdroppers alter the quantum state of the system and so are detected. Developed by Brassard and Bennett, only small laboratory demonstrations have been made.

Quantum Dot Nanocrystals: Quantum Dot Nanocrystals is Used to tag biological molecules, and "measuring between five and ten nanometres across, are made up of three components. Their cores contain paired clusters of atoms such as cadmium and selenium that combine to create a semiconductor. This releases light of a specific colour when stimulated by ultraviolet of a wide range of frequencies. These clusters are surrounded by a shell made of an inorganic substance, to protect them. The whole thing is then coated with an organic surface, to allow the attachment of proteins or DNA molecules. By varying the number of atoms in the core, QDNs can be made to emit light of different colours."

Quantum Mechanics: A largely computational physical theory explaining the behaviour of quantum phenomena, which incorporates the theory of special relativity. Despite dilignet attempts, general relativity has not been sucessfully incorporated into quantum mechanics. A physical model of chemical and optical phenomena, as well as the behaviour of matter in general on a small scale. Quantum mechanics describes a system of particles in terms of a wave function defined over the configuration space of the system. Although the concept of particles having distinct locations is implicit in the potential energy function that determines the wave function (e.g., of a ground-state system), the observable dynamics of the system cannot be described in terms of the motion of such particles from point to point. In describing the energies, distributions, and behaviours of electrons in nanometer-scale structures, quantum mechanical methods are necessary. Electron wave functions help determine the potential energy surface of a molecular system, which in turn is the basis for classical descriptions of molecular motion. Nanomechanical systems can almost always be described in terms of classical mechanics, with occasional quantum mechanical corrections applied within the framework of a classical model.

Quantum Physics: Describes fundamental electronic and optical properties of matter at microscopic level and wave and interference phenomena in particular. Quantum electronics, quantum optics and optoelectronics are important areas of application—atomic clocks, lasers, light emitting diodes, optical fibers, tunnel diodes and superconducting systems are important and well-known examples. The trend towards faster and more complicated microprocessors and microelectronics has resulted in electronic components now approaching the domains of quantum physics at research level. In about 20 years time, miniaturisation will also be halted in commercial applications. It will then probably utilize electronic wave phenomena and single- electron effects in semiconductor components and systems and also to create more complicated transistor components connected in more complicated ways where quantum phenomena, cooperative phenomena and even superconductivity may be important for function. "QNANO"—*quantum nanophysics* - will probably provide a target for research and development in semiconductor physics, molecular electronics and bioelectronics in the foreseeable future.

Quantum Teleportation: 'Teleportation' (or copying) of the quantum state from one particle to another. Not really spectacular as in Star Trek serial where macroscopic objects (humans, weapons and similar) are 'teleported'. Present technology can 'teleport' quantum objects such as photons, electrons ans similar. It is important to note that the object itself is not teleported during this process. What actually happens is that the quantum state (e.g. spin number) is exactly copied from one object to another, both spatially and temporally present *prior* to 'teleportation'. The technique used is based on the Einstein-Podolsky-Rosen effect (EPR) or paradox, which relies on the so-called nonlocality of the wavefunction. An EPR source produces pair of entangled particles, one of which serves as a 'carbon copy'. The third particle (original, source particle to be copied, introduced besides the EPR pair) is 'entangled' with one of the particles from the EPR pair which enables the copying of its state to another EPR particle that has never interacted with the original, yet it becomes its copy. The experimental setup requires one measuring apparatus that entangles the original with one of the EPR particles and another apparatus that acts on another EPR particle which becomes the copy. The experimentalists operating the two apparatuses must be in contact i.e. the one that acts on the copy particle must know the results of the entanglement measurement on the source (original) particle and another EPR particle which is 'waisted' in the process. It should also be noted here that none of the experimentalists does not know the quantum state of the original particle, even when the whole process finishes, neither is it measured during the process.

Rational Drug Design: The input of biocomputing in drug discovery is twofold: firstly the computer may help to optimise the pharmacological profile of existing drugs by guiding the synthesis of new and "better" compounds. Secondly, as more and more structural information on possible protein targets and their biochemical role in the cell becomes available, completely new therapeutic concepts can be developed. The computer helps in both steps: to find out about possible biological functions of a protein by comparing its amino acid sequence to databases of proteins with known function, and to understand the molecular workings of a given protein structure. Understanding the biological or biochemical mechanism of a disease then often suggests the types of molecules needed for new drugs

Reactant: A substance that is consumed in the course of a chemical reaction. It is sometimes known, especially in the older literature, as a reagent, but this term is better used in a more specialized sense as a test substance that is added to a system in order to bring about a reaction or to see whether a reaction occurs (e.g. an analytical reagent).

Reagent Device: A large reagent structure (or a large structure that binds a smaller reagent) serving as a component of a mechanochemical system. A reagent device exists chiefly to hold, position, and manipulate the environment of a reagent moiety.

Recessive: Refers to the member of a pair of alleles that fails to be expressed in the phenotype of the organism when the dominant member is present. Also refers to the phenotype of an individual that has only the recessive allele.

Recombinant DNA: Any DNA molecule formed by joining DNA segments from different sources. Recombinant DNAs are widely used in the cloning of genes, in the genetic modification of organisms, and in molecular biology generally.

Recombination: Process in which chromosomes or DNA molecules are broken and the fragments are rejoined in new combinations. Can occur in the living cell—for example, through crossing-over during meiosis—or in the test tube using purified DNA and enzymes that break and ligate DNA strands.

Reconnaisance Satellite: Also called a spy satellite because it is used to spy on other countries. It can provide intelligence information on military activities, detect missile launches or nuclear explosions, and pick up and record radio and radar transmissions while passing over a country. It can also be used as an orbital weapon by placing warheads on a low orbit satellite to be launched at a ground target, but this is not a recommended or frequent use of satellites.

Remote Sensing Satellite: A type of satellite that performs remote sensing from space. Remote sensing satellites generally monitor important resources for humans. For example, they might be used to track animal migration, locate mineral deposits, watch agricultural crops for weather damage, or see how fast the forests are being cut down. Because they are in space, remote sensing satellites are ideal for monitoring areas with harsh climates or difficult terrains.

Reproducibility: A property of an experiment or process where one tends to receive consistent results from following a specific procedure.

Resolution: The smallest measurable change in input that will produce a small but noticeable change in the output. In the context of chemical separations, defines the completeness of separation.

Resonant Frequency: The frequency at which a moving member or a circuit has a maximum output for a given input.

Respirocrit: In medical nanorobotics, the volume-fraction or bloodstream concentration of respirocyte nanorobots, expressed as a percentage.

Saturated: An organic molecule is described as saturated if it is a closed shell species lacking double or triple bonds; forming a new bond to a saturated molecule requires the cleavage of an existing bond.

Sealed Assembler Laboratory: A work space, containing assemblers, encapsulated in a way that allows information to flow in and out but does not allow the escape of assemblers or their products.

Secretory Vesicle: Membrane-bounded organelle in which molecules destined for secretion are stored prior to release. Sometimes called secretory granule because darkly staining contents make the organelle visible as a small solid object.

Self-Assembler: A specific type of assembler that makes use of self-assembly such that the assembly process would theoretically not require external energy or information input.

Sensors: An equipment which detects, and may indicate, and/or record objects and activities by means of energy or particles emitted, reflected, or modified by objects.

Signal to Noise: Ratio which can interfere with detection. Can also refer to data analysis. Biological data is often very "noisy". This is particularly seen when trying to look at low abundance biomolecules.

Single Crystal: A crystalline solid for which the periodic and repeated atomic pattern extends throughout its entirety without interruption.

Sky Hook: A long, very strong, cable in orbit around a planet which rotates around its center of mass in such a way that when one end is closest to the ground, its relative velocity is almost zero. It would function as a kind of space elevator; shuttle craft would anchor to the end and then be lifted into orbit where they would be released. It is closely related to the idea of a beanstalk.

Slip Casting: A forming technique used to shape ceramic materials. A slip or suspension of solid particles in water is poured into a porous mold. A solid layer forms on the inside wall as water is absorbed by the mold, leaving a shell (or a solid piece) in the shape of the mold.

Small Molecule Therapeutics: Low molecular-weight drug. Compared to larger molecular weight pharmaceuticals such as proteins, peptides, and carbohydrates, small molecules can more easily penetrate cell membranes and the blood brain barrier. Can be delivered orally or intravenously. These molecules tend to incur lower process development and manufacturing costs.

Smart Materials and Products: Here, materials and products capable of relatively complex behaviour due to the incorporation of nanocomputers and nanomachines. Also used for products having some ability to respond to the environment.

Spinning: Fiber forming process: a multitude of fibers are spun as molten material is forced through many small orifices.

Step Coverage: If a layer of material is to be deposited on top of a thin strip of material, it is the ratio of film thickness along the walls of the step to the film thickness at the bottom of the step. LPCVD and PECVD offer excellent step coverage.

Strain: In mechanical engineering, strain is a measure of the deformation resulting from stress (that is, force per unit area); the displacement of one point with respect to another, divided by their equilibrium separation in the absence of stress. In chemistry, a molecular fragment generally has some equilibrium geometry (bond lengths, interbond angles, etc.) when the rest of the molecular structure does not impose special constraints (e.g., bending bonds to form a small ring). Deviations from this equilibrium geometry are described as strain, and increase the energy of the molecule. Strain in the mechanical engineering sense causes strain in the chemical sense.

Stress: Force per unit area applied by one part of an object to another. Pressure is an isotropic compressive stress. Suspending a mass from a fiber places it in tensile stress. Gluing a layer of rubber between two plates and then sliding one over the other (while holding their separation constant) places the rubber in shear stress.

Structure Activity Relationship (SAR): The relationship between chemical structure and pharmacological activity for a series of compounds. Compounds are often classed together because they have structural characteristics in common including shape, size, stereochemical arrangement, and distribution of functional groups. Other factors contributing to structure- activity relationship include chemical reactivity, electronic effects, resonance, and inductive effects.

Thermal Blanketing: Like insulation in a house, the covering on a satellite that regulates temperature. A material commonly used for thermal blanketing on a satellite is Mylar.

Thermal Noise: The vibration and motion of atoms and molecules caused by the fact that they have a temperature above absolute zero. Once used as an argument on why MNT could not work.

Thermocouple: A temperature-measuring device, which contains a pair of end-joined dissimilar conductors in which an electromotive force is developed by thermoelectric effects when the joined ends and the free ends of the conductors are a different temp.

Thermogenic Limit: In medical nanorobotics, the maximum amount of waste heat that may safely be released by a population of in vivo medical nanorobots that are operating within a given tissue volume.

Thermoplastic Polymer: A substance that when molded to a certain shape under appropriate conditions can later be remelted.

Thioester Bond: High-energy bond formed by a condensation reaction between an acid (acyl) group and a thiol group (-SH); seen, for example, in acetyl CoA and in many enzyme-substrate complexes.

Three-axis Stabilized: Done with an internal gyroscope and thrusters. A gyroscope's stable spin can be used as a sensor to tell the satellite when its attitude is changing. The satellite can then correct the problem using thrusters. This is one form of attitude control.

Thrombin: The enzyme that converts *fibrinogen* to *fibrin* to form a blood clot. Thrombin circulates in the inactive form, prothrombin.

Thruster: A way of controlling a satellite's attitude. Thrusters usually contain compressed gas that when sent out of the end of the thruster will move the satellite in space. The force of the compressed gas (the action) causes the satellite to move in the opposite direction (the reaction). This is Newton's third law of motion—for every action there is an equal and opposite reaction.

Titanium Nanopositioner Stages: If the positioning accuracy must be maintained over a prolonged time and temperatures cannot be maintained accurately, use of alternative materials may provide the necessary positioning stability. Titanium's coefficient of thermal expansion (0.9 x 10^{-5} mm/mm per °C) represents a significant improvement over aluminum alloys and offers the additional benefit of higher strength –improving nanopositioner performance.

Transhuman: Someone actively preparing for becoming posthuman. Someone who is informed enough to see radical future possibilities and plans ahead for them, and who takes every current option for self-enhancement.

Transhumanism: Philosophies of life (such as Extropianism) that seek the continuation and acceleration of the evolution of intelligent life beyond its currently human form and human limitations by means of science and technology, guided by life-promoting values.

Transition State: At the saddle point of a col linking two potential wells, the direction of maximum negative curvature defines the reaction coordinate; the transition state is a hypothetical system of reduced dimensionality, free to move only on a hypersurface perpendicular to the reaction coordinate at its point of maximum energy.

Unsaturated: Describes a molecule that contains one or more double or triple carbon-carbon bonds, such as isoprene or benzene. Possessing double or triple bonds.

Viscosity: The ratio of the magnitude of an applied shear stress to the velocity gradient that it produces; in other words: a measure of a noncrystalline material's resistance to permanent deformation.

Visible Spectrum: Visible light makes up only a small part of the *electromagnetic spectrum*. The visible light spectrum can be divided into different wavelenghts of light. The wavelength of the light determines the color of that light. The light spectrum goes from violet to red where red is the longest wavelength.

Vitrification: During firing of a ceramic body, the formation of a liquid phase that becomes a glass-bonding matrix upon cooling.

1

Basics of Nano Physics and Nanotechnology: An Introduction

Understanding Molecular Nanotechnology, Quantum Dots & Lasers and Nanostructured Materials

Thorough, inexpensive control of the structure of matter based on molecule-by-molecule control of products and byproducts; the products and processes of molecular manufacturing. Technology as we know it is a product of industry, of manufacturing and chemical engineering. Industry as we know it takes things from nature—ore from mountains, trees from forests—and coerces them into forms that someone considers useful. Trees become lumber, then houses. Mountains become rubble, then molten iron, then steel, then cars. Sand becomes a purified gas, then silicon, then chips. And so it goes. Each process is crude, based on cutting, stirring, baking, spraying, etching, grinding, and the like. Trees, though, are not crude: to make wood and leaves, they neither cut, grind, stir, bake, spray, etch, nor grind. Instead, they gather solar energy using molecular electronic devices, the photosynthetic reaction centers of chloroplasts. They use that energy to drive molecular machines—active devices with moving parts of precise, molecular structure—which process carbon dioxide and water into oxygen and molecular building blocks. They use other molecular machines to join these molecular building blocks to form roots, trunks, branches, twigs, solar collectors, and more molecular machinery. Every tree makes leaves, and each leaf is more sophisticated than a spacecraft, more finely patterned than the latest chip from Silicon Valley. They do all this without noise, heat, toxic fumes, or human labor, and they consume pollutants as they go. Viewed this way, trees are high technology. Chips and rockets aren't. Trees give a hint of what molecular nanotechnology will be like, but nanotechnology won't be biotechnology because it won't rely on altering life. Biotechnology is a further stage in the domestication of living things. Like selective breeding, it reshapes the genetic heritage of a species to produce varieties more useful to people. Unlike selective breeding, it inserts new genes. Like biotechnology—or ordinary trees—molecular nanotechnology will use molecular machinery, but unlike biotechnology, it will not rely on genetic meddling.

It will be not an extension of biotechnology, but an alternative or a replacement. Molecular nanotechnology could have been conceived and analysed—though not built—based on scientific knowledge available forty years ago. Even today, as development accelerates, understanding grows slowly because molecular nanotechnology merges fields that have been strangers: the molecular sciences, working at the threshold of the quantum realm, and mechanical engineering, still mired in the grease and crudity of conventional technology. Nanotechnology will be a technology of new molecular machines, of gears and shafts and bearings that move and work with parts shaped in accord with the wave equations at the foundations of natural law. Mechanical engineers don't design molecules. Molecular scientists seldom design machines. Yet a new field will grow—is growing today—in the gap between. That field will replace both chemistry as we know it and mechanical engineering as we know it. And what is manufacturing today, or modern technology itself, but a patchwork of crude chemistry and crude machines?

Next chapters will paint a concrete picture of molecular machines and molecular manufacturing, but for now analogy will serve. Picture an automated factory, full of conveyor belts, computers, rollers, stampers, and swinging robot arms. Now imagine something like that factory, but a million times smaller and working a million times faster, with parts and workpieces of molecular size. In this factory, a "pollutant" would be a loose molecule, like a ricocheting bolt or washer, and loose molecules aren't tolerated. In many ways, the factory is utterly unlike a living cell: not fluid, flexible, adaptable, and fertile, but rigid, preprogrammed and specialised. And yet for all of that, this microscopic molecular factory emulates life in its clean, precise molecular construction.

Advanced molecular manufacturing will be able to make almost anything. Unlike crude mechanical and chemical technologies, molecular manufacturing will work from the bottom up, assembling intricate products from the molecular building blocks that underlie everything in the physical world. Nanotechnology will bring new capabilities, giving us new ways to make things, heal our bodies, and care for the environment. It will also bring unwelcome advances in weaponry and give us yet more ways to foul up the world on an enormous scale. It won't automatically solve our problems: even powerful technologies merely give us more power. As usual, we have a lot of work ahead of us and a lot of hard decisions to make if we hope to harness new developments to good ends. The main reason to pay attention to nanotechnology now, before it exists, is to get a head start on understanding it and what to do about it. Semiconductor lasers are key components in a host of widely used technological products, including compact disk players and laser printers, and they will play critical roles in optical communication schemes. The basis of laser operation depends on the creation of nonequilibrium populations of electrons and holes, and coupling of electrons and holes to an optical field, which will stimulate radiative emission. Calculations carried out in the early 1970s by C. Henry (Dingle and Henry 1976) predicted the advantages of using quantum wells as the active layer in such lasers: the carrier confinement and nature of the electronic density of states should result in more efficient devices operating at lower threshold currents than lasers with "bulk" active layers. In addition, the use of a quantum well, with discrete transition energy levels dependent on the quantum well dimensions (thickness), provides a means of "tuning" the resulting wavelength of the material. The critical feature size—in this case, the thickness of the quantum well—depends on the desired spacing between energy levels. For energy levels of greater than a few tens of millielectron volts (meV, to be compared with room temperature thermal energy of 25 meV), the critical dimension is approximately a few hundred angstroms. Although the first quantum well laser, demonstrated in 1975, was many times less efficient than a conventional laser (van der Ziel et al. 1975), the situation was reversed by 1981 through the use of new materials growth capabilities (molecular beam epitaxy), and optimisation of the heterostructure laser design (Tsang 1982).

Even greater benefits have been predicted for lasers with quantum dot active layers. Arakawa and Sakaki (1982) predicted in the early 1980s that quantum dot lasers should exhibit performance that is less temperaturedependent than existing semiconductor lasers, and that will in particular not degrade at elevated temperatures. Other benefits of quantum dot active layers include further reduction in threshold currents and an increase in differential gain—that is, more efficient laser operation (Asada et al. 1986). The key concepts in the laser operation. Stimulated recombination of electron-hole pairs takes place in the GaAs quantum well region, where the confinement of carriers and of the optical mode enhance the interaction between carriers and radiation. In particular, note the change in the electronic density of states, as a function of the "dimensionality" of the active layer.

The population inversion (creation of electrons and holes) necessary for lasing occurs more efficiently as the active layer material is scaled down from bulk (3-dimensional) to quantum dots (0-dimensional). However, the advantages in operation depend not only on the absolute size of the nanostructures in the active region, but also on the uniformity of size. A broad distribution of sizes "smears" the density of states, producing behaviour similar to that of bulk material. Thus, the challenge in realising quantum dot lasers with operation superior to that shown by quantum well lasers is that of forming high quality, uniform quantum dots in the active layer. Initially, the most widely followed approach to forming quantum dots was through electron beam lithography of suitably small featured patterns (~300 Å) and subsequent dryetch transfer of dots into the substrate material. The problem that plagued these quantum dot arrays was their exceedingly low optical efficiency: high surface-to-volume ratios of these nanostructures and associated high surface recombination rates, together with damage introduced during the fabrication itself, precluded the successful formation of a quantum dot laser. Density of electronic states as a function of structure size. With the demonstration of the high optical efficiency self-assembled formation of quantum dots, formed without need of external processing and having the natural overgrowth of cladding material (which addressed issues of surface recombination), there ensued a marked increase in quantum dot laser research. The first demonstration of a quantum dot laser with high threshold density was reported by Ledentsov and colleagues in 1994. Bimberg et al. (1996) achieved improved operation by increasing the density of the quantum dot structures, stacking successive, strain-aligned rows of quantum dots and therefore achieving vertical as well as lateral coupling of the quantum dots. In addition to utilising their quantum size effects in edge-emitting lasers, self-assembled quantum dots have also been incorporated within vertical cavity surface-emitting lasers. A partial summary of the work and achievements in quantum dot lasers.

As with the demonstration of the advantages of the quantum well laser that preceded it, the full promise of the quantum dot laser must await advances in the understanding of the materials growth and optimisation of the laser structure. Although the self-assembled dots have provided an enormous stimulus to work in this field, there remain a number of critical issues involving their growth and formation: greater uniformity of size, controllable achievement of higher quantum dot density, and closer dot-todot interaction range will further improve laser performance. Better understanding of carrier confinement dynamics and capture times, and better evaluation of loss mechanisms, will further improve device characteristics. It should be noted that the spatial localisation of carriers brought about by the quantum dot confinement may play a role in the "anomalous" optical efficiency of the GaN-based materials, which is exceptional in light of the high concentration of threading dislocations (~ 108 - 1010 cm-2) that currently plague this material system. The localisation imposed by the perhaps natural nanostructure of the GaN materials may make the dislocation largely irrelevant to the purely optical (but not to the electrical) behaviour of the material.

The first synthesis and characterisation of carbon nanotubes were reported by Iijima from NEC in late 1991. The initial theoretical study of their electronic structure was soon followed with the work by Dresselhaus and coworkers at MIT (Dresselhaus et al. 1992; Saito et al. 1992a; Saito et al. 1992b). Since then, the fabrication of nanotubes has been improved by several groups, and methods other than arc discharge have been explored. The main issues are to separate the nanotubes from other forms of carbon also produced in the fabrication process and to increase the yield of singlewalled nanotubes (SWNT) for potential applications. Following Iijima's work, macroscopic quantities of MWNT were produced with an improved arc discharge method by Ebbesen and coworkers at NEC (Tsukuba) (Ebbesen and Ajayan 1992).

In the following year, that same group produced what is considered to be among the best SWNT material generated so far; over 70% of the volume of material was nanotubes bundled together into crystalline ropes of metallic character (Thess et al. 1996). Also in 1996, a group from the Chinese Academy of Science used chemical vapor deposition to produce a 50 mm thick film of nanotubes that were highly aligned perpendicular to the surface (Li et al. 1996). Progress in recent years leads one to predict that it will indeed be possible to produce high quality carbon nanotubes in macroscopic quantities needed for many of the applications outlined below. Nanotube bundles form a low density material and are expected to have high stiffness and axial strength as a result of their seamless cylindrical graphitic structure.

It is therefore predicted that they can be used to fabricate a material with better mechanical properties than the present carbon fiber materials. Information about the mechanical properties of nanotubes has been gathered recently by a study of the thermal vibrations of a single attached to a substrate (Treacy et al. 1996). Ebbesen's group at the Princeton NEC Research Institute found that nanotubes have an exceptionally high Young's modulus (~ 2 x 109 Pa) (Treacy et al. 1996). In order to reach a better understanding of the mechanical properties and intrinsic limitations of nanotubes, Bernholc's group from North Carolina State University theoretically studied the behaviour of nanotubes beyond the linear Hooke's law and the nature of the defects leading to dislocations and fractures (Yakobson et al. 1996; Nardelli et al. 1998). Nanotubes are highly polarizable nanoscale straws, a property that confers on them the capacity to ingest inorganic elements by nanocapillarity (Pederson and Broughton 1992). As a result, it has been conjectured that they could be used as minute molds to shape nanometer-sized quantum wires and as miniature test tubes. Ajayan and coworkers at NEC (Tsukuba) have first shown that lead can be introduced into carbon nanotubes (Ajayan and Iijima 1993).

The efficiency of their process is low, and prior removal of the caps from the ends of the nanotubes is expected to improve the situation (Tsang et al. 1993). More information about the mechanism of NT filling was obtained by Pascard and coworkers from the École Polytechnique in France by studying the propensity to form nanowires for 15 encapsulated metal elements (Guerret-Plécourt et al. 1994). Finally, external decoration of nanotubes with metal atoms has been demonstrated and is predicted to have applications in catalysis (Satishkumar et al. 1996). The primary methods of nanotube fabrication and the institutions engaged in specific methods of nanotube fabrication.

Early theoretical studies already showed that the electronic properties of nanotubes strongly depend on their diameter and their chirality leading to metallic or semiconducting structures (Saito et al. 1992c). It was conjectured that these properties can be used to construct nanoscale electronic devices. While theoretical studies were promptly published, it was only in 1996 that Ebbesen and coworkers (1996) at the Princeton NEC Research Institute presented reliable four-point probe conductivity measurements on MWNT, confirming the theoretical predictions. In 1997, two groups, one at Lawrence Berkeley National Laboratory (Bockrath et al. 1997) and the second at Delft University (Tans et al. 1997) in the Netherlands showed that conductivity through nanotubes is controlled by low dimensional effects such as resonant tunneling and single-electron charging effects. Hall effect measurements at the École Polytechnique Fédérale de Lausanne (EPFL) in Switzerland have shown that hole transport is predominant in electronic conductance (Baumgartner et al. 1997). Despite these and other very recent and encouraging efforts such as those studying the mean free path of carriers in nanotubes, the conduction mechanism is still only partially understood (Petit et al. 1997).

For this reason and possibly others related to quantum confinement effects, nanotubes are expected to form outstanding field-emitting materials. In 1995, the Rice University group showed that nanotubes emit electrons very efficiently when immersed in an electric field and irradiated by a laser to remove their cap (Rinzler et al. 1995). They attribute their observation to the unraveling of an atomic wire of carbon atoms. Efficient field emission was also obtained from carefully aligned nanotubes by de Heer and coworkers (1995) at EPFL, whereas Collins and coworkers from the University of California at Berkeley (Collins and Zettl 1996, 1997), have used randomly oriented nanotubes with similar results. Efficient field-emitting material is highly desirable for the production of field-emission displays and microwave tubes. Recently, a group from Mie University in Japan has built a cathode ray tube (CRT) using nanotube field emitters (Saito et al. n.d.). In this work, the layers of nanotubes were cut out from the soot produced in an arc discharge chamber. This fabrication method is presently not compatible with industrial production requirements, and more progress must be made before this effort can be translated into an industrial product. The electrical and field emission properties of nanotubes, with the representative institutions pursuing these studies. In summary, macroscopic amounts of good quality nanotubes can presently be fabricated by several groups around the world, and the theoretical understanding of the electronic structure and related properties of nanotubes has reached a very good level.

However, despite the fact that many potential applications are mentioned in the literature over and over again, only the outstanding field emission properties of nanotubes have achieved realisation in practical devices. One of the main obstacles simply remains the controlled manipulation of nanoscale objects. In this respect it seems that the generation of self-aligned structures is a path to explore further, especially after the encouraging successes reported in the literature in 1997-98.

Bulk nanostructured materials are defined as bulk solids with nanoscale or partly nanoscale microstructures. This category of nanostructured materials has historical roots going back many decades but has a relatively recent focus due to new discoveries of unique properties of some nanoscale materials. Early in the century, when "microstructures" were revealed primarily with the optical microscope, it was recognised that refined microstructures, for example, small grain sizes, often provided attractive properties such as increased strength and toughness in structural materials. A classic example of property enhancement due to a refined microstructure—with features too small to resolve with the optical microscope—was age hardening of aluminum alloys. The phenomenon, discovered by Alfred Wilm in 1906, was essentially explained by Merica, Waltenberg, and Scott in 1919 (Mehl and Cahn 1983, 18), and the microstructural features responsible were first inferred by the X-ray studies of Guinier and Preston in 1938. With the advent of transmission electron microscopy (TEM) and sophisticated X-ray diffraction methods it is now known that the fine precipitates responsible for age hardening, in Al-4% Cu alloys, for example, are clusters of Cu atoms— Guinier-Preston (GP) zones—and the metastable partially coherent q' precipitate (Silcock et al. 1953-54; Cohen 1992). Maximum hardness is observed with a mixture of GPII (or q") (coarsened GP zones) and q' with the dimensions of the q' plates, typically about 10 nm in thickness by 100 nm in diameter.

Therefore, the important microstructural feature of age-hardened aluminum alloys is nanoscale. There are a number of other examples of nanoscale microstructures providing optimised properties. The critical current density JC of commercial superconducting Nb3Sn is controlled by grain size and is inversely proportional to grain size, with grain sizes of 50-80 nm providing high values of JC (Scanlan et al. 1975). The field of nanocrystalline (or nanostructured, or nanophase) materials as a major identifiable activity in modern materials science results to a large degree from the work in the 1980s of Gleiter and coworkers (Gleiter 1990), who synthesised ultrafine-grained materials by the in situ consolidation of nanoscale atomic clusters. The ultrasmall size (< 100 nm) of the grains in these nanocrystalline materials can result in dramatically improved—or different—properties from conventional grain-size (> 1 μm) polycrystalline or single crystal materials of the same chemical composition. This is the stimulus for the tremendous appeal of these materials. While there are a number of bulk properties that may be dramatically changed when the microstructure is nanoscale, this chapter focuses on those for which the recent work with nanostructured materials has been most extensive. These are

- the mechanical properties of nanostructured materials for a variety of potential structural applications, and
- ferromagnetic materials with nanoscale microstructures for potential applications as soft magnetic materials and permanent magnet materials,

and for other special applications such as information storage, magnetoresistance spin valves, and magnetic nanocomposite refrigerants.

Other bulk applications such as hydrogen storage are discussed briefly. The great interest in the mechanical behaviour of nanostructured materials originates from the unique mechanical properties first observed and/or predicted for the materials prepared by the gas condensation method. Among these early observations/predictions were the following:

- lower elastic moduli than for conventional grain size materials—by as much as 30 - 50%
- very high hardness and strength—hardness values for nanocrystalline pure metals (~ 10 nm grain size) are 2 to 7 times higher than those of larger grained (>1 mm) metals
- a negative Hall-Petch slope, i.e., decreasing hardness with decreasing grain size in the nanoscale grain size regime
- ductility—perhaps superplastic behaviour—at low homologous temperatures in brittle ceramics or intermetallics with nanoscale grain sizes, believed due to diffusional deformation mechanisms

While some of these early observations have been verified by subsequent studies, some have been found to be due to high porosity in the early bulk samples or to otherartifacts introduced by the processing procedures. The following summarizes the author's understanding of the state of the art of the mechanical behaviour of nanostructured materials, as determined from the literature, presentations at the U.S. workshop (Siegel et al. 1998), and the WTEC panel's site visits in Japan and Europe.

Elasticity, Hardness and Strength

Early measurements of the elastic constants on nanocrystalline (nc) materials prepared by the inert gas condensation method gave values, for example for Young's Modulus, E, that were significantly lower than values for conventional grain size materials. While various reasons were given for the lower values of E, it was suggested by Krstic and coworkers (1993) that the presence of extrinsic defects—pores and cracks, for example—was responsible for the low values of E in nc materials compacted from powders. This conclusion was based on the observation that nc NiP produced by electroplating with negligible porosity levels had an E value comparable to fully dense conventional grain size Ni (Wong et al. 1994, 85). Subsequent work on porosity-free materials has supported these conclusions, and it is now believed that the intrinsic elastic moduli of nanostructured materials are essentially the same as those for conventional grain size materials until the grain size becomes very small, e.g., < 5 nm, such that the number of atoms associated with the grain boundaries and triple junctions becomes very large. For nanocrystalline Fe prepared by mechanical attrition and measured by a nano-indentation technique. Thus, for most nanostructured materials (grain size > 10 nm), the elastic moduli are not unique properties and not a "negative."

Hardness and strength of conventional grain size materials (grain diameter, d > 1 mm) is a function of grain size. For ductile polycrystalline materials the empirical Hall-Petch equation has been found to express the grain-size dependence of flow stress at any plastic strain out to ductile fracture. In terms of yield stress, this expression is so = si + kd-1/2, where so = yield stress, si = friction stress opposing dislocation motion, k = constant, and d = grain diameter. Similar results are obtained for hardness, with Ho = Hi + kd-1/2. To explain these empirical observations, several models have been proposed, which involve either dislocation pileups at grain boundaries or grain boundary dislocation networks as dislocation sources. In all cases the Hall-Petch effect is due to dislocation motion/generation in materials that exhibit plastic deformation. Ratio of the Young's (*E*) and shear (G) moduli of nanocrystalline materials to those of conventional grain size materials as a function of grain size. The dashed and solid curves correspond to a grain boundary thickness of 0.5 and 1 nm, respectively (Shen et al. 1995). Most of the mechanical property data on nc materials have pertained to hardness, although some tensile test data are becoming available. Several recent reviews have summarised the mechanical behaviour of these materials (Siegel and Fougere 1994, 233–261; Siegel 1997; Morris and Morris 1997; Weertman and Averback 1996, 323–345). It is clear that as grain size is reduced through the nanoscale regime (< 100 nm), hardness typically increases with decreasing grain size and can be factors of 2 to 7 times harder for pure nc metals (10 nm grain size) than for large-grained (> 1 μm) metals. The experimental results of hardness measurements, summarised previously, show different behaviour for dependence on grain size at the smallest nc grains (< 20 nm), including (a) a positive slope ("normal" Hall-Petch behaviour), (b) essentially no dependence (~ zero slope), and (c) in some cases, a negative slope (Siegel and Fougere 1994, 233–261; Siegel 1997; Morris and Morris 1997; Weertman and Averback 1996, 323–345). Most data that exhibit the negative Hall-Petch effect at the smallest grain sizes have resulted from nc samples that have been annealed to increase their grain size. It is suggested that thermally treating nanophase samples in the as-produced condition may result in such changes in structure as densification, stress relief, phase transformations, or grain boundary structure, all of which may lead to the observed negative Hall-Petch behaviour (Siegel and Fougere 1994, 233–261). Only a few cases of negative Hall-Petch behaviour have been reported for as-produced nanocrystalline samples with a range of grain sizes. These include electrodeposited nc alloys and devitrified nc alloys (Erb et al 1996, 93-110; Alves et al. 1996). Nanocrystalline thin films with grain sizes £ 6 nm are also observed to exhibit a negative Hall-Petch effect (Veprek 1998). While it seems likely that in many cases the observed negative Hall-Petch slopes are due to artifacts of the specimen preparation methods, it is also likely that conventional dislocation-based deformation is not operable in nanocrystalline materials at the smallest grain sizes (< ~30 nm). At these grain sizes, theoretically, mobile dislocations are unlikely to occur; nor have they been observed in *in situ* TEM deformation experiments (Siegel and Fougere 1994, 233–261; Milligan et al 1993; Ke et al. 1995). Thus, the hardness, strength, and deformation behaviour of nanocrystalline materials is unique and not yet well understood.

Ductility, Toughness and Superplasticity

It is well known that grain size has a strong effect on the ductility and toughness of conventional grain size (> 1 mm) materials. For example, the ductile/brittle transition temperature in mild steel can be lowered about 40°C by reducing the grain size by a factor of 5. On a very basic level, mechanical failure, which limits ductility, is an interplay or competition between dislocations and cracks (Thomson 1996, 2208–2291). Nucleation and propagation of cracks can be used as the explanation for the fracture stress dependence on grain size (Nagpal and Baker 1990). Grain size refinement can make crack propagation more difficult and therefore, in conventional grain size materials, increase the apparent fracture toughness. However, the large increases in yield stress (hardness) observed in nc materials suggest that fracture stress can be lower than yield stress and therefore result in reduced ductility. The results of ductility measurements on nc metals are mixed and are sensitive to flaws and porosity, surface finish, and method of testing (e.g., tension or compression testing). In tension, for grain sizes < 30 nm, essentially brittle behaviour has been observed for pure nanocrystalline metals that exhibit significant ductility when the grain size is conventional. Elongation to failure in tension vs. grain size for some nanocrystalline metals and alloys. In some metals, Cu for example, ductile behaviour is observed in compression, along with yield strengths about twice those observed in tension. While it is likely that the flaws and porosity present in many nc samples seriously affect the results of mechanical tests and may be partly responsible for the asymmetry of results in compression compared to tension tests, the nature of the deformation process in terms of shear banding may also be important. The above behaviour is presumably due to the inability of usual dislocation generation and motion to occur at these smallest nc grain sizes. An intriguing suggestion based on early observations of ductile behaviour of brittle nc ceramics at low temperatures is that brittle ceramics or intermetallics might exhibit ductility with nc grain structures (Karch et al. 1987; Bohn et al. 1991). Karch and colleagues (1987) observed apparent plastic behaviour in compression in nc CaF2 at 80°C and nc TiO2 at 180°C. These observations were attributed to enhanced diffusional creep providing the plasticity at these temperatures, where conventional grain-size materials would fail in the elastic regime. It was assumed that diffusional creep was responsible for the plasticity; observations were rationalised, with boundary diffusion dominating the behaviour such that the strain (creep) rate is defined as de dt = BsWDDb d3kT where s is the applied stress, W the atomic volume, d the grain size, k the Boltzmann constant, T the temperature, B a constant, and Db the grain boundary diffusion coefficients. Going from a grain size of 1 μm to 10 nm should increase de/dt by 106 or more if Db is significantly larger for nc materials. However, these results on nc CaF2 and nc TiO2 have not been reproduced, and it is believed that the porous nature of these samples was responsible for the apparent ductile behaviour. In addition, the idea of unusually high creep rates at low temperatures has been refuted. Recent creep measurements of nc Cu, Pd, and Al-Zr at moderate temperatures by Sanders et al. (1997) find creep rates comparable to or lower than corresponding coarse-grain rates. The creep curves at low and moderate homologous temperatures could be fit by the equation for exhaustion (logarithmic) creep. One explanation is that the observed low creep rates are caused by the high fraction of low energy grain boundaries in conjunction with the limitation on dislocation activity by the small grain sizes. In sum, the predicted ductility due to diffusional creep in nc brittle ceramics or intermetallics at temperatures significantly less than 0.5 TM has not been realised.

Superplasticity is the capability of some polycrystalline materials to exhibit very large tensile deformations without necking or fracture. Typically, elongations of 100% to > 1000% are considered the defining features of this phenomenon. As grain size is decreased it is found that the temperature is lowered at which superplasticity occurs, and the strain rate for its occurrence is increased. As discussed previously, Equation 1 suggests that creep rates might be enhanced by many orders of magnitude and superplastic behaviour might be observed in nc materials at temperatures much lower than 0.5 TM. As mentioned above, actual creep experiments have not borne out this prediction, but instead have shown creep rates (Equation 1) comparable to or lower than those in coarse-grained samples of the same material. This is presumably why little enhancement in ductility or superplastic behaviour has been observed for nc materials at temperatures < 0.5 TM. However, there is evidence of enhancement of superplastic behaviour in nc materials at temperatures > 0.5 TM. Superplasticity has been observed at somewhat lower temperatures and at higher strain rates in nc materials. The evidence for tensile superplasticity is limited and observed typically at temperatures greater than 0.5 TM and in materials that exhibit superplasticity in coarser grain sizes (1–10 μm). For example, Mishra et al. (1997) observed superplastic behaviour in nc Pb-62%Sn at 0.64 TM and nc Zn–22%Al at 0.52 to 0.60 TM. However, Salishekev et al (1994) observed superplastic behaviour in submicron—200 nm—Ti and several Ti and Ni base alloys. Here, superplasticity (190% elongation, m = 0.32) was observed in Ti at 0.42 TM. This was at a temperature 50°C lower than for 10 μm grain size Ti. The flow stress for the 200 nm Ti at 550°C was 90 MPa, compared to 120 MPa for 10 μm Ti at 600°C. Very recently, Mishra and Mukherjee (1997) have observed superplastic behaviour in Ni3Al with a 50 nm grain size at temperatures of 0.56 to 0.60 TM to strains of 300 - 600%, but with unusual stress-strain behaviour and significant apparent strain-hardening. These new results suggest very different mechanisms may be causing superplastic behaviour in these nc materials.

Mechanical Properties, Theoretical Needs and Applications

While there are still only limited data on the mechanical behaviour— especially tensile properties—of nc materials, some generalisations may be made regarding the deformation mechanisms. It is likely that for the larger end of the nanoscale grain sizes, about 50 - 100 nm, dislocation activity dominates for test temperatures < 0.5 TM. As grain size decreases, dislocation activity apparently decreases. The essential lack of dislocations at grain sizes below 50 nm is presumably the result of the image forces that act on dislocations near surfaces or interfaces. The lack of dislocations in small, confined spaces such as single-crystal whiskers has been known for many years (Darken 1961). Creation of new dislocations is also made difficult as the grain size reaches the lower end of the nanoscale (< 10 nm). Stresses needed to activate dislocation sources, such as the Frank-Read source, are inversely proportional to the distance between dislocation pinning points. Since nanoscale grains will limit the distance between such pinning points, the stresses to activate dislocation sources can reach the theoretical shear stress of a dislocation-free crystal at the smallest grain sizes (~ 2 nm). Thus, at the smallest grain sizes we may have new phenomena controlling deformation behaviour. It has been suggested that such phenomena may involve grain boundary sliding and/or grain rotation accompanied by short-range diffusion-assisted healing events (Siegel 1997). Several examples of deformation by shear banding have been reported for nc materials. Carsley et al (1997, 183-192) have studied nc Fe–10% Cu alloys with grain sizes ranging from 45 to 1,680 nm. In all cases, deformation in compression proceeds by intense localised shear banding. The stress-strain curves exhibited essentially elastic, perfectly plastic behaviour; that is, no measurable strain hardening was observed. Shear banding is also the deformation mode observed in amorphous metallic alloys and amorphous polymers. The deformation shear banding in nc Fe–10% Cu was compared to that for metallic glasses, amorphous polymers, and coarsegrained polycrystalline metals after significant plasticity and work hardening had taken place. While this suggests a close similarity between deformation in nc materials and amorphous materials, not all tensile data on nc materials exhibit a lack of strain hardening. The Fe–10% Cu samples of Carsley et al. (1997, 183-192) showed shear bands even in their larger grained specimens (i.e., about 1,000 nm).

Central to all of the above discussion is the lack of understanding of the microscopic deformation and fracture mechanisms in nc materials. Clearly, a stronger theoretical effort is needed to guide critical experiments and point the direction for optimising properties. There has been limited work in this area, especially in Russia, in applying disclination theory to grain rotation (Romanov and Vladimirov 1992, 191), for example. However, a much larger theoretical effort is required. Another alternate deformation mechanism that may be important in nc materials is mechanical twinning (Huang et al. 1996). Little theoretical work has been carried out to address this possibility. Another important potential approach to understanding the deformation mechanisms in nc materials is to explore the rich older literature on shear banding—mechanical instabilities that have been observed in, for example, mild steels as grain size decreases; the approach to very low strain hardening (perfectly plastic behaviour) is observed. Amorphous materials exhibit many of the phenomenological characteristics of deformation in nc materials, that is, shear banding, asymmetry between tensile and compressive behaviour, and perfectly plastic behaviour. However, deformation mechanisms are not well understood in amorphous materials either. Recent work on bulk metallic glasses may help clarify this. Of the present or near-term applications for nc materials, the hard material WC/Co is an example of several important trends. Nanostructured WC/Co composites have been prepared that can have the following characteristics:

- nanoscale grain sizes after consolidation of powder
- hardness values about twice that of conventional micrograined WC/Co
- enhanced wear resistance and cutting performance

While this material has increased hardness and strength, preliminary reports from Stevens Institute of Technology in Hoboken, NJ, and the Royal Institute of Technology in Stockholm, Sweden, point to similar or increased fracture toughness values for nanostructured WC/Co. As noted earlier, single-phase nanostructured materials studied to date have exhibited high strength and hardness but brittle behaviour at low homologous temperatures (< 0.5 TM). The results of these studies of WC/Co two-phase nanostructured materials suggest that the combination of high hardness/strength and toughness/ductility may be possible in multiphase nanostructured materials. Other examples that point to this possibility come from the work of Professor A. Inoue at Tohoku University in Japan, whose lab the WTEC panel visited. Inoue and coworkers have synthesised a variety of multiphase Al, Mg, and Ni-base alloys with nanoscale microstructures. Many of these alloys consist of nanocrystallites in an amorphous matrix. Some Al-rich alloys contain nanoscale quasi-crystalline particles surrounded by crystalline face-centered cubic Al. The fascinating properties of these multiphase nanostructured alloys include extremely high strength coupled with some ductility. Ductility is high in compression, but uniform elongation in tension is limited. Again, this behaviour is analogous to that exhibited by ductile amorphous alloys. These results again suggest the possibility for development of nanostructured multiphase composites that combine extremely high hardness and strength with toughness and ductility. Such materials could have many applications as unique structural materials. Multiphase ceramic nanocomposites are the focus of several efforts. One significant effort is underway at the Research Center for Intermaterials of the Institute of Scientific and Industrial Research at Osaka University in Japan. Professor Koichi Niihara has a large effort studying micro-nano composites such as Al2O3/SiC that have enhanced toughness and also a variety of hard matrix/soft dispersion or soft matrix/hard dispersion nanocomposites. The enhanced toughness in some ceramic nanocomposites observed by Prof. Niihara has been verified by parallel studies of Dr. Steve Roberts at University of Oxford in England. While "low temperature" superplasticity has not been realised in nanostructured materials, enhanced superplastic behaviour has been observed at elevated temperatures in terms of a somewhat lower temperature range for superplasticity and, perhaps more significant, a higher strain rate regime. While much needs to be understood about superplastic behaviour of nanostructured materials, the possibility for using more conventional strain rates for forming has major industrial implications. This is of particular interest for near net shape forming of ceramics.

Perhaps the overriding issue for defining the mechanical behaviour of nanostructured materials is the lack of understanding of the mechanisms for plastic deformation and fracture in these materials. An extensive experimental and theoretical research effort should be applied to solve these questions. Only with such understanding can the intelligent design of nanostructured materials with optimum mechanical properties be realised. Since many of the synthesis routes for nanostructured materials involve using powder or particulate products, compaction of such powders is needed to form bulk parts without coarsening the nanoscale microstructure. While there has been ongoing work on compaction and the thermal stability of nanocrystalline microstructures, more basic work is needed in this critical area. If some unique properties are limited to the finest grain sizes, methods must be found to stabilize the grain size while attaining theoretical density and complete particulate bonding. Furthermore, conventional processing methods are desirable if realistic scaleup and economy are to be realised. Near-term opportunities for structural applications of nanostructured materials are in the form of coatings prepared by, for example, thermal spray deposition or electrodeposition. There is a significant effort in the United States on thermal spray deposition of nc materials. Several groups in Canada lead in deposition of nc materials by electrodeposition methods. A spin-off company, Nanometals, formed under the auspices of Parteq Research and Development Innovations, has adopted a commercial electrodeposition method to produce nanostructured metals. Structural applications of nanostructured materials may be viewed as a focused approach to a long standing, well known design of materials with optimum mechanical properties. Use of nanoscale precipitates or dispersoids has been known for decades to improve mechanical behaviour. The "new" emphasis on using nanostructured materials comes from 2 significant factors: 1. the special processing methods to push microstructures to the limits of the nanoscale 2. the unique properties and phenomena—as yet not well understood— when this limit is approached

Magnetic Nanocrystalline Alloys, Permanent Magnet Materials and GMR

The discovery of nanocrystalline Fe-based soft magnetic materials is less than ten years old. The first class of such materials was the melt-spun Fe-Si-B alloys containing small amounts of Nb and Cu (Yoshizawa et al. 1988). The Fe-Si-B-Nb-Cu amorphous phase transforms to a body-centered cubic (bcc) Fe-Si solid solution with grain sizes of about 10 nm during annealing at temperatures above the crystallisation temperature. The presence of small amounts of Cu helps increase the nucleation rate of the bcc phase while Nb retards the grain growth. These "Finemet" alloys provide low core losses (even lower than amorphous soft magnetic alloys such as Co-Fe-Si-B), exhibit saturation induction of about 1.2 T, and exhibit very good properties at high frequencies, comparable to the best Co-based amorphous alloys. These were first developed in Japan and have stimulated a large amount of research and development worldwide to optimize the magnetic properties. There has been relatively little work in the United States in this area, however. While many of the soft magnetic properties of Finemet-type nanocrystalline alloys are superior, they exhibit lower saturation inductions than Fe-metalloid amorphous alloys, mainly because of the lower Fe content to attain amorphisation and because of the addition of Nb and Cu (or other elements to control the nucleation and growth kinetics). In order to remedy this problem, another class of Fe-based nanocrystalline alloys was developed by Inoue and coworkers at Tohoku University (Makino et al. 1997), which is commercialised by Alps Electric Co., Ltd., of Nagaoka, Japan. These "Nanoperm" alloys are based on the Fe-Zr-B system; they contain larger concentrations of Fe (83-89 at.%) compared to the Finemet alloys (~ 74 at.% Fe) and have higher values of saturation induction (~ 1.6-1.7 T). The Nanoperm nc alloys have very low energy losses at power frequencies (60 Hz), making them potentially interesting for electrical power distribution transformers. The issues of composition modification, processing, and the brittle mechanical behaviour of these nanocrystalline/amorphous alloys are discussed by V.R. Ramanan in the first volume of this WTEC study, the proceedings of the May 8-9, 1997 panel workshop on the status of nanostructure science and technology in the United States. While there has been extensive research on these alloys, particularly in Japan and Europe, most of the development has been carried out in Japan. The Finemet family of alloys is marketed by Hitachi Special Metals.

Vacuumschmelze GmbH (Germany) and Impky (France) also market similar alloys. The Nanoperm alloys are being commercialised by Alps Electric Co. (Japan). No extensive research nor any commercialisation of these materials has been carried out in the United States. The small single-domain nanocrystalline Fe particles in the amorphous matrix gives these alloys their unique magnetic behaviour, the most dramatic being the lowest energy losses (narrowest B/H hysteresis loop) of any known materials, along with very high permeabilities. These alloys can also exhibit nearly or exactly zero magnetostriction. To date, these materials have been made by crystallisation of rapidly solidified amorphous ribbons. Other methods that might provide geometrically desirable products should be explored or developed. Electrodeposition is one such method that requires further work. Electrodeposited nc Fe-Ni soft magnetic alloys are being developed in Canada. The brittle nature of these materials is a problem for scaleup and transformer manufacture. The brittleness problem must be solved by finding less brittle materials or applying the handling and processing knowledge that exists for embrittled (after annealing) metallic glasses.

The first attempts to produce nanoscale microstructures to enhance the magnetic properties of the Nb-Fe-B permanent magnetic materials used mechanical alloying of blended elemental powders followed by heat treatment (Schultz et al. 1987). Since the grain structure so obtained does not exhibit any crystallographic texture—and limits the energy product— special processing methods such as die-upsetting were used by Schultz and coworkers (1989) to provide the crystallographic anisotropy. While the coercivities of these nanocrystalline alloys are high, the remanent magnetisation is decreased. Recent approaches to increasing the magnetic induction have utilised exchange coupling in magnetically hard and soft phases. The Fe-rich compositions (e.g., Fe90Nd7B3) result in a mixture of the hard Fe14Nd2B phase and soft aFe phase. The nanoscale two-phase mixtures of a hard magnetic phase and a soft magnetic phase can exhibit values of remanent magnetisation. This "remanence enhancement" is associated with exchange coupling between the hard and soft phases, which forces the magnetisation vector of the soft phase to be rotated to that of the hard phase (Smith et al. 1996). Two important requirements for alloys to exhibit remanence enhancement are a nanocrystalline grain size and a degree of coherence across interphase boundaries sufficient to enable adjacent phases to be exchange-coupled. The significant feature of the exchange coupling is that it allows crystallographically isotropic materials to exhibit remanence values approaching those achieved after full alignment. Such two-phase nanoscale ferromagnetic alloys have been prepared by nonequilibrium methods such as melt-spinning, mechanical alloying, and sputter deposition. Besides the high reduced remanence, the material cost is reduced by reduction in the content of the expensive hard rare earth-containing magnetic phase. The theoretical understanding of remanence enhancement appears to be developed to a degree enabling prediction of magnet performance; however, this performance, while a significant improvement over single-phase isotropic magnets, does not reach predicted values. Work is required on optimising the orientation relationships between the hard and soft phases and the interphase properties (coherency) between them. Research on nanocrystalline hard magnetic alloys has received attention worldwide. The U.S. efforts are summarised in the article by G.C. Hadjipanayis (1998, 107-112). While less research seems to be carried out in the world on these materials compared to the nanocrystalline soft magnetic alloys, some efforts exist in most countries. Notable programmes are those of L. Schultz and coworkers at the Institut für Festköper und Werkstofforschung (IFW) in Dresden and P.G. McCormick and coworkers at the University of Western Australia. While the very low losses of the nc soft magnetic materials (Finemet or Nanoperm) are dependent on grain size for their properties, the hard magnetic nc alloys with remanence enhancement provide flexibility in processing, especially with powder materials. These remanence-enhanced nc hard magnetic alloys may find many applications as permanent magnet components.

The phenomenon of giant magnetoresistance (GMR)—the decrease of electrical resistance of materials when exposed to a magnetic field—was first reported in a number of multilayer ferromagnetic/nonferromagnetic thin film systems (Baibich et al. 1988). More recently, GMR was observed in equiaxed granular nanocrystalline materials (Berkowitz et al. 1992). In particular, GMR systems with low saturation fields offer a wide area for application in magnetoresistive devices. GMR sensors have a higher output than conventional anisotropic magnetoresistive sensors or Hall effect sensors. They can operate at higher magnetic fields than conventional magnetoresistive sensors. In multilayer systems the antiferromagnetic alignment of the ferromagnetic layers in zero field becomes ferromagnetic as the field is applied and causes a decrease in resistance. Granular materials that show GMR consist of small ferromagnetic single-domain particles with randomly oriented magnetic axes in a nonmagnetic matrix. An external field rotates the magnetic axes of all magnetic particles. The rotation towards complete alignment of all magnetic axes again reduces the resistance in a similar way as for multilayers. The GMR in granular systems is isotropic. The explanation for the GMR is spin-dependent scattering of the conduction electrons at the ferromagnetic/nonmagnetic interfaces and, to a lesser extent, within the magnetic grains. The GMR scales inversely with the average particle diameter. There is worldwide research on the GMR effect. U.S. programmes are reviewed by R. Shull and G.C. Hadjipanayis in the proceedings of the WTEC U.S. nanotechnologies workshop (Shull 1998, 43-58; Hadjipanayis 1998, 107-112). The NIST work described by Shull has provided material with the largest GMR values for the smallest switching fields. Japanese research on GMR includes studies in Prof. Fujimori's group at Tohoku University. While the theory for GMR of spin-dependent scattering referred to above has been used as an explanation, other explanations taking into account interaction between magnetic regions have been proposed (El-Hilo et al. 1994). Combined theoretical and experimental studies should help to clarify the mechanism for this effect

Explaining Ferromagnetic Nanocrystalline Materials

Magnetic nanocomposite refrigerants, which have four times the magnetocaloric effects of the best low temperature magnetic refrigerant, were developed by NIST and described by R. Shull (1998, 43-58). The entropy change at a given (low) temperature for a system of magnetic spins is enhanced when the isolated spins are clustered. Shull et al. (1993) have shown that the nanocomposite Gd3Ga5–xFexO12 gives superior magnetocaloric effects, which increase with x up to x = 2.5 and can be extended to higher temperatures than conventional materials. Magnetostrictive materials such as Terfonol-D (Tb0.3Dy0.7Fe2) have been of scientific and technological interest in recent years. It is suggested by G.C. Hadjipanayis (1998, 107-112) that nanostructured magnetostrictive materials can have improved properties, such as lower saturation fields, with reduced anisotropy and in multilayers with alternate layers of magnetostrictive and soft magnetic materials that are exchange-coupled. Hadjipanayis states that most of the research in this area is carried out in Japan and Europe. Nanocrystalline magnetic materials offer perhaps the nearest-term prospect of significant applications of bulk nanostructured materials. The remanence enhancement in two-phase hard/soft magnetic materials can result in excellent energy products—comparable to those from rapid solidification processing routes—in powder composites. The flexibility in manufacturing this allows should provide many possible permanent magnet applications. Nanocrystalline soft magnetic materials have the lowest energy losses of any material. While problems such as mechanical brittleness remain, these materials promise to replace existing transformer core materials for power applications. The U.S. research and industrial effort in these materials lags the efforts in Japan.

Nanocrystalline Hydrogen Storage and Corrosion-Resistant Materials

R.B. Schwarz (1998, 93-95) has pointed out that nanostructured materials offer several potential advantages for hydrogen storage materials. Rapid kinetics of absorption/desorption can be aided by refining the microstructure to the nanoscale. For example, nanoscale inclusions of Mg2Ni in Mg catalyze the decomposition of the molecular hydrogen, increasing the hydrogen absorption/desorption kinetics. Another advantage of the nanoscale microstructure is that the alloy powder does not comminute on repeated charging/discharging with hydrogen. This is not strictly a bulk material, since powder agglomerates or green compacts can be used, thus obviating the need for compaction to theoretical density.

The limited work to date on corrosion resistance of nanocrystalline materials indicates that no generalisations can be made. Superior localised corrosion resistance in HCl was observed for nanocrystalline 304 stainless steel (Fe–18%Cr–8%Ni) prepared by sputter deposition (Inturi and Szklavska-Smialowska 1991 and 1992). This was attributed to the fine grain size and homogeneity of the nc material. However, the average dissolution rate of nc Ni was found to be higher than that for coarse-grained material (Rofagha et al. 1991).

Defining Polymers, Polyolefin, Polymer-Silicate and Dendrimer Nanocomposites

There has been enormous interest in the commercialisation of nanocomposites for a variety of applications, and several of these applications will be successful in the near future. Mineral fillers, metals, and fibers have been added to thermoplastics and thermosets for decades to form composites. Compared to neat resins, these composites have a number of improved properties including tensile strength, heat distortion temperature, and modulus. Thus, for structural applications, composites have become very popular and are sold in billion pound quantities. These filled thermoplastics are sold in even larger volumes than neat thermoplastics. Furthermore, the volume of fillers sold is roughly equal to the volume of thermoplastic resin sold. Clearly, the idea of adding fillers to thermoplastics and thermosets to improve properties, and in some cases decrease costs, has been very successful for many years. More recently, with advances in synthetic techniques and the ability to readily characterize materials on an atomic scale has lead to interest in nanometer-size materials. Since nanometer-size grains, fibers and plates have dramatically increased surface area compared to their conventional-size materials, the chemistry of these nanosized materials is altered compared to conventional materials. Polymer nanocomposites combine these two concepts, i.e., composites and nanometer-size materials. Thermoplastics filled with nanometer-size materials have different properties than thermoplastics filled with conventional materials. Some of the properties of nanocomposites, such as increased tensile strength, may be achieved by using higher conventional filler loading at the expense of increased weight and decreased gloss. Other properties of nanocomposites such as clarity or improved barrier properties cannot be duplicated by filled resins at any loading. Polymer nanocomposites were developed in the late 1980s in both commercial research organisations and academic laboratories. The first company to commercialize these nanocomposites was Toyota, which used nanocomposite parts in one of its popular car models for several years. Following Toyota's lead, a number of other companies also began investigating nanocomposites. Most commercial interest in nanocomposites has focused on thermoplastics. Thermoplastics can be broken into two groups: less expensive commodity resins and more expensive (and higher performance) engineering resins. One of the goals of nanocomposites was to allow substitution of more expensive engineering resins with a less expensive commodity resin nanocomposite. Substituting a nanocomposite commodity resin with equivalent performance as a more expensive engineering resin should yield overall cost savings. By a strict definition of nanocomposites, i.e., any filler submicron in size, there already are significant volumes of nanocomposites being produced. These amount to more than 20 million pounds. However, since these fillers are on the upper end of the nanocomposite size range, most sources have excluded them from consideration. This report covers these products but does not add their volumes into aggregate numbers of nanocomposite markets or production volumes to comply with the established convention. At this point in time, there has been much less commercial interest in thermoset nanocomposites compared to thermoplastics. This neglect may not continue much longer since thermoset nanocomposites have some distinct advantages over neat thermoset resins. This report covers some possible markets for thermoset nanocomposites. There are two types of recognised polymer nanocomposites:

- Thermoplastic nanocomposites: these materials are broken into two major categories, i.e., commodity resins and engineering resins. The potential of polymer nanocomposite commodity resin is covered for each resin. Engineering resin nanocomposites are restricted to resins that have been under development.
- Thermoset nanocomposites: thermoset nanocomposites have received less commercial interest in their development than thermoplastic nanocomposites, but these materials may be relatively straightforward to bring into production. Furthermore, thermoset nanocomposites can offer some significant advantages over conventional thermosets.

Polymer nanocomposite (PNC) is a polymer or copolymer having dispersed in it nanoparticles. These may be of different shape (e.g., platelets, fibers, spheroids), but at least one dimension must be ca. 1 to 50 nm (diameter of pencil lead is about 1 mm, or 1,000,000 nm). PNC with all three types of nanoparticles have been prepared (e.g., polycarbonate with carbon nanotubes, polyamide with iron oxide spheres), but only PNC with clay platelets are on the market for the structural (high volume) applications. These PNC's belong to the category of multi-phase systems (MPS, viz. blends, composites, and foams) that consume ca. 95 per cent of plastics production. The key to industrial success of MPS is the desired and stable morphology. Thus, these systems require controlled mixing/compounding, stabilisation of the achieved dispersion, orientation of the dispersed phase, and optimisation of interactions in the finished product. In spite of differences in the nature of the dispersed phase, the compounding strategies for all MPS, including PNC, are similar.

Mixing is the most important operation in polymer processing. Uniformity of the molecular weight and its distribution, entanglement, composition, as well as the stress history and temperature, is the key to good performance. Dispersive (or intensive) and distributive (or extensive) mixing of molten polymers is carried out within the laminar flow regime. The dispersive action is responsible for the breaking of original domains of the minor component into the desired size, while the distributive provides for their homogeneous distribution within the matrix. Microrheology may be used to describe the dispersive process in the shear or extensional flow field, while distributive mixing may be treated in terms of the laminar flow theory. Both models - microrheology and laminar mixing - indicate superiority of the extensional flow field. The presence of the vorticity component in the shear tensor is responsible for rotation of the flow elements that cause excessive energy consumption, polymer degradation and attrition of anisometric solid particles - shear mixing is inefficient. Mixing in the extensional flow field may be carried out by means of the convergent-divergent flow geometry. A motionless extensional flow mixer (EFM) and its newer, dynamic version (DEFM) are fully adjustable, general-purpose mixers used for homogenisation of reaction products, removal of "fish eyes" from films, alloying, blending or dispersing of nanoparticles. In the search for synergism the industry moves toward progressively complex MPS, e.g., foaming of nanoclay reinforced polymer blends. Here, the difficulty is the presence of several phases with potentially different thermodynamic and rheological character as well as reinforcing and foaming the appropriate blend components. However, once the difficulties are over, the advantages translate into superior performance. The clay-containing polymeric nanocomposites (CPNC) constitute new type of polymeric systems, with a great variety of molecular structures and processing difficulties. The only type of nanoparticles used in commercial production of CPNC is the lamellar, expandable, smectite clay. For the best performance of CPNC the clay platelets (ca. 1 nm thick and 300 nm diameter) must be homogeneously dispersed as individual crystalline lamellae (known as exfoliation). To accomplish this task, a balance between the thermodynamics and mechanical mixing must be achieved. Owing to the large aspect ratio, at a concentration of clay exceeding ca. 1 wt per cent the platelets start to interact, creating 3D structures. In commercial CPNC the clay content is about twice as large, hence there is local ordering. Generation of the most suitable orientation is the key for the optimisation of the performance. Three recent developments in the CPNC science and technology are worth noting.

The use of an elongational flow field greatly accelerates dispersion and orientation, facilitating the processing. Several CPNC types (e.g., with polyethylene terephthalate, PET, polystyrene, PS, or polypropylene, PP, matrix) have been prepared using commercial EFM attached either to single-screw extruder (SSE) or twin-screw extruder (TSE). The results demonstrated significant beneficial role even of the standard EFM (developed for melt homogenisation and polymer blending). For example, PET was extruded with organoclay using a TSE and the same machine, but with attached to it EFM. The presence of EFM increased the PNC properties over that from TSE by (data from EFM Inc.):

(i) Tensile modulus by ca. 67 per cent.
(ii) Tensile strength by ca. 60 per cent.
(iii) Impact strength 80 per cent.

In Union Chemical Laboratory, Industrial Technology Research Institute (UCL/ITRI) in Taiwan, PP was compounded with organoclay and a compatibilizer (maleated polypropylene, PP-MA). Three processing methods were used: #1: TSE; #2: TSE + EFM; and #3: SSE + EFM. The method #1 resulted in mixtures with clay dispersed as micron-sized aggregates. By contrast, process #2 produced well-intercalated CPNC with short stacks of organoclay uniformly dispersed in the molten PP. However, the best dispersion was obtained from #3, i.e., using SSE + EFM. This unexpected result was most likely due to higher thermo-mechanical degradation in TSE. The ammonium-ion intercalated organoclay is the main type used for the production of CPNC. However, these materials start to degrade at temperatures above 180°C by the Hofmann elimination mechanism, leaving the clay surface bare. Numerical simulations were carried out to evaluate the effects of independent parameters on the degree of CPNC dispersion.

The most dramatic worsening of the thermodynamic conditions for exfoliation was obtained in the presence of bare clay platelets - even as small amount as 1 or 2 per cent of clay surface precludes exfoliation. Kato et al. from the Toyota Central R&D Lab proposed a revolutionary method of CPNC production. The method uses un-treated clay, e.g., sodium montmorillonite (Na-MMT), which becomes exfoliated and the morphology stabilised during the mixing/compounding step. The process uses a long (L/D = 77) TSE, divided by a pair of seal rings into three zones. In the first zone PP + PP-MA + Na-MMT were melt mixed together. In zone #2 ca. 10 per cent water was injected into the melt and the pressure of the water vapor was well controlled. Adsorption of water by clay platelets, followed by reaction with compatibilizer (PP-MA), resulted in exfoliation and stabilisation of the dispersion. Finally, in zone #3 devolatilisation took place. The resulting CPNC had almost the same properties as the conventionally prepared one, but at a fraction of cost. The method bypasses the need for the costly intercalation of Na-MMT with ammonium ions, avoids the introduction of readily degradable organoclay to molten polymer (which in turn gets degraded by combination of organoclay degradation and air) and at the same time eliminates the need of costly in time and energy pre-drying step. Evidently, the method is in an early stage of development. Its extension to other polymers (e.g., polyesters or polyamides), optimisation of the compatibilizer, processing equipment and process parameters are required.

The main advantages of CPNC stem from the improved stiffness (by a factor of 2 at 5 wt per cent loading), reduction of gas and vapor permeability (by a factor of 100 at 10 wt per cent loading), and reduced flammability. Since these enhancements are achieved with only a slight increase of density, the transport, and packaging industries are expect to be the first to profit from this new technology. Up to now two principal obstacles hindered progress of CPNC into the market: the cost and reproducibility of performance. The demonstrated by Kato et al. suitability of Na-MMT reduces the cost of the nano-filler from ca. $7 to $2 per kilogram, and at the same time reduces the probability of thermo-mechanical degradation during processing and forming. Many questions remain the foremost being whether the achieved improvement of technology is sufficient to lower the cost and improve stability to satisfy the industry standards.

Evidently, the future of CPNC technology hinges on the intelligent application of the available compounding and processing technologies, adopted for the specific needs of the nano-scale mixing. The subject of hybrids based on layered inorganic compounds such as clays has been studied for a considerable time, but the area is enjoying a resurgence of interest and activity as a result of the exceptional properties which can be realised from such nanocomposites. Materials variables which can be controlled and which can have a profound influence on the nature and properties of the final nanocomposite include the type of clay, the choice of clay pre-treatment, the selection of polymer component and the way in which the polymer is incorporated into the nanocomposite. The last of these may be dictated by the available processing methods and whether the prospective user is an integrated polymer manufacturer or a specialist processor.

Common clays are naturally occurring minerals and are thus subject to natural variability in their constitution. The purity of the clay can affect final nanocomposite properties. Many types of clay are aluminosilicates, which have a sheet-like (layered) structure, and consist of silica SiO_4 tetrahedra bonded to alumina AlO_6 octahedra in a variety of ways. A 2:1 ratio of the tetrahedra to the octahedra results in smectite clays, the most common of which is montmorillonite. Other metals such as magnesium may replace the aluminium in the crystal structure. Depending on the precise chemical composition of the clay, the sheets bear a charge on the surface and edges, this charge being balanced by counter-ions, which reside in part in the inter-layer spacing of the clay. The thickness of the layers (platelets) is of the order of 1 nm and aspect ratios are high, typically 100-1500. The clay platelets are truly nanoparticulate. In the context of nanocomposites, it is important to note that the molecular weight of the platelets (ca. 1.3×10^8) is considerably greater than that of typical commercial polymers, a feature which is often misrepresented in schematic diagrams of clay-based nanocomposites. In addition, platelets are not totally rigid, but have a degree of flexibility. The clays often have very high surface areas, up to hundreds of m^2 per gram. The clays are also characterised by their ion (e.g. cation) exchange capacities, which can vary widely. One important consequence of the charged nature of the clays is that they are generally highly hydrophilic species and therefore naturally incompatible with a wide range of polymer types. A necessary prerequisite for successful formation of polymer-clay nanocomposites is therefore alteration of the clay polarity to make the clay 'organophilic'. An organophilic clay can be produced from a normally hydrophilic clay by ion exchange with an organic cation such as an alkylammonium ion. For example, in montmorillonite, the sodium ions in the clay can be exchanged for an amino acid such as 12-aminododecanoic acid (ADA): Na^+-CLAY + HO_2C-R-$NH_3^+Cl^-$ à.HO_2C-R-NH_3^+-CLAY + NaCl. The way in which this is done has a major effect on the formation of particular nanocomposite product forms and this is discussed further below. Although the organic pre-treatment adds to the cost of the clay, the clays are nonetheless relatively cheap feedstocks with minimal limitation on supply. Montmorillonite is the most common type of clay used for nanocomposite formation; however, other types of clay can also be used depending on the precise properties required from the product. These clays include hectorites (magnesiosilicates), which contain very small platelets, and synthetic clays (e.g. hydrotalcite), which can be produced in a very pure form and can carry a positive charge on the platelets, in contrast to the negative charge found in montmorillonites.

The synthetic route of choice for making a nanocomposite depends on whether the final material is required in the form of an intercalated or exfoliated hybrid. In the case of an intercalate, the organic component is inserted between the layers of the clay such that the inter-layer spacing is expanded, but the layers still bear a well-defined spatial relationship to each other. In an exfoliated structure, the layers of the clay have been completely separated and the individual layers are distributed throughout the organic matrix. A third alternative is dispersion of complete clay particles (tactoids) within the polymer matrix, but this simply represents use of the clay as conventional filler.

In recent years, there has been extensive study of the factors which control whether a particular organo-clay hybrid can be synthesised as an intercalated or exfoliated structure. Since clay nanocomposites can produce dramatic improvements in a variety of properties, it is important to understand the factors which affect delamination of the clay. These factors include the exchange capacity of the clay, the polarity of the reaction medium and the chemical nature of the interlayer cations (e.g. onium ions). By modifying the surface polarity of the clay, onium ions allow thermodynamically favourable penetration of polymer precursors into the interlayer region. The ability of the onium ion to assist in delamination of the clay depends on its chemical nature such as its polarity. The loading of the onium ion on the clay is also crucial for success and it should be borne in mind that a commercial organo-clay may not have the optimum loading for a particular application. For positively charged clays such as hydrotalcite, the onium salt modification is replaced by use of a cheaper anionic surfactant. Other types of clay modification can be used depending on the choice of polymer, including ion-dipole interactions, use of silane coupling agents and use of block copolymers. An example of ion-dipole interactions is the intercalation of a small molecule such as dodecylpyrrolidone into the clay. Entropically-driven displacement of the small molecules then provides a route to introducing polymer molecules. Unfavourable interactions of clay edges with polymers can be overcome by use of silane coupling agents to modify the edges. These can be used in conjunction with the onium ion treated organo-clay. An alternative approach to compatibilising clays with polymers has been introduced by TNO (Netherlands), based on use of block or graft copolymers where one component of the copolymer is compatible with the clay and the other with the polymer matrix. This is similar in concept to compatibilisation of polymer blends. A typical block copolymer would consist of a clay-compatible hydrophilic block and a polymer-compatible hydrophobic block. The block length must be controlled and must not be too long. High degrees of exfoliation are claimed using this approach.

The correct selection of modified clay is essential to ensure effective penetration of the polymer or its precursor into the interlayer spacing of the clay and result in the desired exfoliated or intercalated product. Indeed, further development of compatibiliser chemistry is undoubtedly the key to expansion of this nanocomposite technology beyond the systems where success has been achieved to date. Polymer can be incorporated either as the polymeric species itself or via the monomer, which is polymerised in situ to give the corresponding polymer-clay nanocomposite. The second of these is the most successful approach to date, although it probably limits the ultimate applicability of these systems. Polymers can be introduced either by melt blending, for example extrusion, or by solution blending. Melt blending (compounding) depends on shear to help delaminate the clay and can be less effective than in situ polymerisation in producing an exfoliated nanocomposite. Both thermo sets and thermoplastics have been incorporated into nanocomposites, including:

- Nylons
- Polyolefins, e.g. polypropylene
- Polystyrene
- Ethylene-vinyl acetate (EVA) copolymer
- Epoxy resins
- Polyurethanes
- Polyimides
- Poly(ethylene terephthalate) (PET)

Polyolefin nanocomposites are useful for a wide variety of applications in packaging, engineering and fire retardant markets. These nanocomposites are near molecular blends of montmorillonite nanoclay and a polyolefin matrix. Concentrate trade name. Here performance properties are outlined for typical polyolefins with explanations of options for creating nanocomposites. Polyolefins represent a wide range of homo and copolymers, principally containing propylene and ethylene. Also included are polyolefin/rubber alloys usually called TPO's and polyolefin/EVA blends. Because propylene and ethylene contents as well as polydispersities vary widely, some Nanomer Concentrate screening is required to determine the optimum match between the matrix and nanoclay surface treatment. Polyolefin nanocomposites are produced via extrusion compounding using Concentrates. Refer to data for concentrate specifications and grade selection. Film or bottle grade products require twin screw compounders or intensive mixers. Most engineering and fire retardant applications can be accommodated using single screw technology. A specially designed screw, called a NanoMixer™, must be used. Compounding guidelines are available on Tech Data Sheets P-803 through P-806.

Nanocomposite performance depends on the degree of nanoclay de-agglomeration and dispersion within the polyolefin matrix. The process of de-agglomeration/dispersion is commonly referred to as exfoliation. Three factors influence exfoliation: the nanoclay surface treatment, nanocomposite loading, and the amount of dispersive energy used. The optimum nanoclay surface treatment is readily determined through screening, as described earlier. Nanomer loading will not hinder exfoliation within the 2-8 per cent wt/wt range and is optimised at 6 per cent wt/wt. The proper choice of compounding equipment and use of the guidelines assists in achieving the dispersive energy requirement. However, another factor impacting dispersive energy is matrix resin melt flow. Resin melt flow influences the level of shear resulting from dispersive energy input. The lower the melt flow, the greater the shear on Nanomer particles. It follows that exfoliation is maximised in lower melt flow resins, where shear is the highest.

Polyolefin nanocomposites exhibit mechanical properties similar to highly filled systems, but with the added advantage of low density. The density (and resulting part weight) is just slightly higher than unfilled resin. Nanocompositing also provides a high degree of reinforcement without significant loss of impact. In most cases notched Izod impact remains the same as neat (unfilled) resin. Polyolefin nanocomposites have specific gravities and melt flows nearly the same as neat resin, so they process like neat resin. Part output rates will be significantly higher than conventionally filled composites and machine wear will also improve noticeably. Nanocomposites can be used in conjunction with conventional fillers, i.e. glass fiber or talc, to create synergistic materials with customised property sets. For example a 6 per cent wt/wt nanocomposite containing 10-12 per cent wt/wt glass fiber will deliver a property set similar to 30 per cent wt/wt straight glass fiber, but lower specific gravity and part weight. Polyolefin nanocomposites are potent char formers. In fire retardant (FR) applications they function by insulating the resin substrate from the air-surface interface, starving the flame by reducing the fuel needed to continue the burning process. The result is a dramatically lower peak heat release rate (HRR). Nanocomposites alone do not conform to a UL V0 rating. But in conjunction with reduced levels of conventional FR additives they act synergistically to achieve V0. Often traditional FR additives can be reduced by 50 per cent wt/wt. With less FR additive dilution and the enhanced reinforcing effect of nanocompositing, mechanical properties are restored, usually at lower cost than would be the case without nanocompositing. Another benefit of nanocomposite/FR additive synergy is lower density and part weight, since a lot of high density additive has been removed from the formulation. As polyolefins take an increased role in packaging applications, gas barrier becomes an important requirement. Nanocomposites improve oxygen transmission rates (OTR) between 1.2-2X. Carbon dioxide transmission rates (CO2TR) improve 1.4-1.8X. There is also some improvement in water vapor transmission (WVTR), around 10-15 per cent. Improved gas barrier follows the tortuous path model and is valid for cast, blown and molded products. Present performance data for a variety of polyolefins. Predictably, nanocomposite films possess greater tensile strength and modulus, although elongation-atbreak is reduced. Simultaneous improvement in tensile properties and gas barrier provides the formulator with opportunities to downguage films and containers. Tear strengths in both the machine and transverse directions remain the same as neat polyolefin films of the same mil thickness. Tear strength of downguaged nanocomposite film drops in proportion to the reduction in film thickness. The earliest example of the in situ polymerisation method was work by Toyota on synthesis of clay-nylon nanocomposites and this remains probably the most studied system, including work by Bayer and Ube. In a typical synthesis ADA-modified clay is dispersed in the monomer caprolactam, which is polymerised to form the nylon-6-clay hybrid as an exfoliated composite. Complete exfoliation may be preceded by intercalation of the monomer in the clay. Generally, low concentration of clay (a few per cent) are incorporated in these nanocomposites, partly because this is often sufficient to modify the desired properties significantly, but also because higher levels of clay can adversely increase the system viscosity leading to poor processability, although the viscosity increase is shear rate dependent. Other nylons and copolyamides (e.g. nylon-6/6,6) have also been incorporated in clay nanocomposites. Functionality such as hydroxyl groups can be introduced into the onium salt modifiers to improve compatibility with the nylon via hydrogen bonding – this can lead to an enhancement of desirable nanocomposite properties. Similar modification of the chemistry may be required for successful exfoliation with other types of polymer system. In the case of ethylene-vinyl alcohol (EVOH) copolymers, use of hydroxylated quaternary ammonium ions improves compatibility between the clay and the EVOH by introducing favourable hydroxyl group interactions. In polypropylene (PP) nanocomposites, maleic anhydride grafted PP is used as a compatibiliser. Polymerisation initiators can be anchored to the clay platelet surface and this approach has been extended to living free radical polymerisation of styrene, where an initiating species bonded to the TEMPO free radical is attached to the surface of the clay platelets via ion exchange.

For preparation of nanocomposites from high temperature engineering thermoplastics, a major limitation in the use of conventional onium ion modified clays is the thermal instability of the alkylammonium species during processing. The block copolymer route developed by TNO (vide supra) offers one potential solution to this problem. Imidazolium salts such as that depicted. A further alternative is the use of phosphonium salts in place of ammonium salts—this can lead to an increase in the degradation temperature of the organo-clay from 200-300°C to >300°C. Wholly synthetic organo-clays clays, on the other hand, can exhibit thermal stability to beyond 400°C. By using these last two approaches, Triton Systems has succeeded in producing nanocomposites from high temperature resins such as polyetherimide (PEI). Forming the nanocomposite provides a potential route to producing 'speciality' engineering resin performance from cheaper 'standard' engineering resins. Future goals include the extension of this nanotechnology to additional types of polymer system, where the development of new compatibilisation strategies is likely to be a prerequisite. Production of PVC-based systems is still some way off and challenges remain to be solved in PET nanocomposites. Additional reinforcement of clay nanocomposites by glass fibre is currently being investigated. There is also interest in the development of electrically conducting clay nanocomposites. Polymer nanocomposites, particularly polymer-silicate nanocomposites, represent an emerging class of specialty plastic materials.

The melt-processing route would be a powerful approach in the manufacture of these compounds. Techniques such as transmission electron microscopy and x-ray diffraction were used to determine that nylon-6 nanocomposites could be produced via melt compounding. As a result of the "nano"-sized dispersion, reinforcement efficiency is realised at low loading levels (3-5 per cent) which is a small cost in terms of specific gravity. Dramatic enhancements are observed in heat distortion temperature and stiffness. Besides mechanical properties, nanocomposites also have much lower gas permeability than the neat resin. Other potential benefits include good surface appearance, dimensional stability, and lower heat release. Nanocomposites, particularly polymer-silicate nanocomposites, are materials that have been given much attention recently and represent an emerging class of specialty plastic compounds. One of the reasons for this attention is that due to the "nano"-scale dispersion, which results in high aspect ratio and surface area, the reinforcement efficiency of polymer-silicate nanocomposites can be significantly better than conventional mineral fillers. The potential benefits of polymersilicate nanocomposites include higher heat resistance, increased flexural modulus, lower gas permeation rate, easy flow, dimensional stability, good surface appearance, and lower heat release rate. Typically, the type of silicate used in polymersilicate nanocomposites is montmorillonite which is a 2:1 layered smectite clay. This clay has a platy structure with individual platelets having thicknesses of one nanometer and surface lengths on the order of 100 to 1000 nanometers. Montmorillonite is hydrophillic which makes proper exfoliation and dispersion into conventional polymers difficult. Thus, montmorillonite is usually modified through substitution of sodium ions with organic onium ions. The result is an organo-clay complex.

The intercalation procedure expands the gallery and improves the compatibility of the complex with the polymer so that individual platelets can be more easily separated. Possible morphologies of polymers and organo-clay complexes.There are several potential approaches for the production of the polymer-silicate nanocomposite. One procedure involves the synthesis of the compound in the polymerisation stage with the complex *in-situ* with the liquid monomer. Also, nanocomposites could be prepared from solvents in which both polymer and complex are dissolved. Another method involves the preparation of polymer-silicate nanocomposites via melt processing. In this process, the application of shear during compounding assists exfoliation and dispersion. The melt-compounding route is a powerful approach in the production of polymersilicate nanocomposites since existing technologies and equipment could be utilised and easily scaled to commercial quantities. Additionally, this approach allows specialty materials such as those with custom colours, static dissipation, or wear resistance to be readily manufactured. Polymer-silicate compounds were prepared by compounding various organo-clay complexes at different loading levels with commercially available nylon-6 using a twin-screw extruder. Neat nylon-6 and the compounds were dried at 80°C under vacuum prior to compounding, molding, and rheometry. The extent of intercalation, exfoliation, and dispersion of the complexes in the compounds was determined. This was accomplished through wideangle x-ray diffraction (XRD) for preliminary determinations.

Transmission electron microscopy (TEM) was used to confirm the morphology development established by XRD. Test specimens were prepared by injection molding. The mechanical and electrical properties of the specimens were evaluated according to ASTM test methods: D-638 for tensile, D-790 flexural, D-256 for Izod impact, D-648 for deflection temperature, D-955 for shrinkage, and D-570 for water absorption. The barrier capability of these materials was explored by measuring oxygen transmission rates (OTR) with a Mocan apparatus. OTR measurements were performed at constant temperature and humidity for 200 hours on specimens thermoformed from injection molded plaques. Compound melt flow behaviour was analysed with a capillary rheometer. The effect of shear rate at 235°C and melt flow stability at 235°C and 1000 sec^{-1} were determined.

Suitable morphology development in a nanocomposite via the compounding route consists of wetout, intercalation, exfoliation, and dispersion. Therefore, the selection of an appropriate complex for the matrix polymer plays a critical role in determining the potential degree of exfoliation and dispersion in a polymer-silicate compound. One procedure that can be used to assess the extent of intercalation and exfoliation is to analyze x-ray diffraction patterns. XRD patterns for nylon-6 compounds containing two different types of complexes. Finally, not only does this show that XRD patterns can be used to evaluate nanocomposite development but also that nylon-6 nanocomposites are indeed possible via melt compounding. The properties of a typical nylon-6 nanocomposite are given and compared to the unfilled matrix nylon-6. At almost no cost in specific gravity, significant improvements in certain properties are realised. For instance, flexural modulus is increased by almost 50 per cent and deflection temperature raised by over 35°C. Also, tensile strength is increased by over a third. For comparison purposes, the properties of typical mineral and glass fiber filled nylon-6 compounds are given. Reinforcement efficiencies of nanocomposites compared to conventional mineral fillers on a weight basis. Besides mechanical and heat resistance property improvements, nanocomposite compounds can exhibit dramatically enhanced barrier properties as represented by lower oxygen transmission rates. Nylon-6 nanocomposites can achieve an OTR almost four times lower than the unfilled nylon-6. The reduction in OTR is typically attributed to the increase in effective diffusion length as solutes must travel a tortuous path around well-dispersed platelets of high aspect ratio. It is also important to point out that as a result of the nanometer length scale that excellent transparency is retained in sheets and films formed from nylon-6 nanocomposites.

There has been some concern about the ability to process nanocomposites due to the high surface area of the nanoclay upon exfoliation. For instance, the effect of other high surface area additives such as high structure carbon black is to dramatically increase the viscosity. The viscosity of a typical nylon-6 nanocomposite is compared to the viscosity of the neat polymer as a function of shear rate. Contrary to the concerns, the viscosity is observed to be lower than the neat resin at the shear rates of most interest to potential processors. As a result, nylon-6 nanocomposites may be processed as easily as the neat resin. There are several potential explanations for the shear thinning at higher shear rates. One possible reason could be due to orientation of the platelets upon shear. Another reason could be that the polymer has been degraded during compounding with the nanoclay.

However, degradation was not obvious in the nylon-6 nanocomposite samples. The melt flow stability of a nylon-6 nanocomposite was determined. The viscosity was found to not significantly decrease with time. Additionally, the viscosity is equivalent to the viscosity measured under similar conditions during prior testing of the effect of shear rate. Therefore, the degradation if present, would not be expected to continue on further processing. The dimensional stability of nylon-6 nanocomposites is compared. The unfilled resin and other typical reinforced compounds. As seen, the water absorption of the nanocomposite is dramatically reduced by as much as 55 per cent. This is comparable to a filled compound which contains much higher content. This is due to the barrier properties since resin replacement cannot entirely account for all of the change as is the case for conventional mineral and glass compounds. Also observed in table is that the shrinkage of nylon-6 nanocomposites is more similar to the base resin due to the lower filler content. Additionally, the shrinkage in the transverse direction does not differ significantly from the flow direction. This gives a warpage factor much closer to one as shown.

Polymer-clay nanocomposites have been developed via the compounding route in nylon-6. XRD and TEM analysis confirmed that good exfoliation and dispersion was attained. Notable property improvements were observed in tensile strength, flexural modulus, heat deflection temperature, and oxygen transmission rate. Proposed application areas include those requiring structural properties. For structural applications, nylon-6 nanocomposites exhibit equivalent or better properties than typical 20-30 wt per cent mineral filled compounds depending on the property being investigated. It is important to note that these gains come at almost no cost in specific gravity from the unfilled nylon-6. However, the properties of nylon-6 nanocomposites have yet to approach those of glass fiber reinforced compounds. Even so nylon-6 nanocomposites are excellent candidates when improved heat deflection temperature, stiffness, or strength is required especially when glass fiber reinforcement cannot be considered either due to specific gravity, mold shrinkage/ warpage, or wear. Another potential application area is barrier. A significant four-fold improvement was observed in the oxygen transmission rate. This may allow nylon-6 nanocomposites to be utilised in certain packaging and film applications.

Dendrimer based nanocomposites 1-5 are novel organic-inorganic hybrid nanoparticles synthesised by dispersing very small inorganic domains in nanoscopic size polymeric networks. Due to the molecular level mixing of their components they display unique material properties in addition to properties represented by their individual components. Dendrimers are often used as templates in making these particles because they can be made uniform, and using these well-controlled and uniform templates leads to well controlled and uniform composite particles. Apart from obvious materials science applications, such as catalysis and photonic materials, dendrimer nanocomposites (DNC) have a great potential in biomedical applications due to their controlled composition, predetermined size, shape and optional surface functionalities. They may be used as drug delivery vehicles to deliver bioactive guests such as silver or for an incorporated radioactive isotope. Radioactive dendrimer nanocomposites can be delivered directly to the tumor by injecting nanoparticle solutions directly into the tumor microvasculature.6 If a short-lived isotope is used, a procedure like that may serve as an alternate method to direct irradiation. It has also been envisioned that if the surface of such a nanoparticle had been equipped by appropriate targeting moieties the nanoscopic device could find its way selectively to cancerous cells anywhere in the body. A relatively new type of hybrid is based on the silscsquioxane cage structure, which can be regarded as one of the smallest forms of silica or even as 'molecular silica'. This approach has been developed by Hybrid Plastics (USA) and is called polyhedral oligomeric silsesquioxane (POSS) nanotechnology. The particle diameter of the POSS molecules is in the range 0.7-30 Å, on average much smaller that the diameter of, for example, colloidal silica particles. Polymers containing the POSS molecules are therefore true nanocomposites. The physical form of the materials varies from liquids through waxes to crystalline solids. The cost of the POSS chemicals is projected to fall between that of fumed silica and the silicones. The POSS structures can be functionalised with a wide variety of groups, leading to a range of POSS monomers. The nature of the functional group determines compatibility with the polymer matrix. The R group can be simple alkyl, cycloalkyl or aryl, or reactive/ polymerisable groups such as acrylic, a-olefin, styrene, epoxide, carboxylic acid, isocyanate, amine, alcohol and silane. Using such functionalisation, the POSS structures can be either copolymerised with a range of monomers, or grafted onto polymer chains. A variety of phase morphologies is possible, ranging from phase separated to phase inverted to single phase, truly molecular level reinforcement. The POSS compounds can be readily incorporated into polymers using standard processing methods. Interestingly, and in contrast to clay nanocomposites, incorporation of POSS can lead to a reduction in the polymer melt viscosity, which should represent a distinct advantage for many applications. A major effect of POSS incorporation, e.g. in polypropylene, is retention of modulus above the glass transition temperature (Tg) of the unreinforced polymer. Increasing the concentration of the POSS structures in copolymers can have a dramatic effect in increasing the Tg of the polymer. POSS nanocomposites occupy the performance region indicated as 'Hybrids'. Flammability behaviour of the polymer is also claimed to be improved by incorporating POSS.

POSS silanols can be used as coupling agents for fillers and offer a number of advantages over conventional silane coupling agents, including moisture stability, low volatility, no requirement for water addition and no formation of volatile by-products. Surface coverage by POSS cages is also very high. Use of POSS surface treatment can enhance the reinforcing effect of the filler. POSS products are attracting interest in a wide range of materials markets, including aerospace, sporting goods, medical applications and electronics.

Analysizing Nanoclusters, Nanoparticles and Nanocrystals

Nanoscience has taken scientists around the world by storm. It claims to revolutionize the world we live in with radical breakthroughs in areas such as materials and manufacturing, electronics, medicine and healthcare, environment and energy, chemical and pharmaceutical, biotechnology and agriculture, computation and information technology. Nanoparticles are small clusters of atoms about 1 to 100 nanometers long. 'Nano' derives from the Greek word "*nanos*", which means dwarf or extremely small. It can be used as a prefix for any unit like a second or a liter to mean a billionth of that unit. A nanosecond is a billionth of a second. A nanoliter is a billionth of a liter. And therefore a nanometer is a billionth of a meter or 10^{-9} m.

- If a baseball is the size of Earth, a nanoparticle would be the size of an apple.
- We can also compare it with things in the natural world.
- The smallest synthetic nanotechnology made so far are the individual components of an Intel Pentium III microprocessor and they span about 200 nanometers.

Nanoparticles are larger than individual atoms and molecules but are smaller than bulk solid. Hence, they obey neither absolute quantum chemistry nor laws of classical physics and have properties that differ markedly from those expected. There are two major phenomenons that are responsible for these differences. First is the high dispersity of nanocrystalline systems. As the size of a crystal is reduced, the number of atoms at the surface of the crystal compared to the number of atoms in the crystal itself, increases. For example, a 4 nm diameter CdS nanoparticle has about 1500 atoms, of which about a third are on the surface. Properties, which are usually determined by the molecular structure of the bulk lattice, now become increasingly dominated by the defect structure of the surface. The second phenomenon occurs noticeably only in metals and semiconductors. It is called size quantisation and arises because the size of a nanoparticle is comparable to the de Broglie wavelength of its charge carriers (*i.e.* electrons and holes). Due to the spatial confinement of the charge carriers, the edge of the valance and conduction bands split into discrete, quantised, electronic levels. These electronic levels are similar to those in atoms and molecules. The spacing of the electronic levels and the bandgap increases with decreasing particle size. This is because the electron hole pairs are now much closer together and the Coulombic interaction between them can no longer be neglected giving an overall higher kinetic energy. This increase in bandgap can be observed experimentally by the blue-shift in the absorption spectrum or sometimes even visually by the colour of the samples. A larger bandgap means that more energy is required to excite an electron from the valance band to the conduction band and hence light of a higher frequency and lower wavelength would be absorbed. The increase in molar absorption coefficient shows that the oscillator strength also increases as particle size decreases. This is due to the strong overlapping of the wave functions of confined charge carriers. Experiments carried out in the past have found the absorption spectrum to be potential dependent, with decreasing wavelengths as applied potential decreases. This change is fully reversible before the threshold potential.

The first generation of nanofillers is emerging from R&D into commercial application in composites. Federal funding for nanotechnology research and development in USA has increased six-fold, from $116 million (USD) in 1997 to an estimated $961 million in 2004. In the U.S. alone, the FY 2005 budget request that President George W. Bush has sent to Congress calls for a total National Nanotechnology Initiative (NNI) budget of $982 million, which represents a 2 per cent increase. With all this research, what has come out of the R&D pipeline ready for commercial application? As it turns out, plenty. The following are new and, surprisingly, not-so-new applications of two types of nanotechnology: ceramic nanoparticles and carbon nanotubes used as highly functional fillers/additives in polymer products of interest to composites manufacturers. Introduction of these materials can result in the improvement of the polymer's mechanical properties in molded parts and coatings on a wide variety of fronts, including increased part strength, ductility and dimensional stability as well as resistance to damage by abrasion and exposure to caustic chemicals and/or extreme heat. Nanotechnology involves the control and manipulation of materials at the nanoscale—particle sizes from 1 to 100 nanometers (nm), where 1 nm equals 1 *billionth* of a meter— to create new materials and structures that have novel properties due to their small size. Conventional multifunctional materials, such as fiber-reinforced or filled polymers, require a trade-off between various mechanical and physical properties, cost and processability. However, combinations of properties derived at the nanoscale provide opportunities to circumvent these traditional trade-offs. The transition from microparticles to nanoparticles leads to a number of changes in physical properties. One of the major factors is the increase in the ratio of surface area to volume. As the surface area of a particle increases, the portion of its constituent atoms at or near the surface increases exponentially, creating more sites for bonding, catalysis or reaction with surrounding materials, resulting in improved properties such as increased strength or chemical or heat resistance. Additionally, the fact that nanoparticles have dimensions below the critical wavelength of light can render them transparent, which has implications for part aesthetics. One of the most common groups of the new generation of commercial nanotechnologies is ceramics. Ceramics can be categorised into two subgroups: silicates (or silicon oxide), generally in the form of nanoscale flakes of clay, and metal oxide ceramics such as titanium, zinc, aluminum and iron oxides.

Nanoclays—The most common type is montmorillonite (MMT), a layered aluminosilicate in the smectite family of clays. MMT is a component of bentonite clay, formed when volcanic ash settled into ancient brine seas, and now mined in

Wyoming, western Europe, China, Australia, Iran and elsewhere worldwide. Unlike clay minerals such as talc and mica that have been used as fillers for years, MMT can be delaminated and dispersed into individual layers only one nanometer thick by about 70 nm to 150 nm across. The result is a radical increase in the surface area-to-volume ratio: With a surface area of 750 m2/g, 20g of MMT platelets could cover a football field! Southern Clay Products Inc. (Austin, Texas), a manufacturer of functional clay-based additives, began MMT product development in the early 1990s. The company's Cloisite additives, commercially available since 1999, consist of organically modified, nanometer-sized MMT platelets. Unseparated MMT particles are often referred to as tactoids. When tactoids are partially separated in a polymer, they are referred to as being intercalated. When thoroughly separated, individual platelets are said to be exfoliated. To form a nanocomposite, the clay must first be modified with an appropriate organic surface treatment to enable complete dispersion into and miscibility with the polymer matrix. Then the tactoids must be exfoliated into platelets either through the shear stresses exerted during melt compounding or via chemical processes during polymerisation of the matrix. As Karl Kamena, Southern Clay's commercial manager of Cloisite nanoclays, describes it, "The incorporation of single-digit percentages of high surface-area nanoclays into host polymer systems improves the performance of the specific polymer matrix without significant deterioration of properties." The numerous benefits that Cloisite can confer on the host polymer include increased mechanical properties, such as modulus, melt strength and scratch resistance, as well as increased heat resistance, dimensional stability and flame retardation.

The extent of enhancement is related directly to the degree of exfoliation. Currently, about 40 per cent of the MMT platelets can be fully exfoliated, with the remainder forming tactoids of 2 to 10 platelets. Southern Clay continues to improve exfoliation methods, anticipating that, eventually, remaining tactoids will contain as few as one to three platelets. In 1998, Southern Clay began working with General Motors (GM, Detroit, Mich.) and resin manufacturer Basell (Hoofddorp, The Netherlands) to develop a nanoclay-reinforced thermoplastic olefin (TPO) that would outperform the talc-filled TPO then in use. An 8-micron Cloisite clay, each particle of which contains one million platelets, was selected as the filler. Work carried out jointly by the companies allowed the new material to be formulated to different shrinkage rates and to match colour with conventionally filled molded-in-colour parts. Early on, the 8-micron particles had a tendency during the extrusion stage (when the clay was mixed into the TPO) to aggregate into 200-micron clumps rather than exfoliate, causing surface defects and painting problems in the injection molded parts. The team then proved that this issue could be resolved by making adjustments to the extrusion temperature, the extruder screw rotation speed, the material feed rate and other variables during the nanocomposite manufacturing process. GM first introduced a 3 per cent nanoclay-filled TPO on the step assists for the 2002 GMC *Safari* and

Chevrolet *Astro* vans, which had been using a 20 per cent talc-filled TPO. The same material moved into a higher volume application in the body side trim moldings for the 2004 and 2005 Chevrolet *Impala*, molded at Blackhawk Automotive Plastics (Troy, Mich.). Will Rodgers, GM staff research scientist, says GM saw weight savings of 3 to 21 per cent on these redesigned parts. With increased confidence in the material, GM now is using 7 lb of molded-in-colour nanocomposite parts to make this year's *Hummer H2 SUT* cargo bed trim (the center bridge, the sail panels and the box rail protectors), all molded by SportRack Automotive (Sterling Heights, Mich.). These are the first parts that GM has designed as nanocomposites from conception, choosing the material primarily because of its low coefficient of thermal expansion (CTE) and improved appearance. For example, by varying the surface finish of the tool, three levels of gloss were molded into the box rail protectors to match adjoining parts, something that was never possible with talc-filled TPO. GM also reports better scratch/mar performance, increased cohesive strength between the paint and the part, greater ductility at low temperatures, lower flammability, a larger injection molding process window, sharper feature lines, improved knit line appearance and the elimination of tiger striping, (Tiger striping is caused when a filled material groups itself into lines of higher and lower fill weight or viscosity as it flows into a mold, giving the part a striped appearance.) Although the MMT-filled TPO is about an order of magnitude more expensive than the talc-filled material (about $3/lb vs. $0.35/lb), the improvements in part performance and appearance apparently justify the cost.

Metal oxides—Scratch resistance is a highly sought quality in coatings, such as automotive finishes. However, use of silica or clay fillers in quantities sufficient to improve scratch resistance tends to make the coating hazy and reduce its viscosity. Alumina particles are harder than silica but also lead to haziness. This is particularly troublesome in clear coats. Nanophase Technologies Corp. (Romeoville, Ill.) and BYK-Chemie (Wesel, Germany) have teamed to address this problem with two new additives marketed under the tradenames Nanobyk-3600 and Nanobyk-3601. Both additives contain Nanophase's 37-nm diameter Nanodur aluminum oxide particles in concentrations of 30 to 50 per cent by weight, and are intended for use with aqueous (Nanobyk-3600) and nonaqueous UV-curable coatings (Nanobyk 3601). Nanophase Technologies produces its aluminum oxide nanoparticles via physical vapor synthesis: Arc energy is applied to a metallic precursor to generate a vapor at a high temperature. A reactant gas is introduced to the vapor and then the mixture is cooled at a controlled rate so that it condenses and forms nanoparticles, which then are coated with a thin polymeric shell that ensures their compatibility with various polymers and fluids. When the additives are blended into resins and coatings at very low 1.5 to 6 per cent concentrations, scratch resistance reportedly goes up dramatically. In Nanophase tests, particle loadings of 1 per cent resulted in five- to ten-fold increases in scratch resistance, depending on the resin formulation tested, with less than 0.5 percent haze. (At greater than 1 per cent haze, or about a 2 per cent loading of Nanodur alumina particles, a resin is no longer considered "clear.")

Exploration of carbon nanotubes is relatively new. Single wall nanotubes (SWNTs) are long, thin cylinders made of carbon atoms arranged in a graphitic lattice structure that may or may not have closed ends. Multiwall nanotubes (MWNTs) are concentric cylinders of carbon in a similar graphite structure, held together by weak intermolecular-level forces. SWNTs and MWNTs have length-to-diameter ratios that can range from 100:1 to 10,000:1. Their large surface-area-to-volume ratios mean that carbon nanotubes (CNTs) affect matrix properties far beyond what would be predicted from rule-of-mixtures calculations. Nanotubes are currently being commercialised in applications where electrical conductivity and increased mechanical performance are primary design drivers.

Electrical properties. SWNTs can possess electrical conductivity of 1 billion amps/cm2, Young's modulus of more than 1 TeraPascal and a maximum strain of 10 per cent. MWNTs have lesser properties but are easier to produce, though they also are less well understood, due to their greater complexity and variety. Orders of magnitude less costly than SWNTs, MWNTs are more attractive for commercial applications. While they have been used for the last 10 years in thermoplastics as a direct replacement for carbon black in electrical applications, recent developments are enabling their use in thermosets, and significant efforts are underway to exploit their mechanical properties as well. Carbon black and carbon nanotubes are both pure graphitic carbon and so have the same electrical properties. Carbon black particles, however, are spherical, and the rod-like CNTs have a higher aspect ratio, which means far fewer CNT particles are needed to form a conductive "bridge" through a nonconductive polymer. Typical carbon black loading in polymers is 7 to 18 percent (depending on the polymer and the level of conductivity needed), which will make the polymer brittle. CNT's achieve the same conductivity in 1 to 3 percent loadings. Moreover, the CNT's conductivity-versus-loading curve is more gradual and more easily tailored to a specific level of conductivity than carbon black's curve.

In 1984, Hyperion Catalysis (Cambridge, Mass.) was the first company to file a patent on a nanotube manufacturing method. Hyperion's MWNTs have 8 to 12 walls and are about 10 nm wide by 10 micrometers long, an aspect ratio of 1,000:1. They are made through a process which Hyperion's sales and marketing manager Pat Collins describes as "continuous, catalysed vapor phase decomposition and recombination, using light hydrocarbon feedstock." When an atomised metallic catalyst comes in contact with a hydrocarbon gas in a reactor under appropriate time, pressure, and temperature conditions, nanotubes will grow on the catalyst. Each nanotube has a tiny remnant of catalyst inside it. The MWNTs, as formed, tend to clump together in aggregates, attracted to each other by intermolecular Van der

Waals forces. The high shear forces which are necessary to disperse the CNTs into the resin are obtained by using a high-viscosity resin and a twin-screw extruder typically used to compound thermoplastics. The disentangled and dispersed nanotube/resin mixture is formulated in concentrations of up to 20 per cent. A wide variety of nanotube-filled thermoplastics can be purchased, either ready to use with low CNT loadings or as a master batch with higher nanotube concentrations, which then can be let down (diluted) to the appropriate usage level. The ready-to-use resins are offered in antistatic, electrostatic dissipative and conductive grades, with volume resistivities ranging from 1010 ohm/cm for antistatic grades to 1 ohm/cm for conductive grades (as compared to 1.68 x 10-6 ohm/cm for copper). Because the nanotubes are in such low concentrations and are so small, their addition has little effect on the base resin's properties, allowing the material to be processed on the injection molder's existing equipment under conditions similar to those used to process the base resin. Hyperion's continuous process runs 24 hours a day, seven days a week, producing tons of pure (no byproducts, such as amorphous carbon) nanotubes per week. Automotive fuel systems have to be designed for electrostatic discharge. The turbulent flow of fast-moving fuel creates a static charge which must be dissipated. If charge is allowed to build up it can cause pinholes to develop in the fuel system and may damage electronic sensors. On rare occasions, sparks may even discharge during refueling.

The fuel system, which includes the filler neck, inlet check valves, fuel module (pump/sender unit) inside the tank, the fuel lines and the fuel filter, must all, therefore, be conductive. (The Society of Automotive Engineers (SAE) will soon issue its newly revised *Recommended Practice, SAE J1645*, which will outline how all materials that touch fast-moving fuel must be grounded.) Fuel systems static dissipation is a rapidly growing segment of Hyperion's business, according to Collins. BASF (Ludwigshafen, Germany) has already commercialised a conductive polyoxymethylene (POM) with Hyperion nanotubes, specifically for automotive fuel systems. CNT-reinforced nylon 6, manufactured by Degussa (Parsippany, N.J.), is used for the tubing in the flexible portion of the fuel line in 70 percent of cars made in North America. O-rings in the system also must be conductive or they will build unwanted capacitance. Precix (New Bedford, Mass.) recently introduced its F98 series of O-rings, made by incorporating Hyperion's multiwall nanotubes into Precix's own proprietary flouroelastomer resin system. Compared to the company's carbon black-filled O-rings, the F98 compound is said to have three to five times greater resistance to fuel vapor permeation, with significantly reduced fuel swell (the expansion of part dimensions due to absorption of the fuel) as well as a durometer of 79 versus 88, allowing for much lower insertion pressure when assembling parts. Hyperion also supplies various conductive thermoplastics to manufacturers such as GE Plastics (Pittsfield, Mass) for use as injection molding material for electrostatically painted automotive body parts, such as mirror housings, door handles and fenders. According to a *CT* web search, the Renault *Clio* and *Mégane* currently sport fenders made from a nanotube-filled polyamide/polyphenylene ether (PA/PPE) blend. Since CNT loadings are typically 2 to 3 percent rather than the 12 to 15 per cent necessary with carbon black, the plastics do not become brittle, thus negating the need for impact modifiers which would produce the undesirable side effect of reducing the plastic's glass transition temperature (Tg). The CNT-reinforced parts need no primer and can be painted at the same temperature and on the same electrostatic line used with metal parts, reducing cost and improving colour matching. Because the CNT concentrations are low, the surface finish also is smoother than plastics with other types of conductive fillers. Collins adds that Hyperion now has branched out beyond thermoplastics into thermosets, with nanotube-filled versions of Resolution Performance Products' (Ennis, Texas) Epon 1009 and Epon 828 resins. Epon 1009 is a high molecular weight, solid pellet form of bisphenol-A, which can be compounded into liquid epoxies or vinyl esters.

Hyperion sells the Epon 828 nanocomposite in paste form, which is made by grinding the Epon 1009 nanocomposite pellets into a powder and blending the powder into the 828 with a multishaft double planetary mixer in a liquid dispersion process. Collins also notes that Hyperion will introduce easier-to-use liquid master batches of other nanotube-filled epoxies in the near future, primarily for electrical applications, such as conductive epoxies, potting compounds and connectors. Although the nanotubes may add \$5/lb to \$6/lb to the final polymer mix, Collins says there are many applications where the cost is justified either through cost savings in the final product manufacturing process or because the necessary properties cannot be obtained any other way.

Mechanical properties. Zyvex Corp. (Richardson, Texas) approaches the market from a different angle, seeking to use specially modified carbon nanotubes to increase the mechanical performance in sporting goods, aerospace and defense applications. Zyvex does not make its own nanotubes, but instead uses SWNTs and MWNTs from other suppliers, and certifies them for quality and batch-to-batch consistency using a series of standard analytical tests. Zyvex uses a patented process to functionalize the surface of the nanomaterials with a non-covalently bonded "Kentera" molecule which non-covalently bonds the nanomaterial surface via pi-pi bonding and also adheres well to the host matrix. Pi-pi bonding, which involves a slightly stronger molecular force than the intermolecular forces that cause the nanotubes to clump together, enables the Kentera molecule to slip between the nanomaterials and disperse them. Zyvex offers its Nanosolve functionalised nanotubes and nanofibers as dispersions in solvents or water additives, or in solvent-free concentrates, which can be used in thermoplastics and a subset of thermosets comprising polyurethanes and epoxies. This includes concentrates in Resolution Performance Products' Epon

826, 828 and 862 at customised weight loadings specific to an application, whether it is thermal, electrical or mechanical. Marni Rutkofsky, Nanosolve chemist and technical sales representative, is overseeing third-party ASTM physical and mechanical testing on several materials. She says preliminary results at less than 5 per cent loading of MWNTs in thermoplastic polycarbonate show increases of 80, 50 and 150 per cent, respectively, in tensile strength, elongation and Young's modulus. Results at similar loading in polyurethane show a four-fold increase in thermal conductivity. Data on epoxy is expected later this year.

One beneficiary of Zyvex research is Easton Sports (Van Nuys, Calif.) a leading manufacturer of bicycling, hockey, baseball, softball and archery equipment. Three years ago, Easton embarked on an ambitious project to explore the advantages that nanotubes could bring to bicycle components bladder molded at its production facility in Dongguan City, Guangdong Province, China. Easton drum winds its own prepreg and planned to incorporate CNTs into the resin before introducing the resin to the fiber. The challenge was to find a nanotube chemistry that would remain well dispersed in the resin during the course of Easton's manufacturing process and yield improved properties at a reasonable cost. Zyvex and Easton tried several iterations of the resin filled with the CNT/Kentera chemistry before developing a formulation that met the challenge. "We were looking more for strength increase than weight decrease," says John Harrington, VP of Easton's Bicycle Div. "Our products are already some of the lightest on the market, and we want to be conservative with weight reductions when we are using a new material." However, Easton has shaved 10g to 20g from most of the CNT-filled prepreg components, while achieving an average 15 per cent increase in product flexural strength. Road, mountain, and triathlon handlebars, forks, seatposts and OEM seatstays and chainstays are already on the market. The CNT Technology logo distinguishes components which have incorporated nanotubes, and most of the rest of Easton's bicycle components will follow suit. Retail prices on the logoed components have gone up an average of only 10 per cent—an increase that Harrington says consumers are willing to pay for the increase in performance. While nanosized fillers aren't an overnight revolution, it is clear that there is significant activity on many fronts, with increasing use on all. Beyond the applications highlighted here, nanofillers have been incorporated into in golf clubs and golf balls, tennis rackets and a number of other products, with many more to come. The main barriers to nanomaterials use now seem to be development cost and the end-user's ability to obtain proper dispersion in and compatibility with resins. However, all of the nanosuppliers interviewed above are focused on making materials less expensive and more accessible by scaling up production, standardising products, issuing test data and offering them in user-friendly forms.

You must be wondering by now after all the 'over-the-head' theory about how different nanoparticles are from bulk and atomic material, what this really mean to us average everyday individuals. A lot!! Due to the two unique phenomenon which occurs in nanoparticles, their properties (electrical, optical, chemical, mechanical, magnetic, etc.) can be selectively controlled by engineering the size, morphology, and composition of the particles. These new substances will have enhanced or entirely different properties from their parent materials. Nanoparticles can make metals stronger and harder, give ceramics enhanced ductility and formability, enable normally insulating materials to conduct heat or electricity, and make protective coatings transparent. Hence, industries can re-engineer many existing products and design novel new products/processes that function at unprecedented levels. Use the menu bar on the top to read about more about some examples of the possible applications.

Solar cells are photovoltaic devices, which are devices that can convert light into electricity. They are much in demand as they offer an inexhaustible and environmentally benign energy source. Most solar cells are made of amorphous silicon. The problem with this is that the silicon must be of a very high purity and have a near perfect crystal structure. This makes it very expensive to produce. The efficiency of such a cell is also very small, typically converting only 13-18 per cent of sunlight to electricity. However, low efficiency wouldn't matter if huge arrays of cells could be produced cheaply. After all, nature's solar cells, chloroplasts in plants are less than 1 per cent efficient. Most solar powered devices rely on the same principle: a photon of sunlight boosts an electron in the material into a mobile state so that it can be used to generate electricity. The problem with this simple mechanism is that the electrons are negatively charged and will leave a positive charge. These opposite charges attract one another and therefore will tend to recombine, squandering the absorbed energy as heat or as re-emitted light. Silicon solar cells use an electric field to push the negatively charged electrons and positive charges apart. While chloroplasts adopt a more subtle approach of separate charges by making a distinction between the units that generate the electron and those that transport it away. Deciding to copy nature's trick, Michael Gratzel and Brian O'Regan at the Swiss Federal Institute of Technology began research and produced the Grätzel cell in 1991. The Grätzel cell uses intensely coloured organic dye molecules to capture light energy to inject an electron from the dye into a semiconductor such as titanium dioxide (TiO_2). This remarkably efficient charge separation reaction initiates current flow and the output of electrical energy by the cell.

The ideal material used in the cell must have a high surface area for light absorption and charge separation. Nanoparticles, having a comparable surface area to volume ratio, provides for just that. Titanium dioxide nanoparticles are used to make nanoporous thin film supported upon a glass substrate. The material obtained has optical transparency, excellent stability and good electrical conductivity. The benefit of these novel photoelectrical solar cells is that they can be fabricated from cheap, low purity materials by simple and low cost procedures. Contrary to expectation, some of the new devices also have strikingly high conversion efficiency. The size-tunable bandgaps of the semiconductor nanoparticles, due to size quantisation, also means more efficient solar cells can be produced for photovoltaics (electricity production) and water splitting (hydrogen production) processes.

Electrochromism is defined as a reversible colour change of a material caused by the application of an electrical current or potential. It has potential applications in making smart windows for the modulation of incoming light, as well as displays, a field in which competition with liquid crystal displays (LCD) is possible.

There used to be only two types of electrochromic devices. In the first type, the electrodes are made of conducting glass and covered with an organic or inorganic polymer. The two materials would usually display complementary electrochromism and hence produce the same colour change when one is oxidised while the other is reduced. This type of construction is used in electrochromic windows. It is bistable, which means that once the colour change has occurred, the state of the device remains even in absence of applied voltage. The limitation of this typ e of electrochromic devicesis .the slowness of the colour change, due to thelow migration rate of the counterions in the bulk polymer. It is also difficult to obtain strong colour changes or bright colours. In the second type, two complementary electrochromic molecules are dissolved in a solvent. This type of system is very simple to build, reacts very fast and can produce dark or bright colours. The drawback of this type is that an electrical current is needed to maintain the coloured state because the two types of coloured molecules diffuse through the system and react with each other to restore the bleached states. It therefore cannot be used for large area devices or for battery-powered displays.

We therefore need a system which is both bistable and changes colour rapidly. This can be done by attaching a suitable molecule that is colourless in the oxidised state and coloured in the reduced state onto the surface of a colourless semiconductor on conducting glass. When a sufficiently negative potential is applied, electrons are injected from the conducting glass into the conduction band of the semiconductor and reduce the adsorbed molecules. The reverse process occurs when a positive potential is applied and the molecules get bleached. The advantage of this system is that combines immobility of the electrochromic material with the rapidity and colouration efficiency of molecular systems. However, a single molecular monolayer does not absorb a perceptible amount of light. This is where nanoparticles come in!

By using nanocrystalline semiconductor films, it was possible to amplify the light absorption of the molecular monolayer to obtain visible colour changes. The nanocrystalline layer are highly porous and usually the real surface of a 5 μm thick layer is about 500 times the projected area. Therefore, as light passes through the layer, it crosses several hundreds of monolayers of coloured molecules giving a strong absorption.

Some commercial products have already been produced! The most successful is the automatic dimming rear-view mirror sense ambient light and glare from behind, and reduce annoying glare from bright lights. While the company NTera has produced paper quality displays. These displays are bi-stable, not requiring much power keep to a pixel coloured, and reflective, not needing power for backlighting, and therefore can operate with very little power. Coupled with the fast switching, (*i.e.* changing of colour) due to amplification with nanoparticles, this technology could in most flat panel displays.

The invention of the transistor in 1948 by John Bardeen and William Shockley, triggered a new era in electronics. The trend since then has been to create smaller and smaller product using fewer chips of greater complexity and smaller 'feature' sizes. The development of integrated circuits and storage devices have continued to progress at an exponential rate. At present it takes two or three years for each successive halving of component size. However, the technologies currently in use have fundamental limits below, which the devices no longer function in a predictable manner. For instance, oxide layers used in Complimentary Metal Oxide Semiconductors (CMOS) devices are becoming so thin that leakage currents are conducted quantum mechanically by electron tunneling. Therefore, replacement technology to advance miniaturisation even further down, eventually to the dimensions of single atoms and molecules, is keenly sought. In 1985, Dmitri Averin and Konstantin Likharev, proposed the idea of a new three-terminal device called a single electron tunneling (SET) transistor. Two years later Theodore Fulton and Gerald Dolan at Bell Labs in the US fabricated such a device and demonstrated how it operates. Since then much research has gone into this area and better single electron devices has been fabricated. These devices usual have a molecule or atom as a functional unit and operate based on the quantum effect. Due to the small dimensions of the device (a couple of nanometers and hence is a nanoparticle), there are repulsive Coulombic interactions between the electrons and energy is required to add an extra electron. This energy cost can be tuned to zero by applying a voltage to the gate electrode. The electrostatic potential at this gate will be such that an extra electron can hop from the source onto the molecule. However, due to Coulomb repulsion, a second, extra, electron cannot hop on at the same time. The first electron must leave the molecule, moving into the drain before the next electron can enter. This one-by-one electron motion, is known as the single electron quantum tunneling. The creation an ultrafast, single electron transistor would lead to the development of "quantum" computers with supercomputer powers and the size of a thumbtack. However, there is still a long road ahead before atomic or molecular transistors can be assembled into viable, dense, fast logic-circuits. Currently these single electron devices are still no competition for silicon transistors. But they will serve both scientifically, for studying electron motion through nanoscale objects, and technologically, for developing chemical techniques with which to fabricate electronic devices on single molecules.

Nanoart comes in two flavors:

(i) Works created by artists envisioning our nanotechnological future

(ii) Creations and designs produced by nanotechnologists

In 1997, researchers at Cornell University created the world's smallest guitar. Carved out of crystalline silicon it measures only 10 micrometers long, about the size of a single cell, and has six strings, each about 50 nanometers wide. Using an atomic force microscope (AFM) the nanoguitar can be played, but it cannot be heard as its strings resonate at inaudible frequencies.Its creation demonstrates a new technology that could have a variety of uses in fiber optics, displays, sensors and electronics leading to a new generation of microelectromechanical devices. In 1998, writing and printing techniques were miniaturised by a research team led by Chad Mirkin and Charles and Emma Morrison of Northwestern's Center for Nanotechnology when they created the world's smallest pen, utilised by an automated four-colour nanoplotter. The nanolithography had previously allowed us to draw tiny lines with only a single ink or type of molecule. Now, with the nanoplotter, multiple inks, or different kinds of molecules can be placed side by side with such accuracy that the chemical purity of each line can be retained. By 2001, the Northwestern University researchers Seunghun Hong and Mirkinin had invented a nanoplotter that can draw multiple copies of a nanometre-sized pattern: at assorted places on a surface, simultaneously. Their eight-pen nanoplotter could be adapted to use a thousand pens!

The examples so far has focused only on small dimensions of nanoparticles. There are other properties that can be exploited.

Nanoclusters

A nanocluster is a fragment of solid comprising somewhere between a few atoms to a few tens of thousands of atoms. Nanoclusters are therefore a novel state of matter with properties that are neither those of a bulk crystal nor those of individual atoms and molecules. Over the past 10 years huge advances have been made both in the synthesis of size-tunable, monodisperse (i.e. size-selected) nanoclusters of various chemical compositions and in the development of techniques for their assembly into nanostructured solids (facilitating the synthesis of what have been termed "designer materials"). Clusters with sizes in the order of several nanometers can have physical properties with are very different from the bulk material. In just the same way are the (electronic) properties of a cluster very different from those of the individual atoms which make up the cluster. This makes the nanoclusters a very attractive research area in the solid state physics. One of the most characterising properties of these clusters is the very large fraction of surface atoms. This makes the cluster very sensitive to changes at the border of the clusters. Influences from the outside of the clusters are therefore having a great impact on the properties of these clusters. The study of these influences at the border is the main topic of our study. In our research we try to investigate in which way the vibrational state of a cluster is influenced by the surrounding matrix. At the same time we want to get a detailed picture of how these clusters are being formed from a (over)saturated solution, formed by ion implantation. We focus on two different cluster systems. On one hand we study Sn clusters in SiO_2, which is a soft metal in a hard matrix and on the other hand we study Co clusters in Ag which is a hard metal in a soft matrix. These two systems are so to say each others counterpart in the dynamical behaviour we expect. Nanoclusters have a very high zeta potential (negative electrical charge) and are small enough to easily penetrate any cell membrane in the body. Zeta potential represents a basic law of nature. It plays a vital role in the health of all forms of plant and animal life. It is the force that maintains the discreteness of the trillions of circulating cells that nourish an organism. If zeta potential is low, toxins can not be suspended for elimination and nutrients can not be suspended for transportation to the cells. The whole system becomes clogged. Nanoclusters, with their high zeta potential, may increase the absorption of nutrients and the elimination of toxins by helping to maintain the proper polar attraction of negatively charged nutrients to the positively charged intracellular fluid. When these two electrical charges are in the appropriate balance, nutritional absorption and cellular metabolism are at optimum levels for maximum energy and an overall sense of well being.

The properties of silica clusters at temperatures from 1500 to 2800 K are investigated using classical molecular dynamics simulations for particles containing up to 1152 atoms. It was found that atoms in the cluster were arranged in a shell-like structure at low temperatures, and that the radial density profile peaked near the outer edge of the particle. Smaller clusters have much higher pressures, with the magnitudes corresponding quite well to those obtained from the Laplace–Young equation, when evaluated in a self-consistent manner using our derived surface tension. The computed surface tension did not show any significant size-dependent behaviour in contrast to the decreasing surface tension observed for Lennard–Jones liquid clusters. Finally the computed diffusion coefficients in the liquid state are seen to be larger than bulk computed diffusivities. One of the most import methods of synthesising nanoparticles is through vapor phase nucleation and growth. Several methods are used, including combustion, plasmas, thermal reactors, and evaporation-condensation. The evolution of an aerosol undergoing a gas-to-particle growth process is described by a master equation, the "aerosol general dynamic equation" (GDE). The GDE is a population balance equation for the behaviour in time and space of the particle size distribution function, which can include all the driving forces for particle growth (nucleation, surface growth, coagulation/coalescence, transport, etc.). Less conventionally, but of interest for nanoparticle synthesis research, the GDE can be extended to track the degree of coalescence of particle aggregates. One of the biggest challenges in the implementation of the GDE is knowledge of the rates of these individual processes. For nanoparticle synthesis, one of the most fundamental and important kinetic properties of interest in nanoparticle growth is the sintering or coalescence rate between particles during vapor-phase growth. A knowledge of these rates and their dependence on process parameters ultimately impacts the ability to control primary particle and agglomerate growth, which are of critical importance to nanoscale particles whose properties depend strongly on size, morphology, and crystal structure.

Ehrman *et al.* discussed the possible causes of the enhanced rate of silica particle sintering observed in an experiment, including the presence of impurities such as alkali metals which could result in a lower viscosity of silica. On the other hand, in theoretical considerations kinetic coefficients such as the diffusion coefficient and the viscosity, as well as the surface tension, were taken from the properties of bulk material. However, we know that what makes the nanoparticles so interesting is that their fundamental properties are intimately related to the physical size of the component. For an example, a 10-nm-diameter iron particle has almost 20 per cent of its atoms on the surface. This high fraction of surface atoms alters properties such as the melting point and presumably the surface tension14 and accounts for the high reactivity of nanoparticles both toward each other and also toward other materials and gases. One of the interesting aspects of nanoparticles properties is the variation of the internal pressure with particle size.

This implies that as the particle becomes smaller, the pressure will approach infinity as the radius becomes vanishingly small. Of course, this assumes that the surface tension is size independent, but in fact, at some point, the surface tension will tend toward zero. Nevertheless, we should expect that the pressure of a particle can be many hundreds of atmospheres, and may impact the chemical, phase and crystalline structure of particles. For an example, the pressure inside a 10-nm-diameter TiO_2 particle s = 0.5 J/m^2 is on the order of 2000 atmospheres. Such elevated pressures may have a significant impact on properties such as diffusion coefficients and the viscosity, which in turn affect the rates of particle sintering Given these constraints, data obtained at large grain sizes when extrapolated to smaller dimensions may lead to considerable errors. Experimental studies are certainly desirable for investigating these characteristics; however, experiments aimed at determining the fundamental transport and thermodynamic properties of fine aerosols are not easily obtained, because of the ensemble nature of most studies.

Here we study the nanoscale silica clusters of different sizes (N=72, 288, 576, and 1152) and at different temperatures (1500–2800 K) using a classical moleculardynamics (MD) simulation to determine the structural properties, internal pressure, diffusion coefficients, and surface tension. We will particularly focus on the particle size dependence of these properties, and its influence on coalescence/sintering modeling. The clusters consist of N classical atoms which interact through a pair model potential developed by Tsuneyuki *et al.*,which takes into account the steric repulsion due to atomic sizes, screened Coulomb interaction from charge transfer, charge-dipole interactions due to the large electronic polarizability of the anions, and dipole-dipole interactions to mimic the short- and intermediate-order interaction. For higher temperature the potential was modified by adding a Lennard-Jones term, in order to avoid the occasional overriding of the O-O repulsion barrier at a shorter interatomic distance at a higher temperature.

The modified interatomic interaction is referred to as the TTAMm potential with the form $V_{ij}* = V_{ij} + 4e_{ij}\ [(s_{ij}/r_{ij})^{18} - (s_{ij}/r_{ij})^{6}]$, where e_{ij} and s_{ij} are the usual Lennard–Jones parameters. The equations of motion for each atom was solved using the standard velocity version of the Verlet algorithm. In our calculations we constructed a silica cluster with an appropriate number of atoms, and placed it in the center of a spherical cavity of radius R*, which comprised the calculational cell. The center of the calculation cell (r50) coincides with the center of mass of the cluster. The radius of the cell was kept at 2 nm for all clusters. In order to conserve mass at higher temperatures, we force any cluster fragment which escapes the cluster due to evaporation, and reaches the boundary of the cell, to be elastically reflected. All clusters were initially heated to a high temperature liquid state at T=4000 K. Several clusters were then prepared at a variety of temperatures between 1500 and 2800 K by slow numerical cooling. Clusters were allowed to anneal for 35000 time steps (with a time step dt = 1 – 4 fs), followed by a test at constant energy to evaluate if the system had reached an equilibrium state. Production runs for cluster properties were performed at a fixed temperature, and typically required about 1.5×10^{5} time steps to obtain a reasonable statistical averaging of cluster characteristics.

In order to verify the equilibration procedure for each prepared cluster, we gradually decreased the temperature from 4000 to 1500 K, followed by a reversal of this procedure to the desired target temperature. A comparison of cluster properties such as the radial density distribution obtained by cooling and then reverse heating, shows good agreement. The total angular momentum of cluster motion is always reset to zero by the transformation of velocities in order to evaluate transport properties such as diffusion coefficients.

It is not conducive to analyze the cluster behaviour at higher temperatures T>3000 K, where a significant dissociation of fragments takes place during the time interval of the calculation (t μ 0.15 ns). This is because this interatomic potential cannot realistically describe the vaporisation processes of silica. For instance, in the calculations we observe the vaporisation of SiO_2 molecules, while it is well known that silica vaporizes by decomposition to SiO and O_2. At lower temperature T = 1500 K in a solid "glass" state, all clusters exhibit an oscillating density distribution. With increasing temperature there seems to be a tendency to dampen out the radial density gradient, though some peaks still remain. The radial dependence of the mass densities of different-sized clusters at a temperature T=2080 K. Note that the interpeak space is about the mean Si-Si interatom distance which is 0.3175 nm.

To explore the effect of density, we show the radial mass density distribution for a 576-atom cluster at three temperatures. It is clear that the radius of the SiO_2 nanoparticles and, consequently, the averaged density, are effectively insensitive to temperature over quite a large temperature range (1500–2800 K). This is in accordance with the fact that bulk silica has an extremely low thermal expansion coefficient. Over this temperature interval a cluster with N=288 atoms has a radius R=0.92 nm and an average density of p=53.2 g/cm^3. The 576-atom cluster has a radius R=1.16 nm and a density p=52.9 g/cu cm and the cluster of N=1152 atoms has R=1.5 nm and p=2.8 g/cu cm. The radius of a cluster is defined as the distance where the density drops to 0.5 of its interior value. Note that smaller clusters have a higher density, but the values are in close agreement with that calculated for bulk silica in the coesite phase, with a density of 2.9 g/cu cm.

A similar behaviour was observed by Roder *el al* using a BKS interaction potential. They obtained a density of 2.3 g/cu cm. This implies that clusters of different sizes have similar shell structures. Moreover, the cluster surface has the same structure and width regardless of size. The rapid density decrease extends over a distance which is more than a Si-O bond length equal to 0.162 nm. A careful temporal observation makes it clearer that the cluster can be separated into "core"and surface components. On the surface both silicon and oxygen have dangling bonds, with the oxygen atom tending toward the surface. These fragments residing on the surface tend to be highly mobile, so that when averages of the density are taken we see a smooth monotonic decrease in the density profile at the edge of the cluster corresponding to about 0.2–0.3 nm. In general, then, we observe an enhanced density for smaller clusters, small thermal expansion, and an oscillating density profile.

In order to obtain further insight into this oscillating radial density behaviour, we consider the structure of the clusters more closely. A cross-section slice for two different clusters (N=288 and 576 atoms) at T=500 K. The slice is taken through the center of a cluster of thickness 0.328 nm (which is slightly larger than two of nominal Si-O bond lengths) What is being shown are Si and O atom positions following averaging over 75 ps. The slice map of smaller cluster looks like a network of atoms organised in rings. The larger cluster has apparently a spherically symmetric shell structure, and the external shell incorporates both O and Si atoms. The defects in the smaller cluster are denoted by atom trajectories that wander.

The distribution of oxygen and silicon atoms as a function of radial position is presented for the cluster prepared with 576 atoms (192 Si atoms and 384 O atoms). It clearly indicates that oxygen is preferentially found on the surface layer of the cluster under all conditions. Note that the change of temperature has a pronounced effect on the internal arrangement of the particles. In particular, at T=1600 K the peak in the Si concentration corresponds to a minimum in the O atom density. With increasing temperature to T=1680 K, we find that a thermally induced structural transition takes place.

In order to understand the change in the cluster structure we analyze the angular distribution, varying the temperature. The O-Si-O and Si-O-Si angular distributions are shown for a cluster with 576 atoms. At a temperature of T=1600 K, the Si-O-Si angle distribution has two maxima which are at 97.7° and 142°. The former denotes the presence of four member rings, and the latter refers to six member rings. As the temperature increases to T=1680 K, the Si-O-Si angular distribution has only one maximum at 142°.

The results agree well with the angular distributions in the amorphous and molten bulk SiO_2 studied by Vashishta *et al.* who used a three-body potential. For example, at T=2500 K the O-Si-O and Si-O-Si angular distributions have maxima at 110.1° and 141° for the cluster with 576 atoms, and at 109.5° and at 142.5°, respectively, for bulk material. The angular distributions are, however, much broader for the clusters compared to the bulk because of the spherical surface and shell structure. A similar broadening was observed by Roder *et al.* The partial pair correlation functions were calculated for the O-O, Si-O, and Si-Si bonds to understand the enhanced density of smaller particles and mechanism of cluster expansion at higher temperature. It turns out that the first and second peaks of the O-O and Si-O pair correlation functions have the same positions for all clusters.

In contrast to this fact the different clusters have various Si-Si pair correlation functions. One can see that the smaller cluster (N=288 atoms) has a closer Si-Si packing than the largest one (N=576 atoms at T=1500 K. With increasing temperature the peaks of the O-O and Si-O pair correlation function become broader, and their positions remain independent of temperature, whereas the Si-Si pair correlation function is sensitive to the temperature.

Its first peak at r = 0.317 nm does not respond to temperature, but a second peak shifts to the larger coordinate. Thus, we can conclude that the variation of the density with the cluster size as well as the expansion of a cluster with temperature are explained by the change in the Si-Si atom packing due to the strong repulsive nature of the Si-Si interaction.

Semiconductor nanocrystals (also called nanocrystals or quantum dots) can be described as a state of matter intermediate between molecular and bulk crystalline. Their small size and high optical activity make them a very interesting material for study in disciplines ranging from opto-electronics (integrating optical and electronic signals on a single substrate, i.e. crystalline Si) to fluorescence microscopy (as a bright, photostable and minimally perturbative fluorescent marker). This work will concentrate on experiments designed to optically characterize silicon nanocrystals as fluorescent beacons in solution since in biology fluorescent labels are invariably used to image structures or track/probe particles/biomolecules in the liquid state.

Images of single immobilised Si nanocrystals on quartz substrates will also be presented. Just as bulk semiconductors are characterised by the structure of their conduction and valence bands as being direct or indirect, so are semiconductor nanocrystals.

Direct semiconductor nanocrystals (e.g. CdSe, CdS) are characterised by having the minimum transition energy to promote an electron from the valence band to the conduction band without a change in the electron momentum (the energy of this energy separation is known as the band-gap). For indirect semiconductors however excitation at the band gap energy must be accompanied by a change in the electron's momentum (supplied by a phonon). While both types of semiconductors are efficient absorbers of light, it has only been the direct semiconductors which have been emissive enough to be used as an optical source (e.g. in semiconductor GaAs quantum well lasers, Saleh and Teich, 1991) or as fluorescent labels (CdSe) (Chan and Nie, 1998, Bruchez et al, 1998). The discovery of porous silicon as an efficient luminescent material (Canham, 1990) has created great excitement as a prospective candidate for opto-electronic circuit integration. [Integrating direct semiconductor structures on silicon has been unsuccessful due to lattice mismatch]. Though the waited for advances have not materialised, it has been the starting point for experimental and theoretical investigations on nanocrystalline silicon.

Nanocrystalline materials are build from nanoparticles. These particles are also called quantum dots or artificial atoms. These nanoparticles are usually spherical colloids of a semi-conducting material with a diameter of nanometers. Because these colloids are so small new material properties exist which are not present in bulk materials. An important property is that the energy difference between the energy bands, the valence and conducting bands in bulk materials, increases and separate energy levels can be formed. These changes in the energy levels and the small size of the particles result in a new optical and electrical properties which can be used in the devices which were mentioned before. Due to their novel properties and varied potential applications, nanocrystalline materials, with typical grain sizes <100 nm, are attracting increasing attention from researchers all over the world. These materials are traditionally classified as one-dimensional (1-D), or layered structures, two-dimensional (2-D), or rod- or wire-like, and three-dimensional (3-D), or equiaxed, with the 1-D layered and 2-D wire-like structures being extensively investigated for coating and electronic applications and the 3-D equiaxed structures exploited for potential structural or bulk applications. Because of the small grain size of these materials, and consequently the large volume fraction of atoms in or near the grain boundaries, these *materials* exhibit properties that are often superior, and sometimes completely new, in comparison to those of conventional coarse-grained *materials*. However, it should be realised that a full understanding of the structure, thermal stability, and properties of these novel materials is not yet available. A number of issues and concerns still need to be solved.

Nanocrystalline materials can be synthesised by a number of techniques with the starting material existing either in the vapor, liquid, or solid state. Inert-gas condensation, in which the evaporated material is quenched onto a cold substrate, was the first technique used to synthesize nanocrystalline metals and alloys. Subsequently, plasma processing and other physical and chemical vapor deposition techniques have also been used. Electrodeposition and rapid solidification processing use the liquid state as the starting material. Mechanical alloying, friction stir welding, severe plastic deformation, spark erosion, sliding wear, and repeated cold rolling produce nanocrystalline materials in the solid state. While some of these techniques are commercially in use for large-scale production purposes (e.g., inert-gas condensation, electrode-position, and mechanical alloying), many of the other techniques are essentially used for laboratory investigations. The choice of technique for the synthesis of nanocrystalline materials is determined by the ease of the process, its economic viability, its scalability, and the purity of the end product desired. Most of the techniques referred to previously produce the nanocrystalline material in the form of powder. The application of nanocrystalline materials requires that these powders are consolidated to full density with minimal or no porosity and this is not a trivial matter. Full densification with complete inter-particle bonding requires exposure of the powder to high temperatures and pressures for extended periods of time, which is likely to coarsen the micro structural features. Retention of the material in the nanocrystalline state, however, warrants that the powder is not taken to high temperatures for long periods of time; this may not result in complete densification. Thus, successful consolidation to full density requires innovative methods of consolidation. However, since nanocrystalline materials have a large fraction of the atoms in the grain boundaries, the (effective) diffusivity of nanocrystalline materials is much higher than that of a coarse-grained material of the same composition. This will facilitate achieving full consolidation of nanocrystalline materials at temperatures 300°C to 400°C lower than those required for the coarse-grained materials. Successful consolidation of nanocrystalline powders has been achieved by electro-discharge compaction, plasma-activated sintering, shock (explosive) consolidation, hot isostatic pressing (HIP), Ceracon processing, hydrostatic extrusion, strained powder rolling, and sinter forging. Consolidation is not required if the powder could be used as-is (e.g., in slurries, as catalysts, or for coating purposes).

Global Commercial Applications of Nanomaterials

This quote from "The Next Big Thing Is Really Small" by Jack Uldrich and Deb Newberry sums it up nicely: This is not to say that nanotechnology is a far-off, fuzzy, futuristic technology. It is not. It has already established a beachhead in the economy. The clothing industry is starting to feel the effects of nanotech. Eddie Bauer, for example, is currently using embedded nanoparticles to create stain-repellent khakis. This seemingly simple innovation will impact not only khaki-wearers, but dry cleaners, who will find their business declining; detergent makers, who will find less of their product moving off the shelf; and stain-removal makers, who will experience a sharp decrease in customers. This modest, fairly low-tech application of nanotechnology is just the small tip of a vast iceberg—an iceberg that threatens to sink even the "unsinkable" companies." Researchers at Pacific Northwest National Laboratory have developed a coating process to make sponge-like silica latch onto toxic metals in water. Self-Assembled Monolayers on Mesoporous Supports easily captures such metals as lead and mercury, which are then recovered for reuse or contained in-place forever. PNNL One example of a SAMMS nanocomposite (Self-Assembled Monolayers on Mesoporous Supports). An hexagonally close-packed cluster of tubular pores (end view) is shown in the foreground. A single pore, in this case coated with a mercaptopropylsiloxy monolayer, is shown in the background. A model of one surfactant molcule is also shown. A plastic nanocomposite is being used for "step assists" in the GM Safari and Astro Vans. It is scratch-resistant, light-weight, and rust-proof, and generates improvements in strength and reductions in weight, which lead to fuel savings and increased longevity. And in 2001, Toyota started using nanocomposites in a bumper that makes it 60 per cent lighter and twice as resistant to denting and scratching. Will likely be used on other GM and Toyota models soon, and in other areas of their vehicles, as well as the other auto manufactures, lowering weight, increasing milage, and creating longer-lasting autos. Likely to impact repair shops (fewer repairs needed) and auto insurance companies (fewer claims). Will also likely soon be seen everywhere weight, weather-proofing, durability, and strength are important factors. Expect NASA, the ESA, and other space-faring organisations to take a serious look, soon, which will eventually result in lower lift costs, which will result in more material being lifted into space. Electron micrograph of typical silicon nanocomposite cross section showing uniform distribution of conductive carbon nanotube network. Photo courtesy of U.S. Air Force.

Nanocrystals are an ideal light harvester in photovoltaic devices. (They) absorb sunlight more strongly than dye molecules or bulk semiconductor material, therefore high optical densities can be achieved while maintaining the requirement of thin films. Perfectly crystalline CdSe nanocrystals are also an artificial reaction center, separating the electron hole pair on a femtosecond timescale. Fluorescent nanocrystals have several advantages over organic dye molecules as fluorescent markers in biology. They are incredibly bright and do not photodegrade. Drug-conjugated nanocrystals attach to the protein in an extracellular fashion, enabling movies of protein trafficking. (They) also form the basis of a high-throughput fluorescence assay for drug discovery. Stain-repellent Eddie Bauer Nano-Care khakis, with surface fibers of 10 to 100 nanometers, uses a process that coats each fiber of fabric with "nano-whiskers. Developed by Nano-Tex, a Burlington Industries subsidiary. Dockers also makes khakis, a dress shirt and even a tie treated with what they call "Stain Defender", another example of the same nanoscale cloth treatment.

Dry cleaners, detergent and stain-removal makers, carpet and furniture makers, window covering makers. BASF's annual sales of aqueous polymer dispersion products amount to around $1.65 billion. All of them contain polymer particles ranging from ten to several hundred nanometers in size. Polymer dispersions are found in exterior paints, coatings and adhesives, or are used in the finishing of paper, textiles and leather. Nanotechnology also has applications in the food sector. Many vitamins and their precursors, such as carotinoids, are insoluble in water. However, when skillfully produced and formulated as nanoparticles, these substances can easily be mixed with cold water, and their bioavailability in the human body also increases. Many lemonades and fruit juices contain these specially formulated additives, which often also provide an attractive colour. In the cosmetics sector, BASF has for several years been among the leading suppliers of UV absorbers based on nanoparticulate zinc oxide. Incorporated in sun creams, the small particles filter the high-energy radiation out of sunlight. Because of their tiny size, they remain invisible to the naked eye and so the cream is transparent on the skin. Sunscreens are utilising nanoparticles that are extremely effective at absorbing light, especially in the ultra-violet (UV) range. Due to the particle size, they spread more easily, cover better, and save money since you use less. And they are transparent, unlike traditional screens which are white. These sunscreens are so successful that by 2001 they had captured 60 per cent of the Australian sunscreen market.

Makers of sunscreen have to convert to using nanoparticles. And other product manufactures, like packaging makers, will find ways to incorporate them into packages to reduces UV exposure and subsequent spoilage. The $480B packaging and $300B plastics industries will be directly effected. Using aluminum nanoparticles, Argonide has created rocket propellants that burn at double the rate. They also produce copper nanoparticles that are incorporated into automotive lubricant to reduce engine wear. AngstroMedica has produced a nanoparticulate-based synthetic bone. Human bone is made of a calcium and phosphate composite called Hydroxyapatite. By manipulation calcium and phosphate at the molecular level, we have created a patented material that is identical in structure and composition to natural bone. This novel synthetic bone can be used in areas where natural bone is damaged or removed, such as in the in the treatment of fractures and soft tissue injuries.

Biological molecules and systems have a number of attributes that make them highly suitable for nanotechnology applications. For example, proteins fold into precisely defined three-dimensional shapes, and nucleic acids assemble according to well-understood rules. Antibodies are highly specific in recognising and binding their ligands, and biological assemblies such as molecular motors can perform transport operations. Because of these and other favourable properties, biomolecules, biophysics, and biology are themes that run through all of the topics of this report. Although very promising, the bio-related aspects of nanoparticles, nanostructured materials, and nanodevices, are, for the most part, not as well developed as the nonbiological ones. However, a number of recent workshops (e.g., the U.S./EC Workshop on Nanobiotechnology), symposia (e.g., the Conference on Molecular Nanotechnology, and the Symposium on Bio-Nano Electronics), and books attest to the fact that many of the novel developments in this field are poised for rapid expansion.

This study is organised along the lines of the main report. It first puts into perspective current research directed toward biological synthesis and assembly as it pertains to the building blocks of nanotechnology. It then focuses on the current state of the art in biological aspects of dispersions and coatings, high surface area materials, and functional "building blocks" atoms nanoparticles layers nanostructures assembly dispersions and coatings high surface area materials functional nanostructures consolidated materials synthesis large biological component less of a biological component. Organisation of the WTEC study; sections with large biological content are indicated. nanostructures. Finally, it ends with a section on the challenges and opportunities for the biological aspects of nanotechnology.

Understanding Molecular Manufacturing (MM) and Quantum Dots

Manufacturing today works by cutting or deforming large chunks of matter, then fastening together the remaining pieces into products. Molecular manufacturing plans to be more efficient and make better products by assembling products directly from the smallest pieces: atoms and molecules. The basic idea is to develop a small set of chemical reactions that can be applied repeatedly to build large molecules, then control the sequence and/or position of the reactions by computer in order to build engineered molecular systems. Molecular manufacturing is different from biology in that biological systems are not engineered. The functional properties of a cell, or even a protein, are complex and hard to predict. However, the process of building protein molecules from small molecular fragments is quite programmable, and scientists are developing the ability to design and synthesize proteins with desired properties. This will allow protein chemistry to be used in an engineering, rather than a biological, context. This would be one approach to molecular manufacturing. Other approaches using different kinds of chemistry may also work, producing better materials. Eric Drexler has proposed two kinds of chemistry that might work. The first, in 1981, was protein chemistry as described above. The second, in 1992, was three-dimensional carbon lattice, similar to diamond. This would be synthesised by direct robotic operation in vacuum. Although the required chemical reactions have not yet been demonstrated, some of them have been simulated in detail. Despite this, some scientists assert that this kind of chemistry can't possibly work as planned. Discussion of molecular manufacturing has been distorted by several factors. From the beginning, it has been associated with "gray goo" (runaway biosphere-eating self-replicators), leading to excessive fear and attempts to counteract that fear. It has also been associated with with extreme science-fictional projections-though it can be hard to tell fantasy from sober calculation, because the calculations predict some pretty amazing capabilities. In reaction to these factors, some mainstream scientists have attempted to shut down discussion entirely by declaring that molecular manufacturing is impossible or that its major proponents are not credible. Discussion has been further distorted by a variety of widespread conceptual confusions.

The position that molecular manufacturing is impossible is not supportable. Living organisms are not an example of molecular manufacturing, because they are not based on engineering but rather on interlocking complex systems. However, the biochemistry of life could be adapted almost unchanged to a molecular manufacturing system. A few prominent scientists have nonetheless claimed that molecular manufacturing is impossible, and others have echoed them. However, study of their objections shows that the arguments are weak, based on intuition rather than calculation. Some of the arguments elevate engineering difficulties to the status of fundamental limitations. Others are built on basic misunderstandings of the proposals.

Although there are some practical questions remaining to be answered, there is no scientific study demonstrating a limitation or problem with molecular manufacturing theory. Indeed, the chemical mechanisms of life demonstrate that machines of a sort can be constructed out of molecules and can build more machines. Thus the scientists who dismiss the entire field out of hand have not bothered to study it in any depth. An intermediate position is that some form of molecular manufacturing will be possible, but will be limited. This is usually based on the assumption that engineering cannot improve on biology. This may be true for complex problems, but not for the relatively simple tasks that are tackled by engineering. Perhaps because of the opposition of these scientists, not many people are working on molecular manufacturing in the lab; most of the work has been theoretical study, which is easier to do without funding. For now, it remains a controversial theory. But after more than a decade, no one has managed to find a real problem with the theory, and evidence is accumulating in favour of it. Some predicted phenomena, such as exceptionally low friction in certain cases, have now been observed. A single atom has been mechanically removed from a crystal and put back in the same place. Although these were not intended as demonstrations of molecular manufacturing theory, they indicate that at least some of the predictions are valid. Separating science from science fiction, and alarmism from serious policy concerns, will not be an easy task. Most nanotechnology researchers have not paid enough attention to molecular manufacturing to be able to give informed commentary. The section on *"Molecular Manufacturing Myths"* lists some of the common misconceptions about molecular manufacturing.

According to its proponents, molecular manufacturing could be revolutionary. Because chemistry is very precise and repeatable, the manufacturing operations should be reliable enough to allow complete automation. Because the chemical steps are simple, a single fabricator could build a wide variety of products by changing the control programme. Furthermore, machines built this way could have very precise features. This is a good thing, because building with molecules would be quite slow, and small machines can be built more quickly. In fact, if the calculations are right, a complete nanoscale manufacturing system could build a complete copy of itself in a few hours-and this remains true if many small manufacturing systems are combined into one large one, so that even human-scale factories and products can be built quite rapidly. Finally, it appears that the materials produced by the manufacturing system may be extremely high-performance, allowing products to be far more compact and powerful. The section on *"Risks and Benefits of Molecular Manufacturing"* discusses some implications of the technology. Exponential general-purpose molecular manufacturing — that's a mouthful, but what does it mean? Let's take the phrase apart to see why it is so important.

The ability to make products, in this case ranging from clothing, to electronics, to medical devices, to books, to building materials, and much more. The automated building of products from the bottom up, molecule by molecule, with atomic precision. This will make products that are extremely lightweight, flexible, durable, and potentially very 'smart'. A manufacturing technology that will find many applications across many segments of society. Its extreme flexibility, precision, high capacity, and low cost will cause rapid adoption almost everywhere, and therefore will have disruptive effects in many industries. The word exponential refers to the rapid pace — probably unprecedented — at which this technology may be deployed. A compact automated molecular manufacturing system will be able to make more manufacturing systems. We're talking about factories that can build duplicate factories — and probably do it in less than a single day. The math is simple: if one factory makes two, and two factories make four, then within ten days you could have one thousand factories, in ten more days a million factories, and ten days after that a billion factories. Within the span of just a few weeks, in theory, every household in the world could have one of their own, to make most of the products they need, at just the cost of raw materials.

Exponential general-purpose molecular manufacturing means a manufacturing system capable of making a wide range of technologically advanced products, far superior to what we have today, much cheaper, much faster, and able to multiply its own source of production exponentially. From *Responsible Nanotechnology*. The risks and benefits of molecular manufacturing depend heavily on several factors: what it is capable of, when it will be developed, and how it is used. There is substantial disagreement over the first two questions, and almost no information about the third.

As explained above, the idea that molecular manufacturing is impossible is not credible. A separate question is how difficult it will be to develop and whether it will be worth doing. Other technologies are advancing so rapidly that if molecular manufacturing is not developed by 2050 or so, it may not have a very large impact. If it is limited to biochemical materials such as protein, its products may be limited in strength (though still extremely compact compared to today's computers and MEMS). Despite the weight of rhetoric indicating that molecular manufacturing will have little or no impact, the weight of calculation indicates that it will. In the absence of demonstrated problems with the theory, this section is written assuming product performance consistent with carbon-lattice construction-diamond, graphite, and carbon nanotubes, also called buckytubes or fullerenes. The high speed of small systems implies that manufacturing systems based on molecular machines will probably be able to produce their own mass in an hour or so. (Bacteria can do this, and bacteria spend only a fraction of their energy on reproduction.) Building complicated shapes from simple chemistry (under direct computer control), there is no apparent reason why manufacturing systems cannot duplicate themselves.

This implies that the systems would become extremely cheap, and manufacturing capacity could increase to any desired level almost immediately. The products of molecular manufacturing would be very high performance. Even with biochemistry, computers would be millions of times more powerful than today's. With carbon lattice, motors would also be millions of times more powerful. Computation and actuation would be a miniscule (and virtually free) fraction of any human-scale product. And carbon-based materials may be hundreds of times as strong as plastics or even metals. With molecular-scale features, very low friction, and very strong materials, the main limitation on products would be the skill of the designers. Cheap manufacturing of advanced products would allow inefficient infrastructures to be replaced-including our entire energy infrastructure, and much that is wasteful about our agriculture. This would greatly reduce our ecological footprint. Medical research and treatment would be greatly accelerated by the ability to design medical equipment, including massive sensor arrays and tiny robotics, and have them built instantly at negligible cost. Medical research would also be significantly helped by advanced computers. Cheap local self-contained manufacturing offers the possibility of eliminating material poverty. Advanced materials, extremely compact avionics, and rapid, inexpensive manufacturing would allow remarkably cheap access to space and its resources. However, these capabilities also have downsides. Advanced weapons could also be developed quickly. An arms race based on this technology would be far less stable than the nuclear arms race: weapons would be easier to develop and build in secret, easier to use and control, and harder to stockpile (since they would be almost immediately obsolete) than nuclear bombs. An unstable arms race leading to a devastating war is probably the largest as well as the earliest danger posed by this technology. Criminal technologies, from weapons to spy systems to communication to smuggling, would also get a boost from the ability to fabricate advanced products as needed. It remains to be seen whether defensive or offensive technologies will be stronger. Even if defensive technologies are adequate, they will probably be used unevenly. This implies that at least some criminal and terrorist acts using the technology will be successful. And this brings us to the next risk: oppression. For a variety of reasons (including a desire to maintain artificial scarcity), established government and business interests will want to retain control of molecular manufacturing and its effects. A technology that lets each individual design and build whatever they want does not appear compatible with centralised control. But if the technology is too restricted, a thriving black market will arise to supply citizens with restricted and valuable products. Unfortunately, this would also supply the criminals and perhaps even lead to loss of military control, leading to a spiral of further repression. Nearly free manufacturing could lead to an orgy of conspicuous consumption that could damage the ecosystem. And advanced products, up to and including brain and body modification, could shake up society.

Quantum Dots

While the integration of quantum dots into computers threatens to make Pentium processors as obsolete as vacuum tubes, such objects are nothing new to biological systems. Take for instance the *electron transport chain* used by nearly every living organism for *energy conversion*. Numerous proteins in this chain, as well as some smaller organic electron carriers all act as quantum dots. The image above is a model of part of the Iron Sulfur Protein (ISP), which uses its Iron Sulfur cluster to shuttle electrons from one quantum dot (quinone) to the next (Cytochrome C_1). As it does so, *conformation* changes throughout the multi-enzyme complex that it is a part of pump protons across a membrane resulting in a molecular battery. *Single electron transfers* abound in biological systems. Catalases are enzymes that use quantum dots to convert the damaging H_2O_2 into water. The photosynthetic machinery of plants use numerous quantum dots to absorb light and pass their energy and electrons to the complex "photosystems" that convert this energy into nanomechanical fuel, otherwise known as sugar. In the world of computer science, quantum dots are viewed from a different perspective than in biochemistry. While in biochemistry they are generally referred to as *redox groups*, chip designers refer to them as *qubits* [Collins]. Researchers such as Yasunobu Nakamura at the NEC Fundamental Research Laboratories in Tsukuba, Japan, have combined a *superconducting* phase with a quantum dot in an attempt to build a *quantum computer*. Such a computer would use qubits much as a modern computer uses bits. However, qubits may have an advantage over macroscale bits due to the *quantum mechanical* principle of superposition, or the ability to be in two states at once. This principle may allow a single qubit to be used in more than one computation simultaneously. Numerous quirks of quantum mechanics are being found to be advantageous to *quantum computing*. For instance, the "quantum mirage" discovered at IBM provides evidence that physical wires may not be necessary in some nanoelectronic circuits. A promising tool for the organisation of quantum dots into nanoelectronic circuits is *self-assembly*. Living systems have mastered the art of using self-assembly to generate useful structures. For instance, Prof. Dr. Dietmar Pum at the Center for Ultrastructure Research has demonstrated that *S-layer proteins* (Crystalline bacterial cell surface layers) can be effectively used to organise quantum dots (5 to 15 nm in diameter), regularly spaced up to 30 nm with oblique, square or hexagonal lattice symmetries. Such well ordered surfaces organise themselves, and thus, require a minimal amount of effort to prepare. The S-layers are able to self-assemble on a variety of solid supports, including *semiconductors*. Thus, it may not be too long before these nanoscopic arrays are integrated into memory chips, nanoprocessors, and optical devices.

Explaining Things Smaller Than Light and Mechanical VLSI

The primary sense that scientists use for observation is vision. Of course, vision depends on light (electromagnetic waves) bouncing off of the object and into the scientist's eye. Microscopes have been used for centuries to gather light so that scientists can see smaller and smaller objects. For some time, it seemed that improvement on this method (termed "optical microscopy") would allow us to see smaller and smaller objects without limit. That concept may still be true; however, due to the delocalised nature of light, traditional optical microscopy encounters a *far-field diffraction limit*, which limits the resolution to half of the wavelength of light being used. Since the smallest frequencies of light that humans can see is around 400 nm, 200nm is the theoretical limit of simply focusing light into a human eye or detector (the practical limit is even larger). Thus, in order to observe and manipulate nanoscale objects, it is necessary to cross the diffraction barrier.

Scanning Probe Microscopy (SPM)

A recently developed family of techniques that can be used to produce images of nanoscale surfaces with resolution reaching down to the *sub-angstrom level*. Each of these methods has its own unique strengths and weaknesses.

(i) Scanning Tunneling Microscopy (STM)
(ii) Scanning Force Microscopy (SFM)
(iii) Scanning Nearfield Optical Microscopy (SNOM)
(iv) Magnetic Force Microscopy (MFM)
(v) Scanning Capacitance Microscopy (SCM)

The miniaturisation of mechanical devices (MicroElectroMechanical Systems—MEMS) is gaining increasing attention due to numerous commercial applications. Clocks, accelerometers, ink jet printer heads, color projection displays, *scanning probe microscopy*, pressure, temperature, chemical and vibration sensors, light reflectors, switches, vehicle control, pacemakers and data storage are only a few examples. Each of these applications uses MEMS in a unique way, but they are all based on fundamental micromechanical components.

Two fundamental components in MEMS devices are actuators and cantilevers. A mechanical actuator is a device that can convert an electrical signal into mechanical motion. For instance, if one applies a voltage to a quartz crystal, it will change its size in a very precise and predictable way. Such an actuator (known as a *piezoelectric* material) is the heart of a digital watch. A cantilever is basically a small diving board. Cantilevers (and diving boards) bend when pressure is applied to them and oscillate in a way similar to a spring when given the chance (e.g. if they are "driven" at their resonance frequency by an actuator). Such a micromechanical component is what makes atomic force microscopy sensitive enough to see single molecules. The miniaturisation of electronic components is more established than for mechanical components. Fabricating smaller diodes and transistors allows more circuits to be integrated onto a single chip. Combinations of basic components such as diodes and transistors results in new, larger scale components such as a logic gate, which can carry out a single step of logical computation. Over the years, the integration of these logic gates has increased from "small-scale integration" (SSI, less than 10 gates) to medium-scale integration (MSI, 10-100 gates) to large-scale integration (LSI, 100-5000 gates) to today's very-large scale integration (VLSI, with more than 5000 gates). The first steps towards integrating mechanical systems into a single chip have been taken by researchers at IBM. The device known as the *'Millipede'* is an array of cantilevers integrated on a single chip. Each cantilever is sensitive enough to read and write data as nanometer scale indentations in a polymer film, giving it a capacity of 400-500 Gb/in^2. While a single cantilever would be to slow to read all of the pits, integrating them significantly increases the speed, making the "Millipede" a realistic data storage device. Numerous variations on this theme can be conceived. For instance, researchers at Northwestern University have demonstrated that coating the cantilevers with ink allows one to write nanoscale patterns, and have named the method *Dip Pen Nanolithography*. Since this method can be operated in a parallel fashion, it may not be long before arrays of AFM cantilevers coated with ink begin to lay down nanoelectronic circuits.

Since these mechanical systems are controlled electronically, they could conceptually be added as "drop-in" components in current chips. Indeed, most micro-machines are fabricated by the same semiconductor fabrication methods used to make computer chips. Since these methods are approaching the nanoscale, it is reasonable to expect that the basic units of both electronic and mechanical integration will eventually reach atomic and molecular dimensions. Since atoms and molecules are the fundamental building blocks of all matter, we can expect revolutionary capabilities from such NanoElectroMechanical Systems (*NEMS*), also known as *nanomachines*.

Electrons, Alchemy and Atomic Precision

The electron is a most convenient particle (or wave) for microscopy and lithography, not to mention electronic devices such as your computer. The mass of this subatomic particle is so small that only a negative exponent can do it justice: about 10^{-27} g. That is about one billionth of a billionth of a billionth of a gram. See what I mean? Too small for words. Compared to this super-small mass, it has a quantised negative charge of (elementary charge = $1.6*10^{-19}$ Coulomb). As this charge is exactly the same for all electrons, it is possible to precisely predict and control the movement of electrons. This tendency for very small objects such as electrons to be identical to one another is one of the motivating factors behind nanotechnology. That is, unlike components used in larger scale manufacturing, there need not be an inspector to check the mass and charge of each electron in a nanodevice. The small and uniform *mass to charge ratio* of an electron makes its momentum extremely sensitive to electric fields, which exert a force on charged particles. Such fields can be used to accelerate, direct, and focus electrons in the form of an *electron beam*. Scanning an electron beam over a surface, and measuring its interactions with the surface is a method known as *Scanning Electron Microscopy (SEM)*, which allows nanoscale resolution of conducting surfaces.

Furthermore, if a surface is chosen that reacts in response to exposure to electrons then an electron beam can be used as a writing tool for the rapid patterning of surfaces with ultra-high resolution, a method known as *Electron Beam Lithography (EBL)*. Now, consider the wave properties of this beam. The *De Broglie wavelength* of a particle can be related to its momentum by the relatively simple formula $l = h/p$, where l is the wavelength, h is Plank's constant and p is the momentum of the particle. Electrons differ from photons in that their mass is much larger, which results in a larger momentum and smaller wavelength. While a photon has a wavelength much larger than the size of one atom, the wavelength of an electron is far smaller than an atom. Thus, unlike optical microscopy or lithography, the *far-field diffraction limit* is not a barrier to the resolution of electron microscopy or lithography. Rather, other difficulties must be (and are being) overcome, such as our ability to focus the beam on a spot so small as a single atom. It is remarkable how prophetic Feynman was in his speech, seeing as this exact feat was was accomplished 22 years later with the invention of *Scanning Tunneling Microscopy (STM)*. STM is indeed invaluable for determining atomic precision within conducting samples and even selectively positioning individual atoms. However, even with its sub-atomic resolution, STM relies on slow mechanical actuators and is not currently useful for industrial *nanolithography*. Thus, it is still entirely possible that continued improvements in the focusing of electron beams (or other particle beams) might result in more rapid atomic resolution.

Nanotechnology can be qualitatively understood as an approach to alchemy that is based on the current model of matter. For more than 2000 years, the Aristotelian model of matter went relatively unquestioned in the western world. In this model, matter consisted of four elements: earth, air, fire and water. The variety of materials in existence was explained by the idea that the ratios of these elements within a material determined the form that the matter took. Based on this theory, the idea that one might rearrange the elements to create a desired material naturally followed. The arts of alchemy were early attempts to achieve this goal.

In the last centuries, alchemy became more of a philosophy while its techniques were divided between chemistry, nuclear physics and various other disciplines. In the 18th and 19th century, a far more accurate and productive model of the elements was finally developed. Lavoisier started the change with the discovery of a material that he called "oxygen" and began to arrange the other known elements into four groups. Dalton then proposed the importance of atomic masses, after which Mendeleyev noticed a pattern that resulted in the periodic table of the elements. Early in the 20th century, Pauling provided an accurate model of how these elements *bond* together to form molecules. These molecules can in turn interact by means of *weak interactions*. The chemical bonds and molecular interactions seem to be flawlessly explained by *quantum mechanics*, making our model of matter relatively complete. With the new model being far more complex than the original one, the alchemist's goal of arbitrarily rearranging the elements became too confusing to pursue directly, until recently.

During the last century, physicists and mathematicians have been ironing out the kinks in quantum mechanics. Chemists have found numerous ways to synthesize and characterize a plethora of molecules. Biologists have made enormous progress in understanding the molecular mechanisms and weak interactions that allow life to function. Meanwhile, the industrial revolution has provided us with machines, computers and other tools enabling further progress.

Our technology has now reached a point where the idea to more or less arbitrarily arrange the elements to create materials ranging from *nanomachines* to food is beginning to seem achievable. In 1959, Feynman translated the goals of alchemy into more modern terms, proposing that we miniaturize our tools for observing and manipulating matter until atomic precision is reached. In 1991, Drexler coined the term *molecular nanotechnology* referring to thorough, inexpensive control of the structure of matter. In the last decade, the concept of *nanotechnology* has been accepted by the scientific community and numerous organisations have been formed. Unlike alchemy, nanotechnology is based on a more modern model of matter. Nanotechnology does not try to transmute elements as it is far cheaper to extract and purify elements or molecules from existing materials. Arranging the atoms in a material to get the desired elements where we want them is a far more economical approach but still requires significant effort and cooperation. Thus, nanotechnology is an *interdisciplinary* subject drawing from biology, chemistry, physics, engineering and computer science, to name a few. Each discipline contains an important piece of the puzzle, and as the puzzle is assembled we are witnessing remarkable advancements in our ability to economically control the structure of matter.

2

Integrating Physics, Computers and Nanotechnology: An Interdisciplinary Approach

Introduction to Nanosystems, Nanoscale and Nano Devices

The Waitt Family Foundation and its related institutes are very forward-thinking, and Battelle is one of the best research organisations in the world. With their involvement, we will be able to identify the large gap between the basic nanostructured materials` being manufactured today, and the potential of productive nanosystems. This`roadmap initiative will chart the steps required to get from here to there.

Waitt Family Foundation was established by Gateway Founder Ted Waitt in 1993 and is headquartered in La Jolla, California. The Foundation is affiliated with three institutes'formed by Waitt, including the Institute for Historical Discovery, the Institute for Violence Prevention, and the Institute for Scientific Breakthroughs. The Foundation is led by its President, John Heubusch. For more information about The Waitt Family Foundation and its related institutes:

Battelle is'a global leader in science and technology. Headquartered in Columbus, Ohiol it develops and commercializes technology and manages laboratories for customers. Battelle, with the national labs it manages or co-manages, oversees 19,000 staff members and conducts $3 billion in annual research and development. Battelle innovations include the development of the office copier machine (Xerox), pioneering work on compact disc technology, medical technology advancements, and fiber optic technologies.

Nanoelectronics refers to electronics at the sub-micron scale. Many chip components in production already feature sizes at the nanoscale. However, nanoelectronics also encompass what is known as molecular electronics, which uses individual molecules in electronics. Other advances in nanotechnology with applications in electronics include carbon nanotubes that can be used in both chips and displays, and nanomaterials that can be used in films that make smaller, more flexible displays and improved hard disks. Nanoelectronics in Japan and Korea tends to focus on next generation semiconductor devices and single electron devices (SED). Below is a list of individuals who have been instrumental in advancing the state of the art in the field of nanoelectronics and who are among the world's experts on this next-generation technology. The list is organised according to areas of expertise.

- Dr. Jane Alexander—Dr. Alexander is the initiator of the DARPA ULTRA Programme for nanoelectronics research.
- Larry Cooper—Larry Cooper is a long-time advocate and programme manager for a broad range of U.S. Navy programmes in advanced electronics.
- Dr. Gerry Iafrate—Dr. Iafrate is a visionary programme manager and an active research investigator who is widely acknowledged as having been one of the initiators of the U.S. research effort in nanoelectronics during the early 1980s.
- Lt. Col. Gernot Pomrenke—Lt. Col. Pomrenke is the Programme Manager of the the DARPA ULTRA Programme for nanoelectronics research.
- Dr. William Tolles—Dr. Tolles is the author of two major reports on nanotechnology, including one on Nanotechnology in Europe.

They are performing among the most important and comprehensive work in the field of pure nanoelectronics. Lent and Porod have proposed and simulated successfully a wireless design for an electronic nanocomputer based upon quantum cellular automata and a two-state nanoelectronic device they term a "quantum dot cell". Bernstein is attempting to fabricate such devices and the logic based upon them. Their quantum cellular automata are a very important new idea that is stimulating important new research into potential designs and architectures for electronic nanocomputers.

- Brosl Hasslacher and Ronnie Mainieri

  Drs. Hasslacher and Mainieri are on the staff at the Los Alamos National Laboratory. They are leading experts on the limits of nanoscale, low-power electronic devices and the difficult technical issues that must be overcome to build a nanocomputer. They have performed extensive theoretical work and calculations in this area.

- Terry Fountain

  Professor Fountain performs research on computer architecture at the University College London, UK. He is a source of many innovative ideas about how to organise a densely integrated computer system with nanoscale components.

- Rolf Landauer

Dr. Landauer works for IBM where he is a well-known expert on the theory and physics of computation, especially in its nanoscale limits.

- John Storrs-Hall

  Dr. Storrs-Hall is a senior scientist in the Laboratory for Computer Science Research (LCSR) at Rutgers University. He is an expert on the issue of reversibility in nanocomputers. He is also a noted proponent of alternative, nanomechanical approaches for the design and fabrication of nanocomputers.

- Tomasso Toffoli

  Professor Toffoli, of Boston University, is an expert on cellular automaton architectures for nanocomputers and on quantum computation.

- Konstantin Likharev

  Professor Likharev is on the faculty at the State University of New York at Stony Brook. In collaboration with Alexander Korotkov, he helped originate a cellular automaton proposal designed particularly to alleviate some of the weaknesses in the Lent-Porod quantum dot cell approach (which also has been improved). Likharev also is one of the world's preeminent experts on single-electron devices.

- Gary Frazier and Alan Seabaugh

  Drs. Frazier and Seabaugh are on staff at Texas Instruments Corporation (TI). Gary Frazier is the Manager of the Texas Instruments Nanoelectronics Research and Development Group, while Alan Seabaugh is the Technical Leader. TI is among the leaders in the field of nanoelectronics, in large part due to the efforts of Drs. Frazier and Seabaugh, who are involved in many different aspects of the field. Their work in hybrid nanoelectronics—microelectronics technology, as well as their efforts in silicon nanoelectronics research, and in the modeling of nanoelectronic devices are of particular importance. This work could lead to the introduction of the first mass-manufactured nanoelectronics products.

- Robert Bate

  Bob Bate now is a senior investigator at Texas A&M. He is one of the godfathers of nanoelectronics and an early visionary about the potential of this area. He is also the author of a number of fundamental papers on nanoelectronics. In the early 1980s, Bate started and helped to build the Nanoelectronics Group at Texas Instruments prior to his retirement from TI several years ago.

- Federico Capasso

  Capasso works at AT&T where he performed some of the earliest work on "resonance tunneling devices," the class of quantum effect devices that are viewed as among the the more likely candidates for employment as a nanometer-scale transistor.

- Herb Goronkin

Dr. Goronkin is a senior scientist at the Motorola Corporation. A leading investigator in the area of solid-state quantum effect devices, he is an expert and proponent for using methods of chemical self-assembly for the fabrication of next-generation nanoelectronic devices.

- Tom McGill

Professor McGill is on the faculty at the California Institute of Technology, in Pasadena, CA, where he has performed innovative fabrication and experimentation with quantum effect devices.

- Professor Yasuaki Masumoto

Director of the Single Quantum Dot Project in Japan's ERATO Initiative and Professor at the University of Tsukuba in Tsukuba, Japan.

- Professor James Meindl

Professor Meindl is on the faculty of the Georgia Institute of Technology, where he has done comprehensive work on the constraints that apply in scaling down solid-state devices (i.e., transistors) to nanometer-dimensions. He is noted for his extremely clear, comprehensive lectures on this subject, based upon years of research that are summarised in a number of fundamental articles.

- Mark Reed

Professor Reed is Chairman of the Electrical Engineering Department at Yale University and Director of the Yale Center for Microelectronic Materials and Structures. An acknowledged expert on quantum dots, with much experience fabricating and experimenting with solid-state "dots", Prof. Reed concluded some years ago that they must get much smaller in order to operate at room temperature. Thus, Professor Reed now is performing very sensitive experiments on prototype molecular-scale electronic devices, along with Professor James Tour of the University of South Carolina. They are hoping to produce a molecular-scale quantum dot.

- Barry Gilbert

Gilbert is on the Staff of the Mayo Clinic, where he leads the Mayo Special Purpose Processor Development Group. His research on design and fabrication of solid-state quantum effect devices for signal processing is widely recognised as being among the most advanced in the nanoelectronics field.

- James Tour

  Tour is a Professor in the Department of Chemistry and Biochemistry at the University of South Carolina, where he is performing pioneering research on molecular electronics by synthesising the molecules on which Mark Reed of Yale is performing experiments. Recently, Professor Tour and his collaborators successfully fabricated and tested the first functioning molecular wire.

- Ari Aviram

  A long-time proponent of molecular-scale electronic devices, Dr. Aviram is a senior scientist at the IBM Watson Laboratory. He is the author of a series of very influential papers on the subject, dating back to the 1974 classic paper that he co-authored with Professor Mark Ratner. A more recent talk by Dr. Aviram served as the catalyst for the chemical nanofabrication research of Professor James Tour, who has successfully fabricated a conducting molecular wire.

- Mark Ratner

  Professor Ratner, now at Northwestern University in Evanston, Illinois, is one of the originators of the idea of molecular electronics, back in the 1970s. Along with Ari Aviram, Professor Ratner published the seminal technical paper on this topic. He continues to publish innovative research results on the dynamics and modeling of molecular devices, such as those that are being experimented upon by James Tour and Mark Reed.

- Yasuo Wada

  Wada is a senior member of the research technical staff at Hitachi Corp. in Saitama, Japan. He is an expert on nanoprobes. He also has originated an innovative approach for molecular-scale electrical switches.

- Calvin Quate

  Quate is Leland T. Edwards Professor of Electrical Engineering and Applied Physics at Stanford University and the inventor of Atomic Force Microscopy (AFM), a primary tool for the manipulation and investigation of nanostructures. More recently, has been focusing his research on using the techniques for fabricating microelectromechanical systems (MEMS) to build an array of a large number of AFM tips on a single silicon "chip." The comments made above for Noel MacDonald apply here, except Professor Quate is focused on using his arrays for nanofabrication.

- Phaedron Avouris

  Avouris is a member of the research staff at IBM. He is an expert on nanoprobes for nanofabrication.

- Noel MacDonald

  Noel MacDonald is Professor in the School of Electrical Engineering at Cornell University where his conducting research in the Cornell University Nanofabrication Facility Professor MacDonald's research focuses on fabricating an array of scanning tunneling electron microscopes (STMs) on a chip using the techniques for fabricating microelectromechanical systems (MEMS). This work is a groundbreaking engineering achievement and it is potentially very important for a variety of applications in nanofabrication, dense molecular memories, and the exploration of chemical reactions and surface properties. For now, he is focused primarily considering applications of STM arrays to dense memories.

- George Whitesides

  George Whitesides is Mallinckrodt Professor of Chemistry at Harvard University. Professor Whitesides is the one of the world's leading investigators in the area of chemical self-assembly for nanofabrication. He has pioneered in the technique of applying self-assembling monolayers (SAMs) for this purpose. Research in the Whitesides Group covers a wide range of disciplines, including chemistry, biology, physics, and electrical engineering. The work of Professor Whitesides and his collaborators on SAMs and contact printing for nanoelectronic circuitry is an important step toward the fabrication of nanocomputers.

- Steven Chou

  Professor Chou is leading a wide-ranging research programme in the Nanostructure Laboratory (NSL) in the Department of Electrical Engineering at the University of Minnesota where, among his other achievements, he has pioneered highly refined techniques for nanoscale "printing." He has applied this technique to fabricate an ultra-dense magnetic information storage structure. He is working on at least a half dozen projects in VLSI and ULSI (Ultra-Large-Scale Intergration).

- Tom Mallouck

  Professor Mallouck is on the faculty of the Chemistry Department at Penn State University, where he has led experimental efforts that resulted in the chemical self-assembly of a single-electron transistor.

- Harold Craighead

  Professor Craighead works at Cornell University Nanofabrication Facility and is the head of the Cornell Nanotechnology Lab and an expert on self-assembly.

- James Harris

  Professor of Electrical Engineering at Stanford University and leader of the Harris Research Group there. His group specializes in nanofabrication by Molecular Beam Epitaxy (MBE), as well as in studies of the dynamics of quantum dots.

- James Ellenbogen

  As Lead Scientist of the Nanosystems Group at The MITRE Corporation, Dr. Ellenbogen has been performing research to develop faster and more accurate electron-density-based methods for modeling nanoelectronic devices.

- Kevin Empson

  Programme Manager of the U.S. Government's Nanoelectronic Modeling (NEMO) Project.

- Gerhard Klimeck

  A software engineer in the Nanoelectronics Research Group at the Texas Instruments Corporation in Dallas, TX, Gerhard is the driving force behind the innovative design of the software for the Nanoelectronic Modeling (NEMO) software.

- David K. Ferry

  Dave Ferry is Professor of Electrical Engineering and Computer Science at Arizona State University and Director of the Nanostructures Research Group at ASU. He is an authority on the physics and modeling of solid-state nanoelectronic devices, having written two text books and many technical articles on the subject.

- William Frensley

  Bill Frensley is Professor and Head of the Department of Electrical Engineering at the University of Texas, Dallas. He has used his expertise and experience in the simulation of solid-state devices and quantum-effect nanoelectronic devices to develop deep insight into which device designs can work and which probably won't, and why. Also, he has developed a clear, unified picture of device designs, their evolution and relationships.

- Harold Grubin

  As President of SRA, Inc., Hal Grubin is an active participant in a diverse range of modeling and simulation activities at the forefront of solid-state nanoelectronics research.

At first there were only a few of them, but recently, their numbers have multiplied wildly. Newspaper headlines, magazine articles, journal papers, even television commercials now are loaded with those big "nano" words: nanometer, nanoscale, nanosecond, and nanotechnology, to name a few. And it seems that every week some organisation is announcing yet another "nanoconference." Why all the excitement? Because in recent years, as scientists have begun to get a handle on controlling matter at the nanometer scale, they've recognised that these skills can be used to make new materials with unique and useful properties—perhaps leading to a range of commercial applications in sensing, electronics, and other areas.

Impressive laboratory demonstrations of nanoscale dexterity have been widely reported in the scientific and popular press, thereby drawing the attention of more scientists, companies, and technology investors. Nanoelectronics—a developing field in which circuitry is composed of nanometer-sized electronic components—is one topic that's attracting a lot of interest in scientific and nonscientific corners alike. Researchers are driven to explore the field because further miniaturising today's already small electronic circuits will lead to faster, more sophisticated, and more portable devices. Yet it's widely believed that in a decade or so, silicon-based circuitry will have been shrunk as small as is physically possible. And so the search is on for alternative materials from which nanoscale circuits can be constructed. Carbon nanotubes and nanowires of materials such as gallium arsenide are being studied for just that purpose by several research groups. Other investigators, whose work is often labeled "molecular electronics," focus on the electrical behaviour of individual molecules or small numbers of molecules. Scientists who work in these fields aim to design and build new types of computers in which the nanosized entities lie at the heart of logic circuits. Skeptics say that the proposed applications are exaggerated and that reports are full of "hype." But researchers around the globe are exploring and discovering nanometer-scale phenomena, and they are rapidly turning out large numbers of scientific papers. "These are exciting and heady times," says Mark A. Ratner, surveying the combined topics of molecular and nanoelectronics. "There are wonderful experiments being reported these days," he says. "The excitement is justified." A nearly 30-year-old paper by the Northwestern University chemistry professor is often cited as the beginning of modern molecular electronics. In that theoretical study, Ratner and Arieh Aviram, who was at that time a graduate student, proposed that under certain conditions, individual organic molecules could function as p-n junction diodes. These simple electronic devices typically are formed at the interface between positive charge-carrying (p-type) and negative charge-carrying (n-type) semiconductors and can be used as rectifiers to convert AC current to DC current.

Ratner notes that few advances were made in molecular electronics until scanning probe microscopes became available in the late 1980s. The new tools gave the field a much-needed boost by making it possible to measure a whole range of phenomena, including the type Ratner had described in the 1970s. He adds that progress in other areas, for example, in methods for preparing self-assembled monolayers of molecules, was also instrumental in jump-starting nanoscale electronics. "Now the advances are coming thick and fast," he says.

> "Thick and fast" aptly describes the flurry of research being carried out on carbon nanotubes. According to Phaedon Avouris, manager of nanoscale science and technology at IBM's T.J. Watson Research Center, Yorktown Heights, N.Y., the strawlike all-carbon structures with nanometer-sized diameters are endowed with unique properties that make them excellent candidates for nanoelectronic applications.

Nanotubes can be metals or semiconductors, depending on their chirality, Avouris notes. And because of their strong chemical bonds and satisfied valencies, the materials boast high thermal, mechanical, and chemical stability. In addition, carbon nanotubes can be efficient conductors as a result of their tiny diameters, long lengths, and defect-free structures that make them ideal one-dimensional systems. Some years ago, Avouris' research group and, independently, scientists working with Cees Dekker, a professor of applied physics at Delft University of Technology, in the Netherlands, demonstrated that field-effect transistors (FETs) could be fashioned from carbon nanotubes. FETs are a type of switch in which a semiconducting channel bridges two electrodes designated "source" and "drain." Current flow between these electrodes is controlled by a third electrode known as a "gate." By applying a voltage to the gate, the semiconductor's state can be changed—reversibly—from insulating to conducting, thereby switching the transistor on or off. In conventional FETs, the bridging channel is made of silicon. In nanotube devices, the channel is a single carbon nanotube. Today's sophisticated silicon-based integrated circuits owe much of their success to advances in FET technology. Manufacturers pack tens of millions of the tiny transistors into the postage-stamp-sized computer chips used in modern microprocessors. For that reason, the IBM research group and others have been working on improving the properties of nanotube FETs. Much of the work focuses on understanding and controlling electrical conduction through the tiny tubes.

One problem that plagues researchers looking to fashion circuit components from nanotubes is separating metallic tubes from the ones that are semiconducting. Common synthesis procedures produce spaghetti-like mixtures of nanotube ropes that are unusable for semiconductor applications because they contain both types of tubes. But last year, Avouris and coworkers came up with a way to sort through the tangle. After fashioning electrodes around nanotube bundles via lithography methods, the team applied certain voltages to the gate electrodes to switch off the semiconducting tubes, converting them to insulators. Then, by applying high voltage to the circuit, the IBM group oxidised the metallic tubes, causing them to break down without affecting the semiconducting tubes. The group used the method to prepare FET arrays from single-walled nanotube ropes and to peel apart multiwalled nanotubes shell-by-shell. Just a few months later, the IBM group showed that elementary computing circuits known as logic gates—which typically are constructed from combinations of FETs—can be fashioned from a single nanotube bundle. Shortly thereafter, Delft's Dekker reported similar advances in constructing multi-FET nanotube logic circuits.

Complex combinations of AND, OR, and NOT logic gates make up the core of digital processors. To make NOT gates using nanotubes, Avouris and coworkers needed to come up with a procedure for preparing n-type nanotubes. NOT gates require both types of transistors. But the synthesis techniques available at that time, produced only p-type tubes. Hunting for a solution, Avouris and IBM coworkers Vincent Derycke, Richard Martel, and Joerg Appenzeller found that p-type nanotubes can be converted to n-type simply by annealing (heating) in vacuum. Armed with the new preparation trick, the team prepared separate n- and p-type transistors and fabricated a NOT gate. Then they went a step further and fabricated a NOT gate using just one nanotube. For that feat, the team selectively converted a short length of a nanotube to n-type by doping it with potassium through a tiny window in a protective polymer coating while leaving the unexposed portion p-type.

The single tube with p- and n-type segments was then used to construct an intramolecular NOT gate. Further advances have come from transistor design modifications. Earlier this year, for example, the IBM group reported that their so-called top-gated nanotube FETs outperform state-of-the-art silicon FETs in terms of switching rate and the amount of current they can carry per width of conductor. One key difference between the latest design and earlier designs is that the silicon wafers that support the FETs do not function as gates. Instead, the gate is fabricated above the nanotube—allowing all FETs in contact with the silicon wafer to be switched independently.

In addition, the new FET design benefits from switching voltages that are an order of magnitude lower than those needed to switch older FETs Just recently, the IBM team developed a catalyst-free procedure for preparing single-walled carbon nanotubes. Conventional methods for preparing single-walled tubes, such as discharge and vapor deposition techniques, rely on particles of nickel and cobalt or other catalytic transition metals. A problem with those procedures is that the metal particles left behind in the products disturb the nanotubes' electrical properties, forcing scientists to take complex purification measures. But now Avouris and coworkers have shown that single-walled nanotubes can be prepared from SiC in the absence of metal catalysts at temperatures above 1500°C. The method produces nanotubes that are 1.2 to 1.6 nm wide and are aligned along the face of the support. Other researchers have reported techniques for growing vertically aligned products, which they dub nanotube forests. But nanotubes that grow horizontally are amenable to standard vapor deposition methods and may be connected in parallel to lower their electrical resistance Despite the recent advances, Avouris asserts that nanotube FETs remain "far from optimised." Improvements could be made by using thinner insulators with higher dielectric constants, he suggests. "But what's really needed is a better understanding of the mechanism of electrical switching in nanotube FETs." Carbon nanotubes aren't the only game in town.

Nanowires made of semiconductors such as silicon, gallium arsenide, and indium phosphide are being investigated as candidate materials for nanoelectronics. The field is advancing rapidly as researchers are making fast progress in synthesis, device fabrication, and testing. Although Harvard University chemistry professor Charles M. Lieber has published widely on carbon nanotubes, semiconducting nanowires figure prominently into his research programme. The tiny wires are versatile building blocks that can be used in a bottom-up approach to constructing nanoelectronic circuits, Lieber notes, because the size, structure, and functional properties of nanowires can be controlled readily via synthesis procedures.

In 1998, Lieber and coworkers described a vapor-liquid-solid synthesis method in which laser light is used to ablate nanometer-sized metal clusters that serve as nucleation centers and catalysts for nanowire growth. The Harvard researchers used the technique to prepare uniform single-crystalline nanowires of silicon and germanium with diameters as small as 3 nm and lengths up to 30 mm. Since that time, the procedure has been used to prepare nanowires with even smaller diameters, and it's been extended to a wide variety of materials. Examples include III-V semiconductors such as GaAs and II-VI semiconductors such as ZnSe. Recently, a number of research groups boosted the complexity of materials that can be prepared by the cluster-nucleation method.

The teams prepared modulated structures—nanowires composed of dissimilar segments. The two-tone materials are made by turning the supplies of reactants on and off during synthesis with pulsed lasers or by other methods. These "heterostructured" products open the door to new sophisticated applications, such as terahertz-frequency photon emitters. Lieber, graduate student Mark S. Gudiksen, and coworkers used modulation methods to prepare nanowires with 21 alternating segments of GaAs and GaP. And they prepared Si and InP nanowires with modulated doping, such that the wires were endowed with alternating p- and n-type regions. Using similar methods, Peidong Yang, an assistant chemistry professor at the University of California, Berkeley, prepared Si-SiGe nanowires.

And Lars Samuelson of Lund University, in Sweden, synthesised InAs-InP nanowires. To evaluate the usefulness of the tiny wires in nanoscale electronics, the

Harvard researchers fashioned the wires into test circuits that were fabricated using lithography methods and then studied the circuits' electrical properties. According to Lieber, Si nanowire FETs, for example, perform remarkably well. The devices exhibit large on/off ratios (>10^5)—a requirement for effective switches—and charge-carrier mobilities that exceed those measured in conventional silicon devices. In addition to constructing FETs from nanowires, Lieber's team has built and tested photodetectors, biological and chemical sensors, and more complex devices, including light-emitting diodes (LEDs) and logic gates. Most of the papers in which the results are reported were published in just the past 18 months. In the chemical sensing study, Lieber, graduate students Yi Cui and Qingqiao Wei, and Harvard assistant chemistry professor Hongkun Park found that by functionalising boron-doped Si nanowires, they could prepare efficient sensors. Modifying the wires with an amino-silane compound, for example, produced pH sensors that are linear over a large dynamic range.

And they observed that biotin-modified wires can be used to detect picomolar levels of streptavidin. In addition, the Harvard chemists found that the wires could be used to detect proteins reversibly in real time. By crossing an n-type and a p-type InP nanowire and then attaching electrodes to the wires lithographically, Lieber and coworkers made a nanometer-sized junction—at the point of overlap—that behaved as a current rectifier and an LED. Based on the study, Lieber says that more complex devices can be constructed from semiconducting nanowires simply by crossing them and forming tic-tac-toe patterns. But taking advantage of the crossed-nanowire idea requires developing a method for preparing the patterns efficiently and reproducibly. By combining surface patterning techniques and microfluidic alignment methods, the Harvard group developed the needed procedure.

Their idea is to flow nanowires from a suspension into tiny parallel channels in a mold, in a layer-by-layer fashion, while rotating the mold orientation—and hence, nanowire flow direction—between layers. With the crossed-nanowire method for making p-n junctions firmly in hand, Lieber and coworkers were well on their way toward making computational logic circuits. Near the end of 2001, the group published a study in Science showing that their procedures could be used to make arrays of nanowire FETs configured as AND, OR, and NOR logic gates. Lieber stresses that in crossed-nanowire structures, such as the Harvard logic gates, the width and length of the FET channels are controlled by synthesis and assembly—not lithography—unlike today's microelectronic components. These fabrication concepts make it possible "to envision a straightforward path to nanometer-scale electronics of the future," Lieber says. His vision, which is supported by experiments conducted by his group in the past several months, includes a nanoscale logic device known as a decoder that can address individual elements in crossed-wire arrays. Using a handful of molecules—or even just one of them—to make nanoscale switches or other devices is an idea that's gathered momentum in recent times. In 1999, researchers working with chemistry professors James R. Heath and J. Fraser Stoddart of the University of California, Los Angeles, demonstrated that simple circuits—fuses—could be made from a layer of rotaxane molecules. Then a team headed by Rice University chemistry professor James M. Tour and Yale University electrical engineer Mark A. Reed reported success in making reversible switches using benzene thiol molecules.

Other groups have also been investigating the possibilities of single-molecule systems. For example, a team at UC Berkeley including Park (now at Harvard), Paul L. McEuen (now at Cornell), and A. Paul Alivisatos, trapped a single C_{60} molecule in a tiny gap between a pair of electrodes to study electronic conduction mechanisms. And just recently, a single molecule of a transition metal-organic complex—containing either one Co atom or two V atoms—was turned into a single-molecule transistor in independent studies by Park and McEuen. Before molecular devices can be wired up and tested, organic chemists need to synthesize the molecules that drive them. In the UCLA collaboration, Stoddart's group prepares the compounds and Heath's group turns them into circuit components. To make the molecules switchable, Stoddart's group designs them so that one segment of a molecule can move (rotate or translate) relative to another segment. Stoddart notes that, contrary to remarks made by many researchers, the molecules are not difficult to prepare nowadays. But it took some years to build up the necessary expertise in molecular recognition concepts, p-donor/p-acceptor interactions, and other subjects.

He adds that his group has synthesised "hundreds, if not thousands" of catenane and rotaxane analogs since 1989. In 2000, the UCLA team demonstrated a reusable switch based on a catenane—a molecule with a pair of interlocked rings. The switching state of the redox-active molecule was shown to change reversibly, depending on an applied potential. Electrostatic repulsion drives rotation of one ring relative to the other between open and closed states. Looking to improve device performance by tailoring the molecules, Jan O. Jeppesen, a visiting graduate student in Stoddart's lab, and postdoctoral associate Hsian-Rong Tseng prepared various rotaxane analogs of the catenane that brought success in 2000. The series of amphiphilic bistable rotaxanes are dumbbell-shaped molecules capped with large stoppers that are threaded through a cyclophane ring.

As with the catenanes, the ring's position relative to the other piece of the molecule is controlled by redox interactions. Heath and coworkers constructed circuits with the new molecules using a layer-by-layer deposition method to fabricate tic-tac-toe structures with a few thousand molecules situated at each intersection. The team used this fabrication method to prepare 64-bit random-access memory circuits. And Hewlett-Packard Labs announced earlier this month that R. Stanley Williams, Yong Chen, and coworkers have also succeeded in using the UCLA rotaxane molecules to fabricate a 64-bit memory circuit. Only a few years ago, many scientists sat through molecular electronics presentations thinking that the far-fetched nanocomputing goals would never be reached. But the flurry of demonstration results in just the past two years appears to have decreased the number of nonbelievers. Yet some naysayers remain. Just recently, a California professor was furious about a C&EN story on molecular electronics and wrote to the editor, exclaiming, "In reality, there is no such topic!" The professor has a few surprises in store. In reality, molecular and nanoelectronics are bustling fields. Large numbers of commercial products may not be available in the near future, but as the vibrant research activity reveals, they're coming.

In this study, we discuss about nanometer-scale, quantum effect alternatives to micron-scale, and bulk effect transistors in digital circuits. The status of R&D and recent important advances are reviewed briefly. For the past forty years, electronic computers have grown more powerful as their basic sub-unit, the transistor, has shrunk. However, the laws of quantum mechanics, plus the limitations of materials and fabrication techniques soon are likely to inhibit further reduction in the minimum size of today's bulk effect semiconductor transistors. Researchers have projected that as the overall size of the bulk-effect, semiconductor transistor is aggressively miniaturised to approximately 0.1 micron (i.e., 100 nanometers) and beyond, the devices may no longer function as well. Thus, in order to continue the miniaturisation of integrated circuits well into the next century, it is likely that present day, micron-scale or microelectronic device designs will be replaced with new designs for devices that take advantage of the quantum mechanical effects that dominate on the much smaller, nanometer scale. (Note that 1 nanometer, abbreviated nm, is one billionth of a meter or 10 atomic diameters.) Here we briefly survey such quantum effect, nanometer-scale alternatives to micron-scale, bulk-effect transistors for use in electronic digital circuits. The topic of nanoelectronic devices subdivides into two broad areas, as follows:

(i) Solid-state quantum effect nanoelectronic devices.
(ii) Molecular electronic devices.

Solid-state quantum-effect nanoelectronic devices are the nearer-term alternative for continuing to increase the density and speed of information processing. This technology, which changes the operating principles, but not the solid fabrication medium for integrated circuits, might extend Moore's Law of electronics miniaturisation to produce devices down to 25 nm in length, beyond the domain of the bulk effect transistor. Present-day field effect transistors (FETs) on commercial integrated circuits are approximately 1 micron or 1000 nm across, and even the most optimistic proponents of aggressively miniaturised bulk effect FETs predict that they will cease to function effectively when their gate lengths dip below 25 nm, which corresponds to an overall device length of approximately 100 nm. By contrast, quantum effect switching devices tend to function better when they are made smaller. Solid-state quantum-effect nanoelectronic devices might be made as small as approximately 12 to 25 nanometers across and still function effectively if they could be fabricated reliably and uniformly on that scale. Presently, mass fabrication on this small scale does present a set of formidable challenges, no matter what the device design. However, a number of solid-state nanoelectronic switching devices have been fabricated and demonstrated, and prototype solid-state nanoelectronic processors already are operational, as well. Moreover, the drastic decrease in size that could result from the wide-scale introduction of such nanoelectronic devices might produce ultrafast, low power integrated circuits with as many as 100 billion or even 1 trillion switching devices on a single CPU chip, as well as Terabyte nonevanescent memories on a chip. Such great advances would require, however, the reliable, uniform mass fabrication of features only 5 to 10 nanometers wide in solids. This is at least twenty-five times as small as is achievable using UV lithography, the present industrial method of choice. More modest, near term goals do not require such extreme precision, but take advantage of other very desirable properties of solid-state quantum-effect devices, such as low power consumption and high speed.

As is indicated, there are 3 basic categories of solid-state nanoelectronic devices:

(1) Quantum Dots (or "artificial atoms")
(2) Resonant Tunneling Devices
(3) Single Electron Transistors (SETs)

Quantum Dots—A small "island" composed of semiconductor or metal in which electrons may be confined is the essential structural feature that all these quantum-effect, solid-state nanoelectronic devices have in common. The island in a nanoelectronic device assumes a role analogous to that of the channel that forms beneath the gate in a familiar microelectronic field-effect transistor when it is switched on. The composition, shape, and size of the island gives the different types of solid state nanoelectronic devices their distinct properties.

Controlling these factors permits the designer of the device to employ quantum effects in different ways to control the passage of electrons on to and off of the island.

Resonant-tunneling devices—Here, we focus primarily on explaining the operation of resonant tunneling devices, because they employ quantum effects in their simplest form. Presently, these devices usually are fabricated from layers of two different III/V semiconductor alloys, such as the pair GaAs and AlAs. The simplest type of resonant tunneling device is the resonant tunneling diode (RTD). A resonant-tunneling diode is made by placing two insulating barriers in a semiconductor, creating between them an island or potential well where electrons can reside. Resonant tunneling diodes are made with center islands approximately 10 nanometers in width. Whenever electrons are confined between two such closely spaced barriers, quantum mechanics restricts their energies to one of a finite number of discrete "quantised" levels. This energy quantisation is the basis for the operation of the resonant-tunneling diode. The only way for electrons to pass through the device is to "tunnel," quantum mechanically, through the two barriers. The probability that the electrons can tunnel is dependent on the energy of the incoming electrons compared to the energy levels on the island of the device. If the energy of the incoming electrons differs from the energy levels allowed inside the potential well on the island, then current does not flow. The device is switched "off." However, when the energy of the incoming electrons aligns with that of one of the internal energy levels, as shown. By adding a small gate electrode over the island of an RTD one may construct a somewhat more complex resonant tunneling device called a resonant tunneling transistor (RTT). In this three-terminal configuration, a small gate voltage can control a large current across the device. Because a very small voltage to the gate can result in a relatively large current and voltage across the device, amplification or "gain" is achieved. Thus, an RTT can perform as both switch and amplifier, just like the conventional bulk effect transistor.

Unlike conventional bulk effect transistors which usually have only two switching states, "on" and "off," resonant tunneling devices like RTDs and RTTs can have several switching states. This occurs because the quantum well on the island may exhibit several possible energy levels.

Thus, as the voltage bias on an RTD is increased from zero, where it is "off," the device initially switches "on" when the first energy level comes in resonance with the incident electrons. Then, it switches off again as the bias is increased further, past resonance. However, if there is a second energy level in the quantum well, then the device can switch on again as the bias voltage of the RTD (or the gate voltage of an RTT) is increased still further. This multistate switching behaviour permits each device to "count higher" and represent more logic states.

"Hybrid" Microelectronic—Nanoelectronic Resonant Tunneling Transistors—Purely nanometer-scale RTTs tend to be difficult to fabricate with sufficient uniformity in large quantities because of their relatively complex structure, small size, and the sensitivity of the effects they employ. However, there are also nanoelectronic devices which are "hybrids" of solid-state quantum-effect devices combined with micron-scale transistors. One such hybrid transistor-like device, the RTD-FET, is constructed by building tiny, nanometer-scale quantum-effect RTDs into the drain (or source) of a bulk effect micron-scale FET. A schematic of such a device.

Such a hybrid RTT can exhibit multistate switching behaviour of the type described in the preceding section for purely nanometer-scale RTTs. For this reason, the hybrid device can represent more logic states than a pure bulk effect, microelectronic FET with the same "footprint" on the integrated circuit. Thus, the density of the logic can be increased using the multistate switching characteristics of hybrid micro-nanoelectronic RTTs, without appreciably increasing the density of the devices on an integrated circuit. Also, these hybrid RTTs share the advantages of low power and high speed exhibited by the purely nanoelectronic RTTs. Most important, fabricating circuits with this relatively large, hybrid type of three-terminal RTT is easier than fabricating circuits with the much tinier, complex structures for purely nanoelectronic RTTs. For this reason a number of groups are experimenting with such devices.

In recent months, robust prototype circuits based upon such hybrid devices have been demonstrated. Hybrid logic can be viewed as an important, practical engineering step on an evolutionary path toward nanoelectronics. It should accelerate the availability of quantum-effect devices in integrated circuits with very dense functionality. Ultimately, however, it will be desirable to build ultra dense, low-power nanoelectronic circuits made from purely nanometer-scale switching devices and wires. Molecular electronics—using individual covalently bonded molecules to act as wires and switching devices—is the longer-term alternative for achieving this increase in density and for continuing Moore's Law down to the nanometer scale. Individual molecular switching devices could be as small as 1.5 nanometers across, with densities of approximately 1012 devices per sq. cm.

This decrease in size could result in Terabyte memories on a chip and in excess of one trillion switching-devices on a single CPU chip. A primary advantage of molecular electronics is that molecules are natural nanometer-scale structures that can be made absolutely identical in vast quantities (approximately 10^{23} at one time). Also, synthetic organic chemistry—the chemistry of carbon-based compounds—offers more options for designing and fabricating these intrinsically nanometer-scale devices than the technology presently available for producing solid-state chips with nanometer scale features. It has been shown during the past few years that small organic molecules can conduct and control electricity, that large organic bio-molecules are conductive, and that individual molecules of the newly discovered molecular forms of pure carbon called buckminsterfullerenes ("buckyballs" and "buckytubes") have interesting and useful electrical properties. Great progress is being made in understanding and harnessing the electrical properties of individual organic molecules. There has been particular experimental progress in demonstrating that individual molecules can conduct electrical current to function as wires and that they also can act as switches to turn this current on and off. No logic circuits using these molecular switching devices have been fabricated yet. However, designs for such circuits have been proposed recently To some extent, the research and development of molecular electronic switches is recapitulating that for the corresponding solid-state devices, except on a much smaller scale. Thus, it is not surprising, that molecular electronic switching proposals and prototypes may be categorised into molecular analogs of solid-state diodes and analogs for solid-state three-terminal FETs. A molecular analog of the solid state RTD that is discussed above is shown. The hexagonally shaped components are known as benzene "rings." Chains of these rings, such as appear at the right and left ends of the molecule in the figure have been shown to function as conductive "molecular wires". The "CH_2" groups (termed "methylene" groups), on the other hand, act as insulators or "barriers" to electron flow. They can trap electrons on the single benzene ring between them, which forms an island. This molecular sandwich-like structure, with the island in the middle, produces a potential energy well like that sketched at the bottom.. Despite enormous recent progress in the fabrication and demonstration of nanoelectronic devices, many challenges remain.

For solid state nanoelectronics, one of the most important challenges is to be able to produce reliably and uniformly in silicon the characteristic nanometer-scale features required for nanoelectronics: nanometer-scale islands, barriers, and "heterojunctions" between islands and barriers. Up to now, solid state nanoelectronic devices largely have been fabricated in III/V semiconductor compounds such as gallium arsenide and aluminum arsenide. It is believed, however, that silicon nanoelectronics would go a long way toward permitting the inexpensive mass manufacture of nanoelectronic devices. For molecular electronics, one of the great challenges is to develop two-terminal and three-terminal devices than can be incorporated in circuits. Then, it must be demonstrated that such molecular switches can be used to make reliable logic, such as has been proposed recently. Using either the solid-state or the molecular electronic approach, in order to fulfill the promise of nanoelectronics it will be necessary to refine greatly both the fabrication techniques and the architectural concepts to permit the useful assembly of small, low-power computing structures that contain trillions of nanoelectronic switching devices.

- "With this book, Drexler has`established the field of molecular nanotechnology. The detailed analyses show quantum chemists and synthetic` chemists how`to build upon their knowledge of bonds and molecules to`develop the manufacturing systems of`nanotechnlogy, and show`physicists and engineers how`to scale down their concepts of macroscopic`systems to the level of molecules."
- "The most talked about technical book of the year." Bob Schwbach, United Press International.
- William A. Goddard III, Professor of Chemistry and Applied Physicsl Director Materials and Molecular Simulation Center, California Institute of Technology.

- "Devices enormously smaller than before will remodel engineering, chemistry, medicine, and computer technology. How can we understand machines` that are so small?`*Nanosystems* covers`it all: power and strength, friction and wear, thermal noise and quantum uncertainty. This`is *the* book for starting the next century of engineering."
- Marvin Minsky, Professor of Electrical Engineering and Computer Science, Toshiba Professor of Media Arts'and Sciences, Massachusetts'Institute of Technology.
- "What the computer revolution did for manipulating data, the nanotechnology revolution will do`for manipulating matterl juggling atoms like bits. This multidisciplinary synthesis opens the door to the new field of molecular manufacturing."
- Ralph C. Merkle, Member of Research Staff, Computational Nanotechnology Project, Xerox Palo Alto'Research Center.
- "This`work provides`the scientific and technological foundations for the emerging field of molecular systems engineering... It is essential for anyone contemplating research in this`area...a milestone in the develpment of the technologies that will underpin the final industrial revolution." John Walker, Cofounder of Autodesk, Inc.
- "It is`a scholarly examination of how this`technology works, a reference book for the crafters of the future." Byte.
- "Demonstrates not only that nanotechnology is achievable, but shows how it will happen....takes readers from the fundamental physical principles to advanced designs for molecular components and systems." Japan Times |/ STRONG>.
- Best Computer Science Book of 1992, Association of American Publishers.
- Over 10,000 copies in print.

Nanosystems was reviewed by James B. Lewis in The Journal of the American Chemical Society (JACSi Vol. 115, No. 24, December 1993, pages 11657-11658:

The goal of Drexler's investigations is "building complex structures with atom-by-atom control", which is`also the ultimate goal of synthetic chemistry. Drexler's approach is`distinguished from conventional chemistry in that complex structures are to be made by using programmable "nanoscale mechanical systems to guide the placement of reactive molecules" to about 0.1-nm precision. The objective of the book is to`present a theoretical foundation for "molecular manufacturing", which Drexler also`calls`"molecular nanotechnology". The objective is`*not* to`present a detailed review`of recent experimental progress in the many disciplines that converge on what is` being increasingly termed "nanotechnology"…..

Several alternative pathways from current technology to'molecular manufacturing are considered, at least briefly, guiding chemists and others toward a plethora of interesting problems'to pursue.

Functions of Assemblers and Quantum Dots

The September 2001 issue of *Scientific American* devoted six articles and a great deal of text to various perspectives on nanotechnology. The issue included various attacks on the feasibility of molecular assemblers and the work of K. Eric Drexler and his research associates.

Drexler's textbook *Nanosystems* was published in 1992. It is still being used as a technical reference all over the world, and to date no significant errors have been found. As the reader can well imagine, if there were any significant errors *Scientific American* would have pointed them out. They haven't. The technical claims and conclusions of *Nanosystems* have withstood almost a decade of serious public review.

Besides *Nanosystems*, there is now a very large body of technical articles, books, conferences, newsletters, and technical discussion groups that supports the feasibility of molecular machines in general and molecular assemblers in particular.

Two specific articles in the September *Scientific American* were attempts to cast doubt on the feasibility of nonbiological molecular assemblers.

While there are quite a few questionable statements in the rest of that issue, nowhere else is there a serious attempt to advance a technical argument against the feasibility of molecular nanotechnology.

How is it possible that an otherwise respectable publication would publish these attacks? None of their technical criticisms of molecular assemblers has withstood scrutiny — all have fallen by the wayside when it became obvious that they were incorrect. Some of the reasons are reviewed in "That's impossible! How good scientists reach bad conclusions"

Experts everywhere say that nanotechnology will lead to'smarter, faster, and more efficient electronicsl but they also'admit that researchers face some challenging obstacles along the information superhighway. Particle contamination is'one of the thorniest problems'facing the electronics and computing industries. Even the tiniest particle can destroy a computer chip and cause other damage. Several University researchers are searching for ways'to detect particle contaminants and prevent their formation. A team of researchers headed by mechanical engineering professor Peter McMurry is studying'the formation and growth of contaminants in semiconductor processing equipment. Smaller contaminated particles deposit more quickly than larger ones and can destroy a number of memory chips and processors. The team's goal is'to design new processes or tools for producing contaminant-free devices. "It's important to understand how`to prevent particle deposition, and this`will require an`understanding`of how such particles`are formed and transported," McMurry says.

His team has developed a device called the particle beam mass' spectrometer, the only available instrument for measuring nanoparticles'in low-pressure semiconductor processing equipment. Electrical engineering professor Steve Campbell is among the researchers who are looking'for ways to'create electronic'devices with greater memory. His studies are aimed at detecting and preventing the formation of particle contaminants'in chemical reactors.

Although research in the area is'still relatively new, some early prototypes have been made, says'Campbell. Campbell's Microtechnology Laboratory team has developed a machine to detect and measure particlesn "Now we're trying to` demonstrate that isolating`these particles`and using them to`our advantage is`feasible," he says, adding that his group is seeking funding to continue its research.

"The importance of smaller and faster semiconductor devices is evident from the trends`we've all seen [toward] faster, more powerful, and less costly computersl" says McMurry. "Work that is` being done to fabricate devices at the nanoscale will ensure that this trend will continue for some time to"come." The same is true in recording media. Associate Professor Randall Victora and Professor Jack'Judy of the electrical and computer engineering department received grants'two'years ago from the National Storage Industry Consortium and Seagate Corporation to' develop superlattice magnetic recording media. The focus of their research is magnetic'recording technology for hard drivesl which are vulnerable to thermal fluctuations.

These fluctuations can affect the recording process and destroy vital information stored on a disk'in the drive. "We're preventing "that from happeningl" Victora says. "These superlattices cause the disk to`have a stronger resistance to`these thermal fluctuationsn" Once developed, the technology would be used primarily by'the computing industry. Companies such as Quantum, IBM, and Seagate already have expressed interest in the group's research, Victora says. Other University projects also have implications for the electronics industry, including the study of adhesion in computer chip manufacturingl molecular electronics, and the search for an appropriate nonconductor to replace copper, which is'limited in its ability to transport information between resistors.

Quantum dots are nanometer-scale "boxes" for selectively holding or releasing electrons. Over the past decade or so they have been transformed from laboratory curiosities to the building blocks for a future computer industry. Quantum dots are small devices that contain a tiny droplet of free electrons. They are fabricated in semiconductor materials and have typical dimensions between nanometres to a few microns. The size and shape of these structures and therefore the number of electrons they contain, can be precisely controlled; a quantum dot can have anything from a single electron to a collection of several thousands. Atomic physics progressed rapidly at the beginning of the last century, thanks, in large part, to optical spectroscopy. Quantisation and spin were discovered through optical studies, as were other fundamental atomic properties. With the advent of the laser, physicists learned how to manipulate atomic wave functions by applying coherent optical fields. More discoveries followed.

Now, at the beginning of the new century, optical techniques are being used to explore a new scientific frontier: the atom-like entities known as quantum dots (QDs). Measuring 1-100 nm across, QDs are semiconductor structures in which the electron wave function is confined in all three dimensions by the potential energy barriers that form the QD's boundaries. A QD's electronic response, like that of a single atom, is manifest in its discrete energy spectrum, which appears when electron hole pairs are excited. Although the wave function of a QD electron, and its corresponding hole, extends over many thousands of lattice atoms, the pair— termed an 'exciton'—behaves in a quantised and coherent fashion. The coherence is relatively easy to detect and control optically—for two reasons. *First,* the superposition of the ground and excited states dephases more slowly in QDs than in higher-dimensional semiconductor structures. *Second,* QDs have large dipole moments (50-100 times larger than those of atoms). Thanks to these advantages, it is possible to probe and manipulate the wave function of a single QD. QDs possess another attractive property. Their size, shape, and composition can all be tailored to create a variety of desired properties. These "artificial atoms" can, in turn, be positioned and assembled into complexes that serve as new materials. Researchers who work on QDs anticipate that a host of complex, customised QD-based materials will become available. The physics of quantum dots shows many parallels with the behaviour of naturally occurring quantum systems in atomic and nuclear physics.

As in an atom, the energy levels in a quantum dot become quantised due to the confinement of electrons. Unlike atoms however, quantum dots can be easily connected to electrodes and are therefore excellent tools to study atomic-like properties. There is a wealth of interesting phenomena that have been measured in quantum dot structures over the past decade. This page shows a few examples from our group. The next paragraph first discusses briefly the parallels between atoms and quantum dots. The three-dimensional (3D) spherically symmetric potential around atoms yields degeneracies known as shells, 1s, 2s, 3s, 3p... Each shell can hold a specific number of electrons. The electronic configuration is particularly stable when these shells are completely filled wih electrons, occurring at 'magic' atomic numbers 2, 10, 18, 36,... In a similar way, the symmetry of a two-dimensional (2D), disk-shaped quantum dot leads to a shell structure with magic numbers 2, 6, 12, 20,...

The lower degree of symmetry in 2D results in a different sequence of magic numbers than in 3D. By measuring electron transport through quantum dots, a periodic table of artificial 2D elements can be obtained. For this purpose, dots are connected via potential barriers to source and drain contacts. If the barriers are thick enough enough, the number of electrons on the dot, N, is a well defined integer. This number changes when electrons tunnel to and from the dot. However, due to Coulomb repulsion between electrons, the energy of a dot containing N+1 electrons is larger than when it contains N electrons. Extra energy is therefore needed to add an electron to the dot. Consequently, no current can flow which is known as the Coulomb blockade. The blockade can be lifted by means of a third electrode closeby, known as the gate contact. A negative voltage applied to this gate is used to supply the extra energy and thereby change the number of free electrons on the dot. This makes it possible to record the current flow between source and drain as the number of electrons on the dot, and hence its energy, is varied. The Coulomb blockade leads to a series of sharp peaks in the measured current. At any given peak, the number of electrons on the dot alternates between N and N+1. Between the peaks, the current is zero and N remains constant. The distance between consecutive peaks is proportional to the so-called addition energy, which is the difference in energy between dots with N+1 and N electrons. The magic numbers can be identified because significantly higher voltages are needed to add the 2nd, 6th and 12th electron. Quantum dots are 2D analogies for real atoms. But since they have much larger dimensions they are suitable for experiments that can not be carried out in atomic physics. It is especially interesting to observe the effect of a magnetic fieldd, B, on the atom-like properties. A magnetic flux-quantum in an atom requires typically a B-field as high as 10^6 T, whereas for dots this is of the order 1 T, which is experimentally accessible. A quantum dot is a box that holds a discrete number of electrons. This number can be changed by adjusting electric fields in the neighborhood of the dot, for example by applying a voltage to a nearby metal gate. In the following, I will refer to quantum dots that have either zero or one electron in them. Since quantum dots are fabricated in solids, not in vacuum, of course there are many electrons in them. However, almost all of these are tightly bound to atoms in the solid. The few electrons that I will speak of are extra ones beyond those that are tightly bound.

These extra electrons could roam free in a solid were they not confined in a quantum dot. We arrange that two electrons reside in each cell. They may hop (tunnel) from dot to dot within the cell to find a favourable, low-energy configuration. Of course, because electrons are charged, they repel each other, so the two electrons will not both choose to sit in the same dot within the 4-dot cell. In fact, they will choose to occupy two opposite corners of the cell, since this gives them their maximum possible separation. This is nice, because it gives us two possible "polarisation" states for a cell, which we can associate with a binary zero and one. We can use a voltage on an external gate to force a cell into one of its two a priori equally good alternative states. This is a signal, analogous to a high or low voltage at a point within a conventional integrated circuit. A conventional integrated circuit uses two tools to transmit a voltage signal over long distances: an aluminum or silicon wire (to carry signals from place to place) and a transistor used as an amplifier (to refresh weakened signals so that they may travel farther without corruption). In the quantum dot circuits proposed by Lent, both of these tasks are accomplished simply by arranging as many cells as necessary in a row. This, extended to a row of many cells, is the task normally carried out by a wire. In fact, even if the polarisation of the first dot is between zero and one (the electrons are not fully localised on one diagonal), as more dots are added to the chain, the polarisation of later dots will rapidly approach a perfect zero or one. This is the task of amplification or refreshing signals, the trademark of digital circuitry. Although we will not illustrate them, one can also build a corner, fan-out, or inverter (some of the building blocks of digital circuits). "Wires" carrying signals can even cross each other in the plane.

So far, we could make a simple translation from objects in standard integrated circuits to objects in quantum dot circuits. However, the cornerstone of digital circuitry, logic gates, are not so simply translated into this new architecture. A standard logic gate has two inputs and an output determined by the inputs. For example, an AND gate puts out a high output only if both input. One can build any ordinary logic using these gates (most clumsily, one can make an OR gate by fixing input 3 to have polarisation 1, an AND gate by forcing it to have polarisation 0). More impressively, if the center-to-center distance between dots within a cell in 20 nm (extremely aggressive though not impossible with today's electron-beam lithography), a single bit full adder will fit in an area of 1 micron^2. By contrast, the semiconductor industry projects that in 2010 the maximum packing density for transistors will be approximately 1 micron^2. Forty transistors would be needed to build an adder! This dramatic difference in density is explained by two facts:

1. Much of the space in a conventional integrated circuit is taken up by wiring, which is unnecessary here.
2. The minimum feature size required to pack 1 transistor/micron^2 would be 70 nm, and relative alignment between features produced in different lithographic steps generally needs to be good to 10% of this minimum feature size, or 7 nm. Only one stage of quantum dot manufacture need involve small features, so no sensitive alignment is required. This means that the minimum feature size could probably be made smaller than 70

nm even abandoning electron beams in favour of optical techniques favoured by industry for their high throughput. In addition, the diameter of a single cell is only a few times the minimum feature size, as compared to 1 micron/70 nm which is approximately 14 times.

In short, for given lithographic capabilities, quantum dot cells can be packed far more densely than can transistors. Lent's quantum dot cells can be switched from one state to another adiabatically (reversibly). This is accomplished by lowering the barriers between dots within each cell, and removing the global inputs to the circuit, hence relaxing the polarisation of each cell, then applying new global inputs and raising the barriers again. This produces very little heat, an important quality in densely integrated circuits (a DEC Alpha already has a fan mounted on top of it, continually blowing cold air on it so that it doesn't melt). Readout could be accomplished by passing a current through a very narrow constriction near a corner of a cell. An electron in that corner of the cell could shut off current. This was demonstrated by Field et al. This whole discussion has assumed that when new boundary conditions (inputs) are imposed, the system rapidly and reliably finds its ground (or lowest energy) state. This would not be true if the ambient temperature were comparable to the difference in energy between a "correct" and an "incorrect" polarisation of a cell. For the quoted 20 nm dot-to-dot distance, this energy associated with an error in one cell's polarisation would be 1 millielectron volt, or one twentieth of thermal energies (kT) at room temperature, so this is a major worry. Even if the error energy could be made many times larger, which could be accomplished by making the individual dots smaller and closer together, entropy would limit the size of an array of cells to about 20 thousand at error energy = 10 kT. Finally, no one knows how long an array of cells would take to reach its ground state, though Lent estimates 30 picoseconds per cell in a "wire". Stray charges within 1 cell's length of a "wire" would lock a nearby cell permanently into one polarisation state, stymieing computation. Charges are always sprinkled liberally on surfaces of semiconductors, so this would seem to be a major barrier to dense integration using this architecture. Perhaps most seriously, an attendee at the talk pointed out that the ground state of a large array may not be the obvious one determined by propagating signals "correctly" from inputs to outputs (though Lent has showed by modeling that his adder would not suffer from this problem). What length scales are required to produce a quantum dot? In a quantum well the wavelike properties of the electrons only become apparent for layer widths of about a hundredth of a micron. To produce a quantum dot we need to achieve similar length scales along the other two dimensions. Using present techniques it is relatively easy to reduce the lateral dimensions of the islands to about a tenth of a micron. With great effort further reductions can be achieved, but producing a structure one hundredth of a micron across presents an immense technical challenge.

Fortunately nature is on our side. Suppose that we etch two parallel channels through the layered materials to leave a ridge approximately one-tenth of a micron across. In the etching process the atoms on the vertical surfaces of the ridge are disturbed so that the properties of the material are changed considerably. In fact, surfaces in general have quite different properties to bulk material. Let us consider an example using a crystal of silicon. We know that each silicon atom tends to bond with four neighboring atoms to obtain a full complement of electrons in its outer shell. However, atoms at a surface cannot always find another four atoms with which to bond. This means that surface atoms have a tendency to hold on to any nearby free electrons in order to obtain a full shell. This process alone tends to remove conduction electrons from the surface layers. In addition, the accumulation of negative charges trapped by these surface atoms creates an electric field which repels other electrons from this region. As a result we find that the only free electrons in our tiny ridge-like structure are confined to an even smaller channel in the center.

It is rather like a coaxial cable in which the central copper core is surrounded by an insulating layer. For a ridge which is a tenth of a micron across, the electrons are confined to a region which is only about a hundredth of a micron wide. This is just the right length scale required to observe the wavelike nature of the electron. So although at first sight the dimensions appear to be far too large, we can observe lateral quantum confinement in these structures. In this case we have created a quantum wire, since the electrons are still free to travel along the length of the ridge. From this stage the construction of a quantum dot is relatively straightforward. If we etch another set of similarly spaced grooves at right angles to the first set, then the surface effects in these directions are sufficient to cause the electrons to be trapped in a tiny box. Using these techniques it is possible to pattern a relatively large area to produce a regular grid of quantum dots. Low-dimensional nanometer-sized systems have defined a new research area in condensed-matter physics within the last 20 years. Modern semiconductor processing techniques allowed the artificial creation of quantum confinement of only a few electrons. Such finite fermion systems have much in common with atoms, yet they are man-made structures, designed and fabricated in the laboratory. Usually they are called "quantum dots", referring to their quantum confinement in all three spatial dimensions.

A common way to fabricate quantum dots is to restrict the two-dimensional electron gas in a semiconductor heterostructure laterally by electrostatic gates, or vertically by etching techniques. This creates a bowl-like potential in which the conduction electrons are trapped. In addition to the many possible technological applications, what makes the study of these "artificial atoms" or "designer atoms" interesting are the far-reaching analogies to systems that exist in nature and have defined paradigms of many-body physics: atoms, nuclei, and, more recently, metallic clusters or trapped atomic gases. Quantum dots added another such paradigm. Their properties can be changed in a controlled way by electrostatic gates, changes in the dot geometry, or applied magnetic fields. Their technological realisation gave access to quantum effects in finite low- dimensional systems that were largely unexplored. After the initial success in the fabrication and control of mesoscopic semiconductor structures, which are typically about one hundred nanometers in size and confine several hundred electrons, many groups focused on the further miniaturisation of such devices. A breakthrough to the "atomic" regime was achieved with the experimental discovery of shell structure in fluctuations of the charging energy spectra of small, vertical quantum dots: the borderline between the physics of bulk condensed matter and few-body quantum systems was crossed. Much of the many-body physics that was developed for the understanding of atoms or nuclei could be applied.

In turn, measurements on artificial atoms yielded a wealth of data from which a fundamental insight into the many-body physics of low-dimensional finite fermion systems was obtained. With further progress in experimental techniques, artificial atoms will continue to be a rich source of information on many-body physics and undoubtedly will hold a few surprises." The "Kondo effect" is a large anomalous increase in the resistance of certain dilute alloys of magnetic materials in nonmagnetic hosts as the temperature is lowered. In general, the Kondo effect occurs when an impurity atom with an unpaired electron is placed in a metal, producing an interaction of localised electrons with delocalised electrons. The "Kondo temperature" is the temperature at which the Kondo effect predominates. In general, in this context, the term "Coulomb blockade" refers to an effective blockade of quantum mechanical tunneling produced by specific energy barrier constraints. When an individual molecule nanocrystal nanotube, or lithographically defined quantum dot is attached to metallic electrodes via tunnel barriers, electron transport is dominated by single-electron charging and energy-level quantisation. As the coupling to the electrodes increases, higher-order tunneling and correlated electron motion give rise to new phenomena, including the Kondo resonance. To date, all of the studies of Kondo phenomena in quantum dots have been performed on systems where precise control over the spin degrees of freedom is difficult. Molecules incorporating transition-metal atoms provide powerful new systems in this regard, because the spin and orbital degrees of freedom can be controlled through well-defined chemistry. Devices were prepared by an extension of the methods previously used in constructing single-C60 and single-nanocrystal transistors. Using electron-beam lithography, a narrow gold bridge was fabricated on an aluminum pad with a 3-nm oxide layer serving as a gate electrode.

The electromigration-induced break-junction technique was then used to create two closely spaced gold electrodes. Scanning electron microscope imaging and tunnel current measurements reveal that the narrowest gap between the two electrodes is consistently 1 nm. The observation of the Kondo effect in single-molecule transistors, where an individual divanadium molecule serves as a spin impurity. The authors find that the Kondo resonance can be tuned reversibly using the gate voltage to alter the charge and spin state of the molecule. The resonance persists at temperatures up to 30 K and when the energy separation between the molecular state and the Fermi level of the metal exceeds 100 meV. The authors suggest the present study demonstrates that molecules can provide a Kondo system where critical parameters of Kondo physics, such as the spin and orbital degrees of freedom, are defined by chemical synthesis. With the recent advances of synthetic methodology, the preparation of molecular clusters possessing adjustable magnetic properties is becoming feasible. Future investigations of such species are expected to provide detailed insight into electron transport through a molecular system where the spin and orbital degeneracies are precisely controlled.

Quantum dots are small electrically conducting regions, typically less than 1 micron in diameter, that contain from one to a few thousand electrons. Because of the small volume, the electron energies within the dot are quantised, and the behaviour of the quantum dot is intermediate between that of an atom and that of a classical macroscopic object. Such intermediate systems are called "mesoscopic" systems, and in the past several years great attention has been devoted to the physics of such systems, since they apparently can provide insights into quantum systems in general. The electronic states in quantum dots can be probed by transport when a small tunnel coupling is allowed between the dot and nearby source and drain leads. Cronenwett et al report the realisation of a tunable Kondo effect in small quantum dots, with the capability of switching a dot from a Kondo system to non-Kondo system as the number of electrons on the dot is changed from odd to even. The Kondo temperature can be tuned by means of a gate voltage as a single particle energy state nears the Fermi energy. Measurements of the temperature and magnetic field dependence of a Coulomb-blockaded dot show good agreement with prediction of both equilibrium and nonequilibrium Kondo effects. N.B. Zhitenev *et. al.* report a study of the electron localisation- delocalisation transition in quantum dots. The problem of electron localisation has remained a prime focus of experiment and theoretical research over the past 40 years. Single electron capacitance spectroscopy precisely measures the energies required to add individual electrons to a quantum dot. The spatial extent of electronic wave functions was probed by investigating the dependence of these energies on changes in the dot confining potential. For low electron densities, electrons occupy distinct spatial sites localised within the dot. At higher densities, the electrons become delocalised, and all wave functions are spread over the full dot area. Near the delocalisation transition, the last remaining localised states exist at the perimeter of the dot. Unexpectedly, these electrons appear to bind with electrons in the dot center. The motion of electrons through quantum dots is strongly modified by single electron charging and the quantisation of energy levels. Much effort has been directed towards extending studies of electron transport to chemical nanostructures, including molecules, nanocrystals, and nanotubes. The authors report the fabrication of single molecule transistors based on individual C60 molecules connected to gold electrodes.

The coupling is manifest as quantised nano-mechanical oscillations of the C60 molecule against the gold surface, with a frequency of about 1.2 THz. This value is in good agreement with a simple theoretical estimate based on van der Waals and electrostatic interactions between C60 molecules and gold electrodes. Single-C60 transistors were prepared by depositing a dilute toluene solution of C60 onto a pair of connected gold electrodes fabricated using electron-beam lithography. A break-junction technique was then used to create a gap between these electrodes by the process of electromigration. The typical lateral size of the fabricated electrodes was of the order of 100 nm at the point of the gap formation, and the height of the electrodes was 15 nm. Scanning electron microscope images of fabricated electrodes reveal that the gap between two electrodes is not uniform and that the narrowest gap is formed only between small protrusions (10 nm) of two gold electrodes. Current voltage measurements of these electrodes at cryogenic temperatures without deposited C60 molecules show that the size of the gap is consistently about 1 nm. In a significant fraction of the C60 devices, the conductance of the junction after initial breaking is substantially enhanced compared to devices with no C60 deposited, indicating that C60 molecules reside in the junction. The entire structure was defined on a SiO2 insulating layer on top of a degenerately doped silicon wafer that serves as a gate electrode that modulates the electrostatic potential of C60. It was suggested by the researchers that the transport measurements presented demonstrate that single electron-tunneling events can be used both to excite and probe the motion of a molecule: indeed, the single-C60 transistor behaves as a high-frequency nanomechanical oscillator. Furthermore, the oscillations of the C60 molecule must be treated in a quantised fashion, showing that this is a true quantum "mechanical" system. The authors expect that the coupling between the quantised electronic and mechanical degrees of freedom will be generically important in electron transport through nanomolecular systems. The techniques of colloidal chemistry permit the routine creation of semiconductor nanocrystals whose dimensions are much smaller than those that can be realised using lithographic techniques. The sizes of such nanocrystals can be varied systematically to study quantum size effects or to make novel electronic or optical materials with tailored properties. Preliminary studies of both the electrical and optical properties of individual nanocrystals have been performed recently. These studies show clearly that a single excess charge on a nanocrystal can markedly influence its properties.

The researchers present measurements of electrical transport in a single electron transistor made from a colloidal nanocrystal of cadmium selenide. This device structure enables the number of charge carriers on the nanocrystal to be tuned directly, and so permits the measurement of the energy required for adding successive charge carriers. Such measurements are invaluable in understanding the energy-level spectra of small electronic systems, as has been shown by similar studies of lithographically patterned quantum dots and small metallic grains. The researchers suggest their work represents a new type of spectroscopy for single nanocrystals. Unlike optical measurements, where electron hole pairs are created, these measurements probe the energy for adding a single type of charge carrier. The authors hope that these results will stimulate new calculations of the hole addition energies for a CdSe nanocrystal that include the effects of exchange, correlations, and screening by the metallic electrodes. Future measurements will investigate how the ground-state and excited-state properties vary with the size, shape and composition of nanocrystals as well as with the composition of the leads. Semiconductor quantum dots, due to their small size, mark the transition between molecular and solid-state regimes, and are often described as "artificial atoms".

This analogy originates from the early work on quantum confinement effects in semiconductor nanocrystals, where the electronic wavefunctions are predicted to exhibit atomic-like symmetries. Spectroscopic studies of quantum dots have demonstrated discrete energy level structures and narrow transition linewidths, but the symmetry of the discrete states could be inferred only indirectly. The researchers report they use cryogenic scanning tunneling spectroscopy to identify directly atomic-like electronic states with s and p character in a series of indium arsenide nanocrystals. These states are manifest in tunneling current voltage measurements as two- and six-fold single electron-charging multiplets respectively, and they follow an atom-like Aufbau principle of sequential energy level occupation. The researchers suggest that the good correlation between optical transitions and the spacing of levels observed in the tunneling spectra indicates that the charging did not have a significant effect on the QD level structure. The tunneling IV curves are thus well described as an algebraic sum of single electron charging energies and the QD level spacings. Single-photon sources have recently been demonstrated using a variety of devices, including molecules, mesoscopic quantum wells, color centres, trapped ions and semiconductor quantum dots. Compared with a *Poisson*-distributed source of the same intensity, these sources rarely emit two or more photons in the same pulse. Numerous applications for single-photon sources have been proposed in the field of quantum information, but mostly including linear-optical quantum computation—also require consecutive photons to have identical wave packets.

For a source based on a single quantum emitter, the emitter must therefore be excited in a rapid or deterministic way, and interact little with its surrounding environment. When identical single photons enter a 50-50 beam splitter from opposite sides, quantum mechanics predicts that both photons must leave in the same direction, if their wave packets overlap perfectly. This two-photon interference effect originates from the Bose Einstein statistics of photons. This bunching effect was first observed using pairs of highly correlated photons produced by parametric down coversion, but it should also occur with single, independently generated photons. Most proposed applications for single-photon sources in the field of quantum information (with the notable exception of quantum cryptography) involve two-photon interference. Such applications include quantum teleportation, post-selective production of polarisation-entangled photons, and linear-optics quantum computation. It is therefore important to demonstrate that consecutive photons emitted by a single-photon source are identical and exhibit mutual two-photon interference effects. The researchers report a test of the indistinguishability of photons emitted by a semiconductor quantum dot in a microcavity through a Hong Ou Mandel type two-photon interference experiment. The authors find that consecutive photons are largely indistinguishable, with a mean wave-packet overlap as large as 0.81, making this source useful in a variety of experiments in quantum optics and quantum information. Electroluminescence has taken a new turn: S. Coe et al reports the fabrication of high-efficiency organic light-emitting diodes (LEDs) in which the light-emitting centers are cadmium-selenium (CdSe) nanocrystals, or quantum dots. Organic light-emitting diodes (LEDs) bring with them the advantages of robust fabrication technique and high performance, which, when coupled with the excellent luminescent properties of nanocrystals, offer exciting prospects for real, workable devices. Mobile phones with small colour displays using organic LEDs are already commercially available. The images are generated through fluorescence, as electrons make transitions between orbital states of pi-conjugated organic molecules (the pi-bond arises from the overlap of the 2p orbitals of electrons in carbon atoms).

As well as having high quantum efficiency for electron-to-photon conversion, pi-conjugated molecules in organic LEDs have the advantage of colour tunability, so that they can be used to build full-color displays of red green blue (RGB) emitters. But there is also a drawback: the emission spectra of pi-conjugated molecules are very broad, typically spanning 50 to 100 nm (the "full width at half-maximum", or FWHM). This range of molecular fluorescence is caused by the vibrational and rotational motion of atoms inside the pi-conjugated molecules. So with an organic LED, it is difficult to get, for example, pure red light emitted with high quantum efficiency.

Nevertheless, sharp RGB emission has been demonstrated from LEDs based on some specific materials, such as europium chelates, aggregated structures of cyanine dyes or layered inorganic organic perovskite compounds. But these LEDs have not achieved the emission efficiency or device durability needed for practical display applications. Some inorganic nanocrystals emit visible light with sharp emission spectra that are less than 30 nm FWHM. The nanocrystals are in effect quantum dots and confine charge so well within their small volume that a high quantum efficiency, exceeding 50 per cent, is possible. It might be expected, then, that quantum dots incorporated in organic LEDs would make excellent emission centres 4. In fact, electroluminescence has been observed by simply mixing inorganic nanocrystals with -conjugated polymers, but the emission efficiency was far lower than that of conventional polymer LEDs.

Coe et al have fabricated an organic LED with a single layer of CdSe quantum dots sandwiched between organic thin films. Remarkably, the efficiency of their device is about 25 times higher than that achieved so far with quantum-dot LEDs. Thinking ahead to the design of device architectures, there are two noteworthy aspects here: the structure of this quantum-dot LED is already close to that for an optimum device and the process of fabricating a layer of quantum dots is simple.

An electron in a quantum dot can be described by a quantum wave function that is similar to that used for an electron in a single atom, although the energy of the electron in the quantum dot is spread in a coordinated way (spread "coherently") over the lattice of atomic nuclei. The electronic wave functions of quantum dots are often labeled with atomic notation, but quantum dots are very much solid-state nanostructures that can be tailored into different shapes. Recent studies describe the optical behaviour of individual quantum dots and quantum rings, and such behaviour is of considerable interest because quantum dots that emit light are expected to form the basis of a new generation of lasers. In these optical studies of quantum dots, the semiconductor dots and rings are made from indium arsenide embedded in gallium arsenide, and the structures were grown using techniques developed within the past decade that allow much smaller nanostructures to be created than were previously possible. In these new experiments, electrons are introduced one by one into individual quantum dots while the optical emission of the dots is measured with great precision. These studies provide new perspectives on the internal quantum mechanical workings of quantum dots; the ultimate goal is to create useful electronic and optical nanomaterials that have been quantum mechanically engineered by tailoring the shape, size, composition, and position of various quantum dots. Concerning the physics of quantum dots, adding even a single electron to such a system requires a significant amount of extra energy because of the repulsion between the negatively charged electrons as they are forced into a smaller volume.

One result of this, called the "Coulomb blockade", is to make possible a greater laboratory control of the number of electrons in a quantum dot, i.e., researchers can tune the number of electrons by manipulating input energy. In general, optical excitation of a semiconductor leads to the creation of a quasi-particle known as an "exciton"-a negatively charged electron bound together with a positively charged "hole". In contrast to the Coulomb blockade resulting from electrical injection of electrons into a quantum dot, such dots remain neutrally charged following optical excitation, and the quantum dot exciton has been studied in detail by measuring the light emitted when the hole and electron recombine. The researcher concludes: "Quantum dots have great flexibility because their properties can be artificially engineered, but this comes at a price. Nature has given us atoms; scientists must make quantum dots. Further advances in this exciting field of science and technology will depend heavily on the creativity of physicists, chemists, and materials scientists who make these tiny structures." Physicists are actively seeking to transfer the weird phenomena of the quantum world from their laboratories to the real world. In particular, the prospect of quantum information processing has attracted considerable attention for its potential to improve the speed and reliability of data handling.

Information would be encoded in "quantum bits", and the search is on for a physical system that could form a reliable, controllable quantum bit. In contrast to classical bits, which can be in either state 0 or state 1, quantum bits exist as a combination (a linear superposition) of two quantum logic states, represented as $|0>$ and $|1>$. In a quantum computer, the quantum bits first have to be controlled individually in order to initialize the quantum register in which information is stored. Then a controllable interaction between the quantum bits must be established so that the quantum states become entangled. It is this "entanglement" that is the key to a quantum computer's power: in effect, a rigid coupling is introduced between the quantum bits, which can then no longer be considered individually but are affected simultaneously by a calculational operation. It is thanks to this capacity for parallel processing that a quantum computer should be able to perform calculations much faster than a classical computer.

The original proposals for quantum computers were based on atomic systems(1), such as atoms held in traps, where the quantum bit is formed by two energy levels between which an atomic electron can make transitions. Now that semiconductor quantum dots have been synthesised, it opens up the possibility of mimicking these approaches in a solid-state environment. Quantum dots, tiny clusters of semiconductor material, are often called "artificial atoms", because the charge carriers in these systems (electrons or holes) can only occupy a restricted set of energy levels, just like the electrons in an atom. In fact, quantum dots offer a variety of two level systems, based on charge or spin (or both). One such two level system is a coupled electron hole pair-an exciton. The absence (equivalent to the state $|0>$) and presence (state $|1>$) of an exciton in a semiconductor quantum dot could represent the two levels of a quantum bit. Present-day information technology is based mainly on incoherent processes in conventional semiconductor devices. To realise concepts for future quantum information technologies, which are based on coherent phenomena, a new type of "hardware" is required. Semiconductor quantum dots are promising candidates for the basic device units for quantum information processing. One approach is to exploit optical excitations (excitons) in quantum dots. It has already been demonstrated that coherent manipulation between two excitonic energy levels—via so called Rabi oscillations can be achieved in single quantum dots by applying electromagnetic fields.

The researchers report they make use of this effect by placing an InGaAs quantum dot in a photodiode, which essentially connects it to an electric circuit. The researchers demonstrate that coherent optical excitations in the quantum dot two level system can be converted into deterministic photocurrents. The researchers find that this device can function as an optically triggered single-electron turnstile. A coherently driven single-QD photodiode is an optically triggered single-electron source, which offers, just like the related single electron turnstile device, single electrons (or holes) on demand. Both devices have the potential to deliver frequency-controlled currents according to I = fe. With circularly polarised excitation, a single-QD photodiode has the potential to provide deterministic streams of spin-polarised charges. More subtle points concerning the absolute accuracy of this device, like the influence of incomplete tunneling or possible avalanche multiplication, will certainly be the subject of future work. By means of single-QD photodiodes, it becomes possible to transfer optical excitations of single quantum systems into deterministic electric currents. This optoelectronic functionality provides a link between coherent optical excitations and electric currents on the single electron level. The researchers believe that this will allow the electric readout of excitonic quantum gates, and could have a substantial impact on quantum information technology. The strength of radiative transitions in atoms is governed by selection rules that depend on the occupation of atomic orbitals with electrons. Experiments have shown similar electron occupation of the quantised energy levels in semiconductor quantum dots, often described as artificial atoms. But unlike real atoms, the confinement potential of quantum dots is anisotropic, and the electrons can easily couple with phonons of the material. The researchers report electrical pump-and-probe experiments that probe the allowed and "forbidden" transitions between energy levels under phonon emission in quantum dots with one or two electrons (artificial hydrogen and helium atoms).

The forbidden transitions are in fact allowed by higher-order processes where electrons flip their spin. The authors find that the relaxation time is about 200 microseconds for forbidden transitions, 4 to 5 orders of magnitude longer than for allowed transitions. This indicates that the spin degree of freedom is well separated from the orbital degree of freedom, and that the total spin in the quantum dots is an excellent quantum number. The authors suggest this is an encouraging result for potential applications of quantum dots as basic entities for spin-based quantum information storage. The technical and economic difficulties in further miniaturising silicon-based transistors with the present fabrication technologies have motivated a strong effort to develop alternative electronic devices, based, for example, on single molecules. Recently, carbon nanotubes have been successfully used for nanometer-sized devices such as diodes, transistors, and random access memory cells. Such nanotube devices are usually very long compared to silicon-based transistors.

Since the discovery of carbon nanotube a decade ago, researchers have synthesised a variety of forms of carbon nanotubes. Recently, it has been shown experimentally that fullerenes or endohedral etallofullerenes such as Gd encapsulated inside C82 can be inserted into single-wall nanotubes (SWNTs), forming a pea-pod-like structure. When the diameter of the inserted fullerene is smaller than that of the SWNT minus a certain length (0.7 nm, roughly twice the van der Waals bond length), the fullerenes are inserted exothermally. When the diameter of a fullerene slightly exceeds the length described above, the fullerene can be inserted endothermally, but the resultant SWNT is elastically strained. A theoretical study has predicted that the electronic structure is severely modified, including the positions of the van Hove singularities (VHS), when an SWNT is uniaxially strained. Combining these ideas, one can perform "local bandgap engineering" at the site where a fullerene is endothermally inserted. The researchers report a method for dividing a semiconductor nanotube into multiple quantum dots with lengths of about 10 nm by inserting Gd-C endohedral fullerenes. The spatial modulation of the nanotube electronic bandgap is observed with a low-temperature scanning tunneling microscope. The researchers find that a bandgap of 0.5 eV is narrowed down to 0.1 eV at sites where endohedral metallofullerenes are inserted. This change in bandgap can be explained by local elastic strain and charge transfer at metallofullerene sites. The researchers suggest this technique for fabricating an array of quantum dots could be used for nano-electronics and nano-optoelectronics. Single electron charging phenomena are ubiquitous in atoms, molecules, and small electronic devices, and their effects are central to an understanding of the physics and technology of nanoscale systems. Single electron effects arise because the number of electrons residing on a small, quasi-isolated, conducting island is quantised. Adding an additional charge to such a quantum dot costs an electrostatic energy on the order of U = e2/C, where C is the capacitance of the dot and e is the electronic charge.

This charging energy suppresses charge transport when U >> kBT, where kBT is the thermal energy, leading to the Coulomb blockade of charge motion on and off the dot. Although Coulomb blockade phenomena have been studied extensively with transport measurements, such measurements lack the spatial discrimination necessary to probe the interior of a dot or to probe complex multidot systems. An alternative approach is to detect single-charge motion using scanned probe techniques, such as scanned capacitance microscopy, scanned single electron transistors, and atomic force microsopy (AFM). The first two of these have excellent charge sensitivity but are technically very difficult; moreover, they are not easily able to image the topography of the device under study. AFM-based techniques, on the other hand, can be used both to image the sample and to interact with it in a variety of ways. For example, electrostatic force microscopy (EFM), which measures the electrostatic force between a sample and a metallised AFM tip, has been used to detect the motion of single charges during contact electrification of insulating surfaces and to image the potential profile in carbon nanotubes. In addition, scanned gate microsopy (SGM), in which the AFM tip is used to perturb the conducting properties of a sample, has been used to image electron trajectories and scattering centers in two-dimensional electron gases and barriers in carbon nanotubes. The researchers report that an atomic force microscope was used to study single electron motion in nanotube quantum dots. By applying a voltage to the microscope tip, the number of electrons occupying the quantum dot could be changed, causing Coulomb oscillations in the nanotube conductance.

Spatial maps of these oscillations were used to locate individual dots and to study the electrostatic coupling between the dot and the tip. The electrostatic forces associated with single electrons hopping on and off the quantum dot were also measured. These forces changed the amplitude, frequency, and quality factor of the cantilever oscillation, demonstrating how single-electron motion can interact with a mechanical oscillator. The "Kondo effect", first explained by the theoretical physicist Jun Kondo, is observed when magnetic ions occur as dilute impurities in nonmagnetic crystals. If the host crystal is a metal (e.g., copper), the magnetic impurities (e.g., iron) contribute to the electrical resistivity, with the conduction electrons scattered by the magnetic impurity. The scattering is strong at low temperatures, and increases slightly as temperature decreases, an anomalous resistivity-temperature relation.

The effect essentially results from the interactions between the localised spins of the magnetic impurities and the free electrons involved in conduction. There is a characteristic temperature, called the "Kondo temperature", which depends on the impurity and on the host crystal, the resistivity increasing at low temperature, starting near the Kondo temperature. The Kondo temperature of the system iron-in-copper is 24 degrees kelvin. Nearly all transition metal atoms are found as magnetic impurities in copper or gold, and each system has a different Kondo temperature, which varies from 1000 degrees kelvin to a fraction of 1 degree kelvin. Fullerenes are large molecules composed entirely of carbon, with the chemical formula C(n), where n is any even number from 20 to over 100. They apparently have the structure of a hollow spheroidal cage with a surface network of carbon atoms connected in hexagonal and pentagonal rings. Carbon nanotubes are similar to fullerenes, except their shape is tubular. They were first discovered by Sumio Iijima (NEC Laboratories, JP) in 1991, they come in both multi wall and single wall versions, with single wall nanotubes having a diameter of approximately 1 nanometer and multi wall versions having diameters of the order of 10 to 30 nanometers.

There have been rapid developments in understanding the chemistry and physics of carbon nanotubes, and there is much excitement in both the materials science and electronics communities concerning possible applications of these unique structures. In general, the term "quantum dot" refers to an artificial atom. As realised in the laboratory, quantum dots are small electrically conducting regions, typically less than 1 micron in diameter, that contain from one to a few thousand electrons. Because of the small volume, the electron energies within the dot are quantised, and the behaviour of the quantum dot is intermediate between that of an atom and that of a classical macroscopic object. Electron transport in conductors is usually well described by so-called "Fermi-liquid theory", which assumes that the energy states of the electrons near the *Fermi level are not qualitatively altered by Coulomb interactions. In one-dimensional systems, however, even weak Coulomb interactions cause strong perturbations, and the resulting system, known as a "Luttinger liquid" (after Joaquim Luttinger, who did major work on such systems in the 1950s) is predicted to be distinctly different from its 2- and 3-dimensional counterparts. Single walled carbon nanotubes are model systems for the study of 1-dimensional electron transport, and according to theoretical predictions, such systems should be very sensitive to Coulomb repulsion and become insulators at low temperatures.

However, researchers have demonstrated that it is possible to create superconducting junctions when carbon nanotubes are connected to superconducting contacts, a phenomenon called "proximity-induced superconductivity". In the case of nanotubes, the junction properties are apparently related to its highly 1-dimensional character, and such junctions may have potential applications as thermal detectors of weak infrared radiation (bolometers). In this context, the term "ballistic" (as opposed to the term "diffusive") refers to the passage of electrons through a conductor whose length is less than the mean free path of electrons in the conductor, with the result that most of the electrons pass through the conductor without scattering. The researchers point out that the connection of electrical leads to wire-like molecules is a logical step in the development of molecular electronics, but such a system also allows studies of fundamental physics. For example, metallic carbon nanotubes are essentially quantum wires that have been found to act as 1-dimensional quantum dots, Luttinger liquids, proximity-induced superconductors, and ballistic and diffusive 1-dimensional metals. The researchers report that electrically-contacted single-walled carbon nanotubes can serve as powerful probes of Kondo physics, demonstrating the universality of the Kondo effect.

Arising in the prototypical case from the interaction between a localised impurity magnetic moment and delocalised electrons in a metallic host, the Kondo effect has been used to explain enhanced low-temperature scattering from magnetic impurities in metals, and also occurs in transport through semiconductor quantum dots. The far greater tunability of quantum dots (in this case, nanotubes) compared with atomic impurities makes new classes of Kondo-like effects accessible to study. The nanotube devices used by the researchers differ from previous systems in which Kondo effects have been observed in that they are 1-dimensional quantum dots with 3-dimensional metal (gold) reservoirs. This allows observations of Kondo effects for very large electron numbers in the quantum dot, numbers estimated as of the order of tens of thousands. Coupled quantum dots in semiconductor materials are often referred to as artificial molecules, because these electron boxes allow us to precisely control the number of charge carriers and their mutual interaction. For future quantum information processing, entangled quantum states within such mesoscopic systems need to be implemented in electronic circuits.

This requires precise control of the bonds of the artificial molecule in situ. In this context, coupled dots can be used for investigating the probing mechanism of quantum physics; that is, how to attach leads to real molecules. Generally the interaction of quantum systems with their environment induces decoherence and the decay of quantum superpositions. Key questions are how accurately the coupling of two artificial atoms can be adjusted and how coherent coupling can be monitored noninvasively without destroying the molecular quantum states. It has already been shown in transport and microwave spectroscopy that the transition from ionic to covalent binding in quantum dot molecules can be traced. The detection mechanism for those techniques relies on the transmission of a single electron through the coupled dot system. In contrast, the researchers have used a co-tunneling mechanism, involving two electrons tunneling, simultaneously, to probe the coupling between the dots.

Even though the probed quantum states are only weakly coupled to adjacent reservoirs, the dynamics resemble the Kondo effect, in which hybridisation of electron states with source and drain contacts is desired. In summary: The researchers demonstrate how molecular quantum states of coupled semiconductor quantum dots are coherently probed and manipulated in transport experiments. The applied method probes quantum states by the virtual co-tunneling of two electrons and hence resolves the sequences of molecular states simultaneously. This result is achieved by weakly probing the quantum system through parallel contacts to its constituting quantum dots. The overlap of the dots' wave functions and, in turn, the splitting of molecular states are adjusted by the direct influence of coupling electrodes. In a bulk semiconductor, electrons and holes move freely throughout the crystal. However, in a nanocrystal, confinement of the electrons and holes leads to a variety of optical and electronic consequences, including size-dependent molecular-like optical properties, greater electron/hole overlap for enhanced photoluminescence (PL) efficiencies, and discrete single-electron/hole charging. Because of their enormous surface area-to-volume ratios, nanocrystals (NCs) are highly susceptible to heterogeneous redox chemistry with the surrounding environment. Depending on the semiconductor and the surface chemistry, this chemical reactivity can lead to either fatal chemical degradation or new useful properties, such as reversible photocatalytic and electrochromic properties and redox reactivity.

Semiconductor NCs have been prepared with narrow size distributions, controlled surface chemistry, and internal bulk crystal structure, and adsorbed capping ligands are often used to control size and prevent irreversible aggregation. Although the electrochemical properties of monolayer-protected metallic NCs have been well documented, reports concerning the electrochemical properties of semiconductor NCs remain scarce. Difficulties include the limited potential window available in aqueous solutions, the limited solubility of many NCs in nonaqueous solvents, and the need for highly pure, isolated, monodisperse dots. Compound semiconductor NCs, such as CdS, are also chemically unstable upon electron transfer. Reversible electrochemical injection of discrete numbers of electrons into sterically stabilised silicon nanocrystals (NCs) (approximately 2 to 4 nanometers in diameter) was observed by the authors by differential pulse voltammetry (DPV) in N,N'-dimethylformamide and acetonitrile. The electrochemical gap between the onset of electron injection and hole injection —related to the highest occupied and lowest unoccupied molecular orbitals—grew with decreasing nanocrystal size, and the DPV peak potentials above the onset for electron injection roughly correspond to expected Coulomb blockade or quantised double-layer charging energies.

Electron transfer reactions between positively and negatively charged nanocrystals (or between charged nanocrystals and molecular redox-active coreactants) occurred that led to electron and hole annihilation, producing visible light. The electrogenerated chemiluminescence spectra exhibited a peak maximum at 640 nanometers, a significant red shift from the photoluminescence maximum (420 nanometers) of the same silicon NC solution. The authors suggest these results demonstrate that the chemical stability of silicon NCs could enable their use as redox-active macromolecular species with the combined optical and charging properties of semiconductor quantum dots. Quantum dots are small conductive regions in a semiconductor, the regions containing a variable number of electrons (from 1 to 1000) that occupy well-defined, discrete quantum states—for which reason they are often referred to as "artificial atoms". Connecting quantum dots to current and voltage contacts allows the discrete energy spectra of the system to be probed by charge-transport measurements. Two quantum dots can be connected to form an "artificial molecule", and depending on the strength of the inter-dot coupling (which supports quantum-mechanical tunneling of electrons between the dots), the two dots can form "ionic" or "covalent" bonds.

In the ionic bond case, the electrons are localised on individual dots, and in the covalent bond case, the electrons are delocalised over both dots. T.H. Oosterkamp et al report a transition from ionic bonding to covalent bonding in a quantum dot "artificial molecule" probed by microwave excitations. The authors suggest their results demonstrate controllable quantum coherence in single electron devices, an essential requirement for practical applications of quantum dot circuitry in the construction of quantum computers. Nanoscale metal and semiconductor particles are of current interest because they mark a material transition range between quantum and bulk properties. With decreasing particle size, bulk properties are lost as the continuum of electronic states becomes discrete (the quantum size effect) and as the fraction of surface atoms becomes large.

The electronic and magnetic properties of metallic nanoparticles and nanoclusters show new characteristics that can be utilised in novel applications in areas that range from nonlinear optical switching and catalysis to high-density information storage. Numerous methods have been developed to synthesize metal nanoparticles. A major difficulty with scale-up of these methods is that the metal colloid stability is often controlled by electrostatic interactions across the Debye double layer and sterically through adsorption of steric stabilising agents such as polymers and surfactants. As a consequence, such metal colloids are extremely sensitive to their environment. One way to improve the stability of metal nanoparticles is to coat them with silica, which is very resistant to coagulation, even at high volume fractions. It has been reported that particles of noble metals such as Ag and Au can be coated with silica shells, but these procedures usually involve a multi-step process, and only single metal particle cores could be coated.

The researchers report the development of a new method to fabricate nanocomposite silicon dioxide spheres (approximately 100 nanometers in diameter) containing homogeneously dispersed Ag quantum dots (2 to 5 nanometers in diameter). The inclusion morphology is controlled through the timing of the photochemical reduction of silver ions during hydrolysis of tetraethoxysilane in a microemulsion. Nanometer-scale semiconductor crystallites (known as nanocrystals or quantum dots) could dramatically improve the use of fluorescent markers in biological imaging. Because these colloidal particles act as robust broadly tunable nanoemitters that can be excited by a single light source, they could provide distinct advantages over current in vitro and in vivo markers (e.g., organic dyes and fluorescent proteins). However, before nanocrystals can be widely used as biolabels, they must maintain three properties under aqueous biological conditions: efficient fluorescence, colloidal stability, and low nonspecific adsorption. Unfortunately, despite recent advances, these conditions have not been simultaneously satisfied, limiting the development of in vivo applications of nonaggregated (or individual) semiconductor nanocrystals.

The main challenge is that the quantum dots (QDs), as synthesised, have hydrophobic organic ligands coating their surface. To make the QDs water soluble, these organophilic surface species are generally exchanged with more-polar species, and both monolayer and multilayer ligand shells have been pursued. Although the monolayer method is reproducible, rapid, and produces QDs with a regular, well-oriented, thin coating, their colloidal stability is poor. In contrast, the multilayer method yields QDs that are stable in vitro, but the coating process is long and the coating is difficult to control. A more serious concern is that both approaches still produce QDs that tend to aggregate and adsorb nonspecifically. To resolve this problem, researchers have explored two additional coatings. First, the outer ligand shell of the QD has been overcoated with proteins adsorbed through hydrophobic or ionic interactions. Other layers can then be added to allow conjugation with specific biomolecules. Indeed, this method has provided new reagents for fluoroimmunoassays. Second, the outer ligand shell has been overcoated with surfactants or polymers to prevent nonspecific adsorption of biomolecules while still permitting bioconjugation. For example, silica-coated QDs have been further modified with small monomers of poly(ethylene glycol) to reduce nonspecific adsorption. Despite these efforts, nonspecific adsorption and aggregation still occur when QDs are used in biological environments. Studies of cellular uptakes of QDs report large aggregate formation inside the cell. The same aggregation problems are reported when QDs are used for fluorescence in situ hybridisation, or as markers for molecular recognition on cell surfaces. Consequently, the use of QDs in biological applications is still limited and primarily confined to in vitro studies. In summary: Fluorescent semiconductor nanocrystals (quantum dots) have the potential to revolutionize biological imaging, but their use has been limited by difficulties in obtaining nanocrystals that are biocompatible.

To address this problem, the researchers encapsulated individual nanocrystals in phospholipids block-copolymer micelles and demonstrated both in vitro and in vivo imaging. When conjugated to DNA, the nanocrystal-micelles acted as in vitro fluorescent probes to hybridize to specific complementary sequences. Moreover, when injected into Xenopus embryos, the nanocrystal-micelles were stable, nontoxic (<5 z ᵛ 109 nanocrystals per cell), cell autonomous, and slow to photobleach. Nanocrystal fluorescence could be followed to the tadpole stage, allowing lineage-tracing experiments in embryogenesis. Building ordered and defect-free two- and three-dimensional structures on the nanometer scale is essential for the construction of next-generation optical, electronic, and magnetic materials and devices. Traditional assembly approaches have been based on hydrogen bonding, coulombic interactions, and van der Waals forces. Although a bacterial synthetic method was reported to make monodisperse modified polypeptides, it has been difficult to tune the layer spacing and structure of conventional synthetic polymers because of their polydisperse chain lengths. Efforts have been directed toward the use of soft materials to organise inorganic materials at the nanoscale.

Protein cages have been used as templates to synthesize nanoscale materials in capsids. DNA recognition linkers have been successfully used to construct specific gold nanocrystal structures. ZnS and CdS have been nucleated in a lyotropic liquid crystalline medium to make nanowires and nanocrystal superlattice structures by a surfactant assembly pathway. However, these methods have limitations with respect to length scale and type of inorganic material. Monodisperse biomaterials that have an anisotropic shape are promising as components of well-ordered structures. Liquid crystalline structures of wild-type viruses (Fd, M13, and TMV) have been tunable by controlling the solution concentrations, the solution ionic strength, and the external magnetic fields applied to the solutions. The authors recently demonstrated that engineered viruses can recognize specific semiconductor surfaces through the method of selection by combinatorial phage display. These specific recognition properties of the virus can be used to organise inorganic nanocrystals, forming ordered arrays over the length scale defined by liquid crystal formation. In summary: The researchers report that a liquid crystal system was used for the fabrication of a highly ordered composite material from genetically engineered M13 bacteriophage and zinc sulfide (ZnS) nanocrystals. The bacteriophage, which formed the basis of the self-ordering system, were selected to have a specific recognition moiety for ZnS crystal surfaces. The bacteriophage were coupled with ZnS solution precursors and spontaneously evolved a self-supporting hybrid film material that was ordered at the nanoscale and at the micrometer scale into approximately 72-micron domains, which were continuous over a centimeter length scale.

In addition, suspensions were prepared in which the lyotropic liquid crystalline phase behaviour of the hybrid material was controlled by solvent concentration and by the use of a magnetic field. Quantum dots are nanometer-scale "boxes" for selectively holding or releasing electrons. Over the past 10 years they have been transformed from laboratory curiosities to the building blocks for a future computer industry. Preliminary descriptive information about quantum dots (or "artificial atoms") can be found on the Web in the following locations:

- Introduction to the Description of Quantum Dot Cells by David J.K. Goldhaber-Gordon
- A brief functional description of quantum dots is in the first vugraph in the tutorial on Quantum Dot Cells
- A brief structural description of a quantum dot is in the sixth and seventh vugraphs of the tutorial on Quantum Dot Cells
- See also: Discussion of Resonance Tunneling Devices (RTDs)
- Alta Vista's List of Web References on Quantum Dots

Explaining Quantum Dot Nano Materials

Quantum Dots are now available in a new unique ternary core material with a Molecular Plate™ shell, these materials are ready to use in life sciences assays.

Available in deep red colors of 660nm and 680nm, these non-heavy metal quantum dots are potentially less toxic to cells. The deep red colors are ideal for in vivo and NIR imaging.

Basic References on Quantum Dots

- Reed, M. A., "Quantum Dots," *Scientific American*, January 1993, pp. 118-123.
- Kastner, M. A., "Artificial Atoms," *Physics Today*, January 1993, pp. 24-31.
- Turton, R., *The Quantum Dot: A Journey into the Future of Microelectronics*, Oxford U. Press, Oxford, U.K., 1995. (A synopsis of this book may be found on the Oxford University Press Web Site.)
- Akis, R., and Ferry, D. K., "Numerical Modeling of Quantum Dot Structures"

Selected Quantum Dot Research Groups Represented on the World-Wide Web

- Arizona State University's Nanostructures Research Group
- Project Nanostructures of the Institute of Solid State Physics at TU-Berlin
- Stanford University Nanofabrication Facility

Quantum Dots/Nanotech Overview

Quantum dots are semiconductor crystals typically between 1 and 10 nanometers in diameter and have unique properties between that of single molecules and bulk materials. Quantum dots offer tunable optical and electronic properties that can work around natural limits inherent in traditional semiconductors.

Quantum Dots with Molecular Plating for Life Sciences

Do you have an opto-electronic material problem? Need semiconductors with tunable properties to remove nature-imposed limits? Quantum dots offer ways to tune properties through material and size selection. Evident offers the widest array of quantum dot nanomaterials, matrix and composites quantities from milligrams to kilograms. Evident Technologies has introduced the first commercially available non heavy-metal quantum dot product for life science research: Molecular Plate T2-MP EviTags™. T2-MP EviTags feature three significant developments:

- a new ternary (a complex of three semiconductors) core material system,
- new molecular plating shells, and
- a natural lipid coating on the outer layer.

These are the first commercially available bright, long-lasting, quantum dots that do not contain heavy metals – offering a number of benefits covered in the sections below – and to the company's knowledge no other company has reached this level of product development. At the core of the T2-MP EviTags proprietary technology are Evident's Indium Gallium Phosphide (InGaP) quantum dots, which are then coated first with a unique metallic molecular plating (ZnS) and then with a second, natural lipid coating.

Research at Evident, which is focused on advancing product development, led to the first commercially available nanocrystals that comprise a III-V semiconductor material: Indium, Gallium and Phosphide core quantum dots, with a unique molecular metallic platting compound to form the shell or InGaP/ZnS EviDots-MP™.

Evident's Molecular Plate™ covering for InGaP EviDots is lattice matched—with a matched pattern of atoms—to improve its molecular bonding to the core. This results in a quantum dot with high brightness and increased stability for longer lasting fluorescence.

Evident then further envelopes the plated InGaP quantum dots with a natural lipid coating. This coating allows for high stability in water, can be readily functionalised for life science applications, is long lasting in the harsh environments encountered in biological assays, and is less toxic to cells than traditional heavy metal quantum dots.

The new T2-MP EviTags are available in the deep red wavelengths used in life science research – Macoun Red (660nm) and Jonamac Red (680nm) – with carboxyl, amine, biotin and non-functionalised surfaces. Having a biologically compatible quantum dot label like the T2-MP EviTags allows further exploration and development of in vivo assays, following cells inside living organism to get a new understanding of life. In particular, it offers the possibility to create new classes of assays for use where current testing materials could produce cytotoxicity or cell death. Potential applications include live cell and even whole animal imaging, blood cancer assays, and numerous other applications.

Studies have shown that quantum dot toxicity is due to cadmium ions being released into cells. This is exacerbated by oxygen, UV exposure and the large exposed surfaces of the spherical quantum dots. To avoid these problems, surfaces are added to the quantum dots, traditionally consisting of Zinc-Sulfide compounds to create a more benign surface. Polymers are used as well, which have been shown to be stable, but the polymers themselves may kill cells. Studies of quantum dot cytotoxicity are available in: Cytotoxicity of Colloidal CdSe and CdSe/ZnS Nanoparticles Christian Kirchner, Tim Liedl, Stefan Kudera, et al. Evident's new T2-MP EviTags are a potentially more biologically compatible fluorescent tag, and can provide a platform for use in vivo imaging, new forms of assays, new systems and even new markets.

An Overview of Quantum-Dot Cells and Wireless Electronic

On Tuesday, 16 May 1995, I attended a talk at MIT by Professor Craig Lent of Notre Dame. He spoke of a possible architecture for computer logic based on quantum dots. A quantum dot is a box that holds a discrete number of electrons. This number can be changed by adjusting electric fields in the neighborhood of the dot, for example by applying a voltage to a nearby metal gate. In the following, I will refer to quantum dots that have either zero or one electron in them. Since quantum dots are fabricated in solids, not in vacuum, of course there are many electrons in them. However, almost all of these are tightly bound to atoms in the solid. The few electrons that I will speak of are extra ones beyond those that are tightly bound. These extra electrons could roam free in a solid were they not confined in a quantum dot. We arrange that two electrons reside in each cell. They may hop (tunnel) from dot to dot within the cell to find a favourable, low-energy configuration. Of course, because electrons are charged, they repel each other, so the two electrons will not both choose to sit in the same dot within the 4-dot cell. In fact, they will choose to occupy two opposite corners of the cell, since this gives them their maximum possible separation. This is nice, because it gives us two possible "polarisation" states for a cell, which we can associate with a binary zero and one. We can use a voltage on an external gate to force a cell into one of its two a priori equally good alternative states. This is a signal, analogous to a high or low voltage at a point within a conventional integrated circuit.

A conventional integrated circuit uses two tools to transmit a voltage signal over long distances: an aluminum or silicon wire (to carry signals from place to place) and a transistor used as an amplifier (to refresh weakened signals so that they may travel farther without corruption). In the quantum dot circuits proposed by Lent, both of these tasks are accomplished simply by arranging as many cells as necessary in a row. If two cells are placed next to each other, and the polarisation of first is fixed, the second will match its own polarisation. This, extended to a row of many cells, is the task normally carried out by a wire. In fact, even if the polarisation of the first dot is between zero and one (the electrons are not fully localised on one diagonal), as more dots are added to the chain, the polarisation of later dots will rapidly approach a perfect zero or one.

This is the task of amplification or refreshing signals, the trademark of digital circuitry. Although I will not illustrate them, one can also build a corner, fan-out, or inverter (some of the building blocks of digital circuits). "Wires" carrying signals can even cross each other in the plane. So far, we could make a simple translation from objects in standard integrated circuits to objects in quantum dot circuits. However, the cornerstone of digital circuitry, logic gates, are not so simply translated into this new architecture. A standard logic gate has two inputs and an output determined by the inputs. For example, an AND gate puts out a high output only if both input. Here, inputs 1 and 2 vote together and hence determine the middle cell, which in turn determines the output. One can build any ordinary logic using these gates (most clumsily, one can make an OR gate by fixing input 3 to have polarisation 1, an AND gate by forcing it to have polarisation 0). More impressively, if the center-to-center distance between dots within a cell in 20 nm (extremely aggressive though not impossible with today's electron-beam lithography), a single bit full adder will fit in an area of 1 micron^2. By contrast, the semiconductor industry projects that in 2010 the maximum packing density for transistors will be approximately 1 micron^2. Forty transistors would be needed to build an adder! This dramatic difference in density is explained by two facts:

1. Much of the space in a conventional integrated circuit is taken up by wiring, which is unnecessary here.
2. The minimum feature size required to pack 1 transistor/ micron^2 would be 70 nm, and relative alignment between features produced in different lithographic steps generally needs to be good to 10% of this minimum feature size, or 7 nm. Only one stage of quantum dot manufacture need involve small features, so no sensitive alignment is required. This means that the minimum feature size could probably be made smaller than 70 nm even abandoning electron beams in favour of optical techniques favoured by industry for their high throughput. In addition, the diameter of a single cell is only a few times the minimum feature size, as compared to 1 micron/70 nm which is approximately 14 times.

In short, for given lithographic capabilities, quantum dot cells can be packed far more densely than can transistors.

Readout could be accomplished by passing a current through a very narrow constriction near a corner of a cell. An electron in that corner of the cell could shut off current. This was demonstrated by Field et al., PRL 70, 1311 ('93).

This whole discussion has assumed that when new boundary conditions (inputs) are imposed, the system rapidly and reliably finds its ground (or lowest energy) state. This would not be true if the ambient temperature were comparable to the difference in energy between a "correct" and an "incorrect" polarisation of a cell. For the quoted 20 nm dot-to-dot distance, this energy associated with an error in one cell's polarisation would be 1 millielectron volt, or one twentieth of thermal energies (kT) at room temperature, so this is a major worry. Even if the error energy could be made many times larger, which could be accomplished by making the individual dots smaller and closer together, entropy would limit the size of an array of cells to about 20 thousand at error energy = 10 kT. Finally, no one knows how long an array of cells would take to reach its ground state, though Lent estimates 30 picoseconds per cell in a "wire". Stray charges within 1 cell's length of a "wire" would lock a nearby cell permanently into one polarisation state, stymieing computation. Charges are always sprinkled liberally on surfaces of semiconductors, so this would seem to be a major barrier to dense integration using this architecture. Perhaps most seriously, an attendee at the talk pointed out that the ground state of a large array may not be the obvious one determined by propagating signals "correctly" from inputs to outputs (though Lent has showed by modeling that his adder would not suffer from this problem).

Lent's quantum dot cells can be switched from one state to another adiabatically (reversibly). This is accomplished by lowering the barriers between dots within each cell, and removing the global inputs to the circuit, hence relaxing the polarisation of each cell, then applying new global inputs and raising the barriers again. This produces very little heat, an important quality in densely integrated circuits (a DEC Alpha already has a fan mounted on top of it, continually blowing cold air on it so that it doesn't melt).

Explaining Molecular Quantum Dots

In the early 1970's, two farsighted scientists, Ari Aviram and Mark Ratner, began to envision electronic circuit elements made from *single molecules* and to describe in detail how they might function. This was the origin of the field of molecular electronics, now sometimes called "molecular-scale electronics." It has been observed in experiments and shown theoretically that reducing the dimensions of a raises the effective operating temperature of the electron confinement device. Present day quantum dots are large enough (approximately 1-10 microns long and wide) that they require cooling with liquid helium or, at least, liquid nitrogen, to cryogenic temperatures. However, for a practical technology with widespread applications based upon such quantum-effect devices, it will be necessary to achieve room temperature operation. This requirement implies that it is necessary to invent and manufacture molecular-scale quantum dots that are only approximately 1 to 10 nanometers in linear dimension. Such a quantum dot would probably be constructed as a single molecule—i.e., a *molecular quantum dot*. Molecular quantum dots are one example of the next-generation technology. In another example, Professor James Tour of the University of South Carolina and Professor Mark Reed of Rale University are collaborating on the chemical synthesis and testing.

These operate by allowing electrons to move nearly ballistically along the length of a chain of ring-like chemical structures with conjugated pi-orbitals. It has been suggested by Tour and by others, that it may be possible to insert chemical groups of lower conductance into such a molecular wire, creating paired barriers to electron migration through the chain. Such barriers might create a that would function in a fashion similar to solid-state resonance tunneling devices (RTDs) that already have been fabricated, tested, and applied in prototype quantum-effect logic. A group at Purdue University has applied wires of the type invented by Tour as one element of a much more complex, heterogeneous extended structure that functions as a molecular electronic circuit array.

Towards a Potential Two-State Devices

Making much smaller, nanometer-scale electronic digital computers requires making nanometer-scale two-state devices—i.e., nanoswitches or nanotransistors with the two states ON and OFF or 1 and 0. These devices will be used to record and manipulate the bits (1s and 0s) on the nanometer scale. On this page are organised information and links that describe some of the technologies for nanoelectronic devices and mechanisms that have been proposed and investigated for that purpose.

An Introductory Overview of Computational Nanotechnology

The concept of a Universal Computer, a device able to compute anything computable, dates back to Babbage in the early part of the 19th century, and to Turing, Church, and von Neumann in the 20th century. The concept of a "Universal Constructor" is perhaps more recent. The concept was well understood by von Neumann in the 1940's, who defined a "Universal Constructor" in a two-dimensional cellular automata world. Such a model is a mathematical abstraction something like an infinite checkerboard, with several different types of "checkers" that might be on each square. The different pieces spontanesouly change and move about depending on what pieces occupy neighboring squares, in accordance with a pre-defined set of rules. Von Neumann used the concept of a Universal Constructor in conjunction with a Universal Computer as the core components in a self-replicating system.

The possibility of fabricating structures by putting "...the atoms down where the chemist says..." was recognised by Feynman in 1959. Drexler recognised the value of the "assembler" in 1977. (Note: Drexler said "Though assemblers will be powerful... they will not be able to build everything that could exist.") The assembler is analogous to von Neumann's Universal Constructor, but operates in the normal three dimensional world and can build a large, atomically precise structures by manipulating atoms and small clusters of atoms. He published the concept in 1981 and in subsequent work has proposed increasingly detailed designs for such devices. The basic design of Drexler's assembler consists of: (i) a molecular computer, (ii) one or more molecular positioning devices (which might resemble very small robotic arms), and (iii) a well defined set of chemical reactions (perhaps one or two dozen) that take place at the tip of the arm and are able to fabricate a wide range of structures using site-specific chemical reactions.

It is now common for forecasts of future technical abilities to include the ability to fabricate molecular devices and molecular machines with atomic precision. While there continues to be debate about the exact time frame, it is becoming increasingly accepted that we will, eventually, develop the ability to economically fabricate a truly wide range of structures with atomic precision. This will be of major economic value. Most obviously a molecular manufacturing capability will be a prerequisite to the construction of molecular logic devices. The continuation of present trends in computer hardware depends on the ability to fabricate ever smaller and ever more precise logic devices at ever decreasing costs. The limit of this trend is the ability to fabricate molecular logic devices and to connect them in complex patterns at the molecular level.

The manufacturing technology needed will, almost of necessity, be able to economically manufacture large structures (computers) with atomic precision (molecular logic elements). This capability will also permit the economical manufacture of materials with properties that border on the limits imposed by natural law. The strength of materials, in particular, will approach or even exceed that of diamond. Given the broad range of manufactured products that devote substantial mass to load-bearing members, such a development by itself will have a significant impact. A broad range of other manufactured products will also benefit from a manufacturing process that offers atomic precision at low cost. Given the promise of such remarkably high payoffs it is natural to ask exactly what such systems will look like, exactly how they will work, and exactly how we will go about building them.

One might also enquire as to the reasons for confidence that such an enterprise is feasible, and why one should further expect that our current understanding of chemistry and physics (embodied in a number of computational chemistry packages) should be sufficient to explain the operating principles of such systems. It is here that the value of computational nanotechnology can be most clearly seen. Molecular machine proposals, provided that they are specified in atomic detail (and are of a size that can be dealt with by current software and hardware), can be modeled using the tools of computational chemistry. There are two modeling techniques of particular utility. The first is molecular mechanics, which utilizes empirical force fields to model the forces acting between nuclei. The second is higher order ab initio calculations, which will be discussed in a few paragraphs. *A few links to internet computational chemistry resources* are provided here to provide those new to the area with some feel for what's available.

Understanding Molecular Mechanics and Ab Initio Methods

Molecular mechanics allows computational modeling of the positions and trajectories of the nuclei of individual atoms without an undue computational load. Current packages available on personal computers can readily do energy minimisations on systems with thousands of atoms, while supercomputers can handle systems with hundreds of thousands of atoms or more. More complex analyses, particularly analyses that involve searching through large configuration spaces, can limit the size of system that can be effectively handled. As will be discussed later the need to search through large configuration spaces (as when determining the native folded structure of an arbitrary protein) can be avoided by the use of relatively rigid structures (which differ from relatively floppy proteins and have few possible configurations). The modeling of machine components in vacuum reduces the need to model solvation effects, which can also involve significant computational effort. In molecular mechanics the individual nuclei are usually treated as point masses. While quantum mechanics dictates that there must be a certain degree of positional uncertainty associated with each nucleus, this positional uncertainty is normally significantly smaller than the typical internuclear distance. Bowen and Allinger provide a good recent overview of the subject. While the nuclei can reasonably be approximated as point masses, the electron cloud must be dealt with in quantum mechanical terms. However, if we are content to know only the positions of the nuclei and are willing to forego a detailed understanding of the electronic structure, then we can effectively eliminate the quantum mechanics. For example, the H_2 molecule involves two nuclei. While it would be possible to solve Schrodinger's equation to determine the wave function for the electrons, if we are content simply to know the potential energy contributed by the electrons (and do not enquire about the electron distribution) then we need only know the electronic energy as a function of the distance between the nuclei. That is, in many systems the only significant impact that the electrons have on nuclear position is to make a contribution to the potential energy E of the system.

In the case of H_2, E is a simple function of the internuclear distance r. The function E(r) summarizes and replaces the more complex and more difficult to determine wave function for the electrons, as well as taking into account the inter-nuclear repulsion and the interactions between the electrons and the nuclei. The two hydrogen nuclei will adopt a position that minimizes E(r). As r becomes larger, the potential energy of the system increases and the nuclei experience a restoring force that returns them to their original distance. Similarly, as r becomes smaller and the two nuclei are pushed closer together, we also find that a restoring force pushes them farther apart, again restoring them to an equilibrium distance. More generally, if we know the positions r_1, r_2, ... r_N of N nuclei, then $E(r_1, r_2, ... r_N)$ gives the potential energy of the system. Knowing the potential energy as a function of the nuclear positions, we can readily determine the forces acting on the individual nuclei and therefore can compute the evolution of their position over time.

The function E is a newtonian potential energy function (not quantum mechanical), despite the fact that the particular value of E at a particular point could be computed from Schrodinger's equation. That is, the potential energy E is a newtonian concept, but the particular values of E at particular points are determined by Schrodinger's equation. Often called the Born-Oppenheimer approximation, this approach allows a great conceptual and practical simplification in the modeling of molecular systems. While it would in principle be possible to determine E by solving Schrodinger's equation, in practice it is usual to use available experimental data and to infer the nature and shape of E by interpolation. This approach, in which empirically derived potential energy functions are created by interpolation from experimental data, has spawned a wide range of potential energy functions, many of which are sold commercially. Because the gradient of the potential energy function E defines a conservative force field F, molecular mechanics methods are also called "force field" methods.

While it is common to refer to "empirical force field" methods, the more recent use of ab initio methods to provide data points to aid in the design of the force fields makes this term somewhat inaccurate, though still widely used. The utility of molecular mechanics depends crucially on the development of accurate force fields. Good quality force fields have been developed for a fairly broad range of compounds including many compounds of interest in biochemistry. While we will not attempt to survey the wide range of force fields that are available, one particular subset of compounds for which good quality force fields are available involve H, C, N, O, F, Si, P, S, Cl (and perhaps a few others) when they are restricted to form chemically uncomplicated structures (e.g., bond strain is not too great, dangling bonds are few or absent, etc).

Many atomically precise structures which should be useful in nanotechnology fall in this class and can be modeled with an accuracy adequate to determine the behaviour of molecular machines. A potential energy function of particular utility in modeling diamondoid structures was described by Brenner (*Empirical potential for hydrocarbons for use in simulating the chemical vapor deposition of diamond films*, Donald W. Brenner, Phys Rev B, Vol. 42, No. 15, November 15 1990, pages 9458-9471). Brenner's potential energy function, though limited to the elements carbon and hydrogen, has the great advantage that it will handle transition states, unstable structures, and the like.

Thus, given essentially arbitrary coordinates of the carbon and hydrogen atoms in a system, Brenner's potential will return the energy of the system. This permits molecular dynamical modeling of arbitrary hydrocarbon systems, including systems which use synthetic reactions involved in the synthesis of diamond. A particular reaction of interest. See *Surface patterning by atomically-controlled chemical forces: molecular dynamics simulations* by Sinnott et al., Surface Science 316 (1994), L1055-L1060; see also the work of for an illustration of the use of Brenner's potential in modeling the behaviour of graphitic gears, including failure modes). Brenner has commented on the utility of this potential for modeling proposed molecular machine systems. Of course, the "accuracy" of the force fields depends on the application. A force field which was accurate to (say) 10 kcal/ mole would be unable to correctly predict many properties of interest in biochemistry. For example, such a force field would lead to serious errors in predicting the correct three-dimensional structure of a protein. Given the linear sequence of amino acids in a protein, the protein folding problem is to determine how it will fold in three dimensions when put in solution. Often, the correct configuration will have an energy that differs from other (incorrect) configurations by a relatively modest amount, and so a force field of high accuracy is required. Unlike the protein folding problem, where an astronomical range of configurations of similar energy are feasible, the bearing basically has only one configuration: a bearing. While the protein has many unconstrained torsions, there are no unconstrained torsions in the bearing. To significantly change any torsion angle in the bearing would involve ripping apart bonds.

Thus, the same force field which is of marginal utility in dealing with strands of floppy protein is quite adequate for solid blocks of stiff diamondoid material. Not only will small errors in the force field still result in an accurate prediction of the global minima, (which in this case will be a single large basin in the potential energy surface) but also the range of possible structures is so sharply limited that little or no computational effort need be spent comparing the energies of different configurations. Long computational runs to evaluate the statistical properties of ensembles of configurations are thus eliminated. This style of design has been called, only half in jest, "molecular bridge building" because bridges are also designed with large safety margins. This observation, that the same force field that is inadequate for one class of structures is quite adequate for the design and modeling of a different class of structures, leads to a more general principle. Computational experiments generally provide an answer with some error distribution. If the errors produced by the model are of a similar size to the errors that would result in incorrect device function, then the model is unreliable. On the other hand, if the errors in the model are small compared with the errors that will produce incorrect device function, then the results of the model are likely to be reliable.

If a device design falls in the former category, i.e., small errors in the model will produce conflicting forecasts about device function, then the conservative course of action is to reject the proposed design and keep looking. That is, *we can deliberately design structures which are indeed adequately handled by our computational tools.* As illustrated by the bearing, there are a wide range of basically mechanical structures whose stability is sufficiently clear and whose interactions with other structures is sufficiently simple that their behaviour can be adequately modeled with currently available force fields. An even stronger (although somewhat more subtle) statement is possible. The strain in the axle and the sleeve is proportional to the diameter of the bearing. By increasing the diameter, we can reduce the strain. Thus, we can design bearings in this broad class in which the strain can be reduced to whatever level we desire. Because we are dealing with what amounts to a strained block of diamond, the fact that we can reduce strain arbitrarily means that we can design a bearing whose local structure bears as close a resemblance to an unstrained block of diamond as might be wished. We can therefore be very confident indeed that some member of this class will perform the desired function (that of a bearing) and will work in accordance with our expectations. Given that we have developed software tools that are capable of creating and modeling most of the members of a broad class of devices, then we can investigate many individual class members. This investigation then lets us make rather confident statements about the functionality that members of this class can provide, even if there might be residual doubts about individual class members. A moments reflection will show that the class of objects which are chemically reasonably inert, relatively stiff (no free torsions), and which interact via simple repulsive forces that occur on contact; can describe a truly vast class of machines. Indeed, it is possible to design computers, robotic arms and a wide range of other devices using molecular parts drawn from this class.

While empirical force fields are sufficiently accurate to model the behaviour of chemically stable stiff structures interacting with other chemically stable stiff structures, they do not (at present) provide sufficient accuracy to deal with chemical transitions. Thus, if we wish to model the manufacture of a molecular part then we must use higher order ab initio techniques (another paper in this issue illustrates what is meant by this in detail). These techniques impose severe constraints on the number of atoms that can be modeled (perhaps one or two dozen heavy atoms, depending on the hardware, software, and specific type of modeling being attempted), but can provide an accuracy sufficient to analyze the chemical reactions that must necessarily take place during the synthesis of large, atomically precise structures. More generally, higher order ab initio techniques are sufficient to analyze the addition or removal of a small number of atoms from a specific site on a work piece. Synthesis of a large object would then consist of repeated site specific applications of a small number of basic operations, where each basic operation changed the chemical structure of only a small number of atoms at a time. Provided that we reject reaction mechanisms where the result predicted depends on errors that are smaller than can reasonably be modeled, the analysis of these basic operations can be satisfactorily carried out with current methods and hardware. Several reactions mechanisms have been discussed, and a great many more are feasible.

Knowing Sufficiency of Current Modeling Methods and Molecular Compilers

It is quite possible to adequately model the behaviour of molecular machines that satisfy two constraints: (i) they are built from parts that are sufficiently stable that small errors in the empirical force fields do not raise significant questions about the shape or stability of the parts, and (ii) the synthesis of the parts is done by using positionally controlled reactions, where the actual chemical reactions involve a relatively small number of atoms whose behaviour can be adequately modeled with higher order ab initio methods.

Clearly, not all molecular machines satisfy these constraints. Is the range of molecular machines which do satisfy these constraints sufficiently large to justify the effort of designing and modeling them? And in particular, can we satisfactorily model Drexler's assembler within these constraints? The fundamental purpose of an assembler is to position atoms. To this end, it is imperative that we have models that let us determine atomic positions, and this is precisely what molecular mechanics provides. Robotic arms or other positioning devices are basically mechanical in nature, and will allow us to position molecular parts during the assembly process. Molecular mechanics provides us with an excellent tool for modeling the behaviour of such devices. The second fundamental requirement is the ability to make and break bonds at specific sites. While molecular mechanics provides an excellent tool for telling us where the tip of the assembler arm is located, current force fields are not adequate to model the specific chemical reactions that must then take place at the tip/work-piece interface involved in building an atomically precise part (though this statement must be modified in light of the work done with Brenner's potential, discussed above).

For this, higher order ab initio calculations are sufficient. The glaring omission in this discussion is the modeling of the kind of electronic behaviour that occurs in switching devices. Clearly, it is possible to model electronic behaviour with some degree of accuracy, and equally clearly molecular machines that are basically electronic in nature will be extremely useful. It will therefore be desirable to extend the range of computational models discussed here to include such devices. For the moment, however, the relatively modest inclusion of electrostatic motors as a power source is probably sufficient to provide us with adequate "design space" to design and model an assembler.. While it might at first glance appear that electronics will be required for the computational element in the assembler, in fact molecular mechanical logic elements will be sufficient.

Babbage's proposal from the 1800's clearly demonstrates the feasibility of mechanical computation (it is interesting to note that a working model of 's difference engine has been built and is on display at the British Museum of Science. They were careful to use parts machined no more accurately than the parts available to Babbage, in order to demonstrate that his ideas could have been implemented in the 1800's). Drexler's analysis of a specific molecular mechanical logic element and further analysis of system design issues also make it clear that molecular mechanical computation is sufficient for the molecular computer required in an assembler. Molecular electronic proposals for computation have not been worked out as clearly at the system level as the molecular mechanical concepts. Thus, at the moment, molecular mechanical proposals are better understood, at least in this particular context. This situation is likely to change, and when it does the electronic designs could be incorporated (where appropriate) into the design and modeling of an assembler. However, it is not entirely obvious that electronic designs will prove superior to molecular mechanical designs, particularly when device parameters such as size and energy dissipation are considered. While it seems virtually certain that electronic devices will prove faster than molecular mechanical devices, it is less obvious that electronic devices must necessarily be either smaller or dissipate less energy (though confer more recent work on *reversible logic*).

Indeed, given the difficulty of localising individual electrons, it seems possible that electronic devices will prove to be inherently larger than molecular mechanical devices. This might provide a long term role for molecular mechanical devices as high density memory elements, or as logic elements when space (or atom count) is particularly constrained. Whether the assembler is designed and built with an electronic or mechanical computer is less significant than designing and building it. By way of example, a stored programme computer can be built in many different ways. Vacuum tubes, transistors, moving mechanical parts, fluidics, and other methods are all entirely feasible.

Delaying the development of the Eniac because transistors are better than vacuum tubes would have been a most unwise course of action. Similarly, as we consider the design, modeling, and construction of an assembler, we should not hesitate to use simpler methods that we can understand in favour of better methods that are not yet fully in hand. The methods of computational chemistry available today allow us to model a wide range of molecular machines with an accuracy sufficient in many cases to determine how well they will work. We can deliberately confine ourselves to the subset of devices where our modeling tools are good enough to give us confidence that the proposals should work. This range includes bearings, computers, robotic arms, site-specific chemical reactions utilised in the synthesis of complex structures, and more complex systems built up from these and related components. Drexler's assembler and related devices can be modeled using our current approaches and methods.

Computational nanotechnology includes not only the tools and techniques required to model proposed molecular machines, it must also include the tools required to specify such machines. Molecular machine proposals that would require millions or even billions of atoms have been made. The total atom count of an assember might be roughly a billion atoms. While commerically available molecular modeling packages provide facilities to specify arbitrary structures, it is usually necessary to "point and click" for each atom involved. This is obviously unattractive for a device as complex as an assembler with its roughly one billion atoms. It should be clear that molecular CAD tools will be required if we are to specify such complex structures in atomic detail with a reasonable amount of effort. An essential development is that of "molecular compilers" which accept, as input, a high level description of an object and produce, as output, the atomic coordinates, atom types, and bonding structure of the object. The first line, "scale 0.9," is simply an instruction to shrink the size of the axle by 10% compared with the normal size. This allows the axle to be positioned inside the sleeve before joint minimisation of the axle and sleeve is done (using MSI's PolyGraf, a commercially available molecular mechanics package that implements both MM2 and the Dreiding II force field). If the axle w were not done, then the atoms in the axle and the sleeve would be comingled, and minimisation would produce meaningless results. The second line begins with the "tube" specifier. This tells the programme to produce a tubular structure, rather than a "block" or a "ring." the "ring" specifier is used to produce toroidal structures, e.g., tubes that have been further bent into donuts. The first triplet of numbers following "tube," "0 0.75 0.75," specifies the offset to be used for the crystal lattice. The second triplet, "1.75 0 0" specifies that the surface of the tube is (100), and that the thickness of the tube wall is 1.75 lattice spacings.

The third triplet, "0 17 -17," specifies the direction and length of the circumference of the tube. The direction (or [0 1 -1]) is at right angles to [1 0 0], the direction of the tube surface. The circumference of the tube is simply the length of the [0 17 -17] vector. Finally, "0 5.25 5.25" gives the direction of the axis and the length of the tube. This does not fully specify the axle of the bearing, for the 100 surface has been cut circumferentially to produce grooves. The next command, "grid 0 0.5 -0.5 0 1 1" specifies that a grid is to be laid down on the surface. The grid is specified by two vectors, "0 0.5 -0.5" and "0 1 1,"" which give the directions and lengths of the edges of the unit parallelogram from which the grid is composed.

The next command, "delete 1.25 0 0"" specifies that the point at coordinates "1.25 0 0" is to be deleted, and further, that all points in the same location with respect to any unit parallelogram of the grid are also to be deleted. Thus, this "grid .... delete ...." cuts out the grooves that are visible on the outer surface of the axle. A similar set of commands was used to specify the sleeve of the bearing. It is easy to specify bearings with different surfaces, surface orientations, circumferences, lengths, etc.

To select a (111), (110), (312), or any other surface, it is sufficient to change the vector that specifies the surface, and the vectors that specify the tangent to the surface and the axis of the tube (which must be at right angles to the surface vector).

Thus, complex structures involving many atoms can be generated quickly and easily with a few lines of input specification. The C source code is available at URL ftp://ftp.parc.xerox.com/pub/nano/tube.c. The programme will generate output in either PolyGraf format or in Brookhaven format, so the output should be readable by most computational chemistry packages. The software required to design and model complex molecular machines is either already available, or can be readily developed over the next few years. Many of the modeling issues can be dealt with by existing commercially available computational chemistry packages. The molecular compilers and other molecular CAD tools needed for this work can be implemented using generally understood techniques and methods from computer science. Using this approach, it will be possible to substantially reduce the development time for complex molecular machines, including Drexler's assembler. This approach is similar in spirit to the computer aided design and modeling used to speed the development of many products today.

This included the hardware, software, compilers, operating system, etc. During this process, extensive use was made of computational models to verify each level of the design. The operating system was written in Pascal, and checked out on another computer. The compilers were written in Pascal, and also checked out on another computer. The code produced by the compilers was checked out on an instruction set simulator. The microcode was checked out on a micro-instruction simulator. The logic design was checked out with logic level simulation tools, and circuit simulation packages were used to verify the detailed electronic design. All this work was done at the same time. The software was written and debugged even though the machine on which it would eventually be executed didn't exist. The microcode was written and debugged before the hardware was available. When the hardware was finally made available, system integration went relatively rapidly. Imagine, for a moment, how long it would have taken to develop this system had we carried out the development in the obvious sequential fashion. First, we would have implemented the hardware and only then begun work on the microengine.

Later, with the hardware and microengine fully checked out and working, we could have developed and debugged the microcode. With this firmly in hand, we could then have written the compilers and verified the code they produced. Finally, we could have implemented and checked out the operating system. Needless to say, such a strategy would have been very slow and tedious. Doing things in the simple and most obvious way often takes a lot longer than is needed. If we were to approach the design and construction of an assembler using the simple serial method, it would take a great deal longer than if we systematically attacked and simultanesouly solved the problems that arise at all levels of the design at one and the same time. That is, by using methods similar to those used to design a modern computer, including intensive computational modeling of individual components and sub-systems, we can greatly shorten the time required to design and build complex molecular machines. This can be further illustrated by considering the traditional manner of growth of our synthetic capability over time. Today, we find we are able to synthesize a certain range of compounds and structures. As time goes by, we will be able to synthesize an ever larger range of structures.

This growth in our ability will proceed on a broad front, and reflects the efforts of a broad range of researchers who are each pursuing individual goals without concern about the larger picture into which they might fit. Sufficient time we will eventually be able to synthesize complex molecular machines simply because we will eventually develop the ability to synthesize just about anything. However, if we wish to develop a particular kind of device, e.g., an assembler, then we can speed the process up by conducting computational experiments designed to clarify the objective and to specify more precisely the path from our current range of synthetic capabilities to the objective. Such computational experiments are inexpensive, can provide very detailed information, are possible for any structure (whether we can or cannot synthesize it) and will, in general, reduce the "time to market" for the selected product. Computational experiments let us examine structures quickly and easily, rejecting those which have obvious defects. This kind of examination of the "design space" is impossible with physical experiments today, but is easily done with computational experiments. Computational experiments also provide more information. For example, molecular dynamics can literally provide information about the position of each individual atom over time, information which would usually be inaccessible in a physical experiment. Of course, the major advantage of computational experiments over physical experiments in the current context is the simple fact that physical experiments aren't possible for molecular machines that we can't make with today's technology. By using computational models derived from the wealth of experimental data that is available today, we can (within certain accuracy bounds) describe the behaviour of proposed systems that we plan to build in the future. If we deliberately design systems that are sufficiently robust that we are confident they will work regardless of the small errors that must be incurred in the modeling process, we can design systems today that we will not be able to build for some years, and yet still have reasonable confidence that they will work. By fully utilising the experience that has been developed in the rapid design and development of complex systems we can dramatically reduce the development time for molecular manufacturing systems.

It is possible to debate how long it will be before we achieve a robust molecular manufacturing capability. However, it is very clear that we'll get there sooner if we develop and make intelligent use of molecular design tools and computational models. These will let us design and check the blueprints for the new molecular manufacturing technologies that we now see on the horizon, and will let us chart a more rapid and more certain path to their development.

Elaborating on Nanosystem and Resonant Tunneling Devices (RTDs

Here we talk about the future nanoelectronic devices using quantum effects employed in semiconductor devices for fabrication of computer devices and high frequency oscillators. The underlying principles used for defining the perpendicular transport of electrons in low dimensional structures up to room temperature and its current and voltage characteristics are explained using energy band diagrams. MOSFETS with its special characteristic features serves two purposes. First case its acts as an excellent conductor device even with its scaling properties and even with small changes in its bias voltage results in large changes in its conductivity, hence its acts as an amplifier device. Though it functions as a two state device there some factors need to be considered for its implementation. The obstacles associated with MOSFET are high electric field, heat power dissipation, vanishing bulk properties, shrinkage of depletion layers and unevenness of the depletion layer. Corrections have to be applied for these devices. Once the devices approach the nanoelectric properties, the bulk properties of semiconductor devices can be replaced by quantum physics which depend on the electron properties, energy quantisation effects and tunneling come into effect. When quantum effects are applied to semiconductor devices the consideration of doped materials no longer considered as major issue. These devices are of great importance for high speed electronics. Here we discuss the quantum properties, general used terms such a quantum dot, tunneling effect, island, subband and electron behaviour in quantised energy levels to deal with the resonant tunneling devices and the application of the quantum properties in various devices such as Resonant Tunneling Diodes (RTDs), Resonant Tunneling Transistors (RTTs). Here we explain the theory related to tunneling and the mathematical derivation for transmission probability of the particle in the double barrier junction.

Resonant Tunneling Devices, Island, Potential Well and Quantisation Effects

An island confining electrons with one or two classical degrees of freedom. These solid state nanoelectronic devices are fabricated by using semiconductor materials. The V group element in the periodic table Si/Ge are generally used, but in current nanoelectronic devices III – IV group elements in periodic table such as GaAs Gallium Arsenide and AlGaAs Aluminium Gallium arsenide are used because of its excellent mobility of electrons and easier to fabricate defect free junctions than it is for IV group elements S/Ge. It's a quantum heterostructure exhibiting quantum perpendicular transport of electrons up to room temperature. An example for such a property is resonant electron transmission through a double barrier heterostructure. If more than one material is used in a device is referred as heterostructure. A heterojunction is formed when two different materials are joined together at a junction. The pairs of semiconductor materials that form the heterostructures are GaAs/AlGaAs, GaInAs/InP, GaInAs/AlInAs, GaSb/AlSb, GaN/AlN, InN/GaN etc.

A Quantum Well is potential structure which spatially confines electrons. Molecular Beam Epitaxy method is for growing the hetrostructures provide one dimensional confinement along the superlattice axis. The basic theory of Quantum Mechanics illustrates that an electron subjected to potential confinement has its energy quantised and a discrete energy spectrum is expected for the electron system. The quantum wells with the smallest dimension of the island in the quantum devices ranges from approximately 5 to 100nm. This island is a layer or region different from the surrounding material; otherwise edges of the island create electric fields from small electrodes patterned in the shape of the desired island. The island is surrounded by two narrow walls called tunneling barriers or potential barriers of other material or an insulating defect zone in the substrate. The island surrounded by the potential barrier impedes the flow of electrons in and out of the island region. But within the island mobile electrons tend to form puddles which are smaller than the dimensions of the island. Thus fabrication of the device becomes complicated for defining confinement of the electrons in the island. There are two basic concepts for defining the confinement of the electron in quantum mechanical Effects for nanometer scale islands between closely spaced potential barriers. Firstly, quantum mechanics restricts electrons energy to one of the finite number of one electron energy levels. The smaller the distance between the barriers the smaller the island, the more widely spaced energy levels for the electrons in the potential well between the barriers. The energy space between the energy levels in the potential well. If the potential barriers are thin approximately 5 to 10 nm or less depends on the height of the barrier, electrons occupying the energy levels lower than the height barrier has a finite probability of "Tunneling "through the barrier to get on or off the island. But for the electron of a given energy to tunnel through a barrier empty state must exist on one other side of barrier. Energy quantisation and tunneling strongly influence the flow of electrons through a nano-electronic device.

Analysing Quantum Theory

The familiar Schrodinger wave equation for the electron in the free space is given by

$$-h^2/2m + \_\ \nabla^2 =$$

$$-h^2/2m\ (\partial^2\_ / x^2 + \partial^2\_ / y^2 + \partial^2\_ / z^2)$$

where '*m*' is the mass of free electron. The wave function of the electron or particle _ *k* (*r*) = 1/ (2_)3 (e^{ikr}) are phase waves.

The wave vector $k = (kx, ky, kz)$

Its energy given by

$$= \_^2 k^2/2m = \_2/2m\ (kx^2 + ky^2 + ky^2)$$

Vector components of "K" are the quantum numbers for the free motion of the electron, for each degree of freedom.

For a K space, the number of states in a volume *dk*, $dk = dkx, dky, dkz$.

Tunneling Transport and Bound Particles in a Quantum Well

Electron is a free particle with potential $V(x) = 0$. The familiar time dependent Schrodinger equation is independent of potential given in (2) for a free particle. Electron as wave packet moves with group velocity and phase velocity and the wavelets inside the wave move with a phase velocity.

$$(-h^2/2m)\ d^2\ \_/dx^2 = E\ \_\ (x)$$

E is the energy of the particle. Electron as a wave packet moves with the group velocity and the wavelets inside the wave move with a phase velocity Vph.

The movement of the electron is restricted in a finite region of space rather than to propagate in space is defined as" Quantum well or a particle in a box". It's an infinitely deep square well and has zero potential energy in the region 0 < x < a, and infinitely high potential barrier preventing to flow above the region. The width 'a 'is the full width and is sometimes studied as half width. The Schrodinger equation of motion of particle is same for the particle inside the well as there is no potential energy. The three solutions for

$$E = h^2K^2/2m$$

Are Exp(ikx), exp(–ikx), Sinkx or Coskx E is +ve for k is imaginary and waves are real exponential. E is –ve for Hyperbolic function. The wave functions and energies are given by:

$$\_{n}\ (x) = A_n\ \text{Sin}\ (n\_x)/a\ (4)$$

$$\_n = \_o\ (kn)$$

$$= \_^2K^2\ n/2m = \_^2n^2.\_^2/2m\ a^2$$

The integer n = 1, 2, 3. ..... is a quantum number indicates the energy states in the well. Quantum wells and barriers are the layered structures for the heterostructure devices for the electrons in the conduction band. The quantum well GaAs is sandwiched between two thick layers of AlGaAs and the well exists for conduction and valence bonds for the heterojunction devices. The conduction band has its complications for the Alx Ga1-x Asx with x > 0.45. The tunneling barrier is provided by a layer of AlGaAs surrounded by GaAs. Viewing classically electron cannot cross the barrier unless it has sufficient kinetic energy, but in quantum mechanics the transmission coefficient increases rapidly with the energy in the tunneling area. Electron selects high energy states to cross the tunneling area. The opposite of the barrier is a quantum well in which a thin layer of GaAs sandwiched between two thick layers of AlGaAs.

Resonant Tunneling and Resonant Tunneling Transistor (RTT)

Quantum well with thin barriers where an electron can "leak out" arising to "Quasibound or Resonant State" instead of three bound state. This double barrier structure provides the active region of Resonant -Tunneling Diode. Let us consider electrons that impinge on the barrier from the left of energy "E". Their probability of crossing the barrier is given by the transmission coefficient T. The transmission coefficient is defined as the product of transmission coefficients of the two individual barriers (i.e Left and right barriers TL and TR) respectively. T ˜ TL * TR, TL, TR are small for typical barriers. For energies near resonance T value increases to higher than TL and TR. Resonant tunneling diodes are studied using this concept at extremely high frequencies.

For our study of resonant tunneling devices we consider a resonant tunneling diode is shown below. The sequential layer structure of the RT Diode is shown below in Fig 6.3. The thickness of the barriers, wells and spacers can be varied as desired. The layer structure of GaAs/AlGaAs with two tunnel barriers of AlGaAs sandwiched by a quantum well of GaAs. The quantum well is thin enough for the bottom of conduction band to reach the quantum well in order to form 1 or more 2D energy levels in the quantum level. The barriers AlGaAs are thin enough electrons are capable of crossing the barrier through the quantum well and current flows between the heavily doped contacts. The well consist of numerous quantised states these form the quasi–bound states, as there is minimum but finite probability with which the electrons can tunnel through the well. This tunneling is key aspect for the finite lifetime of the electrons on these levels leading to some broadening of the quantum wells. The resonant tunneling diode is a system with two contacts containing three dimensional electrons and quantum well containing two dimensional electrons (if any). These subsystems are weakly coupled through tunneling. The Energy Band diagram of this RTD device can be explained under three voltage bias conditions.

Condition 1: The equilibrium condition exists for the structure when no voltage is applied, but in the well there exists a single level of energy e1 which is lower of the bottom of the two- dimensional sub-band. The parameters of the diode are chosen in such a way even in the nonbiased state energy el exists, which lies above the Fermi–energy level.

Condition 2: By applying a voltage bias in the contacts produces a downward shift of the level in the quantum level making it possible for the electrons to tunnel through the double barrier. The structure is designed in to prevent flow of electrons during thermal agitation process over the barriers. The condition for which electrons flow this quantum structure is when the quasi –bound level lies below the fermi energy but above the conduction band bottoms of the contacts. For this case the electrons from the emitter (left contact) whose kinetic energy of the perpendicular motion is given coincides with the e1 are transmitted through the structure with finite probability. This is called resonant tunneling process which forms an important case study exhibiting negative differential resistance. When resonant condition is reached the transmission coefficient and the current through the diode increases sharply as the voltage bias is increased, and the resonant level in the well falls below the level of the conduction band and there are no electrons to tunnel through.

Condition 3: As there are no electrons to tunnel resonantly with the applied bias, the current decreases despite of the increased voltage, the current voltage characteristics is identified by the negative differential resistance. The I–V characteristics of Resonant Tunneling Diode. If more than one subband exists in the well contributes to series of NDR peaks as it passes to each subband in the quantum well. The device acts as switch. There is always a background current due to electrons tunneling at energies away from resonant peaks, these peaks dominate the transmission coefficient. For approximation to calculate the current in a resosnant tunneling diode we consider only a single peak and use equation for T(E) to calculate the current. The effect of bias on a resonant tunneling structure across the diode is small, Epk is above sea of the incoming electrons and current flows. As the applied bias increases this decreases the energy of the resonant level, so the sea of electrons on the left can pass through it and large current flows. In three dimensions the current increases linearly as Epeak approaches the bottom of the electrons on the left as as shown in Energy band diagram of Fig 6.6(c). A further increase of the bias decreases the energy of the resonant level down and no longer available to electrons and the current decreases abruptly. The result is a current–voltage characteristics that shows Differential Conductivity is used for the application of amplifier or oscillator.

The integration is carried over E from – infinity to + infinity. The current remains constant at this value while the peak remains well inside the range of incoming energies and falls to zer over a width of about as Epk passes outside the range.

Based on the characteristics of the diode the RTT introduces a third terminal called GATE. The energy levels in the potential well relative to the source can be varied by applying voltage on the third (gate) terminal rather than voltage on source. A small gate voltage controls the flow of the current through the device. The barriers in the RTT create potential around the well or island and prevent charge from flowing through the device, even it is under a voltage bias. Potential well is configured in the same way as RTD, until the gate voltage is charged by the voltage bias the energy levels of the well are lowered for all the states and bring them into resonance with the mobile electrons in the conduction band in the source, so that an electron current can be transmitted through the device.

The RTDs and RTTs employ quantisation effects and tunneling property which is used for best amplification and multiple on and off states associated with multiple discrete quantum levels inside the potential well on a very small well or narrow islands. If there is a second energy level in the quantum well, then the device can switch on again as the bias voltage of the RTD (or the gate voltage of an RTT) is increased still further. This multi-state switching behaviour permits each device to "count higher" and represent more logic states. The RTD is expected to operate up to terahertz (1000 GHz). Brown and coworkers fabricated microwave oscillators operating around 712 GHz.

Though the characteristics and behaviour of RTDs seemingly convinced but there are some challenges that are focused, which depends on its fabricating and testing of these nanoelctronic devices. They drawbacks are Multistate quantum effect behaviour RTD's do not turn-off their current completely when they are offresonance. There is a residual current in the valley between the current peaks and leading to the on and off states barely distinguishable. The circuit architectures must be designed keeping in mind its potential sensitivity and peak to valley current ratio as large as possible. Launder pointed out that RTD's and RTTs can simply turn-off if there are any fluctuations in the input voltage. The operation of RTD and RTT was possible to manufacture at room temperature but at temperatures below absolute scale the SET exhibited the oscillation for which the random thermal motion often provide electrons with small additional energy to get to the island, this could increase the relevant energy separation U for states in the island, resulting switching off instead of switching on.

This limit is only possible in molecular electrons. Device fabrication is concentrated on III–V semiconductors because of its excellent purity and easy manufacturing. But these materials contaminated easily when they are exposed to mobile charge carriers leads to instability of the device, also they cannot withstand electric field. While silicon, SiO_2 is amorphous crystalline is the best insulating material can acts as barrier as little as 5nm can be grown uniformly and prevents electrons to cross the barrier.

The new trend is to make use of Si/Ge in the heterostructure devices particularly sandwiches of Si/Ge. The random background charges tend to accumulate in semiconductor for a small operating quantum- effect or single electron device. This can be overcome by the use of Silicon-On-Insulator (SOI) technology. SOI could be used in molecular electronics for its relatively non-conductive, nonpolarizable organic compound. The exponential sensitivity of the tunneling current to width of potential barrier device operability depends on the nanoscale barriers which are made with extreme precision and uniform width. Sufficient reliability and precision, predictable behaviour of large number of devices is very difficult to achieve on a several-nanometer scale in solids.

Even though it is possible in hetero-structure Devices such as RTDFETs accuracy is achieved in one of the three dimensions. Rapid scaling on microelectronic devices, employing quantum effects in nanoscale and using high resolution techniques for fabrication devices has put some challenges that to be worked on for its efficient functions. For solid state nanoelectronics, one of the most important challenges is to be able to produce reliably and uniformly in silicon the characteristic nanometer-scale features required for nanoelectronics: nanometer-scale islands, barriers, and "heterojunctions" between islands and barriers. Up to now, solid state nanoelectronic devices largely have been fabricated in III/V semiconductor compounds such as gallium arsenide and aluminum arsenide. It is believed, however, that silicon nanoelectronics would go a long way toward permitting the inexpensive mass manufacture of nanoelectronic devices. During the past decades the scaling of electronic devices and the development of data-processing systems such as microprocessors, memories and more recently complete systems on chip are the driving forces behind the progress in semiconductor industry. Today, the combination of lithography and heterostructure growth techniques with atomic layer precision enables the fabrication of devices so small that quantum mechanical effects are observable. Among solid-state electronics the most promising devices are resonant tunneling and single electron devices.

Energy quantisation due to coulombic repulsion of single charges and quantum confinement are their underlying physical phenomena to control the charge transport through this nanoscale structures. Both device categories have in common a current voltage characteristics with several, in the case of single electron devices periodical, negative-differential resistance (NDR) regions. Handling these NDR characteristics and exploiting them in a logic family or memory cell is a widely accepted approach to nanoscale circuits, especially as a method to improve the computational functionality of basic circuit components. From the technological point of view resonant tunneling devices are presently the most mature type of quantum effect device because their interfacing with conventional transistors has reached an advanced level and even in silicon room temperature operation of NDR-devices is achievable. Besides the high-speed operation in the GHz regime the combination of the NDR with electronic amplification is attractive to reduce circuit complexity, that is the number of devices and the wiring. This has motivated the development of different three terminal devices. Significant examples are resonant hot electron transistors, gated resonant tunneling diodes and two dimensional electron gas tunneling field effect transistors. Here a monolithic integration of a resonant tunneling diode (RTD) and a heterostructure field effect transistor (HFET) is used.

This approach has proven to be large-scale integration capable on the InP material system and is the basis for different resonant tunneling device circuits investigated by industrial as well as academic research groups. In addition to the technologically oriented research a further prerequisite for nano-scale integration is the investigation of suitable logic families, architectures and the development of a design framework. Recapitulating some milestones, the first approach to build a logic circuit with RTDs is a 1-bit full adder based on XNOR gates proposed by Capasso in 1989. By taking advantage from the multistate behaviour of the RTD the circuit complexity of the adder is reduced.

Other applications following this idea are different kinds of multiple-valued logic gates as well as resonant tunneling diode memory cells and analog-digital mixed circuits. For large scale integration of nano-scale devices an essential aspect is the question if these functionally integrated circuits could be combined with existing VLSI algorithms for bit-level computations. Independent of the specific technology and the operating principles of the different categories of nano-scale devices, following aspects have to be considered:

- The use of quantum effects to reduce the logic depth of a circuit.
- A flexible and intuitive methodology to transfer a give logic functionality into a circuit design. The impact of fabrication tolerances and physics related parameter variations on the circuit functionality and error rate.
- The capability to fabricate logic and memory circuit components with one technology.
- A regular layout with a small number of different circuit modules to enable the a library based design.
- Local interconnections on the circuit and the system level to solve the wiring problem.
- The impact of increasing clocking frequencies on the circuit design indicated by a continuing decrease of gate delays per clock cycle which limits the number of logic circuit stages between two synchronising registers.
- Additional circuit overhead due to clocking and temporal data storage in latches.

- The implementation of concurrent VLSI algorithms to achieve a low latency and a high data throughput for basic low-level operations such as addition, multiplication, and accumulation.
- The enormous design complexity ranging from device and technology to the system level. In this context sever problems arouse from the pure increase of the number of gates, devices, and wires (horizontal complexity) as well as from themore complicated modeling of these components (vertical complexity).

For clocked logic families various high speed circuits based on the monostable-bistable transition logic element (MOBILE) have been proposed and operated at clocking frequencies of several GHz. The MOBILE is a rising edge triggered, current controlled gate. It consists of two RTDs which are switched into a monostable or bistable state depending on an oscillating bias voltage VCLK. In the sense of a pulsed power supply the bias voltage also serves as clock to synchronize the gates. The basic idea of implementing logic gates by means of such a monostable-bistable transition in a symmetric configuration of two tunneling diodes has been investigated first by Goto in 1960. Since field effect transistors were not available at that time to isolate inputs and outputs these circuits have not played an important role in microelectronics. Today, their comeback in nanoelectronics is primarily due to the technological progress in the field of quantum effect devices and themore profound understanding of nonlinear dynamic phenomena.

In this study, it was shown that the monostable-bistable behaviour of MOBILE circuits strongly depends on the RTD peak voltage. Accordingly it is also to expect that the shape of the resonance and the exponentially increasing thermionic current will influence the equilibrium points of the circuit. The stable (solid lines) and unstable equilibrium points (dashed lines) of three MOBILEs composed of RTDs with different I-V characteristics. A reduced peak to valley ratio significantly decreases the output voltage swing. In contrast to this the RTD has small peak voltage of V_P = 0.2 V, an non-symmetric resonance peak and a large valley region due to the slowly increasing thermionic current.

As a result a distinct tristable region appears for clock voltages of 0.7 V < V_{CLK} < 1.25 V. In digital logic circuits especially the central stable state in the tristable region is a potential error source if the amplitude of the clock voltage exceeds 0.7 V and the output remains at $V_{OUT} = V_{CLK}/2 = 2$ instead of switching to the logic high level. For the first RTD, this tristable region is less pronounced. These investigations illustrate that the use of RTDs in MOBILE circuits requires a careful optimisation of the device parameters and a precise control of the double barrier heterostructure. Since the RTD depicted.Among the different implementations of the MOBILE input stage mentioned above, the RTD-HFET input stage is primarily advantageous to increase the fan-in of logic families withmultiple input terminals and has been applied to a bit-level pipelined threshold logic full adder. A programmable NAND/NOR gate as a simple extension of the MOBILE inverter. The NAND/NOR gate has two input RTD-HFETs T_1 and T_{22} and a third input transistor T_3 to modify the logic function. Adding the third RTD-HFET to the NAND gate effectively increases the input current so that the ratios of RTD peak currents correspond to the NOR function if T_3 is on. The inputs (X, Y) are sensed during the monostable-bistable transition and compared to the current of the load RTD D_2 (dashed curve of the load RTD V_{CLK} = 0.62 V). As it is depicted a change of the input voltages increases or decreases the input current sum $I_P^{tot} = I_P(T_1) + I_P(T_2) + I_P(D_1)$ in equidistant steps of the RTD peak current $I_P = A_{RTD}$ jP. Here A_{RTD} is the minimum RTD anode area and jP is the peak current density. The load line diagram is obtained by HSPICE-circuit simulation using a semi-empirical RTD large signal equivalent circuit which has been implemented as SPICE-macro model with voltage-controlled current sources. From the viewpoint of the design methodology this current controlled switching of the input RTD/HFETs enables the introduction of a design parameter l which describes the ratios of the RTD peak currents according to $I_P = l\, jP\, A_{RTD}$. For a given RTD-HFET technology the current density, peak voltage and the anode area as well as the HFET device parameters are fixed.

Thus, transferring a Boolean logic equation into a circuit design is straightforward by selecting appropriate design parameters for the RTDs in the input and the latch stage. For the inverting logic family this is illustrated. Here, the only difference between the NAND and NOR function is the increased anode area of load RTD D_2. Compared to previous RTD logic families with different devices in the input stage and the RTD latch this is a simple methodology to scale the circuits. The RTD-HFET is amonolithically integrated, 3-dimensional InP based heterostructure where the double barrier quantum well of the RTD is grown on the HFET drain contact layer. The resulting I-Voutput characteristics exhibits a gate controllable NDR-like behaviour if the HFET channel current I_{DS} is larger than the RTD peak current I_P. In this case the RTD acts as current limiting device and the peak current is independent of the gate voltage if a logic high level of $V_H > 0.3$ V is applied.

For gate voltages in the region of the threshold voltage the device shows the usual amplifying HFET characteristics. The RTD structure is composed of an n-InGaAs cathode layer, an i-InGaAs spacer and an InGaAs quantum well which is enclose by two AlAs-barriers. At the boundary between the RTD an HFET structure an InAlAs etch stop is included. The HFET has alloyed Ni/Ge/Au contacts. Between the InP substrate and the 20 nm InGaAs channel a 115nm InAlAs buffer is inserted. The Schottky barrier of the HFET is made of a Pt/Ti/Pt/Au gate, a 22nm InAlAs/AlAs barrier, and a 4nm InAlAs spacer. Since the logic swing of MOBILE circuits is $DV_{OUT} = V_H—V_L$ the layer structure has been optimised to obtain a low peak voltage of V_P =

0.27 V so that gate currents due to HFET Schottky gate ($V_H < V_{GS,max}$ » 0.8 V) are avoided for logic high input voltages V_H » 0.7 V.

Recent investigations of the homogeneity have shown that a RTD peak voltage and current variation of 16 mV and 0.41 mA over a quarter 2" wafer are achievable without any optimisation. These values are comparable to standard deviations reached with other technologies. A technological advantage of the RTD-HFET input stage are the relaxed precision requirements of the HFET parameters. In the original version of this RTD-based LTG proposed by Yamamoto et al. at NTT the input terminals are single HFETs with different transistor widths. Here, the HFET acts as a switch. Consequently, the gates are more robust against device parameter variations than in the original MOBILE where theHFET input stage has to provide an exactly defined amount of current at a distinct bias voltage. Based on this RTD-HFET technology and the design methodology discussed above a NAND/NOR gate with 3.0 mm^2 mode area and j_p = 9 kAcm^{-2} has been fabricated and tested. The RTD HFET has a gate length of L = 1.0 mm and a gate width of W = 3 mm. RTD anode areas of A_{RTD} = 1.0 mm the circuit configuration is modified. First, the load RTD D_2 is substituted by an depletion type RTD-HFET. The clock V_{CLK} is now applied to the HFET gate and therefore a constant supply voltage of V_{DD} = 0.8 V > $2V_P$ is used instead of the difficult pulsed power supply scheme of the original MOBILE. The monostable-bistable transition of an inverter for different supply voltages from V_{DD} = 0.6 . . . 1.0 V. Compared to the RTD-HFET inverter with pulsed power supply here the voltage levels in the bistable region are more robust against variations of the clock voltage.

Furthermore, by modifying the supply voltage the output voltage swing can be optimised independent of the clock signal amplitude. A further improvement is achieved by inserting a pull down HFET in the output branch which is controlled by the inverse clock V_{CLK}. The pull down HFET speeds up the discharging of the load capacitance and thereby the high-lowtransition (bistable-monostable transition) at the end of a clock cycle is significantly reduced. Since MOBILE gates require a multi-phase clocking scheme, usually a four phase clocking scheme, this inverse clock signal is available and does not lead to an additional circuit overhead. In a second, more stringent approach a further NAND gate has been designed where the driver RTD of the MOBILE is simply omitted. A comparable output behaviour is obtained but the total current is drastically reduced due to the smaller load RTD area.

Based on the device parameters extracted from the fabricated test circuits the novel gates NAND1 and NAND2 have been simulated assuming an improved technology with 200 nm gate length enhancement and depletion mode HFETs. A HSPICE simulation of these scaled MOBILE gates. For both gates the driving capability at the critical logic input combination (V_{in1} = 0, V_{in2} = 1) and (V_{in1} = 1, V_{in2} = 0) has been investigated for load capacitances of C_L = 5.0 . . . 25 fF, corresponding to fan-out values up to 5. The load capacitance C_L is the sum of the RTD-HFET gate-source capacitances of in the subsequent circuit stage (C_{GS} = 4.8 fF per fan-out). The rise time of the clock is 15 ps for a supply voltage of V_{DD} = 0.8 V.

However, the fan-out has a strong impact on the low-high transition time of both gates: t_{LH} = 23 . . . 40 ps for NAND1 and t_{LH} = 19 . . . 36 ps for NAND2. Due to the omitted driver RTD NAND2 is faster than NAND1 for low fan-out values. In contrasts to this NAND1 has a better driving capability. This is reflected by the smaller fan-out loading factor Dt_{LH} = 4 per fan-out. For NAND2 a larger value of Dt_{LH} = 5.8 ps per fan-out is obtained. Furthermore, if the load capacitance is larger than C_L = 20 fF NAND2 switches to the logic low level. This is a general phenomenon of MOBILE gates if the RTD speed index $S = I_P/2C_L$ » 10 . . . 100 V/ns, that is the ratio of the switching current difference $I_P/2$ and the sum of the load capacitances, is insufficient to provide the necessary current to charge the output node during the rising clock edge. Here, the trade-off between speed and driving capability improves the fan-out at the cost of an increased clock rise time to 25 ps. In this case our simulations indicate that the delay time and low high transition time are slightly affected only. Due to the pull-down HFET the high-low transition time t_{HL} is comparable to the clock fall time and less fan-out dependent than the low-high transition. Therefore, total clock periods of T = 140 . . . 150 ps and multi GHz operation will be possible with scaled RTD-HFET technology. Concerning the power dissipation NAND2 performs better because the static power dissipation as well as the switching current during the monostable-bistable transition are depending on the area of the load RTD. Before the design of larger threshold logic circuits is presented we will discuss the question of an appropriate clocking scheme for MOBILE-type gates.

To operate cascaded threshold gates in circuits composed of several stages it has to be ensured that the evaluation of one stage only starts after the previous stage has finished the computation. A possible solution is a four phase overlapping clocking scheme with a phase delay between the clocks of two adjacent stages. Since MOBILE circuits are self-latching as soon as they have reached the bistable state this four phase clocking scheme leads to a bit-level pipelined operation, also known as fine-grain pipelining, bit-level systolic operation and nano-pipelining.

After the evaluation phase has finished the gate holds the result as long as the supply voltage remains constant at $V_{CLK} = 3V_P$. During this hold phase the output is valid and the subsequent circuit stage starts with the evaluation phase. In the reset phase, that is the falling edge of the clock, the load capacitance is discharged and the gate returns to the initial, monostable state. Since this reset operation is a bistable monostable transition the switching process is symmetric for equal rise and fall times.

The clock cycle is finished by an inactive phase. During this phase the inputs of the gate are changed according to the results of the preceding pipeline stage. The clock signals of the four phase clocking scheme have slowly rising and falling clock edges (t_{rise} = t_{fall} = $T/4$) which are frequently used in adiabatic circuit architectures. From the perspective of the power dissipation a result of such an adiabatic operation is a reduction of the dynamic switching energy by slowly charging and discharging the load capacitances. First estimations for MOBILE-type circuits consisting of laterally scaled resonant tunneling devices with minimum feature sizes below 200 nm have shown that the dynamic power dissipation is the major source of power dissipation. Compared with a non-adiabatic operation (t_{rise}, t_{fall} << $T/4$) the symmetric adiabatic clocking scheme decreases the total power dissipation by about one order of magnitude.

To extend the computational capabilities of RTD logic gates the current controlled switching of the MOBILE in connection with the RTD-HFET input stage allows a functional implementation of linear threshold gates (LTGs). The characteristic feature of LTGs is the parallel processing of multiple inputs. A LTG calculates the weighted sum c of the digital inputs x_k, $k = 1, \ldots, N$. The weighted sum is converted into a digital output y by comparing c with a given threshold value q. If the weights w_k and the threshold value q are selected properly the LTG computes any linear separable Boolean function of the N inputs. Compared to a Boolean logic gate, a threshold gate combines an internal multiple-valued computation of the weighted sum with digital encoded input and output states. Actually, this capability of processing multiple input signals enables it to design circuits with bit-level parallelism and reduced complexity. Thus, our approach shows a certain similarity to multiple valued logic, but differs in regard to the digital input and output logic states. Although not explicitly mentioned the programmable NAND/NOR gate of the previous subsections is basically a LTG with two negative weighted inputs and a modifiable threshold value.

An RTD-based threshold gate composed of two serially connected RTDs and four RTD-HFETs with two positive weighted inputs (x_1, x_2) and two negative weighted inputs (x_3, x_4). The threshold value is implemented by modifying the RTD anode area of the driver RTD. According to the current controlled switching the weighted sum of a LTG is then given by the total input current at the output node according Kirchhoff's current law. Addition is the most frequently used operation in general purpose and application specific processors. Therefore, the design of an efficient adder is essential for every emerging circuit technology and is the motivation to investigate two threshold logic full adder circuits as basic components of a n-bit ripple carry adder (RCA). As a further improvement a pipelined threshold logic version of a parallel n-bit carry look ahead adder (CLA) is presented using a logarithmic depth tree structure. In connection with the bit-level pipelined operation of the RTD-HFET threshold logic gates our approach is interesting for future digital signal processing applications where data throughput is of great relevance. Due to the inherent bistability of RTD-HFET threshold logic gates, an advantage of the proposed bit-level pipelined adders is that the area and latency increasing methodology of introducing D-latches between two different logic stages in pipelined arithmetic circuits can be avoided.

Bit-level Pipelined Addition—To operate a threshold logic full adder in a bit-level pipelined fashion the first design has to be modified in such a way that the input operands (a_i, b_i, c_{i-1}) are not required to compute the sum bit si in the subsequent stage. An alternative adder design is shown where the complete adder cell comprises four threshold gates which are arranged in two circuit stages. The underlying idea of this second threshold logic algorithm is to exploit the periodical relationship between the sum bit *si* and the operand sum $c = a_i + b_i + c_{i-1}$. The first stage of the full adder contains three gates having the threshold values $Q_i = 1,2,3$ and classifies the operand sum according to the intervals shown. Finally, the second stage detects whether the operand sum lies in one of the two high intervals c = [1, 2] and c = [3] and or not. Since the inputs of the second circuit stage are the intermediate results (x_i; y_i) and the new carry ci, the input operands of the first stage (a_i; b_i; c_{i-1}) are not required in the second clock cycle. Therefore, at the cost of two additional gates the modified full adder can be fully pipelined and is ideally suited for a pipelined n-bit adder.

The most simple hardware algorithm for parallel addition of two n-bit numbers is a ripple carry adder (RCA) composed of pipelined full adder cells. The carry bits are propagating in diagonal direction so that the logic depth comprises $d = n + 1$ pipeline stages. This carry-propagation path is the critical path of the ripple carry adder. Since the four phase clocking scheme leads to four gate delays per clock cycle the overall delay (latency) of an n-bit ripple carry adder is $T_{add} = (n+1)\ T/4$. Apart from n RTD-HFET full adders the RCA also contains two triangular arrays of delay elements which are typical for bit-level pipelined circuits to consider the propagation or the carries along the diagonal direction. This ensures that the operands a_i and b_i are arriving simultaneously with the carry c_{i-1} at the corresponding full adder. The delay elements are implemented as RTD-HFET threshold gate with a single input similar to the RTD-HFET inverter. The input registers where the n-bit operands (a_0, ..., a_{n-1}) and (b_0, ..., b_{n-1}) are stored before they are loaded into the first pipeline stage. The final sum bits (s_0; ..., s_n) are stored in the output register. Both, the input and output registers are updated synchronously with the full adder array at each clock cycle and thus providing a high data throughput.

This motivates a further lateral scaling of resonant tunneling devices. RTD anode areas, to achieve power delay products below 1 fJ for MOBILE based logic gates. It has been shown that the increased functionality of monolithic integrated nanoelectronic devices can be transfered to the circuit level to reduce the complexity of the designs by using threshold logic gates. The advantages of a threshold logic design have been illustrated by the fast carry computation in two full adder designs and a pipelined. The proposed circuits demonstrate that the use of threshold logic as a novel design methodology for resonant tunneling devices offers a significant performance improvement for basic arithmetic components. Above was a comprehensive analysis of the design and application of RTD/HFET-based circuit architectures. Considering the present state of technology the impact of these quantum effect devices is investigated for different aspects of the circuit design, such as signal timing, clocking scheme and design methodology. After investigating the impact to the RTD I-V characteristics on the circuit behaviour a high speed logic family for three terminal resonant tunneling devices with drastically reduced design parameter sensitivity (RTD area only) is presented.

Experimental data prove the functionality at ultra-low voltages V_{DD} = 0.8 V, V_H = 0.7 V, V_L = 0.1 V). HSPICE simulations of two novel scaled NAND gates based on experimental results indicate multi-GHz operation with a fan-out up to 5, a gate delay below 10 ps, and a clock rise time of 15 ps. The total power dissipation of these gates is below 2 mW and 1:2 mW, respectively.

Explaining the Applications of Nanodevices, Nanoelectronics and Nanosensors

In the broadest sense, nanodevices are the critical enablers that will allow mankind to exploit the ultimate technological capabilities of electronic, magnetic, mechanical, and biological systems. While the best examples of nanodevices at present are clearly associated with the information technology industry, the potential for such devices is much broader. Nanodevices will ultimately have an enormous impact on our ability to enhance energy conversion, control pollution, produce food, and improve human health and longevity.

In the past decade, our ability to manipulate matter from the top down, combined with advances and in some cases unexpected discoveries in the synthesis and assembly of nanometer-scale structures, has resulted in advances in a number of areas. Particularly striking examples include the following:

Current Scientific Advances: The unexpected discovery and subsequently more controlled preparation of carbon nanotubes and the use of proximal probe and lithographic schemes to fabricate individual electronic devices from these materials:

- The ability in only the last one or two years to begin to place carefully engineered individual molecules onto appropriate electrical contacts and measure transport through the molecules.

- The explosion in the availability of proximal probe techniques and their use to manipulate matter and thereby fabricate nanostructures.
- The development of chemical synthetic methods to prepare nanocrystals, and methods to further assemble these nanocrystals into a variety of larger organised structures.
- The introduction of biomolecules and supermolecular structures into the field of nanodevices.
- The isolation of biological motors, and their incorporation into nonbiological environments.

A number of examples of devices in the microelectronics and telecommunications industries rely on nanometer-scale phenomena for their operation. These devices are, in a sense, "one-dimensional" nanotechnologies, because they are micrometer-scale objects that have thin film layers with thicknesses in the nanometer range. These kinds of systems are widely referred to in the physics and electronics literature as two-dimensional systems, because they have two classical or "normal" dimensions and one quantum or nanoscale dimension. In this scheme, nanowires are referred to as one-dimensional objects and quantum dots as zero-dimensional. In this document, and at the risk of introducing some confusion, we have chosen to categorize nanodevices by their main feature nanodimensions rather than by their large-scale dimensions. Thus, two dimensional systems such as two-dimensional electron gases and quantum wells in our notation are one-dimensional nanotechnologies, nanowires are two-dimensional nanotechnologies, and quantum dots are three-dimensional nanotechnologies. Examples include high electron mobility transistors, heterojunction bipolar transistors, resonant tunneling diodes, and quantum well optoelectronic devices such as lasers and detectors. The most recent success story in this category is that of giant magnetoresistance (GMR) structures. These structures can act as extremely sensitive magnetic field sensors. GMR structures used for this purpose consist of layers of magnetic and nonmagnetic metal films.

The critical layers in this structure have thicknesses in the nanometer range. The transport of spin-polarised electrons that occurs between the magnetic layers on the nanometer length scale is responsible for the ability of the structure to sense magnetic fields such as the magnetic bits stored on computer disks. GMR structures are currently revolutionising the hard disk drive magnetic storage industry worth $30-40 billion/year. Our ability to control materials in one dimension to build nanometer-scale structures with atomic scale precision comes from a decade of basic and applied research on thin film growth, surfaces, and interfaces. The extension from one nanodimension to two or three is not straightforward, but the payoffs can be enormous. Breakthroughs in attempting to produce three-dimensional nanodevices include the following:

- Demonstration of Coulomb blockade, quantum effect, and single electron memory and logic elements operating at room temperature.
- Integration of scanning probe tips into sizeable arrays for lithographic and mechanical information storage applications.
- Fabrication of photonic band-gap structures.
- Integration of nanoparticles into sensitive gas sensors.

In order to exploit nanometer-scale phenomena in devices, we must have a better understanding of the electronic, magnetic, and photonic interactions that occur on and are unique to this size scale. This will be achieved through experiment, theory, and modeling over the next decade. In addition, new methods to image and analyze devices and device components will be developed. These might include three-dimensional electron microscopies and improved atomic scale spectroscopic techniques. Over the same time period, we believe that it will become possible to integrate semiconductor, magnetic, and photonic nanodevices as well as molecular nanodevices into functional circuits and chips. The techniques now being developed in biotechnology will merge with those from nanoelectronics and nanodevices. Nanodevices will have biological components.

Biological systems will be probed, measured, and controlled efficiently with nanoelectronic devices and nanoprobes and sensors. There will be significant progress in nanomechanical and nanobiomechanical systems, which will exhibit properties that are fundamentally different from their macroscopic counterparts. There are important applications for instruments that will fly into space: nanocomponents are needed to achieve overall instrument sizes in the micron or millimeter range. Some of the same issues apply to battlefield sensors for situational awareness. Finally, a significant goal is the development of nanometer-scale objects that manipulate and perform work on other nanometer-scale objects, efficiently and economically achieving the same things we currently rely on scanning tunneling microscopy (STM) or atomic force microscopy (AFM) to carry out. A first step towards this goal might be the integration of nanometer-scale control electronics onto micromachines. In the information technology arena, nanodevices will both enable and require fundamentally new information processing architectures. Early examples of possible architectural paradigm shifts are quantum computation, quantum dot cellular automata, molecular electronics, and computation using DNA strands. Such architectures will fundamentally change the types of information technology problems that can be attacked. Effective implementation of these types of architecture will require nanodevices. Other paradigm shifts include the emergence of quantised magnetic disks; single photonic systems that will allow efficient optical communication; nanomechanical systems; a broad class of structures and devices that merge biological and non-biological objects into interacting systems; and use of nanocomponents in the shrinking conventional circuit architectures. Research on nanodevices using nanoscale wiring and molecular logic, as well as new principles for devices such as spin electronics, have made significant inroads in the past year or two. The exploration and fabrication of nanodevices requires access to sophisticated and sometimes expensive tools. More and better access to such equipment as well as rapid prototyping facilities is needed. Of equal importance is the recognition that success in nanodevices will draw upon expertise from a broad range of traditional disciplines. Therefore, it is imperative that programmes be established that facilitate and strengthen cross-fertilisation among diverse disciplines and that allow rapid adoption of new methods across field boundaries. Nanodevices are in some ways the most complicated nanotechnological systems. They require the understanding of fundamental phenomena, the synthesis of appropriate materials, the use of those materials to fabricate functioning devices, and the integration of these devices into working systems. For this reason, success will require a substantial funding level over a long period of time. There is strong sentiment for single investigator funding as well as for structured support of interdisciplinary teams.

Molecular Electronics—If the reduction in size of electronic devices continues at its present exponential pace, the size of entire devices will approach that of molecules within a few decades. However, well before this happens, both the physics upon which electronic devices are based and the manufacturing procedures used to produce them will have to change dramatically. This is because current electronics are based primarily on classical mechanics, but at the scale of molecules, electrons are quantum mechanical objects. Also, the cost of building the factories for fabricating electronic devices, or fabs, is increasing at a rate that is much larger than the market for electronics; therefore, much less expensive manufacturing process will need to be invented.

Organic Nanostructures: The Electrical Conductivity of a Single Molecule—By combining chemical self-assembly with a mechanical device that allows them to break a thin gold wire with nanometer scale control, researchers have succeeded in creating a "wire" consisting of a single molecule that can connect two gold leads. Using this structure, they have been able to begin to measure and study the electrical conductivity of a single molecule.

Thus, an extremely important area of research is molecular electronics, for which molecules that are quantum electronic devices are designed and synthesised using the batch processes of chemistry and then assembled into useful circuits through the processes of self-organisation and self-alignment. If molecular electronics achieves the ultimate goal of using individual molecules as switches and carbon nanotubes as the wires in circuits, we can anticipate nonvolatile memories with one million times the bit area density of today's DRAMs and power efficiency one billion times better than conventional CMOS circuitry. Such memories would be so large and power-efficient that they could change the way in which computation is performed from using processors to calculate on the fly to simply looking up the answer in huge tables. A major limitation of any such process is that chemically fabricated and assembled systems will necessarily contain defective components and connections. By describing a silicon-based computer that was designed to operate perfectly in the presence of huge numbers of manufacturing defects, researchers from Hewlett-Packard (HP) and the University of California–Los Angeles (UCLA) presented an architectural solution to the problem of defects in molecular electronics, as described. In 1999, researchers from HP Labs and UCLA experimentally demonstrated the most crucial aspect for such a system, an electronically addressable molecular switch that operates in a totally "dry" environment. Logic gates were fabricated from an array of configurable molecular switches, each consisting of a monolayer of electrochemically active rotaxane molecules sandwiched between metal electrodes. The operation of the switches. In the "closed" state, current flow is dominated by resonant tunneling through the electronic states of the molecules. The switches are irreversibly opened by applying an oxidising voltage across the device. In this case, since the memory of the molecules is not volatile, only one set of wires is needed to set and read the state of the molecules, and in principle, one molecule can replace seven transistors in a conventional silicon circuit. In the demonstration, several devices were configured together to produce AND and OR logic gates. The high/low current levels of those gates were separated by factors of 15 and 30, respectively, which is a significant enhancement over that for conventional wired-logic gates.

A Field-Effect Transistor Made from a Single-Wall Carbon Nanotube: Several research groups around the world have succeeded in fabricating electrical switches such as the field-effect transistor from single-walled carbon nanotubes. A single-walled carbon nanotube 1.6 nm in diameter was manipulated into place using an atomic force microscope. Once placed on the metal contacts, the semiconducting tube behaved like the channel in a field-effect transistor, turning on or off depending on the applied gate voltage. Nanotubes hold great promise as electronic elements for a variety of different nanostructures. Researchers are just beginning to understand how they conduct electricity and how to place them into appropriate device structures. It is interesting to note that both the atomic force microscope used to fabricate this structure and the carbon nanotubes that form the critical element were developed only in the past decade.

A Commercial IBM Giant Magnetoresistance Read Head: When certain kinds of materials systems are exposed to a magnetic field, their electrical resistance changes. This effect, called the magnetoresistive effect, is useful for sensing magnetic fields such as those in the magnetic bits of data stored on a computer hard drive. In 1988, the giant magnetoresistance effect was discovered in specially prepared layers of nanometer-thick magnetic and nonmagnetic films. By 1991, work at the IBM Almaden Research Center demonstrated that the GMR effect could be observed in easily made samples and that a special kind of GMR structure, a spin valve, could sense very small magnetic fields. This opened the door to the use of GMR in the read heads for magnetic disk drives. IBM first announced a commercial product based on this design in December 1997.

Molecular Logic: Electron transport studies in molecular-scale systems have recently become possible with the utilisation of advanced microfabrication and self-assembly techniques. Investigations are now possible of the electronic conduction through conjugated molecules that are end-bound onto surfaces; these have been demonstrated with a scanning tunneling microscope, with micromachined silicon nanopores, and with proximal probes. Work on the proximal probes demonstrated that 0.1 microamp of current can be transported through a single molecule. However, in all of the past embodiments, the electronic properties exhibit simple diodic behaviour that is unsuitable for potential circuit applications. Researchers recently observed the first large and useful reversible switching behaviour in an electronic device that utilizes molecules as the active component. The essential feature of the fabrication process is the employment of nanoscale device area that gives rise to a small number of self-assembled molecules (ca. 1,000), which eliminates pinhole and other defect mechanisms that hamper through-monolayer electronic transport measurements. This technique has demonstrated good control over the device area and intrinsic contact stability and produces a large number of devices with acceptable yield so that statistically significant results can be produced The active electronic component was made from an organic compound upon exposure to Au. The I-V is fully reversible upon change in bias sweep direction. This is the first observation of robust and large negative differential resistance (NDR) in a device where molecules form the active region with peak-to-valley-ratios (PVRs); and the PVRs here are >1000:1. The performance exceeds that observed in typical solid state quantum well resonant tunneling heterostructures. Therefore, in addition to the obvious size advantages for scaling, the intrinsic device characteristics (i.e., valley current shutoff) may be superior to solid state embodiments; present silicon devices rarely exceed PVRs of 100:1.

Nanoelectronics offers a broad set of opportunities by focusing on quantum devices and addressing their potential for high performance through increases in density (factors of 5 to 100), speed (factors of 10 to 100), and reduced power (factors of more than 50). Resonant tunneling devices are being explored with demonstrated successes in multivalued logic and various logic circuits and memory circuits. SET logic and memory concepts are being explored with focus on memory applications. Molecular electronics and self-assembly approaches have shown a path towards manufacturing alternatives and device options for regimes beyond traditional scaling.

Spin devices in the form of nanomagnetics using the magnetoresistive effect in magnetic multilayers have demonstrated their use for nonvolatile, radiation-hard memory. Quantum cellular automata and coupled quantum dot technology are being explored and their potential assessed for transistorless computing. By exploring Si-based heterojunctions, bandgap engineering, vertical device structures, and quantum devices, inroads are being made into extending CMOS capabilities. Potential applications are in digital radar, electronic support measures (ESM) receivers, ATM data stream processing, wide bandwidth communications, digital image processing, waveform generation, and the broad area of analog to digital (A/D) applications. Demonstrations have shown the efficacy of resonant tunneling devices in various network environments. The long-term vision for nanoelectronics sees the use of quantum devices in other high performance systems especially in telecommunications for signal processors and electronics for A/D converters in detectors. After more than three decades of exploring space, the National Aeronautics and Space Administration (NASA) has completed an initial reconnaissance of our solar system. The next missions will involve sending spacecraft to destinations that are much more difficult to travel to, like the Sun or Pluto. Also, spacecraft will be required to perform more difficult tasks, such as landing on a celestial body, collecting a sample of its material, and returning it to Earth.

To carry out such technically challenging missions at an affordable cost, NASA has created the Deep Space Systems Technology Programme, known as X2000. Every two to three years starting in the year 2000, the programme will develop and deliver advanced spacecraft systems and body structures to missions bound for different areas of the solar system and beyond. In order to achieve reduction in the size of spacecraft, the avionics systems of the spacecraft are being reduced in size with each delivery of X2000, in part by means of integrating nanotechnology with microtechnology. The forecasts of the mass, volume, and power of future avionics systems of spacecraft.

The leftmost column shows the Mars Pathfinder spacecraft, which represents the current state of the art. The first delivery for X2000 is an integrated avionics system that subsumes the functionalities of command and data handling, attitude control, power management and distribution, and science payload interface. Advanced packaging technologies as well as advanced design automation techniques are used to define a highly integrated, modular, building-block architecture for highly reliable and long-term survivable deep-space planetary missions. Advanced low-power techniques and architectures will drastically reduce overall power consumption compared to currently available flight hardware. “System On A Chip” (SOAC) will prototype single chip and multichip module solutions that lead towards an avionics system on a chip. This chip will integrate the avionics system that is being developed for the X2000 avionics deliverable. That is, the chip will include power management, sensor technology, and telecommunications modules, together with CPU and storage technology. To accomplish this, nanotechnology will be needed to miniaturize and integrate the different subsystems.

The goal for the year 2020 is to establish and maintain a world-class programme to research revolutionary computing technologies (RCT) that will not only take us beyond the limits of semiconductor technology scaling but also will enable the vision of a "thinking spacecraft." A thinking spacecraft would be a totally autonomous, highly integrated, extremely capable spacecraft that operates at ultralow power. To achieve this goal, without a doubt, we need to employ nanoscience. In spite of the phenomenal advances in digital computing in recent years and those expected in the near future, even future supercomputers cannot compete with biological systems in performing certain ill-defined tasks such as pattern recognition, sensor fusion, fault-tolerant control, and adaptation to the environment.

- *Biomimetics*, that is, systems or technologies inspired by architectures, functions, mechanisms, and principles found in biological systems, for example:
  - One gram of DNA could possibly store all the data in the Library of Congress.
  - The human brain contains about 1014 interconnects and operates at 1016 operations per second, using ultra low power and imprecise computing elements.
  - Humans are endowed with an immune system that provides recovery from illness—a "self-repair system".
- *Quantum computing*, that is, a joint venture between computer science and quantum physics. Although, the concept of a quantum computer is simple, its realisation is not. Two issues motivate quantum computing:
  - Quantum mechanical concepts must be applied to solve intractable (NP-complete) computing problems.
  - From a computer miniaturisation point of view, the size limit of a bit of information is important. Recently, this issue has attracted increased attention, due to the current development of nanotechnology and the design problems of semiconductor and metal devices that are approaching the quantum size limit. Consequently, the idea of quantum computing, in which the elements that carry the information are atoms, has attracted the attention of many scientists.

As devices become smaller, lighter, and consume less power, NASA will be able to design and fly space probes on missions that are not currently possible.

Nanotechnology on a Chip: A New Paradigm for Total Chemical Analysis Systems, the ability to make chemical and biological information much cheaper and easier to obtain is expected to fundamentally change healthcare, food safety, law enforcement, national security, and many other areas of direct interest to the American public. The vision of broadly available chemical analysis is fueling an international effort to develop "mChemLabTM-on-a-chip" technology. Micro-total analysis systems (m-TAS, as they are often referred to) are distinguished from simple sensors because they conduct a complete analysis; a raw mixture of chemicals goes in and an answer comes out. Sandia National Laboratories is developing a hand-held mChemLabTM demonstrator that will analyze for air-born chemical warfare agents and liquid-based explosives agents. The mChemLabTM development project brings together an interdisciplinary team of about 50 staff members from throughout the laboratory in areas of expertise including microfabrication, chemical sensing, microfluidics, and information sciences.

Although nanotechnology plays an important role in current mTAS efforts, most mTAS approaches use miniaturised versions of conventional architecture and components to achieve system tasks. Small valves, pipes, pumps, separation columns, etc. are patterned after their macroscopic counterparts. Even though we are finding that these miniaturised components can work as well as (and sometimes better than) their macroscopic analogs, they simply will not allow for the vision of chemical laboratories in a grain of sand. Nanotechnology will enable a completely new architecture, or nano-TAS. The ability to build materials with switchable molecular functions could provide completely new approaches to valves, pumps, chemical separations, and detection. For example, fluid streams could be directed by controlling surface energy without the need for a predetermined architecture of physical channels. Switchable molecular membranes and the like could replace mechanical valves. By eliminating the need for complex fluidic networks and micro-scale components used in current m-TAS efforts, nano-TAS is a fundamentally new approach to allow greater function in much smaller, lower power total chemical analysis systems.

A Biological Nanodevice for Drug Delivery: The nanobiological anticancer agent PK1 exploits the enhanced permeability and retention (EPR) effect associated with disease tissues with low integrity vasculature in order to deliver cytotoxin (doxorubicin) to tumors. The synthetic backbone of PKI (N-2-(hydroxypropyl) methacrylimide or HPMA) gives the complex a size (diameter in the mid-nanometer range) that makes it unable to extravasate efficiently into healthy tissues with normal vasculature. Tumor vasculature is abnormally permeable, allowing preferential accumulation of PK1 in tumor tissue. HPMA-bound doxorubicin is non-toxic, limiting toxicity to healthy tissue, and active doxorubicin is released from the complex preferentially in tumor tissues. The labile peptidic linker tethering doxorubicin to HPMA was selected because it is the substrate for a protease known to be over-expressed in the target tumor types. PK1 increases the tolerated doxorubicin dose by more than an order of magnitude by virtue of EPR-based targeting and its engineered tumor-preferred doxorubicin release properties. It is in human clinical trial in Europe. Contemplated embellishments to this and similar polymer therapeutics include use of monodisperse nanopolymers (dendritic polymers) to enhance control of EPR properties, incorporation of protein docking domains that recognize tumor associated antigens to tether the complex following its delivery to the tumor, and incorporation of additional antitumor agents thought to have synergistic effects with cytotoxins, that is, angiostatic agents, among others.

The Development of Useful Nanotech Robotic Systems: It is potentially feasible to manufacture nanorobots that are capable of sophisticated symmetric behaviours, either through independent function or by assembling themselves into collective units. Imagine, for example, high-resolution video screens that can repair themselves simply by having a microscopic robot at each screen element. These "pixelbots" would be capable of producing light, but also be smart enough to remove themselves from the video array should they ever fail. Other pixelbots would sense the vacancy left by any defective device and reorganise themselves to fill the hole. Another example is the incorporation of autonomous "nurse" robots into the human body that are chemically benign but are capable enough to remove cancer cells at the source. Having the ability to discriminate between healthy and cancerous cells, the "nurse-bots" would function independently to continuously heal tissue in ways beyond the current ability of the human body. The same notion can be implemented in self-optimising silicon memories or processors, where a blown transistor would mean one just had to wait for the computer to heal itself.

One key to establishing such capabilities is research into autonomous self-assembled systems. Researchers at Los Alamos National Laboratory are exploring these systems by creating very inexpensive macro-scale robots. These sense and adapt to their environment, including assimilating other robots to execute such tasks as searching for and marking the location of unexploded land mines. The capabilities of these intelligent cellular systems are readily scaled, providing untapped possibilities for large numbers of inexpensive nano-machines to become microscopic building blocks for heretofore unimaginable functions—a form of "nano-Lego" for the new millennium with novel, untapped market potential.

Integrated Nanotechnology in Microsystems: Advances in nanotechnology will have a profound effect on the future of integrated microsystems. The integration of microelectronic, microelectromechanical, optical, and chemical microsensors into "systems on a chip" is an area that may involve mechanical, optical, and/or chemical functions as well. These advances will make possible miniaturised systems that sense, think, talk (communicate), and act. However, these microscale systems will only become a reality if enabled by the control of performance at the nanoscale. Thus, for example, advances in microelectromechanical systems (MEMS) and photonics shown in the figure depend on discoveries in nanoscience and nanoscale fabrication.

Uses of MITRE Nanosystems Research Task

The prospective affordability of any future nanotechnology is a key consideration in attempting to understand its practicality. Thus, one task in the MITRE Nanosystems Modeling and Nanoelectronic Computers research project has been an effort to initiate a prospective economic analysis of nanoelectronics. The initial objective of the Economic Analysis Task was to develop and to explain an approach to a prospective cost-benefit analysis (CBA) for nanoelectronics. As a basis for this effort, the members of the MITRE Nanosystems Group began to assess the past and current expenditures/costs for nanoelectronics research and a development (R&D) in order to make projections of anticipated future costs. While an actual CBA has not yet been conducted, several methods and approaches to such an analysis were identified. In addition to approaches to conducting a CBA, methods for estimation of R&D costs of advanced technologies such as nanoelectronics, were identified. They include:

- Evaluation of R&D investments in electronics industries.
- Use the Cost Estimation Relationships developed at RAND.
- Extend two different cost curves for microelectronics (developed by MITRE and RAND)
- Evaluate cost, benefits, and possibilities for next generation parallel computing hardware.
- Modify cost estimation technique for state-of-the-art technologies.

The project team determined that the first two of these were most applicable to assessing the potential costs of nanoelectronics. Task Leader Michael D. Smith has continued the research for this economic analysis of nanoelectronics in his Masters' Thesis, "Estimation of Future Manufacturing Costs for Nanoelectronics," which he completed in May 1996 in the Department of Industrial Engineering at the Virginia Polytechnic Institute and State University in Blacksburg, VA. Michael D. Smith's thesis has been made available on the World-Wide Web by the Virginia Polytechnic Institute.

Understanding the Distributed Supercomputing Platform

We firmly believe that the converging fields of nanotechnology, evolutionary computing, and pervasive networking together represent both the greatest opportunity and threat for our universe. We think that the safest and best way forward as these fields merge is to provide a virtual environment for self-replicating machines running evolving software to be tested, evaluated and even farmed. This is no trivial task. If we are to avoid the so called 'gray goo' problem, and at the same time harness the power of natural selection, we must find ways of containing self-replicators while the possibility of their 'going postal' exists. To this end, we will be deploying a massively parallel supercomputer based on the "generic workers around a shared jobs bag" pattern. It provides a very simple structure for the distributed comutation of all manner of scenarios. We seek the following intersectable sets of collaborators: To register your interest, please provide your email address and indicate the nature of your interest using the checkboxes below.

- Helpers—People willing to run services of the distributed computing platform on their computers for some, if not all day. You should have installed Java1.2 or better. Mac Users note Java1.2 will not be available until the release of MacOS X. We will make available a a suite of Jini Services as a single installer and simple launch interface.
- Developers—Java programmers familiar with CVS and other Open Source tools for distributed development who are keen to write simulation scenarios, simulation components and various utilities, visualisation tools, high level editors and user interfaces. We will provide a simple upload utility for adding new simulations to the network.
- Gardeners—Bringing up healthy, usefull nanotechology simulation is a lot like gardening. The gardens will need constant weeding, and the more people willing to keep an occasional eye on a crop of networked nanobots the better. We will provide simple browser utilities enabling you to provide critical feedback to active simulations.
- Writers—Like any community, we want to be social, and we want to make sure that everyone who is interested can get the most out of the efforts of the community. We want commentators to comment, critics to critique, and writers to write. By imagining the future, we can help ward off many disasters without even having to simulate them. Speculative fiction is simulation.
- Visual Artists—Visual representations of possible scenarios, "an artist's impression", if you can draw, paint or 3d render we need your help. First up we need a logo which simply illustrates both the collective and the various aspects of the gRobots project.

We are 4 private individuals who have come together to kick start the gRobots project. It is our intention to establish a not-for-profit company structure along the lines of the Foresight Institute. In the meantime we are not a legally incorporated body, but an ensemble of impassioned professionals. We are not soliciting funds of any kind.

(i) Dave Sag

Dave has been writing and programming on the themes of Artificial Life and Nanotechnology for almost 20 years. He is a Senior Associate of the Foresight Institute and Director/Equity partner in a diverse range of commercial and non-commercial online projects. He is internationally regarded as an artist, programmer, business person and portable whole.

(ii) David (DayV) Parry.

In 1984 DayV owned the only copy of Mandelbrot's "the fractal geometry of nature" in Australia. He has been obsessed with self replicating forms, evolutionary comupting and nanotechnolgy almost as a consequence of his earlier obsessions with electronic music and mathematics. He is most recently the author of a handy tutorial on XSP Logic sheets. He is a member of the Eurpoean Technology Forum, and God Emperor King of the virtual nation of Suran.

(iii) Dr Phillip Blythe

Phil is an interdisciplinary scientist, with fingers in many different research pies. Originally an aeronautical engineer, He has since conducted research in adaptive computation, control theory, cognitive psychology, and economics. There have been three central themes that have remained with him through each of these various research projects; evolution, dynamic systems and decision making. He was recently a visiting professor to the Centre for Adaptive Behaviour at the Max Planck Institute for Human Development.

(iv) Anthony Aldridge

Tony is a musician, inventor, programmer and all round chaos junkie. He's worked on virtual interfaces for the blind, recorded servral albums and for a while ran the UK's largest music technology department at University of London. He is also a specialist software security expert and quite a dab hand at opening a beer bottle with a cigarette lighter.

Regarding the possible ecological risks of nanotech, the first quantitative technical analysis of Drexler's "gray goo" scenario, in part as an answer to Bill Joy's recent concerns.

The Original Nanotechnology Weblog of Nanodots

"As part of a five-day business trip here, Mr. Kim met with reporters in southern Seoul yesterday to introduce cutting-edge future technologies and offer predictions on the world's information-technology industry.

"In the next five to 10 years, mobile phone users will be able to detect changes in the facial expressions and even in the smell of the person they are talking to, according to Jeong Kim, president of Bell Labs, the research arm of the U.S. firm Lucent Technololgies.

"Elaborating on the new cell phone technology developed by the laboratory, Mr. Kim said that smell-transmitting sensors, lenses that follow the movement of eyeballs and microphones as narrow as a human hair already exist, and would be on sale within a decade.

"Mr. Kim said nanotechnology — techniques used to create structures as small as one-millionth of a millimeter — had played a key role in the development of the new technologies.

"Mr. Kim predicted that nanotechnology could bring huge changes to the way humans live, citing the example of a new computer that will enable simultaneous translation of telephone conversations.

"If nanotechnology maintains its current pace of development, it will give birth to a computer that has the information processing capacity equivalent to every human brain combined by 2060, he said.

"Mr. Kim said that while communication technologies have so far mostly focused on speed, future developments will look to improve their convenience. Eventually, he added, phones will no longer need to be touched or even spoken to, but will instead respond to mental commands."

Involved Nanotech Ambition and Pessimism

Richard Jones asks: "Why does the molecular manufacturing community seem to have many fewer members in the UK than it does in the USA? I don't think it's fair to say that the dramatic vision of molecular manufacturing is pursued in a contextual vacuum—I think there is quite a well-developed world view that underlies the molecular manufacturing vision. Is there something about that world-view that makes it more attractive in one country than another?...Is this really just a clash between the habitual rainsoaked pessimism of the British, and sunny American optimism and its associated can-do attitude?" An excellent question, Richard. It's hard to miss the difference between U.S. attitudes — not just toward molecular manufacturing, but toward nanotechnologies in general, and technology more broadly — and attitudes in Europe. Why is this? Well, for centuries now, risk-takers have paddled (now, flown) across the oceans to try their luck here. American culture is far more admiring of risk-takers. Perhaps this centuries-long process has led to a difference in national attitudes toward what timeframe is of interest when considering engineering goals. I've visited Sweden and Norway. They're very nice places, but I'm glad my forebears took the boat. I carry on the tradition, in a way, by betting my career on nanotechnology instead of, say, practicing medicine, which would have been a much safer career choice. I'll caricature the pro-vs-anti-MNT debate: The anti's say "Man will never fly, unless it's on the back of a very large, biological bird." The pro's say "The heck with that — let's build an airplane." This can-do, "let's build it" attitude seems quite American to me. It has worked pretty well for us so far, though taking risks doesn't always pay off, and can be...risky. –CP

About Public Database on Nanoparticle Safety and Nanotech in 20 Years

Tired of garbled news articles on nanoparticle safety? I certainly am. Now, the International Council on Nanotechnology (which I serve as an Advisor) has published the first public database of peer-reviewed journal articles examining environmental, health, and safety issues arising from both engineered and incidental nanoparticles (press release in PDF). You can help improve it by using the addition/correction form. It's free, but there's a catch: "Site registration or payment may be required to access the full content of these articles." Sigh — now we just need more open access journals. Titled Nanotechnology – a Key Technology for the Future of Europe (PDF), it includes on pages 26-28 a scenario of the role of nanotech in 2025. The technology will sound familiar, but the political context is quite different from most U.S. nanotech scenarios.

"Rogerio remembers the days when people travelled by car or aeroplane, until droughts, fires, famines, plagues and hurricanes became so frequent that the scientific evidence for an imminent, irreversible ecological global disaster could no longer be ignored. Governments had to then make some very hard-hitting decisions to ensure the survival of the planet, and the continuation of the human race. Travel was forbidden, except in emergency, new forms of energy use were enforced through legislation...

"Surprisingly, also, the bank had done rather well out of the new world order. The use and development of these new technologies had been enforced by law, and this meant a new role for the bank in providing finance. By law also, banks today had to use a high proportion of their surplus to support projects in specified regions of the world at very low interest rates. Not popular to begin with, but the benefits were now apparent. However, it had been sad to see so many businesses such as those involved in import and export of out-of-season fruits, flowers and vegetable, and those needing a high energy input such as glass factories, go to the wall."

Regarding Fun Nano Video and Nanoexplosions-on-a-chip

On Athenaweb (free registration req'd), a new site brought to you by the European Commission, is a five-minute video called The Principle of Nanotechnology with enjoyable nano graphics, worth viewing. That the audio is in German is not a problem, since those of us who are language-challenged can simply enjoy the video portion. If you're curious, though, you can read the transcript in English, which is above average. I especially like "Even the infinitesimal 'arterial submarine', destroying deposits blocking blood circulation, must remain merely an illusion for the moment." For the decade..but let's hope not for too many decades. I'd like to have those things available before I need them.

From the ever-diligent Charles Q. Choi, on work by physicist Shubhra Gangopadhyay at the University of Missouri at Columbia: "The researchers [have] coated devices made of glass with a mixture of nanoparticle fuel such as aluminum and oxidizer such as iron oxide. The nanoparticle quality of the fuel and oxidizer provides far greater surface area for explosive reactions than a conventional mixture would. When ignited, the nanoparticle combination releases a tremendous amount of thermal and mechanical energy...Their device could generate pulses of energy in only fractions of a second, of potential use in defense applications such as initiating explosives. This power could also be stored for later use in portable electronics, much like batteries. The shock waves might also find use in medical devices, to destroy kidney stones perhaps, Gangopadhyay explained." But it sounds like daunting challenges remain: " 'We need to perform a thorough study to reduce the energy needed to initiate the chemical reaction. We also need to design a chip to sustain energy generation for a longer time period and store it properly for later use,' she added." She's starting a company: NEMS & MEMS Works, LLC.

Aggregated Diamond Nanorods and Other Interesting Nnanotech Experiments

"Physicists in Germany have created a material that is harder than diamond. Natalia Dubrovinskaia and colleagues at the University of Bayreuth made the new material by subjecting carbon-60 molecules to immense pressures. The new form of carbon, which is known as aggregated diamond nanorods, is expected to have many industrial applications (App. Phys. Lett. 87 083106).

"The hardness of a material is measured by its isothermal bulk modulus. Aggregated diamond nanorods have a modulus of 491 gigapascals (GPa), compared with 442 GPa for conventional diamond. Dubrovinskaia and two of her co-workers—Leonid Dubrovinky and Falko Langenhorst—have patented the process used to make the new material.

"Diamond derives its hardness from the fact that each carbon atom is connected to four other atoms by strong covalent bonds. The new material is different in that it is made of tiny interlocking diamond rods. Each rod is a crystal that has a diameter of between 5 and 20 nanometres and a length of about 1 micron.

"The group created the ADNRs by compressing the carbon-60 molecules to 20

GPa, which is nearly 200,000 times atmospheric pressure, while simultaneously heating to 2500 Kelvin."

From CORDIS: "Dr Dubrovinskaia and two of her colleagues have patented the process used to make the new material. 'We have developed a concept for innovative technology to produce the novel material in industrial-scale quantities and now we are looking for partners in order to realise our ideas,' she said." Chemist Derek Lowe describes some amazing-sounding work at Caltech: "They're taking electron microscope snapshots, one trillionth of a second at a time. And what is this technique good for? Well, electron microscopy has long been used for imaging all sorts of materials and biological samples. Fast freezing of the samples has revealed an extraordinary amount of information in the past, and [Ahmed] Zewail's new method basically allows this to happen in real time, at room temperature, under normal conditions. The energies required to do it aren't huge, and it's quite likely that we'll be able to get useful data without destroying delicate targets. We could end up with extreme slow-motion movies of molecular processes, imaged at electron-diffraction resolutions. We're actually going to be able to watch nanotechnology experiments as they happen." He speculates that this could be worth a Nobel Prize. (Credit: Instapundit)

Professor David Leigh, of the University of Edinburgh, said the demonstration was one of the first examples to show molecular machines integrating successfully with the macroscopic world... The tiny machines that coat the surface are essentially rod-like structures with rings that, in their normal state, furiously jump up and down because of Brownian motion (the random movement of molecules caused by collisions with molecules around them). But when these structures are stimulated by ultra-violet light, a chemical reaction takes place and the rings all go to one end, changing the nature of the surface under the blob from a repulsive one to an attractive one...

Professor Leigh concedes the work is early proof-of-principle but he is bolstered by the knowledge that nature uses molecular machines to do all of its jobs. He believes it can only be a matter of time before we are tapping into that know-how.

> "Mankind, despite all of our technology in the 21st century, we use molecular machines to do nothing," he told the British Association's Festival of Science here in Dublin, Ireland.
>
> "The properties of every drug, every polymer and every catalyst are just based on their static or non-smart properties. When we learn to do the same as nature, you really will have materials that today just sound like science fiction."
>
> Jenny Hogan blogs for Nature.com: "The historic city of Edinburgh in

Scotland is this week hosting hundreds of scientists and politicians discussing "Nanotechnology and the Health of the EU Citizen in 2020".

"[Using iron oxide nanoparticles to treat cancer], we are told, is the first anti-cancer therapy that uses nanoparticles to have entered clinical trials. It is being developed by a company called Magforce, based in Germany.

"The approach seems very simple: the iron oxide nanoparticles are coated to prevent them irritating the immune system, then injected directly into the tumour. Next the patient is exposed to an alternating magnetic field, and the magnetic nanoparticles become very hot. Because they are localised to the tumour, the heating destroys the cancerous cells without harming the tissue around it.

"And what happens to the nanoparticles? They seem to end up in the spleen, causing no adverse reaction."

In the popular blog Instapundit, Foresight director Glenn Reynolds interviews Foresight advisor Ray Kurzweil on the topic of his new book, *The Singularity is Near: When Humans Transcend Biology*. Some excerpts:

"Achieving the hardware requirement was controversial when my last book on this topic, The Age of Spiritual Machines, came out in 1999, but is now pretty much of a mainstream view among informed observers. Now the controversy is focused on the algorithms...

"Just recently, scanning tools can see individual interneuronal connections, and watch them fire in real time. Already, we have mathematical models and simulations of a couple dozen regions of the brain, including the cerebellum, which comprises more than half the neurons in the brain. IBM is now creating a simulation of about 10,000 cortical neurons, including tens of millions of connections. The first version will simulate the electrical activity, and a future version will also simulate the relevant chemical activity. By the mid 2020s, it's conservative to conclude that we will have effective models for all of the brain...

"By 2030, a thousand dollars of computation will be about a thousand times more powerful than a human brain. Keep in mind also that computers will not be organised as discrete objects as they are today. There will be a web of computing deeply integrated into the environment, our bodies and brains."

Scenarios have been written about how advanced nanotechnology could help victims of natural disasters, and those otherwise in need, in the future. And at the upcoming Foresight Conference we'll be looking at how nanotech can provide clean water cheaply. But what can nanotech do, right now, for those affected by Hurricane Katrina?

Nothing, to my knowledge (send corrections if you know of something). But what the nanotech community can do today is send donations to relief organisations, and this blog encourages us all to do so as part of Foresight director (and Alpha Blogger) Glenn Reynolds' blogburst to promote Flood Aid. Other nanotech bloggers are urged to participate in this blogburst. [UPDATE: I used Amazon to give to American Red Cross, and got a match from FreeMoneyFinance.com — you can too, until they reach their limit.]

Nanotech in U.S and Foresight Nanotech Conference

New research on public attitudes by U. Wisconsin journalism prof Dietram Scheufele:

"Scheufele worked with Cornell University associate professor Bruce Lewenstein on the survey, the results of which will be published in the Journal of Nanoparticle Research and Science Communication.

"The survey showed that about 25 percent of respondents reported never having heard of the issue, even if it was explained to them by the interviewer, and only 16 percent felt at least "somewhat informed" about nanotechnology.

"Survey results also showed that there were significant differences between those who were aware of nanotechnology and those who were not.

"For instance, 59 percent of those who indicated they were aware of nanotechnology expressed overall support of the emerging technology, compared to 28 percent support among those who were not aware.

"The survey also revealed that those who were aware of the technology had a much more optimistic view of the potential positive effects of nanotechnology.

"Two-thirds to three-quarters of respondents who were aware of the issue agreed that nanotechnology has potential benefits in health and medicine, national defense and environmental policy, while only about half of the respondents who were unaware shared this perspective."

Early registration ends tomorrow, Sept 1, for the Foresight Conference and/or Vision Weekend. We have been asked whether those affected by Hurricane Katrina can get an extension: Yes, of course. We've received some registrations covering all five-and-a-half days of this segmented meeting covering Applications, Policy, Research, and Vision. These hearty conference-goers will receive an endurance award at the end of the meeting. As will the Foresight staff! –CP

Nanoinvesting and Productive Nanosystems

A very quick intro to nanotech investing in Wired News by Joanna Glasner includes some very good advice that one hopes anyone contemplating such an investment will pay close attention to, such as: "Beware of marketing labels: If a company has 'nano' in its name, [Matthew Nordan of Lux Research] cautions, that doesn't necessarily mean it's on the cutting edge of nanotechnology. Dig into the technology behind potential investments before you place a bet." Unfortunately the piece also includes this quote: " 'The concept of self-replicating molecular machines is science fiction,' said Darrell Brookstein, an investment adviser and author of the book *Nanotech Fortunes*. 'It's not a matter of waiting 30 years to develop them. They're never going to develop.' " His book seems okay for those making near-term investments, but for technical analysis of what is theoretically possible in the long term, I'll go with Freitas and Merkle on this issue. See their book.

Interview on Technology Roadmap for Productive Nanosystems

Check out the interview of Foresight president Scott Mize over at *Nanotechnology Now* on the "Technology Roadmap for Productive Nanosystems." Excerpt:

"The 'Technology Roadmap for Productive Nanosystems' will chart anticipated developments in nanotechnology which will be required to take us from current abilities to advanced nanosystems. The Roadmap will examine what can be developed in labs today and then map out a step-by-step plan for developing these new nanotechnologies.

"The Roadmap process will bring together and coordinate the thinking and activity of key stakeholders who have a role in developing the next generations of nanotechnology. These stakeholders will include governments, corporations, research institutions, policy professionals, investors, educators and the media. The goal is to provide a common vocabulary and framework that each stakeholder can use as input for their own businesses strategy, investment, or research and development process...

"Our working definition is that 'Productive Nanosystems are molecular-scale systems that make other useful materials and devices that are nanostructured.' However, through the Roadmap process, this definition is likely to evolve."

We try not to let Foresight's activities dominate this blog, but for those of you considering attending the 13th Foresight Conference on Advanced Nanotechnology, Oct. 22-27: the deadline is this coming Thursday, Sept. 1, to take advantage of the early registration fee discount. See the speakers, the programme, the Research page, and the list of organisations attending previous meetings. Late abstracts are still being considered. This is the most comprehensive nanotech conference we've ever done: Applications, Policy, Vision, and Research are all covered, and attendees can customize their registration to select the sessions of most interest to them personally. We have more high-profile speakers than ever before, by far. And because we're non-profit, and rather experienced at holding these meetings, registration fees are lower than most other nano meetings. The Feynman Prize Banquet is an especially exciting part of the meeting.

A sample: "In yet another potentially high impact nanotherapeutic approach, Professor Reza Ghadiri and his group at Scripps Research Institute have developed nanobiotic agents that can kill bacteria with the risk of them developing resistance substantially lower than with conventional antibiotics. Antibiotics often work by latching onto specific molecules in the bacteria's outer membrane. If the bacteria modify these molecules even slightly, they become resistant to the antibiotic and can proceed to infect their host. In Ghadiri's work, self-assembled nanotubes are made from rings of peptides stacked on top of each other. The nanotubes bundle together to form carpet-like aggregates which punch holes through bacterial membranes. The specific peptides are chosen so that they perforate the bacterial membranes without harming the cells of the host. The bacteria die instantly. Modifying a particular molecule would be insufficient for the bacteria to become resistant – they would have to substantially modify their membrane."

Prof. Morinobu Endo of Shinshu University in Japan, widely regarded as a Nobel candidate for his work with carbon nanotubes, speaks out bluntly in an article in RedNova News: " 'Nanotechnology is certain to play a crucial role in creating technological innovations in the 21st century,' says Endo, who is believed to be a Nobel Prize candidate for his discovery of nanotubes and a method to mass produce them. Endo says success in nanotechnology is a 'matter of life and death' for developed nations, because developing nations are rapidly catching up in the mature technologies that led the 20th century, such as antibiotics and semiconductors. 'Without technological innovations, industries in developed nations will be caught in a price competition with developing nations and will consequently suffer a further hollowing-out,' he says.

He also believes nanotech could do much to assist smaller firms: "In Japan, Endo believes that nanotechnology will be a key force in boosting the competitive edge of small and midsize companies, the sector most severely affected by cheap imports from emerging economies. 'Nanotechnologies can be incorporated easily into manufacturing at not only large firms but also at small- and medium-sized local companies,' he says. 'They differ from 20th century-style technologies, which tend to be dominated by large firms because of the large-scale funds and research needed for industrial use.' "

From Roland Piquepaille: Gecko lizards, which can climb any vertical surface and hang from a ceiling with one toe, have fascinated scientists for a long time. Their foot-hairs have a structure which allow them to strongly adhere to [almost] any type and shape of surface. Now, according to this short news release from the National Science Foundation (NSF), researchers from the University of Akron, Ohio, have developed synthetic hairs from multiwalled carbon nanotubes (MWNT) that have adhesion forces 200 times higher than those observed with gecko foot-hairs. This could lead to new dry adhesives used in microelectronics, robotics or space applications.

Analysing Computational Molecular Modeling and Nanotech Applications

Reviews in Computational Chemistry by Lipkowitz and Boyd is an excellent series of books which does just that. Unlike most review series, it includes introductory and tutorial as well as more advanced material. The World-Wide Web Virtual Library: Chemistry is an excellent starting place to look for any chemistry related information. The Computational Chemistry Archive is more specifically focused on issues surrounding computational chemistry. Infochem has many links to Molecular Modelling Software. Chemistry Software and Information Resources of the National HPCC. QCPE (The Quantum Chemistry Programme Exchange) provides a wide range of computational chemistry software, often free. NetSci's list of molecular modeling software and their computational chemistry page A list of chemistry resources on the internet is maintained at Rensselaer. NIH has a page on molecular modeling and has a guide to software. An online-text which discusses aspects of quantum mechanics, the Born-Oppenheimer approximation, and molecular mechanics.

CambridgeSoft provides a package that runs on the Mac and under Windows. It is more limited in its modeling capabilities: it can effectively do energy minimisations on structures with perhaps a hundred atoms (their minimizer does only steepest descent). They provide truly excellent facilities for making black & white pictures of molecules, and equally excellent facilities for drawing journal-quality 2-dimensional chemical structures using standard notation. Gaussian is an ab initio programme that usually costs over $10,000 but which is now available on the PC for under $1,000. While it has an antediluvian user interface (designed for punch cards) it is the de facto standard against which competing companies (like the startup Schrodinger) compare themselves and is well known in the computational chemistry literature. They can be reached by e-mail at info@gaussian.com. While I haven't had direct experience with it, Jay Ponder (who gave an excellent talk on molecular mechanics at the First Foresight conference on Molecular Nanotechnology) has made available a molecular mechanics and dynamics package written in Fortran 77 which is free (including source code). To quote their description—TINKER is a complete package for performing molecular mechanics calculations. It is intended to serve as a platform for algorithm development and parameterisation, while still being efficient enough for most production work. The currently available potentials include MM2 and MM3, AMBER-95, AMBER/OPLS, CHARMM22 and our own TINKER potential. Many other literature potentials can be implemented via external parameter files; parameter sets for ENCAD and MMFF-94 are under development.

Programmes are provided to perform many functions including the following:

- energy minimisation over Cartesian or torsional coordinates via conjugate gradient, variable metric or truncated Newton methods,
- molecular dynamics with periodic boundaries and options to control temperature and pressure,
- normal mode vibrational analysis,
- distance geometry including a new, efficient random atom-pair metrisation,
- simulated annealing with a choice of cooling protocols,
- analysis and breakdown of single point potential energies,
- verification of analytical derivatives of the standard potential and/or user defined functions,
- location of a transition state between two minima,
- free energy perturbation calculations,
- fitting of intermolecular potential parameters to crystal structure and thermodynamic data, and
- diffusion equation method and Gaussian density annealing approaches to optimisation via smoothed potential energy surfaces.

Laboratories throughout the world are rapidly gaining atomically precise control over matter. As this control extends to an ever wider variety of materials, processes and devices, opportunities for applications relevant to NASA's missions will be created. This document surveys a number of future molecular nanotechnology capabilities of aerospace interest. Computer applications, launch vehicle improvements, and active materials appear to be of particular interest. We also list a number of applications for each of NASA's enterprises. If advanced molecular nanotechnology can be developed, almost all of NASA's endeavors will be radically improved. In particular, a sufficiently advanced molecular nanotechnology can arguably bring large scale space colonisation within our grasp. This study describes potential aerospace applications of molecular nanotechnology, defined as the thorough three-dimensional structural control of materials, processes and devices at the atomic scale.

The inspiration for molecular nanotechnology comes from Richard P. Feynman's 1959 visionary talk at Caltech in which he said, "The problems of chemistry and biology can be greatly helped if our ability to see what we are doing, and to do things on an atomic level, is ultimately developed—a development which I think cannot be avoided." Indeed, scanning probe microscopes (SPMs) have already given us this ability in limited domains. Synthetic chemistry, biotechnology, "laser tweezers" and other developments are also bringing atomic precision to our endeavors. An expanded version of Drexler's MIT PhD thesis, examines one vision of molecular nanotechnology in considerable technical detail. It proposes the development of programmable molecular assembler/replicators. These are atomically precise machines that can make and break chemical bonds using mechanosynthesis to produce a wide variety of products under software control, including copies of themselves. Interestingly, living cells exhibit many properties of assembler/replicators. Cells make a wide variety of products, including copies of themselves, and can be programmed with DNA.

Replication is one approach to building large systems, such as human rated launch vehicles, from molecular machines manipulating matter one or a few atoms at a time. Note that biological replication is responsible for systems as large as redwood trees and whales. Another approach to nanotechnology is supramolecular self-assembly, where molecular systems are designed to attract each other in a particular orientation to form larger systems. Hollow spheres large enough to be visible in a standard light microscope have been created this way using self-assembling lipids. There are many other examples and this field is rapidly advancing. Biological systems can do most of what molecular nanotechnology strives to accomplish — atomically precise products, active materials, reproduction, etc.

However, biological systems are extremely complex and molecular nanotechnology seeks simpler systems to understand, control and manufacture. Also, biological systems usually work at fairly mild temperature and pressure conditions in solution — conditions that are not found in most aerospace environments. Today, extremely precise atomic and molecular manipulation is common in many laboratories around the world and our abilities are rapidly approaching Feynman's dream. The implications for aerospace development are profound and ubiquitous. A number of applications are mentioned here and a few are described in some detail with references. From this sample of applications it should be clear that although molecular nanotechnology is a long term, high risk project, the payoff is potentially enormous — vastly superior computers, aerospace transportation, sensors and other technologies; technologies that may enable large scale space exploration and colonisation. This study is organised into two sections. In the first, we examine three technologies—computers, aerospace transportation, and active materials—useful to nearly all NASA missions. In the second, we investigate some potential molecular nanotechnology payoffs for each area identified in NASA's strategic plan.

Some of these applications are under investigation by nanotechnology researchers at NASA Ames. Some of the applications described below have relatively near-term potential and working prototypes may be realised within three to five years. This is certainly not true in other cases. Indeed, many of the possible applications of nanotechnology that we describe here are, at the present time, rather speculative and futuristic. However, each of these ideas have been examined at least cursorily by competent scientists, and as far as we know all of them are within the bounds of known physical laws. We are not suggesting that their achievement will be easy, cheap or near-term. Some may take decades to realise; some other ideas may be scrapped in the coming years as insuperable barriers are identified. But we feel that they are worth mentioning here as illustrations of the potential future impact of nanotechnology. The applicability of manufacturing at an ever smaller scale is nowhere more self-evident than in computer technology.

Indeed, Moore's law (an observation not a physical law) says that computer chip feature size decreases exponentially with time, a trend that predicts atomically precise computers by about 2010-2015. This capability is being approached from many directions. Here we will concentrate on those under development by NASA Ames and her partners. Carbon nanotubes can be viewed as rolled up sheets of graphite from 0.7 to many nanometers in diameter. The smaller tubes are single molecules. Dai placed carbon nanotubes on an SPM tip thus extending our ability to manipulate a single molecule with sub-angstrom accuracy.

Not only are the tips atomically precise, but they should have approximately the same chemistry as C_{60}, and thus be functionalizable with a wide variety of molecular fragments. Functionalising carbon nanotube tips will allow mechanical manipulation of many molecular systems on various surfaces with sub-angstrom accuracy. One particularly intriguing possibility along this line is to utilize a carbon nanotube SPM tip to engrave patterns on a silicon surface. It should be possible to create features a few nanometers across. These would be perhaps 100 times finer than the current state of the art in commercial semiconductor photolithography. Further, in contrast to approaches such as electron microscope lithography for which the speed of operation now appears to be an insuperable obstacle for industrial production, nanotube SPM-based lithography can be accelerated by utilising an array with thousands of SPM tips simultaneously engraving different parts of a silicon surface. Also, nanotube SPM lithography could provide a practical means to explore various futuristic electronic device technology ideas, such as quantum cellular automata, which require exceedingly small feature sizes. Needless to say, if these ideas pan out, they could literally revolutionize computer device technology, paving the way for systems that are many times more powerful and more compact than any available today. For the near term, it should be noted that the semiconductor industry is a major market for SPM products. These are used to examine production equipment. High performance carbon nanotube tips should be of substantial value. NASA Ames is collaborating with Dr. Dai, now at Stanford, to develop these tips. It is possible to store data on long chain molecules (for example, DNA) and it may be possible to read these data with carbon nanotube tipped SPMs. Existing DNA synthesis techniques can be used to write data. If the different DNA base pairs can be distinguished with a carbon nanotube tipped SPM, then the data can be read non-destructively (current techniques allow a destructive read). However, the difference between base pairs is not great.

If the base pairs cannot be distinguished, techniques for attaching modified enzymes to specific base pair sequences could be used. Certain enzymes (DNA (cytosine-5) methyltransferases) attach themselves onto a specific sequence of base pairs with a covalent bond. The enzyme then performs its operation and breaks the bond. Smith modified the enzyme such that the initial covalent bond was formed but the subsequent operation was disrupted. The result is that DNA synthesised with the target base pair sequences at the desired location can force precise placement of the enzymes. The presence of an enzyme could represent 1 and its absence 0. Enzymes are sufficiently large that distinguishing their presence should be straightforward. If the DNA/enzyme approach proves impossible, a wide variety of other polymer systems could be examined. Bauschlicher computationally studied storing data in a pattern of fluorine and hydrogen atoms on the (111) diamond surface. If write-once data could be stored this way, 10^{15} bytes/cm^2 is theoretically possible. By comparison, the new DVD write-once disks now coming on the market hold about 10^8 bytes/cm^2.

Bauschlicher compared the interaction of different probe molecules with a one dimensional model of the diamond surface. This study found some molecules whose interaction energies with H and F are sufficiently different that the force differential should be detectable by an SPM. These studies were extended to include a two dimensional model of the diamond surface and two other systems besides F/H. Other surfaces, such as Si, and other probes, such as those including transition metal atoms, have also been investigated. Among the better probes was C_5H_5N (pyridine). Quantum calculations suggest that pyridine is stable when attached to C_{60} in the orientation necessary for sensing the difference between hydrogen and fluorine. Half of C_{60} can form the end cap of a (9,0) or (5,5) carbon nanotube, and carbon nanotubes have been attached to an SPM tip. Thus, it might be possible using today's technology to build a system to read the diamond memory surface. Avouris has shown that individual hydrogen atoms can be removed from a silicon surface. If this could be accomplished in a gas that donates fluorine to vacancies on a diamond surface, the data storage system could be built. Thummel computationally investigated methods for adding a fluorine at the radical sites where a hydrogen atom had been removed from a diamond surface. As mentioned before, carbon nanotubes can be described as rolled up sheets of graphite. Different tubes can have different helical windings depending on how the graphite sheet is connected to itself. Theory suggests that single-walled carbon nanotubes can have metallic or semiconductor properties depending on the helical winding of the tube. Chico, Han, Menon, and others have computationally examined the properties of some of hypothetical devices that might be made by connecting tubes with different electrical properties.

Such devices are only few nanometers across — 100 times smaller than current computer chip features. Several authors, including Tour have described methods to produce conjugated macromolecules of precise length and composition. This technique was used to produce molecular electronic devices in mole quantities. The resultant single molecular wires were tested experimentally and found to be conducting.

The three and four terminal devices have been examined computationally and look promising. The features of these components are approximately 3 angstroms wide, about 750 times smaller than current silicon technology can produce. Helical logic is a theoretical proposal for a future computing technology using the presence or absence of individual electrons (or holes) to encode 1s and 0s. The electrons are constrained to move along helical paths, driven by a rotating electric field in which the entire circuit is immersed. The electric field remains roughly orthogonal to the major axis of the helix and confines each charge carrier to a fraction of a turn of a single helical loop, moving it like water in an Archimedean screw. Each loop could in principle hold an independent carrier, permitting high information density.

One computationally universal logic operation involves two helices, one of which splits into two "descendant" helices. At the point of divergence, differences in the electrostatic potential resulting from the presence or absence of a carrier in the adjacent helix controls the direction taken by a carrier in the splitting helix. The reverse of this sequence can be used to merge two initially distinct helical paths into a single outgoing helical path without forcing a dissipative transition. Because these operations are both logically and thermodynamically reversible, energy dissipation can be reduced to extremely low levels. ... It is important to note that this proposal permits a single electron to switch another single electron, and does not require that many electrons be used to switch one electron. The energy dissipated per logic operation can likely be reduced to less than 10^{-27} joules at a temperature of 1 Kelvin and a speed of 10 gigahertz, though further analysis is required to confirm this. Irreversible operations, when required, can be easily implemented and should have a dissipation approaching the fundamental limit of $\ln 2 \times kT$. One study not conducted by Ames or partners is particularly worth mentioning since it places a loose lower bound on the computational capabilities of molecular nanotechnology. Drexler designed a number of computer components using small diamondoid rods with knobs that allow or prevent movement to accomplish computation.

While this tiny mechanical Babbage Machine is probably not an optimal computational engine, its calculated performance for a desktop computer is 10^{18} MIPS — about a million times more powerful than the largest supercomputer that exists today. Note that with very fast computation energy use and heat dissipation become a severe problem. One approach to addressing this issue is reversible logic. Drexler proposed a nanotechnology based on diamond and investigated its potential properties. In particular, he examined applications for materials with a strength similar to that of diamond (69 times strength/mass of titanium). This would require a very mature nanotechnology constructing systems by placing atoms on diamond surfaces one or a few at a time in parallel. Assuming diamondoid materials, McKendree predicted the performance of several existing single-stage-to-orbit (SSTO) vehicle designs. The predicted payload to dry mass ratio for these vehicles using titanium as a structural material varied from < 0 (the vehicle won't work) to 36 per cent, i.e., the vehicle weighs substantially more than the payload.

With hypothetical diamondoid materials the ratios varied from 243 per cent to 653 per cent, i.e., the payload weighs far more than the vehicle. Using a very simple cost model ($1000 per vehicle kilogram) sometimes used in the aerospace industry, he estimated the cost per kilogram launched to low-Earth-orbit for diamondoid structured vehicles should be $153-412. This would meet NASA's 2020 launch to orbit cost goals. Estimated costs for titanium structured vehicles varied from $16,000-59,000/kg. Although this cost model is probably adequate for comparison, the absolute costs are suspect. Drexler used a more speculative methodology to estimate that a four passenger SSTO weighing three tons including fuel could be built using a mature nanotechnology. Using McKendree's cost model, such a vehicle would cost about $60,000 to purchase — the cost of today's high-end luxury automobiles. These studies assumed a fairly advanced nanotechnology capable of building diamondoid materials. In the nearer term, it may be possible to develop excellent structural materials using carbon nanotubes. Carbon nanotubes have a Young's modulus of approximately one terapascal— comparable to diamond. Issacs and Pearson proposed a space elevator—a cable extending from the Earth's surface into space with a center of mass at geosynchronous altitude. If such a system could be built, it should be mechanically stable and vehicles could ascend and descend along the cable at almost any reasonable speed using electric power (actually generating power on the way down). The first incredibly difficult problem with building a space elevator is strength of materials. Maximum stress is at geosynchronous altitude so the cable must be thickest there and taper exponentially as it approaches Earth. Any potential material may be characterised by the taper factor—the ratio between the cable's radius at geosynchronous altitude and at the Earth's surface.

Nanotechnology advocates and detractors are often preoccupied with the question "When?" There are three interrelated answers to this question:

1. The time-to-nanotechnology will be measured in decades, not years. While a few applications will become feasible in the next few years, programmable assembler/replicators and swarms will be extremely difficult to develop.
2. Nobody knows. There are far too many variables and unknowns. Beware of those who have excessive confidence in any date.
3. The time-to-nanotechnology is very sensitive to the level of effort expended. Resources allocated to developing nanotechnology are likely to be richly rewarded, particularly in the long term.

For steel the taper factor is tens of thousands—clearly impossible. For diamond, the taper factor is 21.9 including a safety factor. Diamond is, however, brittle. Carbon nanotubes have a strength in tension similar to diamond, but bundles of these nanometer-scale radius tubes shouldn't propagate cracks nearly as well as the diamond tetrahedral lattice. Thus, if the considerable problems of developing a molecular nanotechnology capable of making nearly perfect carbon nanotube systems approximately 70,000 kilometers long can be overcome, the first serious problem of a transportation system capable of truly large scale transfers of mass to orbit can be solved. The next immense problem with space elevators is safety—how to avoid dropping thousands of kilometers of cable on Earth if the cable breaks. Active materials may help by monitoring and repairing small flaws in the cable and/or detecting a major failure and disassembling the cable into small elements. Drexler calculates that lightsails made of 20 nm aluminum in tension should achieve an outward acceleration of ~14 km/s per day at Earth orbit with no payload and minimal structural overhead. For comparison, the delta V from low Earth to geosynchronous orbit is 3.8 km/s.

Lightsails generate thrust by reflecting sunlight. Tension is achieved by rotating the sail. The direction of thrust is normal to the sail and away from the Sun. By directing thrust along or against the velocity vector, orbits can be lowered or raised. This form of transportation requires no reaction mass and generates thrust continuously, although the instantaneous acceleration is small so sails cannot operate in an atmosphere and must be large for even moderate payloads.

Today, the smallest feature size in production systems is about 250 nanometers — the smallest feature size in computer chips. Since atoms are an angstrom or so across and carbon nanotubes have a diameter as small as 0.7 nanometers, atomically precise molecular machines can be smaller than current MEMS devices by two to three orders of magnitude in each dimension, or six to nine orders of magnitude smaller in volume (and mass). For example, the size of the kinesin motor, which transports material in cells, is 12 nm. Han computationally demonstrated that molecular gears fashioned from single-walled carbon nanotubes with benzyne teeth should operate well at 50-100 gigahertz. These gears are about two nanometers across. Han computationally demonstrated cooling the gears with an inert atmosphere. Srivastava simulated powering the gears using alternating electric fields generated by a single simulated laser. In this case, charges were added to opposite sides of the tube to form a dipole. To make active materials, a material might be filled with nano-scale sensors, computers, and actuators so the material can probe its environment, compute a response, and act. Although this document is concerned with relatively simple artificial systems, living tissue may be thought of as an active material. Living tissue is filled with protein machines which gives living tissue properties (adaptability, growth, self-repair, etc.) unimaginable in conventional materials.

Active materials can theoretically be made entirely of machines. These are sometimes called swarms since they consist of large numbers of identical simple machines that grasp and release each other and exchange power and information to achieve complex goals. Swarms change shape and exert force on their environment under software control. Although some physical prototypes have been built, at least one patent issued, and many simulations run, swarm potential capabilities are not well analysed or understood. We briefly discuss some concepts here. Michael proposes brick-shaped machines of various sizes that slide past each other to assume a variety of shapes. He has generated a large number of videos showing computer simulations of simple motions. Although his web site contains rather extravagant claims, this work has received a U.K. patent. Yim built a small swarm with macroscopic (size in inches) components called polypod, built a simulator of polypod, and programmed it to move in various ways to study locomotion. There are two brick shaped components in polypod, one of which has two prismatic joints linked by a revolute joint. The second component is a cubic connector with no mechanical motion. Polypod is programmed by tables for each member of the swarm. Each member is programmed to move at various speeds in each degree of freedom for certain amounts of time. The swarm components are implicitly synchronised so there is no clock signal.

Hall proposes a swarm with 10 micron dodecahedral components each with 12 arms that can move in and out, rotate a little, and grab and release each other. This concept is called the "utility fog." Hall estimates that the utility fog would have a density of 0.2, tensile strength of 1000 psi in action and 100,000 psi in a passive mode, and have a maximum shear rate of 100 km/second/meter. Bishop proposes a swarm consisting of 100 nanometer brick-shaped components that slide past each other to change shape. Globus proposes a swarm with two kinds of components —

edges and nodes. The terms "node" and "edge" are chosen to correspond to those in graph theory. The roughly spherical nodes are capable of attaching to five edges (for a tetrahedral geometry with one free edge per node) and rotating each edge in pitch and yaw. The rod-like edges are capable of changing length, rotating around their long axis, and attaching/detaching to/from nodes. Component design, power distribution and control software are significant challenges for swarm development. Consider that with 10 micron components a cubic meter of swarm would contain about 10^{15} devices, each with an internal computer communicating with its neighbors to accomplish a global task. NASA's mission is divided into five enterprises: Mission to Planet Earth, Aeronautics, Human Exploration and Development of Space, Space Science, and Space Technology. We will examine some potential nanotechnology applications in each area. The Earth Observing System (EOS) will use satellites and other systems to gather data on the Earth's environment. The EOS data system will need to process and archive >terabyte per day for the indefinite future. Simply storing this quantity of data is a significant challenge—each day's data would fill about 1,000 DVD disks. With projected nanocomputer processing speeds of 10^{18} MIPS, a million calculations on each byte of one day's data would take one second on the desktop. Given a mature nanotechnology, it should be possible to build sensors in balloon-borne systems approximately the size of bacteria.

With replication based manufacturing, these should be quite inexpensive. If the serious communication and control problems can be solved, one can imagine spreading billions of tiny lighter-than-air vehicles into the atmosphere to measure wind currents and atmospheric composition. A similar approach might be taken in the oceans—note that the oceans are full of floating microscopic living organisms that can sense and react to their environment. Smart dust might sense the environment, note the location via a GPS-like system, and store that information until close enough to a data-collection point to transfer the data to the outside world. The strength of materials and computational capabilities previously discussed for space transportation should also allow much more advanced aircraft. Stronger, lighter materials can obviously make aircraft with greater lift and range. More powerful computers are invaluable in the design stage and of great utility in advanced avionics. MEMS technology has been used to replace traditional large control structures on aircraft with large numbers of small MEMS controlled surfaces.

This control system was used to operate a model airplane in a windtunnel. Nanotechnology should allow even finer control—finer control than exhibited by birds, some of which can hover in a light breeze with very little wing motion. Nanotechnology should also enable extremely small aircraft.

A reasonably advanced nanotechnology should be able to make simple atomically precise materials under software control. If the control is at the atomic level, then the full range of shapes possible with a given material should be achievable. Aircraft construction requires complex shapes to accommodate aerodynamic requirements. With molecular nanotechnology, strong complex-shaped components might be manufactured by general purpose machines under software control.

The aeronautics mission is responsible for launch vehicle development. Payload handling is an important function. Very efficient payload handling might be accomplished by a very advanced swarm. The sequence begins by placing each payload on a single large swarm located next to the shuttle orbiter. The swarm forms itself around the payloads and then moves them into the payload bay, arranging the payloads to optimize the center of gravity and other considerations. The swarm holds the payload in place during launch and may even damp out some launch vibrations. On orbit, satellites can be launched from the payload bay by having the swarm give them a gentle push. The swarm can then be left in orbit, perhaps at a space station, and used for orbital operations. This scenario requires a very advanced swarm that can operate in an atmosphere and on orbit in a vacuum. Besides the many and obvious difficulties of developing a swarm for a single environment, this provides additional challenges. Note that a simpler swarm might be used for aircraft payload handling. Aerospace vehicles often require complex checkout procedures to insure safety and reliability. This is particularly true of reusable launch vehicles. A very advanced swarm with some special purpose appendages might be placed on a vehicle. It might then spread out over the vehicle and into all crevices to examine the state of the vehicle in great detail. Nanotechnology-enabled Earth-to-orbit transportation has the greatest potential to revolutionize human access to space by dropping the current $10,000 per pound cost of launch, but this was discussed above. Other less dramatic technologies include:

Space structures with a long design life (such as space station modules) need high-reliability materials that do not degrade. Active materials might help. The machines monitor structural integrity at the sub-micrometer scale. When a portion of the material becomes defective, it could be disassembled and then correctly reassembled. It should be noted that bone works somewhat along these lines. It is constantly being removed and added by specialised cells.

To effect timely repairs, space stations require a large store of spare parts and tools that are rarely used. A mature nanotechnology might create a "matter compiler," a machine that converts raw materials into a wide variety of products under software control. Contemporary examples of very limited matter compilers are numerically controlled machines and polypeptide sequencers. With a substantially more capable nanotechnology-based matter compiler, a space station crew could simply make spare parts and tools as needed. The programmes could be stored on-board or on the ground. New tools invented on Earth could be transferred as software to the station for manufacture. Once used, unneeded tools and broken parts could be ionised in a solar furnace, transferred using controlled magnetic fields, and the constituent atoms stored for later manufacture into new products.

An advanced nanotechnology might be able to build filters that dynamically modify themselves to attract the contaminant molecules detected by the air and water quality sensors. Once attached to the filter, the filter could in principle move the offending molecules to a molecular laboratory for modifications to useful or at least inert products. A swarm might implement such an active filter if it was able to dynamically manufacture proteins that could bind contaminant molecules. The protein and bound contaminant might then be manipulated by the swarm for transportation. With a sufficiently advanced nanotechnology it might even be possible to directly generate food by non-biological means. Then agriculture waste in a self-sufficient space colony could be converted directly to useful nutrition. Making this food attractive will be a major challenge.

Sleeping crew members in the shuttle experience considerable pain and sleep disruption when the reaction control system fires and they collide with the cabin walls. If crew members were connected to the walls by a swarm, the swarm could absorb most or all of the force before the crew member struck the wall. The swarm could then gradually return the crew member to center (without the oscillations associated with bungee cords) in preparation for the next firing.

For resupply, spacecraft docking is a frequent necessity in space station operations. When two spacecraft are within a few meters of each other, a swarm could extend from each, meet in the middle, and form a stable connection before gradually drawing the spacecraft together. A swarm could support space-suited astronauts in simulated partial-g environments by holding them up appropriately. The swarm moves in response to the astronaut's motion providing the appropriate simulation of partial or 0 gravities. Tools and other objects are also manipulated by the swarm to simulate non-standard gravity.

Active nanotechnology materials (see active materials) might enable construction of a skin-tight space suit covering the entire body except the head (which is in a more conventional helmet). The material senses the astronaut's motions and changes shape to accommodate it. This should eliminate or substantially reduce the limitations current systems place on astronaut range of motion.

In situ resource utilisation is undoubtedly necessary for large scale colonisation of the solar system. Asteroids are particularly promising for orbital use since many are in near Earth orbits. Moving asteroids into low Earth orbit for utilisation poses a safety problem should the asteroid get out of control and enter the atmosphere. Very small asteroids can cause significant destruction. The 1908 Tunguska explosion, which was calculated to be a 60 meter diameter stony asteroid, leveled 2,200 km^2 of forest. Hills calculated that 4 meter diameter iron asteroids are near the threshold for ground damage. Both these calculations assumed high collision speeds. At a density of 7.7 g/cm^3 Babadzhanov, a 3 meter diameter asteroid should have a mass of about 110 tons. Rabinowitz estimates that there are about one billion ten meter diameter near Earth asteroids and there should be far more smaller objects. For colonisation applications one would ideally provide the same radiation protection available on Earth. Each square meter on Earth is protected by about 10 tons of atmosphere. Therefore, structures orbiting below the van Allen belts would like 10 tons/meter^2 surface area shielding mass. This would dominate the mass requirements of any system and require one small asteroid for each 11 meter^2 of colony exterior surface area. A 10,000 person cylindrical space colony such as Lewis One with a diameter of almost 500 meters and a length of nearly 2000 meters would require a minimum of about 90,000 retrieval missions to provide the shielding mass. The large number of missions required suggests that a fully automated, replicating nanotechnology may be essential to build large low Earth orbit colonies from small asteroids. A nanotechnology swarm along with an atomically precise lightsail is a promising small asteroid retrieval system. Lightsail propulsion insures that no mass will be lost as reaction mass. The swarm can control the lightsail by shifting mass. When a target asteroid is found, the swarm spreads out over the surface to form a bag. The interface to the sail must be active to account for the rotation of the asteroid—which is unlikely to have an axis-of-rotation in the proper direction to apply thrust for the return to Earth orbit. The active interface is simply swarm elements that transfer between each other to allow the sail to stay in the proper orientation. Of course, there are many other possibilities for nanotechnology based retrieval vehicles.

Extraterrestrial materials brought into orbit could be fed into a high-temperature solar furnace and partially ionised. Magnetic fields might then be used to separate the nuclei. These are fed in appropriate quantities to a matter-compiler to build the products desired.

Several authors, including Freitas have speculated that a sufficiently advanced nanotechnology could examine and repair cells at the molecular level. Should this capability become available—presumably driven by terrestrial applications—the small size and advanced capabilities of such systems could be of great utility on long duration space flights and on self-sufficient colonies.

Self-replicating systems permit efforts of great scope to be pursued economically. Adjusting the environment on another planet to suit the tastes of humans is one such undertaking. Heating and cooling can be achieved by (among many other methods) using space-based mirrors. Chemical modifications of the planetary surface and atmosphere can be achieved in relatively short periods by the use of self-replicating systems that absorb sunlight and raw materials, and convert them into the desired products. Much as plants changed the environment of the earth to what we see today, so self-replicating molecular manufacturing systems might more rapidly convert the environments of other planets.

As interstellar trips might last many years, the ability to conserve supplies by maintaining some crew members in a suspended state would be useful. An extremely advanced nanotechnology might use molecular manipulations of each cell to provide (a) better methods of slowing or suspending the metabolic activity of crew members and (b) better methods of restoring metabolic activity to a normal state when the destination is reached.

Molecular manufacturing should enable the creation of very precise mirrors. Unlike lightsail applications, telescope mirrors require a very precise and somewhat complicated shape. A swarm with special purpose appendages capable of bonding to the mirror might be able to achieve and maintain the desired shape.

A very advanced nanotechnology would be capable of imaging and then removing the surface atoms of an extra-terrestrial sample. By removing successive surface layers the location of each atom in the sample might be recorded, destroying the sample in the process. This data could then be sent to Earth. Besides requiring a very advanced nanotechnology, there is a more fundamental—but not necessarily fatal—problem: as the outside layer of atoms is removed the next layer may rearrange itself so the sample is not necessarily perfectly recorded.

Solar Power—Low Earth orbit spacecraft generally depend on solar cells and batteries for power. For energy collection, molecular manufacturing can be used to make solar photovoltaic cells at least as efficient as those made in the laboratory today. Efficiencies can therefore be > 30 per cent. In space applications, a reflective optical concentrator need consist of little more than a curved aluminum shell < 100 nanometers thick (photovoltaic cells operate with higher efficiency at high optical power densities). A metal fin with a thickness of 100 nanometers and a conduction path length of 100 microns can radiate thermal energy at a power density as high as 1000 W/m^2 with a temperature differential from base to tip of < 1 K. Accordingly, solar collectors can consist of arrays of photovoltaic cells several microns in thickness and diameter, each at the focus of a mirror of ~100 micron diameter, the back surface of which serves as a ~100 micron diameter radiator. If the mean thickness of this system is ~1 micron, the mass is ~10^{-3} kg/m^2 and the power per unit mass, at Earth's distance from the Sun, where the solar constant is ~1.4 kW/m^2, is > 10^5 W/kg." By comparison, the U.S. built Photovoltaic Panel Module solar cells currently used on the Mir Space Station and planned for use on the International Space Station generate about 118 W/kg.

Fuel Cells—A critical component in hydrogen/oxygen fuel cells is the PEM (Proton Exchange Membrane). This membrane must (a) permit the passage of protons while (b) blocking everything else. Present membranes do a rather poor job. One group at Ames is designing and computationally testing PEMs to study possible energy mechanisms in early life. While these studies are not meant to design optimal membranes for fuel cell use, the basic knowledge and approach may be of value. Another proposal is to design a diamond membrane a few nanometers thick with "proton pores." The pores might be lined with fluorine, oxygen and nitrogen to create a region with a high proton affinity. In addition, a positionally controlled platinum might be held at the mouth of the pore to verify that H_2 can be catalytically split into H+ and e–, and that the barrier for migration of the H+ into the pore is modest in size. Nanotechnology must provide precise control over the manufacturing process of the diamondoid PEM since the pores must be made very precisely.

Studies of H_2 absorption and packing in carbon nanotubes and nanoropes are in progress at NASA Ames and elsewhere. Nanotubes provide large pore sizes and nanoropes have different pore sizes depending on interstitial and other locations. Dillon estimated that the single walled nanotubes in their sample contained 5 to 10 per cent by weight of H_2. The nanotubes were about 0.1 to 0.2 per cent by weight of the total sample. Computational studies at Ames suggest that to store 7-10 per cent H_2 in single walled nanotubes at room temperature the H_2s must be stored inside the tubes, not merely adsorbed on the walls. This work suggests that carbon nanotubes might be developed into an excellent H_2 storage medium within 3-5 years. Calculations with oxygen suggest that a diamondoid sphere ~0.1 microns in diameter should easily hold oxygen at ~1,000 atmospheres. While higher pressures are feasible, they offer declining returns. At higher pressures, the pressure-volume relationship becomes severely non-linear and the density approaches a limiting value. Other gases might also be stored if diamondoid spheres can be built, but the analysis has not been done. High strength light-weight materials will allow greater efficiency of energy storage as angular momentum.

Many kinds of ultraminiature electromechanical devices have utility on a miniaturised space craft. It has been shown that manipulating carbon nanotubes changes their electrical properties. This might be exploited to build nanometer scale strain devices. This may be achievable within 3-5 years, and simulations along these lines are in progress. Similar results have been achieved experimentally with C_{60}. The electrical properties of a C_{60} molecule were changed by applying pressure to the molecule with an SPM tip.

Smaller, lighter spacecraft are cheaper to launch (current costs are about $10,000/lb) and generally cheaper to build. Diamondoid structural materials can radically reduce structural mass, miniaturised electronics can shrink the avionics and reduce power consumption, and atomically precise materials and components should shrink most other subsystems.

Thermal protection is crucial for atmospheric reentry and other tasks. The carbon nanotubes under investigation at NASA Ames and elsewhere may play a significant role. Most production processes for carbon nanotubes create a tangled mat of nanotubes that has a very low mass-to-volume ratio. Like graphite, the tubes should withstand high temperatures but the tangled mat should prevent them from ablating. This may lead to high temperature applications. Many of the applications discussed here are speculative to say the least. However, they do not appear to violate the laws of physics. Something similar to these applications at these performance levels should be feasible if we can gain complete control of the three-dimensional structure of materials, processes and devices at the atomic scale. How to gain such control is a major, unresolved issue. However, it is clear that computation will play a major role regardless of which approach —positional control with replication, self-assembly, or some other means —is ultimately successful. Computation has already played a major role in many advances in chemistry, SPM manipulation, and biochemistry. As we design and fabricate more complex atomically precise structures, modeling and computer aided design will inevitably play a critical role. Not only is computation critical to all paths to nanotechnology, but for the most part the same or similar computational chemistry software and expertise supports all roads to molecular nanotechnology.

Thus, even if NASA's computational molecular nanotechnology efforts should pursue an unproductive path, the expertise and capabilities can be quickly refocused on more promising avenues as they become apparent. As nanotechnology progresses we may expect applications to become feasible at a slowly increasing rate. However, if and when a general purpose programmable assembler/replicator can be built and operated, we may expect an explosion of applications. From this point, building new devices will become a matter of developing the software to instruct the assembler/replicators. Development of a practical swarm is another potential turning point. Once an operational swarm that can grow and divide has been built, a large number of applications become software projects. It is also important to note that the software for swarms and assembler/replicators can be developed using simulators — even before operational devices are available.

3

Historical Evolution and Development of Nanotubes: Towards Understanding Their Commercial Uses

Explaining Historical Evolution and Development of Nanotubes

The beginning of this story has been told many times. In 1985, a confluence of events led to an unexpected and unplanned experiment with a new kind of microscope resulting in the discovery of a new molecule made purely of carbon – the very element chemists felt there was nothing more to learn about. Buckyballs – sixty carbon atoms arranged in a soccer ball shape – had been discovered and the chemical world, not to mention the physical and material worlds, would never be the same. In fact, what had been discovered was not just a single new molecule but an infinite class of new molecules: the fullerenes. Each fullerene—C_{60}, C_{70}, C_{84}, etc.—possessed the essential characteristic of being a pure carbon cage, each atom bonded to three others as in graphite. Unlike graphite, every fullerene has exactly 12 pentagonal faces with a varying number of hexagonal faces (e.g., buckyball – C_{60} – has 20). Some fullerenes, like C_{60}, were spheroidal in shape, and others, like C_{70}, were oblong like a rugby ball. Dr. Richard Smalley recognised in 1990 that, in principle, a tubular fullerene should be possible, capped at each end, for example, by the two hemispheres of C_{60}, connected by a straight segment of tube, with only hexagonal units in its structure. Millie Dresselhaus, upon hearing of this concept, dubbed these imagined objects "buckytubes." In actuality, however, carbon nanotubes had been discovered 30 years earlier, but had not been fully appreciated at that time. In the late 1950s, Roger Bacon at Union Carbide, found a strange new carbon fibre while studying carbon under conditions near its triple point. He observed straight, hollow tubes of carbon that appeared to consist in graphitic layers of carbon separated by the same spacing as the planar layers of graphite. In the 1970s, Morinobu Endo observed these tubes again, produced by a gas-phase process. Indeed, he even observed some tubes consisting in only a single layer of rolled-up graphite. In 1991, after the discovery and verification of the fullerenes, Sumio Iijima of NEC observed multiwall nanotubes formed in a carbon arc discharge, and two years later, he and Donald Bethune at IBM independently observed single-wall nanotubes – buckytubes. These pure carbon polymers could now be understood in the context of fullerenes, changing the perception of them to molecules, with all that special designation implies. Nanotubes had been fullerenised.

Now, ten years after Iijima's initial observation, we know a great deal about nanotubes and tubular fullerenes. We know that multiwall nanotubes are invariably produced with a high frequency of structural defects. (In comparison with their larger relations, the 5-20 micron-diameter graphite fibers used in aerospace and sporting goods applications, multiwall nanotubes are quite sound structurally; nevertheless, they frequently contain regions of structural imperfection.) As any material scientist knows, it is the occurrence of defects that inevitably degrade the material properties of a substance, such as strength. The intrinsic properties of a material may be world-beater, but typically the actual properties of the bulk material are only a few percent of what the material would exhibit if it were structurally perfect. For example, a structural defect such as a microcrack in a steel wire, will lead to catastrophic failure at 1-2% of the theoretical breaking strength one would predict based on fundamental chemical principals.

In contrast, buckytubes are fullerenes, and are thus molecules: perfect, hollow molecules of pure carbon linked together in a hexagonally bonded network to form the hollow cylinder. The tube is seamless, with either open or capped ends. The diameter of single-wall carbon nanotubes is 0.7 to 2 nm (typically about 1.0 nm) – 100,000 times thinner than a human hair. Buckytube lengths are typically hundreds of times their diameters. The molecular aspect of buckytubes is critical. Every atom is in the right place. This is a profound difference with their larger, defective cousins. To a chemist, a molecule is a very special thing. A molecule is complete, and is usually relatively happy with its identity. When a molecule is confronted with the opportunity to change, that is, to undergo a chemical reaction with other bits of stuff, there is almost always a fairly significant barrier to overcome.

Other, non-molecular "stuff", when confronted with other stuff, is usually quick to change, and add in with the new part to make a larger whole. Metals are like this: a chunk of metal exposed to more metal (this could take place in the molten, gaseous, or solid phase) will accommodate the addition because it does not possess molecular completeness and invariance. One aspect of molecular invariance is the reliability and predictability of chemical change. Molecules can be induced, by various means, to overcome their barriers to change, for example using heat. But by and large the reaction products when dealing with molecules are consistent. This is not true of changes that non-molecular things undergo. No two chunks of metal ever look the same, just as no two snowflakes look the same: snowflakes are not molecules. The effect of molecular invariance on material properties is equally profound. Whatever the intrinsic nature of the material, there are no defects to degrade the properties. You get what you get.

Understanding the Structure, Name and Properties of Nanotubes

The special nature of carbon combines with the molecular perfection of buckytubes (single-wall carbon nanotubes) to endow them with exceptionally high material properties such as electrical and thermal conductivity, strength, stiffness, and toughness. No other element in the periodic table bonds to itself in an extended network with the strength of the carbon-carbon bond. The delocalised pi-electron donated by each atom is free to move about the entire structure, rather than stay home with its donor atom, giving rise to the first molecule with metallic-type electrical conductivity. The high-frequency carbon-carbon bond vibrations provide an intrinsic thermal conductivity higher than even diamond. In most materials, however, the actual observed material properties—strength, electrical conductivity, etc.—are degraded very substantially by the occurrence of defects in their structure. For example, high strength steel typically fails at about 1% of its theoretical breaking strength. Buckytubes, however, achieve values very close to their theoretical limits because of their perfection of structure—their molecular perfection. This aspect is part of the unique story of buckytubes. Buckytubes are an example of true nanotechnology: only a nanometer in diameter, but molecules that can be manipulated chemically and physically. They open incredible applications in materials, electronics, chemical processing and energy management. Buckytubes are single-wall carbon nanotubes, in which a single layer of graphite—graphene—is rolled up into a seamless tube. Graphene consists of a hexagonal structure like chicken wire. If you imagine rolling up graphene or chicken wire into a seamless tube, this can be accomplished in various ways. For example, carbon-carbon bonds (the wires in chicken wire) can be parallel or perpendicular to the tube axis, resulting in a tube where the hexagons circle the tube like a belt, but are oriented differently. Alternatively, the carbon-carbon bonds need not be either parallel or perpendicular, in which case the hexagons will spiral around the tube with a pitch depending on how the tube is wrapped. There is a straightforward labelling convention to distinguish differently wrapped tubes from one another. The mapping specifies the number of unit vectors required to connect two atoms in the planar hexagonal lattice to form a seamless tube. These numbers specify a "vector" for the mapping, commonly expressed as (m,n), where m and n are integers. These numbers constitute a unique "name" for a tube. Any tube "named" (n,0) has carbon-carbon bonds that are parallel to the tube axis, and form, at an open end, a "zig-zag" pattern; these tubes are referred to as "zig-zag" tubes. Tubes named (n,n), where the two integers are equal, have carbon-carbon bonds that are perpendicular to the tube axis, and are often called "armchair" tubes. These two basic types are achiral, meaning they do not have a distinct mirror-image, like left and right hands. All the other tubes, named (m,n), where m does not equal n, and neither is 0, are chiral, and have left-and right-handed variants. In most respects, the properties of tubes of different types are essentially the same. The exception to this is in their electrical conductivity, where these subtle structural differences can have profound effects. For example, all armchair tubes—i.e., where m=n—have truly metallic electrical conductivity. They transport electrons along the tube axis just as metals do, without a single atom of metal in their structure! This behaviour in a molecule is unprecedented. In contrast, the other tubes are intrinsically semiconducting, either with a very small band gap of a few meV, or with moderate band gaps on the order of 1 eV. The rule here is that those tubes where (n-m) is a multiple of 3 are the small-gap type, while the others have medium gaps. Since tubes with different (n,m) are molecularly distinct, there exists the possibility of chemically separating different types, and even of growing only selected tube types, although currently, all production processes produce a random mixture. Another structural aspect of tubes is their self-organisation into "ropes," which consist in many (typically, 10-100) tubes running together along their length in van der Waals contact with one another. Ropes are far longer than any individual tube in them: whereas tubes are typically about 100-1000 nm in length (and about 1 nm in diameter) ropes are essentially endless, branching off from one another, then joining others.

Qualities of Nanotube Flexural Strain Sensing & Carbon Nanotube Film

The sensor/actuator embedded smart materials and systems are essential for structural damage detection and control. There is need for smart structural materials, which can be used to sense strains, and adjust the structural properties so as to achieve desired response. Carbon nanotube composites have the potential to be such a smart structural material. Results (Baughman *et al.*, 1999) have shown that when carbon nanotube films, also called as "buckypapers", are used as actuators, large actuator strains can be achieved by smaller operating voltages as compared with ferroelectric and electrostrictive materials. Single wall carbon nanotubes (SWCNTs) promise to be a reinforcement material for nanocomposites because of their superior mechanical properties. Researchers (Hadjiev *et al.*, 2001; Zhao *et al.*, 2001) have studied the Raman band shift in SWCNT's as a function of stress/strain in carbon nanotubes. Raman studies on multi-wall carbon nanotubes have also been carried out (Schadler *et al.*, 1998; Wagner *et al.*, 1998). Though the experiments are carried at nanoscale, the results illustrate the potential of SWCNT as strain/stress sensor. It has also been shown that the electrical band gap changes as a function of axial compression, tension stretch, torsion, and bending strain (Peng and Cho, 2002; Tombler *et al.*, 2000; Yang and Han, 2000). Strong dependence of SWCNT's band structure on mechanical deformation makes it possible to develop nano-electro-mechanical sensors. Most of the studies to date relate the mechanical deformation with the change in electrical properties at the nanoscale. In this study, change in the electronic property of carbon nanotubes due to strain at macroscale is demonstrated using the experimental results. Such carbon nanotube films have randomly oriented SWCNTs with isotropic strain sensing properties; thus, have multidirectional and multi-location strain sensing capability.

The carbon nanotube films used in this study are produced by mixing SWCNTs with 0.25?mg/ml *N,N*-Dimethylformamide (DMF) and filtering the mixture through a 0.2?mm teflon membrane. Freestanding carbon nanotube film (Buckypaper) is peeled from the filter after rinsing and drying of the remaining material. The carbon nanotube film is further dried for 24?h under vacuum. The SEM picture of a carbon nanotube film with ~15?ìm thickness. To explore the strain sensing potential of the carbon nanotube film, it is attached to a 0.25×2×24?in. brass specimen and subjected to pure bending moment using two-point loading. For insulation, a PVC film is attached between the brass specimen and carbon nanotube film. High strength epoxy and a vacuum-bonding method is used to attach carbon nanotube film and PVC to the brass specimen in order to ensure the perfect strain transfer – between brass, PVC, and the carbon nanotube film. A conventional electrical resistance strain gage is attached next to the carbon nanotube film to measure flexural strain for comparison purposes. Load P is applied in increments using a servo-hydraulic machine. The corresponding change in voltage across the carbon nanotube film is measured using four-point probe measurement (Smits, 1958). Input voltage across two outer probes is kept constant during the measurement and change in voltage across the two inner probes is recorded. Strain measurements are also made using conventional strain gage. The change in voltage is proportional to the change in resistivity of the film (Dharap *et al.*, 2004). The change in voltage due to the change in dimensions of the film is small as compared to the change in voltage due to the change in resistivity of the film (Li *et al.*, 2003).

The change in voltage in the carbon nanotube film as a function of measured flexural strain from the conventional strain gage. As the strain measured by the conventional strain gage increases from 0 to ~600?ìm/m the voltage change measured across the two outer probes increases from 0 to ~200?ìV. A nearly linear trend between the change in voltage and the strain is evident; although, deviation from the linear trend exists. Further study is needed to establish causes for deviation from linear trend such as temperature and gas exposure history (Bezryadin *et al.*, 1998; Collins and Avouris, 2000; Hone *et al.*, 2002). The change in voltage is measured by moving the four-point probe to several parallel locations on a single carbon nanotube film, which also yielded a similar linear trend. It is evident from the experimental study that the carbon nanotube film can be used for measuring flexural strains at the macroscale. Carbon nanotube films are made up of randomly oriented SWCNTs; hence, their electronic properties are independent of direction. Hence taking measurement along different directions provides corresponding strains. Carbon nanotube films can also be integrated into composites to simultaneously act as strong structural material and sensor. Such composites may find application in smart structures.

Towards Nanotube Manufacturing on a Commercial Scale

Hardly a day goes by without someone suggesting a new application for carbon nanotubes. Since their discovery, academics and industrialists have researched or considered their use in brake discs; fuel cells; advanced aerospace composites; co-axial cable; conductive fuel lines; EMI (Electromagnetic Interference) shielding in electronic devices; conductive tyres; conductive inks; and, of course, the hotly disputed space elevator – to name but a few. While some of these applications are clearly closer to reality than others, possibly the most immediate products will be developed in the coatings industry. The launch in April 2004 of the UK's first commercial manufacturing process for high-purity single-wall carbon nanotubes at Thomas Swan & Co. Ltd., has cleared the way for the development of a wide range of advanced coatings applications. Carbon nanotubes offer a range of attractive properties and could enhance many coatings applications. However, the current barriers to mass-market use of these materials include the availability of high-quality materials in commercial quantities. Now a partnership between industry and academia has resulted in a commercial manufacturing process for carbon nanotubes. For those who have not come across them before, carbon nanotubes consist of molecular cylinders of pure, hexagonally-arranged carbon atoms that resemble rolled-up sheets of chicken wire with a diameter measured in a few nanometres (1 nanometre is 1 billionth of a metre) and a length of many microns. They occur in two main types, the single-wall carbon nanotube (SWNT) composed of a single cylinder of carbon, and the multi-wall version (MWNT) consisting of concentric tubes or cylinders of carbon (effectively straws within straws). The ends of the tubes are usually closed off by a carbon end-cap. Other variations on this theme include the double-wall tube, 'herringbone' and 'bamboo' structures. Since their discovery, and Sumio Iijima's watershed paper in 1991, there has been unprecedented academic and industrial interest in carbon nanotubes and their potential use in a wide range of commercial applications.

There is no denying that the basic properties of both single and multi-wall carbon nanotubes are truly remarkable. Recent studies have suggested that single-wall carbon nanotubes have a tensile strength of 50-100 GPa and a modulus of 1-2 TPa. This places them well ahead of steel in both strength and modulus and all for one sixth the weight. Combine this with their high thermal and electrical conductivity and you begin to understand why so many scientists and commercial organisations have sat up and taken notice. The basic properties of the individual carbon nanotube (both single and multi-wall) have been considered in some fascinating potential applications. In essence carbon nanotubes are advanced fillers—highly conductive (thermally and electrically), extremely strong and light-weight. Add to this their apparent ability to absorb electro-magnetic interference and the fascinating photo-acoustic effect (whereby they spontaneously ignite when exposed to an intense light source) and you are left with some intriguing possibilities for advanced coatings. The addition of low concentrations of carbon nanotubes to plastic it makes it electrically conductive. This allows electrostatic painting to be used when coating automotive parts, removing the need for costly primers. The high aspect ratio of carbon nanotubes (length divided by diameter) means that they form a percolating matrix across the non-conductive filler at much lower loadings than traditional fillers. Carbon black typically requires 30 or 40% by volume loading to confer conductivity, at which point the mechanical properties of the composite are often severely degraded. In comparison Cambridge University recently achieved a conductive epoxy using only 0.005 wt% loading of aligned multi-wall carbon nanotubes. It is also likely that modest loadings of single-wall carbon nanotubes will actually enhance the structural properties of the composite. The anti-static properties of carbon nanotube coatings may also have potential applications in packaging. Anti-static coatings are used in the electronics to prevent damage to sensitive electronic components during shipping and storage.

Carbon nanotubes (in their pure and undamaged form) are possibly the best thermally conductive material ever discovered. As such they are being researched for their heat management potential in a number of applications. As the electronics world strives to fit more and more functionality into less and less space, overheating is becoming a major problem. This is forcing greater pressure on materials and coatings involved in electronics manufacturing to be thermally conductive in an attempt to remove the heat from the device as quickly as possible. Another aspect of thermal management is the demand for rapid curing times of polymeric materials such as ink resins and elastomers. The use of low loadings of carbon nanotubes should increase the speed of thermal dissipation from a coating or resin allowing for a much faster curing time. This would clearly be of interest in automated processes where the speed of curing limits the rate of production. Electro-magnetic interference (EMI) shielding in laptops and mobile phones is becoming increasingly important to preventing interference with and from other portable electronic devices. As an extension to the electrostatic dissipation properties of carbon nanotube coatings, a number of companies are developing coatings that are designed to absorb EMI. While there are already some products in the market-place and a whole range of applications in development, the carbon nanotube market has been constrained by three main issues:

(i) A lack of commercially-available material of consistently high quality.

(ii) Cost—until recently single-wall carbon nanotubes were selling for EUR

400/g which is clearly some way from commercial viability as a bulk material.

(iii) Successfully expressing the basic properties of the raw material in the end application – the incredibly small scale of the material poses some interesting challenges for advanced material and coatings science.

Such teething problems are not new. Carbon fibres took many years to be widely accepted in the materials world from both a cost and performance aspect and have only just made a regular appearance in popular sporting markets such as golf equipment. Despite this, some early applications of carbon nanotubes in products such as conductive fuel lines in the automotive industry have already made a successful appearance on the market. In addition, recent advances in the manufacture of carbon nanotubes on an industrial scale will allow vital commercial research and evaluation to be conducted and this will open the way to a wide range of fully industrialised applications. The UK's first commercial manufacturing process for high-purity single-wall carbon nanotubes was announced by Thomas Swan & Co in April 2004. The plant is the result of four years of collaboration between Thomas Swan and Cambridge University's Department of Material Science and Department of Chemistry. The unique collaboration brought the academic expertise of Cambridge University together with over 77 years of chemical manufacturing experience from Thomas Swan. The goal of the project was to supply high-purity single and multi-wall carbon nanotubes in commercial quantities and to back up this initial offering with the ability to scale up the process to keep up with demand as the industrial market expanded. The problem was that at the start of the programme no-one knew which production method would supply both single and multi-wall tubes and be fully scalable. It was this problem that was presented to Cambridge University. Carbon nanotubes can be manufactured using a variety of methods:

- Laser ablation uses a high-power laser to vaporise a graphite source loaded with a metal catalyst. The carbon in the graphite reforms as predominantly single-wall nanotubes on the metal catalyst particles.
- Arc discharge involves an electrical discharge from a carbon-based electrode in a suitable atmosphere to produce both single and multi-wall tubes of high quality but in low quantities.
- Chemical vapour deposition (CVD) is where a hydrocarbon feedstock is reacted with a suitable metal-based catalyst in a hot furnace to 'grow' nanotubes which are subsequently removed from the substrate and catalyst by a simple acid wash.

This concept was then taken to design stage (and patented by Cambridge University). Using a team of chemical engineers and an independent consultant, the plant was built at Swan's site in North East England towards the end of 2003. Over the following months the plant was commissioned and then optimised until the material that was being produced was of consistently high quality. A consistent supply of high-purity material that is also scaleable is a major stepping stone for the developing carbon nanotubes market. It allows companies to conduct meaningful industrial research using commercial quantities of raw material. In addition, this research can now be done with the knowledge that, should an application prove itself, there is a scalable source of the same raw material available to supply any ongoing development. Combine this with lowering prices (Thomas Swan predicts that within a few years the price of single-wall carbon nanotubes with be measured in tens of GBP per kg) and the market should start to take off.

A major problem in achieving a saleable product that is on average 100,000 times smaller than a human hair is that it is rather hard to 'see' what you have made. The advent of nanomaterial manufacturing brings to light the necessity for advanced industrial and cost-effective analytical techniques. While Scanning Electron Microscopes (SEM) provide the ability to view the sample in remarkable detail, it only shows a minute fraction of the product. The analogy often used is like looking at a football pitch through a jam jar. The grass beneath you is green and healthy but you have no idea what it is like on the other side of the pitch.

This will not be an easy problem to solve and while RAMAN techniques (which involve studying the spectrum profile of a reflected laser from your sample) offer some hope, it will be some time before a standard industry protocol for purity measurement is identified and accepted. Thomas Swan currently bases purity measurement on a combination of SEM, (Transmission Electron Microscopy) TEM, RAMAN and Thermal Gravimetry Analysis (TGA) techniques, combined with an experienced technical eye and has so far received very encouraging feedback from early customers. As the company moves to scale the process up, ongoing customer feedback and collaboration will become increasingly vital. Carbon nanotubes are still in the very early stages of industrial development. While they are still expensive compared to a fully commoditised product, such as carbon black, the price will fall as the demand increases. Despite the problems associated with analysis and downstream processing (namely dispersion) the basic properties of the raw material offer a wide range of exciting applications that are beginning to be realised. As the research into these applications moves out of the lab and into industry it is important that the supply of carbon nanotubes is consistent, of high quality and performed under the appropriate regulatory and health and safety protocols. Thomas Swan has taken the first step towards this goal and material is now available to allow meaningful industrial evaluation and development. The next ten years of development should see the creation of a whole new area of material and coatings science and associated commercial applications.

4

Applications of Nanotubes and Buckytubes in Conductive Plastics, Adhesives, Connectors and Other Areas

Nanotubes: An Introductory Overview

A nanotube is a nanometer scale tube-like structure. It may refer to:

- Carbon nanotube
- Inorganic nanotube
- DNA nanotube
- Membrane nanotube - a tubular membrane connection between cells

The term nanotube is normally used to refer to the carbon nanotube, which has received enormous attention from researchers over the last few years and promises, along with close relatives such as the nanohorn, a host of interesting applications. There are many other types of nanotube, from various inorganic kinds, such as those made from boron nitride, to organic ones, such as those made from self-assembling cyclic peptides (protein components) or from naturally-occurring heat shock proteins (extracted from bacteria that thrive in extreme environments). However, carbon nanotubes excite the most interest, promise the greatest variety of applications, and currently appear to have by far the highest commercial potential. Only carbon nanotubes will be covered in this chapter. Carbon nanotubes are often referred to in the press, including the scientific press, as if they were one consistent item. They are in fact a hugely varied range of structures, with similarly huge variations in properties and ease of production. Adding to the confusion is the existence of long, thin, and often hollow, carbon fibers that have been called carbon nanotubes but have a quite different make-up from that of the nanotubes that scientists generally refer to. To distinguish these we will refer to them as carbon nanofibers.

Carbon nanotubes were 'discovered' in 1991 by Sumio Iijima of NEC and are effectively long, thin cylinders of graphite, which you will be familiar with as the material in a pencil or as the basis of some lubricants. Graphite is made up of layers of carbon atoms arranged in a hexagonal lattice, like chicken wire. Though the chicken wire structure itself is very strong, the layers themselves are not chemically bonded to each other but held together by weak forces called Van der Waals. It is the sliding across each other of these layers that gives graphite its lubricating qualities and makes the mark on a piece of paper as you draw your pencil over it.

Now imagine taking one of these sheets of chicken wire and rolling it up into a cylinder and joining the loose wire ends. The result is a tube that was once described by Richard Smalley (who shared the Nobel Prize for the discovery of a related form of carbon called buckminsterfullerene) as "in one direction . . . the strongest damn thing you'll ever make in the universe". In addition to their remarkable strength, which is usually quoted as 100 times that of steel at one-sixth of the weight (this is tensile strength—the ability to withstand a stretching force without breaking), carbon nanotubes have shown a surprising array of other properties. They can conduct heat as efficiently as most diamond (only diamond grown by deposition from a vapour is better), conduct electricity as efficiently as copper, yet also be semiconducting (like the materials that make up the chips in our computers). They can produce streams of electrons very efficiently (field emission), which can be used to create light in displays for televisions or computers, or even in domestic lighting, and they can enhance the fluorescence of materials they are close to. Their electrical properties can be made to change in the presence of certain substances or as a result of mechanical stress. Nanotubes within nanotubes can act like miniature springs and they can even be stuffed with other materials. Nanotubes and their variants hold promise for storing fuels such as hydrogen or methanol for use in fuel cells and they make good supports for catalysts. But let's look at some of the different types of nanotubes, and nanotube pretenders. One of the major classifications of carbon nanotubes is into single-walled varieties (SWNTs), which have a single cylindrical wall, and multi-walled varieties (MWNTs), which have cylinders within cylinders.

The lengths of both types vary greatly, depending on the way they are made, and are generally microscopic rather than nanoscopic, i.e. greater than 100 nanometers (a nanometer is a millionth of a millimeter). The aspect ratio (length divided by diameter) is typically greater than 100 and can be up to 10,000, but recently even this was made to look small. In May 2002 SWNT strands were made in which the SWNTs were claimed to be as long as 20 cm. Even more recently, the same group has made strands of SWNTs as long as 160 cm, but the precise make-up of these strands has not yet been made clear. A group in China has also found, purely by accident, that packs of relatively short carbon nanotubes can be drawn out into a bundle of fibers, making a thread only 0.2 millimeters in diameter but up to 30 centimeters long. The joins between the nanotubes in this thread represent a weakness but heating the thread has been found to increase the strength significantly, presumably through some sort of fusing of the individual tubes. These are the stars of the nanotube world, and somewhat reclusive ones at that, being much harder to make than the multi-walled variety. The oft-quoted amazing properties generally refer to SWNTs. As previously described, they are basically tubes of graphite and are normally capped at the ends, although the caps can be removed. The caps are made by mixing in some pentagons with the hexagons and are the reason that nanotubes are considered close cousins of buckminsterfullerene, a roughly spherical molecule made of sixty carbon atoms, that looks like a soccer ball and is named after the architect Buckminster Fuller (the word fullerene is used to refer to the variety of such molecular cages, some with more carbon atoms than buckminsterfullerene, and some with fewer). The theoretical minimum diameter of a carbon nanotube is around 0.4 nanometers, which is about as long as two silicon atoms side by side, and nanotubes this size have been made. Average diameters tend to be around the 1.2 nanometer mark, depending on the process used to create them. 8 SWNTs are more pliable than their multi-walled counterparts and can be twisted, flattened and bent into small circles or around sharp bends without breaking.

Discussions of the electrical behaviour of carbon nanotubes usually relate to experiments on the single-walled variety. As we have said, they can be conducting, like metal (such nanotubes are often referred to as metallic nanotubes), or semiconducting, which means that the flow of current through them can be stepped up or down by varying an electrical field. The latter property has given rise to dreams of using nanotubes to make extremely dense electronic circuitry and the last year has seen major advances in creating basic electronic structures from nanotubes in the lab, from transistors up to simple logic elements. The gulf between these experiments and commercial nanotube electronics is, however, vast. There are various ways of producing SWNTs, which are briefly discussed later. The detailed mechanisms responsible for nanotube growth are still not fully understood and computer modeling is playing an increasing role in fathoming the complexities. The ambition of SWNT producers is to gain greater control over their diameters, lengths, and other properties, such as chirality (explained below). The volumes of SWNTs produced are currently small and the quality and purity are variable. Carbon Nanotechnologies Inc. of Houston, Texas, are currently ramping up production to a half a kilogram a day, which is actually huge in comparison to amounts of SWNTs that have been made historically. Various companies pursuing specific nanotube applications produce their own material in house.

Imagine again the chicken wire that we roll up to make the nanotube. In fact, imagine a chicken wire fence out of which will be cut a rectangle to roll into a tube. You could cut the rectangle with the sides vertical or at various angles. Additionally, when joining the sides together, you can raise or lower one side. In some cases it will not be possible to make a tube such that the loose ends match and hexagons are formed, but in other cases it will, and these represent the possible permutations of SWNTs. The possibilities are two forms in which a pattern circles around the diameter of the tube, often called zigzag and armchair (not the most intuitive of names, unfortunately, but they are now widely used), and a variety of forms in which the hexagons spiral up or down the tube with varying steepness, these being the chiral forms. There is theoretically an infinite variety of the latter, if you allow for infinite diameters of nanotubes.

Which of these forms a nanotube takes is the major determinant of its electrical properties, i.e. whether the tube is semiconducting or conducting. For a long time, the fact that all known production methods created a mix of types has been considered one of the hurdles to be overcome if the electronic properties are to be exploited. Claims have now been made that it is possible to produce only the semiconducting kind (specifically, the zigzag form). Additionally, there are approaches that can yield only semiconducting nanotubes from a mix of semiconducting and conducting ones. One such approach relies on vaporizing the conducting nanotubes with a strong electric current, leaving only the semiconducting kind behind. A more recent approach is simply to leave the mix of nanotubes lying around for a while—the metallic ones are oxidized and become semiconducting (the process can, of course, be speeded up). In the early part of 2001 there were also reports that nanotubes had been induced to form crystals, each of which contained just one type of nanotube. This would have been a nice separation method but the silence on this approach since then suggests that these results have not been duplicated.

Multi-walled carbon nanotubes are basically like Russian dolls made out of SWNTs—concentric cylindrical graphitic tubes. In these more complex structures, the different SWNTs that form the MWNT may have quite different structures (length and chirality). MWNTs are typically 100 times longer than they are wide and have outer diameters mostly in the tens of nanometers. Although it is easier to produce significant quantities of MWNTs than SWNTs, their structures are less well understood than single-wall nanotubes because of their greater complexity and variety. Multitudes of exotic shapes and arrangements, often with imaginative names such as bamboo-trunks, sea urchins, necklaces or coils, have also been observed under different processing conditions. The variety of forms may be interesting but also has a negative side—MWNTs always (so far) have more defects than SWNTs and these diminish their desirable properties. Many of the nanotube applications now being considered or put into practice involve multi-walled nanotubes, because they are easier to produce in large quantities at a reasonable price and have been available in decent amounts for much longer than SWNTs. In fact one of the major manufacturers of MWNTs at the moment, Hyperion Catalysis, does not even sell the nanotubes directly but only pre-mixed with polymers for composites applications.

The tubes involved typically have 8 to 15 walls and are around 10 nanometres wide and 10 micrometers long. Other companies are moving into this space, notably formidable players like Mitsui, with plans to produce similar types of MWNT in hundreds of tons a year, a quantity that is greater, but not hugely so, than the current production of Hyperion Catalysis. This is an indication that even these less impressive and exotic nanotubes hold promise of representing a sizable market in the near future.

These are single-walled carbon cones with structures similar to those of nanotube caps that have been produced by high temperature treatment of fullerene soot. Sumio Iijima's group at NEC has demonstrated that nanohorns have good adsorptive and catalytic properties (i.e. desired substances stick to them and they enhance chemical reactions), and the company is working on using them in a new generation of fuel cells for personal electronics.

We use this term to refer to hollow and solid carbon fibers with lengths on the order of a few microns and widths varying from some tens of nanometers to around 200 nanometers. These materials have occasionally been referred to as nanotubes. However, they do not have the cylindrical chicken wire structure of SWNTs and MWNTs but instead consist of a mixture of forms of carbon, from layers of graphite stacked at various angles to amorphous carbon (lacking any large-scale regular structure). Because of this variable structure they do not exhibit the strength of pure nanotubes but can still be quite strong and possess other useful properties. The US company Applied Sciences Inc. is already producing tons of such material a year and there are several major producers in the Far East.

Production processes for carbon nanotubes, crudely described, vary from blasting carbon with an electrical arc or a laser to growing them from a vapor, either en masse (usually in tangled bundles) or on nanoparticles, sometimes in predetermined positions. These processes vary considerably with respect to the type of nanotube produced, quality, purity and scalability. Carbon nanotubes are usually created with the aid of a metal catalyst and this ends up as a contaminant with respect to many potential applications, especially in electronics. IBM have very recently, however, grown nanotubes on silicon structures without a metal catalyst. Scalability of production processes is an essential commercial consideration—some of the approaches use equipment that simply cannot be made bigger and the only way to increase production is to make more pieces of equipment, which will not produce the economies of scale required to bring down costs significantly.

By now we hope you have an idea of the different types of nanotubes and how their qualities and ease of production vary. Reports of one company producing tons of the material, alongside reports of researchers not being able to get enough to get meaningful results from their research, should no longer be confusing. It would, of course, be useful if commentators got into the habit of making clear the type of material they are talking about. Understanding these differences is essential for understanding the commercial potential of the various applications of nanotubes and related structures that already exist or have been proposed. The variety of these is vast, and the commercialisation timelines involved vary from now to ten years from now or more. Some of the potential markets are enormous. We will leave you with a taste of the possibilities. The materials markets are already seeing applications for composites based on multi-walled carbon nanotubes and nanofibers. In many ways this is an old market—that of carbon fibers, which have been around commercially for a couple of decades.

The benefits of the new materials in these markets are the same as those of carbon fibers, just better; the main properties to be considered being strength and conductivity. Carbon fibers are quite large, typically about a tenth of a millimeter in diameter, and blacken the material to which they are added. MWNTs can offer the same improvements in strength to a polymer composite without the blackening and often with a smaller amount of added material (called the filler). The greater aspect ratio (i.e. length compared to diameter) of the newer materials can make plastics conducting with a smaller filler load, one significant application being electrostatic painting of composites in products such as car parts. Additionally, the surface of the composite is smoother, which benefits more refined structures such as platens for computer disk drives.

When thinking about structural applications such as these, it should be remembered that, in general, as the fibers get smaller so the number of defects decreases, in a progression towards the perfection of the single-walled nanotube. The inverse progression is seen in terms of ease of manufacturing—the more perfect, and thus more structurally valuable, the material, the harder it is to produce in quantity at a good price. This relationship is not written in stone—there is no reason that near-perfect SWNTs should not be producible cheaply and in large quantities. When this happens, and it might not be too far off, the improvements seen in the strength-to-weight ratios of composite materials could soar, impacting a wide variety of industries from sports equipment to furniture, from the construction industry to kitchenware, and from automobiles to airplanes and spacecraft (the aerospace industry is probably the one that stands to reap the greatest rewards). In fact a carbon nanotube composite has recently been reported that is six times stronger than conventional carbon fiber composites. This is an appropriate point at which to introduce a note of caution, and an area of research worth keeping an eye on. Just because the perfect nanotube is 100 times stronger than steel at a sixth of the weight doesn't mean you are going to be able to achieve those properties in a bulk material containing them. You may remember that the chicken wire arrangement that makes up the layers in graphite does not stick at all well to other materials, which is why graphite is used in lubricants and pencils. The same holds true of nanotubes—they are quite insular in nature, preferring not to interact with other materials. To capitalize on their strength in a composite they need to latch on to the surrounding polymer, which they are not inclined to do (blending a filler in a polymer is difficult even without these issues—it took a decade to perfect this for the new nanoclay polymers now hitting the markets).

One way of making a nanotube interact with something else, such as a surrounding polymer, is to modify it chemically. This is called functionalisation and is being explored not just for composite applications but also for a variety of others, such as biosensors. For structural applications, the problem is that functionalisation can reduce the valuable qualities of the nanotubes that you are trying to capitalize on. This is an issue that should not be underestimated. Of course, in theory you don't need to mix the nanotubes with another material. If you want to make super-strong cables, for example, the best solution would be to use bundles of sufficiently long nanotubes with no other material added. For this reason, one of the dreams of nanotube production is to be able to spin them, like thread, to indefinite lengths. Such a technology would have applications from textiles (the US military is in fact investigating the use of nanotubes for bullet-proof vests) to the 'space elevator'. The space elevator concept, which sounds like something straight out of science fiction (it was, in fact, popularized by Arthur C. Clarke) involves anchoring one end of a huge cable to the earth and another to an object in space. The taut cable so produced could then support an elevator that would take passengers and cargo into orbit for a fraction of the cost of the rockets used today. Sounds too far out? It has, in fact, been established by NASA as feasible in principle, given a material as strong as SWNTs. The engineering challenges, though, are awesome, so don't expect that 'top floor' button to be taking you into orbit any time soon.

The materials market is a big one, and there are others, which we'll come to, but smaller ones exist too. Nanotubes are already being shipped on the tip of atomic force microscope probes to enhance atomic-resolution imaging. Nanotube-based chemical and biosensors should be on the market soon (they face stiff competition from other areas of nanotechnology). The thermal conductivity of nanotubes shows promise in applications from cooling integrated circuits to aerospace materials. Electron beam lithography, which is a method of producing nanoscale patterns in materials, may become considerably cheaper thanks to the field emission properties of carbon nanotubes. Recent developments show promise of the first significant change in X-ray technology in a century. Entering more speculative territory, nanotubes may one day be used as nanoscale needles that can inject substances into, or sample substances from, individual cells, or they could be used as appendages for miniature machines (the tubes in multi-walled nanotubes slide over each other like graphite, but have preferred locations that they tend to spring back to). Big markets, apart from materials, in which nanotubes may make an impact, include flat panel displays (near-term commercialisation is promised here), lighting, fuel cells and electronics. This last is one of the most talked-about areas but one of the farthest from commercialisation, with one exception, this being the promise of huge computer memories (more than a thousand times greater in capacity than what you probably have in your machine now) that could, in theory, put a lot of the $40 billion magnetic disk industry out of business. Companies like to make grand claims, however, and in this area there is not just the technological hurdle to face but the even more daunting economic one, a challenge made harder by a host of competing technologies.

Despite an inevitable element of hype, the versatility of nanotubes does suggest that they might one day rank as one of the most important materials ever discovered. In years to come they could find their way into myriad materials and devices around us and quite probably make some of the leaders in this game quite rich.

Understanding Applications of Carbon Nanotubes, Its Commercial Uses for Mass-Markets

The special nature of carbon combines with the molecular perfection of buckytubes (single wall carbon nanotubes) to endow them with exceptionally high material properties such as electrical and thermal conductivity, strength, stiffness, and toughness. No other element in the periodic table bonds to itself in an extended network with the strength of the carbon-carbon bond. The delocalised pi-electron donated by each atom is free to move about the entire structure, rather than stay home with its donor atom, giving rise to the first molecule with metallic-type electrical conductivity. The high-frequency carbon-carbon bond vibrations provide an intrinsic thermal conductivity higher than even diamond. In most materials, however, the actual observed material properties—strength, electrical conductivity, etc.—are degraded very substantially by the occurrence of defects in their structure. For example, high strength steel typically fails at about 1 per cent of its theoretical breaking strength. Buckytubes, however, achieve values very close to their theoretical limits because of their perfection of structure—their molecular perfection. This aspect is part of the unique story of buckytubes. Buckytubes are an example of true nanotechnology: only a nanometer in diameter, but molecules that can be manipulated chemically and physically. They open incredible applications in materials, electronics, chemical processing and energy management.

Buckytubes have extraordinary electrical conductivity, heat conductivity and mechanical properties. They are probably the best electron field-emitter possible. They are polymers of pure carbon and can be reacted and manipulated using the tremendously rich chemistry of carbon. This provides opportunity to modify the structure and to optimise solubility and dispersion. Very significantly, buckytubes are molecularly perfect, which means that they are free of property-degrading flaws in the nanotube structure. Their material properties can therefore approach closely the very high levels intrinsic to them. These extraordinary characteristics give buckytubes potential in numerous applications.

Buckytubes are the best known field emitters of any material. This is understandable, given their high electrical conductivity, and the unbeatable sharpness of their tip (the sharper the tip, the more concentrated will be an electric field, leading to field emission; this is the same reason lightening rods are sharp). The sharpness of the tip also means that they emit at especially low voltage, an important fact for building electrical devices that utilize this feature. Buckytubes can carry an astonishingly high current density, possibly as high as 1013 A/cm^2. Furthermore, the current is extremely stable. An immediate application of this behaviour receiving considerable interest is in field-emission flat-panel displays. Instead of a single electron gun, as in a traditional cathode ray tube display, here there is a separate electron gun (or many) for each pixel in the display. The high current density, low turn-on and operating voltage, and steady, long-lived behaviour make buckytubes attract field emitters to enable this application. Other applications utilising the field-emission characteristics of buckytubes include: general cold-cathode lighting sources, lightning arrestors, and electron microscope sources.

Much of the history of plastics over the last half century has been as a replacement for metal. For structural applications, plastics have made tremendous headway, but not where electrical conductivity is required, plastics being famously good electrical insulators. This deficiency is overcome by loading plastics up with conductive fillers, such as carbon black and graphite fibres (the larger ones used to make golf clubs and tennis racquets). The loading required to provide the necessary conductivity is typically high, however, resulting in heavy parts, and more importantly, plastic parts whose structural properties are highly degraded. It is well established that the higher aspect ratio of filler, the lower loading required to achieve a given level of conductivity. Buckytubes are ideal in this sense, since they have the highest aspect ratio of any carbon fibre. In addition, their natural tendency to form ropes provides inherently very long conductive pathways even at ultra-low loadings. Applications that exploit this behaviour of buckytubes include EMI/RFI shielding composites and coatings for enclosures, gaskets, and other uses; electrostatic dissipation (ESD), and antistatic materials and (even transparent!) coatings; and radar-absorbing materials.

Buckytubes have the intrinsic characteristics desired in material used as electrodes in batteries and capacitors, two technologies of rapidly increasing importance. Buckytubes have a tremendously high surface area (~1000 m^2/g), good electrical conductivity, and very importantly, their linear geometry makes their surface highly accessible to the electrolyte. Research has shown that buckytubes have the highest reversible capacity of any carbon material for use in lithium-ion batteries. In addition, buckytubes are outstanding materials for supercapacitor electrodes and are now being marketed. Buckytubes also have applications in a variety of fuel cell components. They have a number of properties including high surface area and thermal conductivity that make them useful as electrode catalyst supports in PEM fuel cells. They may also be used in gas diffusion layers as well as current collectors because of their high electrical conductivity. Buckytubes' high strength and toughness to weight characteristics may also prove valuable as part of composite components in fuel cells that are deployed in transport applications where durability is extremely important.

The same issues that make buckytubes attractive as conductive fillers for use in shielding, ESD materials, etc., make them attractive for electronics materials, such as adhesives and other connectors (e.g., solders).

The idea of building electronic circuits out of the essential building blocks of materials—molecules—has seen a revival the past five years, and is a key component of nanotechnology. In any electronic circuit, but particularly as dimensions shrink to the nanoscale, the interconnections between switches and other active devices become increasingly important. Their geometry, electrical conductivity, and ability to be precisely derived, make buckytubes the ideal candidates for the connections in molecular electronics. In addition, they have been demonstrated as switches themselves.

The record-setting anisotropic thermal conductivity of buckytubes is enabling applications where heat needs to move from one place to another. Such an application is electronics, particularly advanced computing, where uncooled chips now routinely reach over 100°C. CNI's technology for creating aligned structures and ribbons of buckytubes is a step toward realising incredibly efficient heat conduits. In addition, composites with buckytubes have been shown to dramatically increase the bulk thermal conductivity at small loadings.

The world-record properties of buckytubes are not limited to electrical and thermal conductivities, but also include mechanical properties, such as stiffness, toughness, and strength. These properties lead to a wealth of applications exploiting them, including advanced composites requiring high values in one or more of these properties.

Fibres spun of pure buckytubes have been demonstrated and are undergoing rapid development, along with buckytube composite fibres. Such super strong fibres will have applications including body and vehicle armour, transmission line cables, woven fabrics and textiles.

Buckytubes have an intrinsically high surface area; in fact, every atom is not just on a surface—each atom is on two surfaces, the inside and outside! Combined with the ability to attach essentially any chemical species to their sidewalls provides an opportunity for unique catalyst supports. Their electrical conductivity may also be exploited in the search for new catalysts and catalytic behaviour. The exploration of buckytubes in biomedical applications is just underway, but has significant potential. Cells have been shown to grow on buckytubes, so they appear to have no toxic effect. The cells also do not adhere to the buckytubes, potentially giving rise to applications such as coatings for prosthetics and anti-fouling coatings for ships. The ability to chemically modify the sidewalls of buckytubes also leads to biomedical applications such as vascular stents, and neuron growth and regeneration. There is a wealth of other potential applications for buckytubes, such as solar collection; nanoporous filters; catalyst supports; and coatings of all sorts. There are almost certainly many unanticipated applications for this remarkable material that will come to light in the years ahead and which may prove to be the most important and valuable of all.

Carbon nanotubes offer a range of attractive properties and could enhance many coatings applications. However, the current barriers to mass-market use of these materials include the availability of high-quality materials in commercial quantities. Now a partnership between industry and academia has resulted in a commercial manufacturing process for carbon nanotubes.

Hardly a day goes by without someone suggesting a new application for carbon nanotubes. Since their discovery, academics and industrialists have researched or considered their use in brake discs; fuel cells; advanced aerospace composites; co-axial cable; conductive fuel lines; EMI (Electromagnetic Interference) shielding in electronic devices; conductive tyres; conductive inks; and, of course, the hotly disputed space elevator – to name but a few. While some of these applications are clearly closer to reality than others, possibly the most immediate products will be developed in the coatings industry. The launch in April 2004 of the UK's first commercial manufacturing process for high-purity single-wall carbon nanotubes at Thomas Swan & Co. Ltd., has cleared the way for the development of a wide range of advanced coatings applications.

Composition and Classification of Carbon Nanotubes—Single-Wall (SWNT) and Multi-Wall (MWNT)

For those who have not come across them before, carbon nanotubes consist of molecular cylinders of pure, hexagonally-arranged carbon atoms that resemble rolled-up sheets of chicken wire with a diameter measured in a few nanometres (1 nanometre is 1 billionth of a metre) and a length of many microns. They occur in two main types, the single-wall carbon nanotube (SWNT) composed of a single cylinder of carbon, and the multi-wall version (MWNT) consisting of concentric tubes or cylinders of carbon (effectively straws within straws). The ends of the tubes are usually closed off by a carbon end-cap. Other variations on this theme include the double-wall tube, 'herringbone' and 'bamboo' structures.

Strength and Properties of Carbon Nanotubes

Since their discovery, and Sumio Iijima's watershed paper in 1991, there has been unprecedented academic and industrial interest in carbon nanotubes and their potential use in a wide range of commercial applications. There is no denying that the basic properties of both single and multi-wall carbon nanotubes are truly remarkable. Recent studies have suggested that single-wall carbon nanotubes have a tensile strength of 50-100 GPa and a modulus of 1-2 TPa. This places them well ahead of steel in both strength and modulus and all for one sixth the weight. Combine this with their high thermal and electrical conductivity and you begin to understand why so many scientists and commercial organisations have sat up and taken notice.

Properties of the Individual Carbon Nanotube

The basic properties of the individual carbon nanotube (both single and multi-wall) have been considered in some fascinating potential applications. In essence carbon nanotubes are advanced fillers—highly conductive (thermally and electrically), extremely strong and light-weight. Add to this their apparent ability to absorb electro-magnetic interference and the fascinating photo-acoustic effect (whereby they spontaneously ignite when exposed to an intense light source) and you are left with some intriguing possibilities for advanced coatings.

Electrical Conductivity of Carbon Nanotubes

The addition of low concentrations of carbon nanotubes to plastic it makes it electrically conductive. This allows electrostatic painting to be used when coating automotive parts, removing the need for costly primers.

Benefits of Modest Loadings of Single-Wall Carbon Nanotubes

The high aspect ratio of carbon nanotubes (length divided by diameter) means that they form a percolating matrix across the non-conductive filler at much lower loadings than traditional fillers. Carbon black typically requires 30 or 40% by volume loading to confer conductivity, at which point the mechanical properties of the composite are often severely degraded. In comparison Cambridge University recently achieved a conductive epoxy using only 0.005 wt% loading of aligned multi-wall carbon nanotubes. It is also likely that modest loadings of single-wall carbon nanotubes will actually enhance the structural properties of the composite.

Anti-Static Properties of Carbon Nanotubes May Be Used in Packaging for Electronics

The anti-static properties of carbon nanotube coatings may also have potential applications in packaging. Anti-static coatings are used in the electronics to prevent damage to sensitive electronic components during shipping and storage.

Thermal Management Potential of Carbon Nanotubes

Carbon nanotubes (in their pure and undamaged form) are possibly the best thermally conductive material ever discovered. As such they are being researched for their heat management potential in a number of applications. As the electronics world strives to fit more and more functionality into less and less space, overheating is becoming a major problem. This is forcing greater pressure on materials and coatings involved in electronics manufacturing to be thermally conductive in an attempt to remove the heat from the device as quickly as possible.

Low Loadings of Carbon Nanotubes Should Speed Up Curing Times

Another aspect of thermal management is the demand for rapid curing times of polymeric materials such as ink resins and elastomers. The use of low loadings of carbon nanotubes should increase the speed of thermal dissipation from a coating or resin allowing for a much faster curing time. This would clearly be of interest in automated processes where the speed of curing limits the rate of production.

Electrostatic Dissipation Properties of Carbon Nanotube Coatings

Electro-magnetic interference (EMI) shielding in laptops and mobile phones is becoming increasingly important to preventing interference with and from other portable electronic devices. As an extension to the electrostatic dissipation properties of carbon nanotube coatings, a number of companies are developing coatings that are designed to absorb EMI.

Using Carbon Nanotubes in Conductive Inks for Spray-On Circuits and Coatings

A significant amount of research is being directed into the use of carbon nanotubes in conductive inks. The ability to spray on conductive ink opens up some intriguing possibilities for 'spray-on' circuits and coatings. A leader in this field is US-based Eikos which was awarded USD 860,000 in May this year from the US Air Force Research Laboratory to develop transparent conductive polymers specifically for military aircraft canopies.

While such contracts indicate that the carbon nanotube market is already up and running in some niche applications, there are still some major barriers to overcome.

The Three Main Restrictions in the Market-Place for Carbon Nanotubes

While there are already some products in the market-place and a whole range of applications in development, the carbon nanotube market has been constrained by three main issues:

1. A lack of commercially-available material of consistently high quality.
2. Cost—until recently single-wall carbon nanotubes were selling for EUR 400/g which is clearly some way from commercial viability as a bulk material.
3. Successfully expressing the basic properties of the raw material in the end application – the incredibly small scale of the material poses some interesting challenges for advanced material and coatings science.

Such teething problems are not new. Carbon fibres took many years to be widely accepted in the materials world from both a cost and performance aspect and have only just made a regular appearance in popular sporting markets such as golf equipment.

Early Applications of Carbon Nanotubes now on the Market

Despite this, some early applications of carbon nanotubes in products such as conductive fuel lines in the automotive industry have already made a successful appearance on the market. In addition, recent advances in the manufacture of carbon nanotubes on an industrial scale will allow vital commercial research and evaluation to be conducted and this will open the way to a wide range of fully industrialised applications.

The UK's First Production Plant for Single-Wall Carbon Nanotubes (SWNTs)

The UK's first commercial manufacturing process for high-purity single-wall carbon nanotubes was announced by Thomas Swan & Co in April 2004. The plant is the result of four years of collaboration between Thomas Swan and Cambridge University's Department of Material Science and Department of Chemistry. The unique collaboration brought the academic expertise of Cambridge University together with over 77 years of chemical manufacturing experience from Thomas Swan.

Problems Encountered When Setting Up this Plant for Carbon Nanotubes

The goal of the project was to supply high-purity single and multi-wall carbon nanotubes in commercial quantities and to back up this initial offering with the ability to scale up the process to keep up with demand as the industrial market expanded. The problem was that at the start of the programme no-one knew which production method would supply both single and multi-wall tubes and be fully scalable. It was this problem that was presented to Cambridge University.

Methods Used for Manufacturing Carbon Nanotubes

Carbon nanotubes can be manufactured using a variety of methods:

- Laser ablation uses a high-power laser to vaporise a graphite source loaded with a metal catalyst. The carbon in the graphite reforms as predominantly single-wall nanotubes on the metal catalyst particles.
- Arc discharge involves an electrical discharge from a carbon-based electrode in a suitable atmosphere to produce both single and multi-wall tubes of high quality but in low quantities.
- Chemical vapour deposition (CVD) is where a hydrocarbon feedstock is reacted with a suitable metal-based catalyst in a hot furnace to 'grow' nanotubes which are subsequently removed from the substrate and catalyst by a simple acid wash.

System Based on Chemical Vapour Deposition (CVD) Growth

The team at Cambridge developed a novel system based on CVD growth, as it enabled the production of both single and multi-wall nanotubes of reasonably high quality and consistency while offering the greatest potential for scale up. This concept was then taken to design stage (and patented by Cambridge University). Using a team of chemical engineers and an independent consultant, the plant was built at Swan's site in North East England towards the end of 2003. Over the following months the plant was commissioned and then optimised until the material that was being produced was of consistently high quality.

A consistent supply of high-purity material that is also scaleable is a major stepping stone for the developing carbon nanotubes market. It allows companies to conduct meaningful industrial research using commercial quantities of raw material. In addition, this research can now be done with the knowledge that, should an application prove itself, there is a scalable source of the same raw material available to supply any ongoing development. Combine this with lowering prices (Thomas Swan predicts that within a few years the price of single-wall carbon nanotubes with be measured in tens of GBP per kg) and the market should start to take off.

Problems in Seeing What has been Manufactured

A major problem in achieving a saleable product that is on average 100,000 times smaller than a human hair is that it is rather hard to 'see' what you have made. The advent of nanomaterial manufacturing brings to light the necessity for advanced industrial and cost-effective analytical techniques. While Scanning Electron Microscopes (SEM) provide the ability to view the sample in remarkable detail, it only shows a minute fraction of the product. The analogy often used is like looking at a football pitch through a jam jar. The grass beneath you is green and healthy but you have no idea what it is like on the other side of the pitch.

Tools and Techniques Used to Help Manufacturers See what they are Making

This will not be an easy problem to solve and while RAMAN techniques (which involve studying the spectrum profile of a reflected laser from your sample) offer some hope, it will be some time before a standard industry protocol for purity measurement is identified and accepted. Thomas Swan currently bases purity measurement on a combination of SEM, (Transmission Electron Microscopy) TEM, RAMAN and Thermal Gravimetry Analysis (TGA) techniques, combined with an experienced technical eye and has so far received very encouraging feedback from early customers. As the company moves to scale the process up, ongoing customer feedback and collaboration will become increasingly vital.

Prices of Carbon Nanotubes Will Fall as Demand Increases

Carbon nanotubes are still in the very early stages of industrial development. While they are still expensive compared to a fully commoditised product, such as carbon black, the price will fall as the demand increases. Despite the problems associated with analysis and downstream processing (namely dispersion) the basic properties of the raw material offer a wide range of exciting applications that are beginning to be realised.

What the Next Ten Years Might Hold for Producers of Carbon Nanotubes

As the research into these applications moves out of the lab and into industry it is important that the supply of carbon nanotubes is consistent, of high quality and performed under the appropriate regulatory and health and safety protocols. Thomas Swan has taken the first step towards this goal and material is now available to allow meaningful industrial evaluation and development. The next ten years of development should see the creation of a whole new area of material and coatings science and associated commercial applications.

Explaining the Theory of Nanotube

Effect of twisting on the properties of a metallic armchair (6,6) nanotube. Three twisted configurations are shown. Twisting is found to transform the metallic nanotube to a semiconducting one with a band-gap that varies with the twist angle as shown.

Variation of the local density of states (LDOS) along the length of a (6,6) nanotube as a function of the bending angle.

Here, we focus on our recent work on *carbon nanotubes*; their structure, properties and uses in nano-electronic devices. Carbon nanotubes are extremely thin (their diameter is about 10,000 times smaller than a human hair), hollow cylinders made of carbon atoms.

This study consists of the following sections (to access them you may also use the navigation bar on the left):

- *Nanotube manipulation:* We can manipulate the nanotube positions, change their shape, cut them and place them on electrodes.
- *Molecular mechanics:* We can simulate the mechanical behaviour of nanotubes by calculating the forces acting between nanotubes and other objects such as the substrate.
- *Nanotube Field-Effect Transistor:* We have successfully used semiconducting single and multi-walled nanotubes as channels of field-effect transistors.
- *Nanotube rings:* While normally nanotubes are straight, we have devised ways to prepare them in a ring form.
- Nanotube theory: Computation and theory of the electrical and mechanical properties

Nanotubes, depending on their structure, can be metals or semiconductors. They are also extremely strong materials and have good thermal conductivity. The above characteristics have generated strong interest in their possible use in nano-electronic and nano-mechanical devices.

For example, they can be used as nano-wires or as active components in electronic devices such as the field-effect transistor shown in this study.

Market

According to market report in 2004, "a total tonnage of 65 tons of nanotubes`and fibers was produced in 2004 which equates at manufacturer's current prices to`• 144 million worth. By 2010 this figure will surpass $ 3 billion. This`represents a compound annual growth rate of well over 60%."

Fundamentals and Applications

The small scale and the one dimensional structure of carbon nanotubes are directly related to their unique properties which find more and more applications as the understanding and progress in synthesis advances.

Carbon nanotubes represent an exemplary system where the bottoms-up approach to synthesis results in perfect structures with sizes less than 10nm, a range which remains inaccessible for advanced projection lithography techniques.

Applications of carbon nanotubes range from reinforcement of composites or conductive plastics to electrodes for batteries or flat screens, field effect transistors, chemical or force sensors and electromechanical memory. These applications have been demonstrated in bulk, on surfaces and on individual tubes. This course is focused on the fundamental properties of carbon nanotubes and provides the basic knowledge needed to follow the important developments in this field.

Couse Objectives

We will show how the helical structure of the tube influences the electronic properties of the tube in a fundamental way. In the first part we will explain the structural peculiarities, give an overview of the main synthesis techniques and then show how, starting from the known electronic properties of graphite, how those of the nanotubes can be described. The second part of the course covers electronic transport, electron emission properties, optical properties, mechanical properties and composites. We will show to what extend analytical techniques can provide information about individual tubes and we will see the importance of the immediate environment of the nanotube surface.

In summary the goal of the course is:

- To give an introduction to the elementary properties of carbon nanotubes.
- To give the context and current status of the scientific research on carbon nanotubes.
- Synthesis Methods
- Characterisation Methods
- CNT Based Materials
- Carbon Nano Industrial Applications

Applications of Carbon Nanotubes (Buckytubes) in Conductive Plastics, Adhesives and Connectors

Much of the history of plastics over the last half century has been as a replacement for metal. For structural applications, plastics have made tremendous headway, but not where electrical conductivity is required, plastics being famously good electrical insulators.

This deficiency is overcome by loading plastics up with conductive fillers, such as carbon black and graphite fibres (the larger ones used to make golf clubs and tennis racquets). The loading required to provide the necessary conductivity is typically high, however, resulting in heavy parts, and more importantly, plastic parts whose structural properties are highly degraded. It is well established that the higher aspect ratio of a filler, the lower loading required to achieve a given level of conductivity.

Buckytubes are ideal in this sense, since they have the highest aspect ratio of any carbon fibre. In addition, their natural tendency to form ropes provides inherently very long conductive pathways even at ultra-low loadings.

Applications that exploit this behaviour of Buckytubes include EMI/RFI shielding composites and coatings for enclosures, gaskets, and other uses; electrostatic dissipation (ESD), and antistatic materials and (even transparent!) coatings; and radar-absorbing materials.

The same issues that make Buckytubes attractive as conductive fillers for use in shielding, ESD materials, etc., make them attractive for electronics materials, such as adhesives and other connectors (e.g., solders).

The special nature of carbon combines with the molecular perfection of buckytubes (single-wall carbon nanotubes) to endow them with exceptionally high material properties such as electrical and thermal conductivity, strength, stiffness, and toughness. No other element in the periodic table bonds to itself in an extended network with the strength of the carbon-carbon bond. The delocalised pi-electron donated by each atom is free to move about the entire structure, rather than stay home with its donor atom, giving rise to the first molecule with metallic-type electrical conductivity.

The high-frequency carbon-carbon bond vibrations provide an intrinsic thermal conductivity higher than even diamond. In most materials, however, the actual observed material properties—strength, electrical conductivity, etc.—are degraded very substantially by the occurrence of defects in their structure. For example, high strength steel typically fails at about 1% of its theoretical breaking strength.

Buckytubes, however, achieve values very close to their theoretical limits because of their perfection of structure—their molecular perfection. This aspect is part of the unique story of buckytubes. Buckytubes are an example of true nanotechnology: only a nanometer in diameter, but molecules that can be manipulated chemically and physically. They open incredible applications in materials, electronics, chemical processing and energy management.

Buckytubes have extraordinary electrical conductivity, heat conductivity and mechanical properties. They are probably the best electron field-emitter possible.

They are polymers of pure carbon and can be reacted and manipulated using the tremendously rich chemistry of carbon. This provides opportunity to modify the structure and to optimise solubility and dispersion. Very significantly, buckytubes are molecularly perfect, which means that they are free of property-degrading flaws in the nanotube structure. Their material properties can therefore approach closely the very high levels intrinsic to them.

These extraordinary characteristics give buckytubes potential in numerous applications.

Buckytubes are the best known field emitters of any material. This is understandable, given their high electrical conductivity, and the unbeatable sharpness of their tip (the sharper the tip, the more concentrated will be an electric field, leading to field emission; this is the same reason lightening rods are sharp). The sharpness of the tip also means that they emit at especially low voltage, an important fact for building electrical devices that utilize this feature. Buckytubes can carry an astonishingly high current density, possibly as high as 1013 A/cm^2. Furthermore, the current is extremely stable [B.Q. Wei, et al. Appl. Phys. Lett. 79 1172 (2001)].

An immediate application of this behaviour receiving considerable interest is in field-emission flat-panel displays. Instead of a single electron gun, as in a traditional cathode ray tube display, here there is a separate electron gun (or many) for each pixel in the display. The high current density, low turn-on and operating voltage, and steady, long-lived behaviour make buckytubes attract field emitters to enable this application.

Other applications utilising the field-emission characteristics of buckytubes include: general cold-cathode lighting sources, lightning arrestors, and electron microscope sources. Much of the history of plastics over the last half century has been as a replacement for metal. For structural applications, plastics have made tremendous headway, but not where electrical conductivity is required, plastics being famously good electrical insulators. This deficiency is overcome by loading plastics up with conductive fillers, such as carbon black and graphite fibres (the larger ones used to make golf clubs and tennis racquets). The loading required to provide the necessary conductivity is typically high, however, resulting in heavy parts, and more importantly, plastic parts whose structural properties are highly degraded. It is well established that the higher aspect ratio of filler, the lower loading required to achieve a given level of conductivity. Buckytubes are ideal in this sense, since they have the highest aspect ratio of any carbon fibre. In addition, their natural tendency to form ropes provides inherently very long conductive pathways even at ultra-low loadings. Applications that exploit this behaviour of buckytubes include EMI/RFI shielding composites and coatings for enclosures, gaskets, and other uses; electrostatic dissipation

(ESD), and antistatic materials and (even transparent!) coatings; and radar-absorbing materials.

Buckytubes have the intrinsic characteristics desired in material used as electrodes in batteries and capacitors, two technologies of rapidly increasing importance. Buckytubes have a tremendously high surface area (~1000 m^2/g), good electrical conductivity, and very importantly, their linear geometry makes their surface highly accessible to the electrolyte. Research has shown that buckytubes have the highest reversible capacity of any carbon material for use in lithium-ion batteries [B. Gao, Chem. Phys. Lett. 327, 69 (2000)]. In addition, buckytubes are outstanding materials for supercapacitor electrodes [R.Z. Ma, et al., Science in China Series E-Technological Sciences 43 178 (2000)] and are now being marketed.

Buckytubes also have applications in a variety of fuel cell components. They have a number of properties including high surface area and thermal conductivity that make them useful as electrode catalyst supports in PEM fuel cells. They may also be used in gas diffusion layers as well as current collectors because of their high electrical conductivity. Buckytubes' high strength and toughness to weight characteristics may also prove valuable as part of composite components in fuel cells that are deployed in transport applications where durability is extremely important.

The same issues that make buckytubes attractive as conductive fillers for use in shielding, ESD materials, etc., make them attractive for electronics materials, such as adhesives and other connectors (e.g., solders).

The idea of building electronic circuits out of the essential building blocks of materials—molecules—has seen a revival the past five years, and is a key component of nanotechnology. In any electronic circuit, but particularly as dimensions shrink to the nanoscale, the interconnections between switches and other active devices become increasingly important. Their geometry, electrical conductivity, and ability to be precisely derived, make buckytubes the ideal candidates for the connections in molecular electronics. In addition, they have been demonstrated as switches themselves.

The record-setting anisotropic thermal conductivity of buckytubes is enabling applications where heat needs to move from one place to another. Such an application is electronics, particularly advanced computing, where uncooled chips now routinely reach over 100°C. CNI's technology for creating aligned structures and ribbons of buckytubes [D.Walters, et al., Chem. Phys. Lett. 338, 14 (2001)] is a step toward realising incredibly efficient heat conduits. In addition, composites with buckytubes have been shown to dramatically increase the bulk thermal conductivity at small loadings.

The world-record properties of buckytubes are not limited to electrical and thermal conductivities, but also include mechanical properties, such as stiffness, toughness, and strength. These properties lead to a wealth of applications exploiting them, including advanced composites requiring high values in one or more of these properties.

Fibres spun of pure buckytubes have recently been demonstrated [R.H. Baughman, Science 290, 1310 (2000)] and are undergoing rapid development, along with buckytube composite fibres. Such super strong fibres will have applications including body and vehicle armour, transmission line cables, woven fabrics and textiles.

Buckytubes have an intrinsically high surface area; in fact, every atom is not just on a surface—each atom is on two surfaces, the inside and outside! Combined with the ability to attach essentially any chemical species to their sidewalls provides an opportunity for unique catalyst supports. Their electrical conductivity may also be exploited in the search for new catalysts and catalytic behaviour.

Biomedical Applications

The exploration of buckytubes in biomedical applications is just underway, but has significant potential. Cells have been shown to grow on buckytubes, so they appear to have no toxic effect. The cells also do not adhere to the buckytubes, potentially giving rise to applications such as coatings for prosthetics and anti-fouling coatings for ships.

The ability to chemically modify the sidewalls of buckytubes also leads to biomedical applications such as vascular stents, and neuron growth and regeneration.

Other Applications

There is a wealth of other potential applications for buckytubes, such as solar collection; nanoporous filters; catalyst supports; and coatings of all sorts. There are almost certainly many unanticipated applications for this remarkable material that will come to light in the years ahead and which may prove to be the most important and valuable of all.

Tracing the History and Development of Carbon Nanotubes

The beginning of this story has been told many times. In 1985, a confluence of events led to an unexpected and unplanned experiment with a new kind of microscope resulting in the discovery of a new molecule made purely of carbon – the very element chemists felt there was nothing more to learn about. Buckyballs – sixty carbon atoms arranged in a soccer ball shape – had been discovered and the chemical world, not to mention the physical and material worlds, would never be the same.

Fullerenes

In fact, what had been discovered was not just a single new molecule but an infinite class of new molecules: the fullerenes. Each fullerene—C_{60}, C_{70}, C_{84}, etc.—possessed the essential characteristic of being a pure carbon cage, each atom bonded to three others as in graphite. Unlike graphite, every fullerene has exactly 12 pentagonal faces with a varying number of hexagonal faces (e.g., buckyball – C_{60} – has 20).

Some fullerenes, like C_{60}, were spheroidal in shape, and others, like C_{70}, were oblong like a rugby ball. Dr. Richard Smalley recognized in 1990 that, in principle, a tubular fullerene should be possible, capped at each end, for example, by the two hemispheres of C_{60}, connected by a straight segment of tube, with only hexagonal units in its structure. Millie Dresselhaus, upon hearing of this concept, dubbed these imagined objects "buckytubes."

Carbon Nanotubes

In actuality, however, carbon nanotubes had been discovered 30 years earlier, but had not been fully appreciated at that time. In the late 1950s, Roger Bacon at Union Carbide, found a strange new carbon fibre while studying carbon under conditions near its triple point. He observed straight, hollow tubes of carbon that appeared to consist in graphitic layers of carbon separated by the same spacing as the planar layers of graphite. In the 1970s, Morinobu Endo observed these tubes again, produced by a gas-phase process. Indeed, he even observed some tubes consisting in only a single layer of rolled-up graphite.

In 1991, after the discovery and verification of the fullerenes, Sumio Iijima of NEC observed multiwall nanotubes formed in a carbon arc discharge, and two years later, he and Donald Bethune at IBM independently observed single-wall nanotubes – buckytubes. These pure carbon polymers could now be understood in the context of fullerenes, changing the perception of them to molecules, with all that special designation implies. Nanotubes had been fullerenized.

Multiwall Nanotubes

Now, ten years after Iijima's initial observation, we know a great deal about nanotubes and tubular fullerenes. We know that multiwall nanotubes are invariably produced with a high frequency of structural defects. (In comparison with their larger relations, the 5-20 micron-diameter graphite fibers used in aerospace and sporting goods applications, multiwall nanotubes are quite sound structurally; nevertheless, they frequently contain regions of structural imperfection.) As any material scientist knows, it is the occurrence of defects that inevitably degrade the material properties of a substance, such as strength. The intrinsic properties of a material may be world-beater, but typically the actual properties of the bulk material are only a few percent of what the material would exhibit if it were structurally perfect. For example, a structural defect such as a microcrack in a steel wire, will lead to catastrophic failure at 1-2% of the theoretical breaking strength one would predict based on fundamental chemical principals.

Buckytubes

In contrast, buckytubes are fullerenes, and are thus molecules: perfect, hollow molecules of pure carbon linked together in a hexagonally bonded network to form the hollow cylinder. The tube is seamless, with either open or capped ends. The diameter of single-wall carbon nanotubes is 0.7 to 2 nm (typically about 1.0 nm) – 100,000 times thinner than a human hair. Buckytube lengths are typically hundreds of times their diameters.

Molecular Considerations

The molecular aspect of buckytubes is critical. Every atom is in the right place. This is a profound difference with their larger, defective cousins. To a chemist, a molecule is a very special thing. A molecule is complete, and is usually relatively happy with its identity. When a molecule is confronted with the opportunity to change, that is, to undergo a chemical reaction with other bits of stuff, there is almost always a fairly significant barrier to overcome. Other, non-molecular "stuff", when confronted with other stuff, is usually quick to change, and add in with the new part to make a larger whole. Metals are like this: a chunk of metal exposed to more metal (this could take place in the molten, gaseous, or solid phase) will accommodate the addition because it does not possess molecular completeness and invariance.

Molecular Invariance

One aspect of molecular invariance is the reliability and predictability of chemical change. Molecules can be induced, by various means, to overcome their barriers to change, for example using heat. But by and large the reaction products when dealing with molecules are consistent. This is not true of changes that non-molecular things undergo. No two chunks of metal ever look the same, just as no two snowflakes look the same: snowflakes are not molecules. The effect of molecular invariance on material properties is equally profound. Whatever the intrinsic nature of the material, there are no defects to degrade the properties. You get what you get.

Understanding the Structure, Naming and Properties of Carbon Nanotubes (Buckytubes)

The special nature of carbon combines with the molecular perfection of buckytubes (single-wall carbon nanotubes) to endow them with exceptionally high material properties such as electrical and thermal conductivity, strength, stiffness, and toughness. No other element in the periodic table bonds to itself in an extended network with the strength of the carbon-carbon bond. The delocalised pi-electron donated by each atom is free to move about the entire structure, rather than stay home with its donor atom, giving rise to the first molecule with metallic-type electrical conductivity. The high-frequency carbon-carbon bond vibrations provide an intrinsic thermal conductivity higher than even diamond. In most materials, however, the actual observed material properties—strength, electrical conductivity, etc.—are degraded very substantially by the occurrence of defects in their structure. For example, high strength steel typically fails at about 1% of its theoretical breaking strength. Buckytubes, however, achieve values very close to their theoretical limits because of their perfection of structure—their molecular perfection. This aspect is part of the unique story of buckytubes. Buckytubes are an example of true nanotechnology: only a nanometer in diameter, but molecules that can be manipulated chemically and physically. They open incredible applications in materials, electronics, chemical processing and energy management.

Basic Structure

Buckytubes are single-wall carbon nanotubes, in which a single layer of graphite—graphene—is rolled up into a seamless tube. Graphene consists of a hexagonal structure like chicken wire. If you imagine rolling up graphene or chicken wire into a seamless tube, this can be accomplished in various ways. For example, carbon-carbon bonds (the wires in chicken wire) can be parallel or perpendicular to the tube axis, resulting in a tube where the hexagons circle the tube like a belt, but are oriented differently. Alternatively, the carbon-carbon bonds need not be either parallel or perpendicular, in which case the hexagons will spiral around the tube with a pitch depending on how the tube is wrapped.

Naming Convention

There is a straightforward labelling convention to distinguish differently wrapped tubes from one another. The mapping specifies the number of unit vectors required to connect two atoms in the planar hexagonal lattice to form a seamless tube. These numbers specify a "vector" for the mapping, commonly expressed as (m,n), where m and n are integers. These numbers constitute a unique "name" for a tube. Any tube "named" (n,0) has carbon-carbon bonds that are parallel to the tube axis, and form, at an open end, a "zig-zag" pattern; these tubes are referred to as "zig-zag" tubes. Tubes named (n,n), where the two integers are equal, have carbon-carbon bonds that are perpendicular to the tube axis, and are often called "armchair" tubes. These two basic types are achiral, meaning they do not have a distinct mirror-image, like left and right hands. All the other tubes, named (m,n), where m does not equal n, and neither is 0, are chiral, and have left-and right-handed variants.

Properties of Different Tube Types

In most respects, the properties of tubes of different types are essentially the same. The exception to this is in their electrical conductivity, where these subtle structural differences can have profound effects. For example, all armchair tubes—i.e., where m=n—have truly metallic electrical conductivity.

They transport electrons along the tube axis just as metals do, without a single atom of metal in their structure! This behaviour in a molecule is unprecedented. In contrast, the other tubes are intrinsically semiconducting, either with a very small band gap of a few meV, or with moderate band gaps on the order of 1 eV. The rule here is that those tubes where (n-m) is a multiple of 3 are the small-gap type, while the others have medium gaps.

Since tubes with different (n,m) are molecularly distinct, there exists the possibility of chemically separating different types, and even of growing only selected tube types, although currently, all production processes produce a random mixture.

Ropes

Another structural aspect of tubes is their self-organisation into "ropes," which consist in many (typically, 10-100) tubes running together along their length in van der Waals contact with one another. Ropes are far longer than any individual tube in them: whereas tubes are typically about 100-1000 nm in length (and about 1 nm in diameter) ropes are essentially endless, branching off from one another, then joining others. These ropes are useful in providing very long conductive pathways well below what would normally be the loading required to reach a "percolation threshold" for conductivity.

Towards Flexural Strain Sensing Using Carbon Nanotube Film

The sensor/actuator embedded smart materials and systems are essential for structural damage detection and control. There is need for smart structural materials, which can be used to sense strains, and adjust the structural properties so as to achieve desired response. Carbon nanotube composites have the potential to be such a smart structural material. Results (Baughman *et al.*, 1999) have shown that when carbon nanotube films, also called as "buckypapers", are used as actuators, large actuator strains can be achieved by smaller operating voltages as compared with ferroelectric and electrostrictive materials. Single wall carbon nanotubes (SWCNTs) promise to be a reinforcement material for nanocomposites because of their superior mechanical properties. Researchers (Hadjiev *et al.*, 2001; Zhao *et al.*, 2001) have studied the Raman band shift in SWCNT's as a function of stress/strain in carbon nanotubes. Raman studies on multi-wall carbon nanotubes have also been carried out (Schadler *et al.*, 1998; Wagner *et al.*, 1998). Though the experiments are carried at nanoscale, the results illustrate the potential of SWCNT as strain/stress sensor. It has also been shown that the electrical band gap changes as a function of axial compression, tension stretch, torsion, and bending strain (Peng and Cho, 2002; Tombler *et al.*, 2000; Yang and Han, 2000). Strong dependence of SWCNT's band structure on mechanical deformation makes it possible to develop nano-electro-mechanical sensors. Most of the studies to date relate the mechanical deformation with the change in electrical properties at the nanoscale. In this study, change in the electronic property of carbon nanotubes due to strain at macroscale is demonstrated using the experimental results. Such carbon nanotube films have randomly oriented SWCNTs with isotropic strain sensing properties; thus, have multidirectional and multi-location strain sensing capability.

The carbon nanotube films used in this study are produced by mixing SWCNTs with 0.25?mg/ml *N,N*-Dimethylformamide (DMF) and filtering the mixture through a 0.2?mm teflon membrane. Freestanding carbon nanotube film (Buckypaper) is peeled from the filter after rinsing and drying of the remaining material. The carbon nanotube film is further dried for 24?h under vacuum. The SEM picture of a carbon nanotube film with ~15?ìm thickness. To explore the strain sensing potential of the carbon nanotube film, it is attached to a 0.25×2×24?in. brass specimen and subjected to pure bending moment using two-point loading. For insulation, a PVC film is attached between the brass specimen and carbon nanotube film. High strength epoxy and a vacuum-bonding method is used to attach carbon nanotube film and PVC to the brass specimen in order to ensure the perfect strain transfer – between brass, PVC, and the carbon nanotube film. A conventional electrical resistance strain gage is attached next to the carbon nanotube film to measure flexural strain for comparison purposes. Load P is applied in increments using a servo-hydraulic machine. The corresponding change in voltage across the carbon nanotube film is measured using four-point probe measurement (Smits, 1958). Input voltage across two outer probes is kept constant during the measurement and change in voltage across the two inner probes is recorded. Strain measurements are also made using conventional strain gage. The change in voltage is proportional to the change in resistivity of the film (Dharap *et al.*, 2004). The change in voltage due to the change in dimensions of the film is small as compared to the change in voltage due to the change in resistivity of the film (Li *et al.*, 2003).

The change in voltage in the carbon nanotube film as a function of measured flexural strain from the conventional strain gage. As the strain measured by the conventional strain gage increases from 0 to ~600?ìm/m the voltage change measured across the two outer probes increases from 0 to ~200?ìV. A nearly linear trend between the change in voltage and the strain is evident; although, deviation from the linear trend exists. Further study is needed to establish causes for deviation from linear trend such as temperature and gas exposure history (Bezryadin *et al.*, 1998; Collins and Avouris, 2000; Hone *et al.*, 2002). The change in voltage is measured by moving the four-point probe to several parallel locations on a single carbon nanotube film, which also yielded a similar linear trend.

It is evident from the experimental study that the carbon nanotube film can be used for measuring flexural strains at the macroscale. Carbon nanotube films are made up of randomly oriented SWCNTs; hence, their electronic properties are independent of direction. Hence taking measurement along different directions provides corresponding strains. Carbon nanotube films can also be integrated into composites to simultaneously act as strong structural material and sensor. Such composites may find application in smart structures.

Analysing Carbon Nanotube Transistors, Rings and Theories

Scientists have long been intrigued by the electrical properties of carbon nanotubes, those tiny straws of pure carbon measuring less than a hair's width across. Now one research team has unveiled an improved design for transistors based on the minuscule tubes, one that could substantially reduce the lower limit on transistor size.

The size of conventional transistors made of metal oxide semiconductors (MOS) has shrunk over the years and can now be as small as about 100 nanometers across. This reduction has allowed for an impressive increase in both the speed and power of a variety of electronics. By employing carbon nanotubes, a team led by Prabhakar R. Bandaru of the University of California at San Diego hopes to shrink transistor dimensions even more--down to just a few nanometers thick. "The small size and dramatic switching behaviour of these nanotubes makes them candidates for a new class of transistor," he explains.

The novel transistors were first grown as regular, straight nanotubes. Using catalyst particles containing iron, the team then coaxed a second branch to grow out of the first tiny straw. The resulting y-shaped nanotubes carry current when attached to electrical contacts, and varying the amount of voltage applied to the stem controls the flow of electrons through the junction. The new approach improves on previous carbon nanotube-based transistors, in which separate components had to be added in. "We can synthesize functionality at the nanoscale, in this case to include the three elements of a circuit—the gate, source and drain," Bandaru says, "and we don't have to go to the trouble of making them separately and assembling them."

The tubes are metallic or quasi, the others are semiconductors with moderate band-gaps. If solid particles become small enough, in the nanometer range, they become electronically comparable to atoms and molecules. Instead of the laws of classical physics, they become quantum systems with discrete energy states. This is the basis for a number of useful applications. Quantum dots are reminiscent of the rhetorical question raised by Greek philosophers in antiquity: consider a sand-heap. If one removes from it a single grain of sand, obviously it is still a sand-heap. One continues doing so. At which stage does the sand-heap become something else? One of the most exciting applications of quantum dots is the single electron tunneling transistor. An electrometer measures electric charge. A single electron transistor is, at present, the most sensitive electrometer. Small alterations of the charge distribution beneath a small isolated island of aluminum atoms markedly affect the conductivity across that island. When a gate voltage is applied to an electron donor material, such as tellurium, placed underneath the island, changes in conductivity can be monitored. Nanocrystals are indeed important in single electron devices operating at room temperature, such as supersensitive electrometers and memory storage. Capped nanocrystals of both metals and semiconductors by virtue of their size possess capacitance in the range of attofarads (1 aF = 10^{-18} F). Charging a nanocrystal with an extra electron perturbs it to such an extent that the next electron requires an appreciable change in the charging potential. This is often seen as a 'Coulomb staircase' in the current-voltage tunneling spectra. Indeed, the charging energy varies linearly with the inverse of the diameter of the nanocrystal. Such sensitivity to single electron charging makes a nanocrystal an ideal candidate for use in single electron transistors and memory devices. There are metallic and there are semiconducting nanoparticles. The former enjoy industrial and medical applications, for their magnetic properties in particular. The latter have uses in electronics and optoelectronics. One can also write with nanoparticles, dip-pen lithography is one of the techniques. More generally, the challenge is to organise nanoparticles in three dimensions.

The natural and the artificial once again meet! Biosystems, such as enzymes, antibodies, and nucleic acids have sizes commensurate with those of man-made nanoparticles. Hence, it becomes tempting to marry and hybridize them into nanocomposites, which is one of the most recent developments. Two of the main thrusts of nanotechnology have been self-assembled ordered arrays of atoms with specific properties (electrical conductivity, magnetism, *etc.*) and well-defined, molecule-like clusters of atoms endowed with specific properties. The keyword here is specificity. The standard account of the birth of nanotechnologies dates it to the lecture by Richard P. Feynman on December 29, 1959 at the annual meeting of the American Physical Society held at Caltech. The books under review rightfully salute Feynman as the Founding Father of the field. As is well-known, the brilliant Caltech physicist ended his talk with a dare, announcing two prizes financed with his own money for the miniaturisation of the pages of a book and for the building of a miniscule electric motor, only 1/64 inch cube. Feynman's lecture makes interesting reading. Its perspective is identical to that in Jules Verne's books, not visionary in the sense of a prophecy, imagining revolutionary developments and devices. The outlook is much more down to earth. Jules Verne and Richard P. Feynman took heed of recent scientific and technological advances to extrapolate to what the future would almost certainly witness, in a prospective. In his lecture Feynman took as his starting point the feasibility of scaling down the contents of the world's libraries. He guesstimated their holdings at about 24 million volumes. He figured that their total information could be contained on an area of about 35 pages of the *Encyclopaedia Britannica*.

The discussions of potential economic, medical, and environmental benefits may have given the false impression that nanotechnology will create a wondrous utopia in which all human problems are solved and we all live happily ever after. This is even more mistaken than the idea that new technologies always cause more problems than they solve. Many of the main constraints and difficulties faced by people are based not on technology or its lack, but instead by the very nature of the world we live in and the essence of our humanness. Increasing affluence based on molecular manufacturing won't end economic problems any more than past increases in affluence have. Wilderness can still be destroyed; people can be oppressed; financial markets can be unstable; trade wars can be waged; inflation can soar; individuals, companies, and nations can go into debt; bureaucracy can stifle innovation; tax levels can become crippling; wars and terrorism can rage. None of these will automatically be stopped by advanced technology. What is more, the potential benefits of new technologies aren't automatic. Nanotechnology could be used to restore the environment, to spread wealth, and to cure most illness. But will it? This depends on human action, working within the limits set by the real world. This chapter first describes some of the limits to what nanotechnology can accomplish, and then some of the adverse side effects of its basically good applications. The next will discuss the problem of accidents, which seems manageable, and then the far greater problem of potential abuse of new capabilities. The world imposes limits on what we can do. Technology in general (and nanotechnology in particular) can provide padding for us as we throw ourselves against these hard, sharp limitations, and can sometimes help us slip past old limits through previously unknown gaps. Eventually, though, we will encounter new limits. In the end, solid constraints will limit human action no matter how much we juggle atoms and molecules, or the bits and bytes of information. Let's look at some of these, starting with the most abstract and long term—the most definite and hardest to avoid—and moving toward the more personal and near term. Many problems differ fundamentally from the material problems of limited matter and energy: they involve information.

Some of the most precious stores of information in the world today are the genetic codes of the biosphere. This information, different for virtually every individual organism, is the product of millions of events that we are incapable of modeling or recreating. When this information is lost, it is lost forever. When the atoms encoding this information are thoroughly scattered, there seems to be no way to retrieve it. With any species, most genetic information is shared in common, found in all members of that species. But the variations in genetic code between individuals are important, both to the individuals themselves and to the health and prospects of the species as a whole. Consider the northern white rhino, whose numbers have dropped to an estimated thirty-two animals, or the California condor, of which only forty remain, all in captivity. Even if biologists succeed in reestablishing these species—eight condors were hatched in 1989—much of the diversity of their genetic information has been lost. Worse yet are extinctions of species for which no tissue samples were saved. The future may see some amazing recoveries: dry skin and bones may yield a complete set of genes when sifted by molecular machinery, and even current techniques have been used to recover genes from an ancient leaf, almost 20 million years old. Our eyes and instruments cannot yet tell us how much information from the past remains, but we do know that genetic information is being lost every day, and once lost, it is irretrievable. People have often been wrong about physical limits, confusing the limits of their technology with the limits of the possible.

As a result, learned men first dismissed the idea of heavier-than-air flight, and then dismissed the idea of flying to the Moon. Yet physical limits are real, and all technology—past, present, and future—will stay within those limits. There is even reason to suspect that some of those limits are where the learned now believe them to be. Nanotechnology will make it possible to push closer to the real limits set by natural law, but it will not change those laws or the limits they set. It will not affect the law of gravity, the gravitational constant, the speed of light, the charge of the electron, the radius of the hydrogen atom, the value of Planck's constant, the effects of the uncertainty principle, the principle of least action, the mass of the proton, the laws of thermodynamics, or the boiling point of water. Nanotechnology won't make energy or matter from nothing. It seems a good bet that no one will build a faster than light spacecraft, or an antigravity machine, or a cable twice as strong as diamond. There are limits. Science today may be wrong about some limits, but scientific knowledge is practically defined to be our best information about how the world works, so it isn't wise to bet against it. There will be claims that nanotechnology will be able to do things that it can't, or that capabilities are around the corner when they aren't. Sometimes these will be innocent errors, sometimes they will be culpably stupid errors, and sometimes they will be what amounts to fraud. Among the problems that nanotechnology cannot solve is that of misguided claims, by people calling themselves "scientists," "engineers," or "businesspeople," that they have a big technical breakthrough worth a fortune. Every interesting new technology, particularly in its early days, is a chaotic mix of competent workers and charlatans.

For every Thomas Edison inventing useful products such as light bulbs or the precursor of movie projectors, there were people promoting electric hairbrushes to cure baldness, and electric shoes, electric belts, electric hats—the list goes on—that authoritatively claimed cures for infertility, overweight, underweight, and all the ills and discomforts of mankind. Today, we laugh at the credulity of our forefathers who bought these gadgets; we shouldn't, unless we laugh at our own times as well. Natural law imposes limits, but so does the nature of human beings. These will continue as long as people do. Reproduction is a deeply ingrained instinct enforced by the march of time, which ruthlessly discards the genetic material of all who neglect it. Many would argue that the Earth is already overpopulated.

While nanotechnology could enable the current population, and even a greatly increased one, to live more lightly on the Earth, there will still be limits to Earth's capacity. The norms of human life are shaped by ancient patterns: high rates of infant or childhood mortality have been facts of life for millennia, and having many, many children has been a way to ensure that one or two will survive to work on the farm, and to care for you in your old age. Large families naturally become traditional. When modern medicine and reliable food supplies change those conditions—as they have, in cultural terms, virtually overnight—behaviour does not shift as quickly. The result is the Third World population boom. In Western countries, where there has been time for behaviour to adapt, a huge family is the exception.

It might seem that our problem is solved. Molecular manufacturing can make everyone wealthy, and wealthy populations today have stable or shrinking populations. The Earth can support more people with advanced technologies, and these will also open up the vast room and resources of the world beyond Earth. Would that this were true. If 99 per cent of the people in a population respond to wealth by reducing childbearing, the population will indeed stabilize or shrink, for a while. But populations are not uniform. What of the 1 per cent, say, who are members of a minority with different values? If that minority has a growth rate of 5 per cent per year, then in ninety-five years they will be the majority, and in one thousand years their population will have grown by a factor of 1,500,000,000,000,000,000,000, if resource limits or genocide haven't intervened. Note that the Hutterites of North America, a reasonably wealthy religious group viewing fertility control as a sin and high fertility as a blessing, have managed an average of ten children per woman. Given enough time, exponential growth of even the smallest population can consume all the resources in reach.

The right to reproduce is often regarded as basic, as illustrated by the outrage at reports of forced abortion in the People's Republic of China. The Hutterites and many others regard it as part of their freedom of religion. But what happens when parents have more children than they can support—does redistribution solve the problem? If reproduction is not forcibly suppressed, and if resources are forcibly and repeatedly redistributed so that each human being has a roughly equal share, then each person's share will steadily shrink. Even given the most optimistic assumptions regarding available resources, with a policy of resource redistribution and unlimited reproduction, the amount per person would eventually be insufficient to sustain life. This policy must be avoided, because if it is followed, it will kill everyone. As soon as we grant that any entity is entitled to certain rights—whether that entity be a human child, an animal, or some future artificial intelligence—the question arises of who is responsible for providing resources to support it when it can't do so for itself.

The above argument indicates that a policy of coercion by some central power to compel the entire population to support an exponentially exploding population of these individuals would lead directly to disaster. Ultimately, this responsibility must rest with the entities' initiator: the designer of the artificial intelligence, the owner of the pet, the parents of the child. No new technology can magically remove the limits imposed by natural law, and thereby lift the burden of human responsibility. Every time a technology solves a problem, it creates new problems. This doesn't mean that the change is neutral, or for the worse, of course. The Salk and Sabin vaccines for polio virtually destroyed the iron-lung industry, and the pocket calculator virtually destroyed the slide-rule industry, but these advances were worth the price of some economic adjustment. Molecular manufacturing and nanotechnology will bring far greater changes, placing far greater strains on our ability to adapt. We shouldn't be surprised when basically beneficial applications make someone miserable. Our lives are largely centered around problems.

If we can solve many of these problems, the centers of our lives will shift, creating fresh problems. This section sketches some of the issues of change and adaptation more to raise questions than to offer solutions. Molecular manufacturing offers the possibility of drastic change, a change in the means of production more fundamental than the introduction of industry, or of agriculture. Our economic and social structures have evolved around assumptions that will no longer be valid. How will we handle the changes in the way we work and live? Nanotechnology will have wide-ranging impact in many areas, including economic, industrial, and social patterns. What do historical patterns in similar circumstances tell us about the future? Any powerful technology with broad applications revolutionizes lives, and nanotechnology will be no exception.

Depending on one's point of view, this may sound exciting or it may sound disturbing, but it most certainly does not sound comfortable. In comparison to many projections of the twenty-first century, though, nanotechnology may lead to comparatively comfortable change. The changes most often projected—for a future not including nanotechnology—have been ecological disaster, resource shortages, economic collapse, and a slide back into misery. The rise of nanotechnology will offer an alternative—green wealth—but that alternative will bring great changes from the patterns of recent decades. Times of rapid technological change are disconcerting. For most of humanity's existence, people lived in a stable pattern. They learned to live as their parents had lived—by hunting and gathering, later by farming—and changes were small and gradual.

A knowledge of the past was a reliable guide to the future. Sudden changes, when they did occur, were apt to be ruinous: invasions or natural disasters. These sudden changes were fought or repaired or survived as best one could. Making major changes by choice was rare, and radical innovations were generally for the worse: the old ways at least ensured the ancestors' survival, the new might not. This made cultures conservative. It is only natural that there be efforts to resist change, but before undertaking such an effort, it makes sense to examine the record of what works and what doesn't.

The only examples of successful change fighters have been communities that have created and maintained barricades to isolate themselves from the outside world socially, culturally, and technologically. For the two centuries before 1854, Japan turned its back on the outside world, following a deliberate policy of seclusion. The leaders of Albania restricted contacts for many years; only recently have they started to open up. Isolation attempts have worked better on a smaller scale, when participation is voluntary rather than decreed by government. Today, within the Hawaiian island chain, the tiny, privately owned island of Niihau, sixteen miles long and six miles wide, is deliberately kept as a preserve of the nineteenth-century Hawaiian lifestyle. Over two hundred full-blooded Hawaiians there speak the Hawaiian language and use no telephones, plumbing, television, and no electricity (except in the school). The Amish of Pennsylvania have no surrounding ocean to help maintain their isolation, but rely instead on tight social, religious, and technological rules aimed at keeping external technology and culture out, and themselves grouped in; those who leave the fold are excluded.

On a national scale, attempts to take only one part of the package—whether social or technological—haven't done well at all. For decades, the Soviet Union and the Eastern bloc nations welcomed Western technology but attempted tight restrictions on the passage of people, ideas, and goods. Yet illegal music, thoughts, literature, and other knowledge still crept in—as they do into the Islamic countries. Fighting technological change in society at large has had little success, where that change gave some large group what it wanted. The most famous fighters of technological change—the Luddites—were unsuccessful.

They smashed "automated" textile machinery that was replacing old hand looms during the early industrial revolution in England, but people wanted affordable clothing, and smashing equipment in one place just moved the business elsewhere. Change has sometimes been postponed, as when a later group, under the banner of "Captain Swing," smashed hundreds of threshing machines in a wide area of southern England in 1830. They succeeded in keeping the old, labor-intensive ways of harvesting for over a generation. In previous centuries, when the world was less tightly connected by international trade, communications, and transportation, delays of years and even decades could be enforced through violence or legal maneuvers such as tariffs, trade barriers, regulations, or outright banning. Attempting to stop or postpone change is less successful today, when technology moves internationally almost as easily as people do—and human travel is so easy that 25 million people cross the Atlantic each year. Change fighters find that the problems they create mount with time.

Products made using the old, high-cost techniques are uncompetitive. There is no way to bring back the "old jobs": they no longer make sense. But old habits die hard, and these same responses to the prospect of technological change continue today—ignoring it, denying it, and opposing it. Societies that have fought change, as Britain did, have fallen behind in a cloud of coal smoke. Why did the Luddites respond violently? Perhaps their response can be attributed to three factors: First, the change in their lives was sudden and radical; second, it affected a large group of people at one time, in one area; and third, in a world unprepared for rapid technological change, there was no safety net to catch the unemployed. While local economies might have been able to absorb a trickle of hungry laid-off workers, they lacked the size and diversity needed to offer other employment options quickly to large numbers of unemployed. In the twentieth century, however, societies have of necessity become somewhat better adapted to change. This has been a matter of necessity, because sluggish communities soon fall behind.

In the ancient days of peasant stability, there was no need for institutions like Consumer Reports to study and rate new products, or regulators like the Environmental Protection Agency to watch over new hazards. We developed the needs, and we developed the institutions. These mechanisms represent important adaptions, not so much to the technologies of the twentieth century, but to the increasing change in technology during the twentieth century. There is great room for improvement, but they can perhaps provide a basis for adapting to the next century as well. Even with the best of institutions to cushion shocks and discourage abuse, there will be problems. The very act of solving problems of production—of increasing wealth—will create problems of economic change. Over centuries, the trend has seemed to be toward centralisation, beginning with the rise of factories and industrial towns. What drove these developments was the high cost of machinery and plant operations, the need to be near power sources, the impracticality of transportation among many small, dispersed sites, and the need for face-to-face communication. Beginning with the first industrial revolution, factories employed large numbers of people in one place, leading to overcrowding and making local economies dependent on one industry and sometimes on a single company.

Costly equipment necessitated central locations for textile production, rather than the cottage industries where a lone woman could earn a livelihood carding wool and creating thread on a spinning wheel (providing the origin of the term *spinster*). By the 1930s, the belief in the virtues of centralisation and central planning—the supposed efficiencies and economies of scale—led to nationwide or continentwide experiments in centralisation. But over the last decade, these large-scale experiments have been dismantled, from Britain's privatisation of nationalised utilities to the beginning of a return to the market system in Eastern European countries. Because the old limits on transportation, energy sources, and communication have fallen, business is now decentralising.

Between 1981 and 1986, the Forbes 500 companies cut their employees by 1.8 million. But during those same years, total civilian jobs went up by 9.2 million. Start-up companies created 14 million jobs; small companies created another 4.5 million. Telecommuting is booming, as are new businesses, independent professionals, and cottage industries. We've also seen the resurgence of small, but highly diverse stores: gourmet food shops, specialty ethnic shops, tea and coffee purveyors, organic and health food stores, bakeries, yogurt shops, gourmet ice-cream stores, convenience stores offering twenty-four-hour access, shops selling packaged food plus snacks. These stores epitomize something fundamental: At some point, what we want is not a standard good at an ever cheaper price, but special things customised to meet our own individual tastes or needs.

The trend for advanced technologies seems to be leading away from centralisation. Will nanotechnology counter or accelerate this trend? By reducing the cost of equipment, by reducing the need for large numbers of people to work on one product, and bringing greater ability to produce the customised goods that people want, nanotechnology will probably continue the twentieth-century trend toward decentralisation. The results, though, will be disruptive to existing businesses. The computer industry perhaps provides a clue to what might happen as costs are lowered by nanotechnology. The computer-software industry is characterised by the garage-shop start-up. When your equipment is cheap—inexpensive PCs built around low-cost chips—and you can make a product by throwing in some ingenuity and human labor, it's possible to start a new industry on a shoestring.

In 1900, when cars were simple, there were many car manufacturers. By the 1980s, if you weren't an industrial giant like General Motors or Ford, Honda or

Nissan, you had to be John De Lorean to even get a shot at acquiring the capital to play in the business. If molecular manufacturing can slash the capital costs for producing cars or other plant-intensive equipment, we will see the equivalent of garage-shop businesses springing up to offer new products, and hiring workers away from the industrial giants of today just as the personal computer has destroyed the dominance of the mainframe. The U.S.A. dream is to be an entrepreneur, and the technological trends of the twentieth century point in that direction. Nanotechnology probably continues it.

In one area, however, the late twentieth-century trend has been toward uniformity. The nations of Western Europe are in the process of uniting under one set of economic rules, and parts of Eastern Europe are anxious to join them. More and more supranational and transnational organisations knit the world together. The growth of trade has motivated economic integration. Molecular manufacturing will work against this trend as well, permitting radical decentralisation in economic terms. This will help groups that wish to step aside from the stream of change, enabling them to be more independent of the turbulent outside world, picking and choosing what technologies they use. But it will also help groups that wish to free themselves from the constraints of the international community. Economic sanctions will have little force against countries that need no imports or exports to maintain a high standard of living. And export restrictions will likewise do little to hamper a military buildup. By weakening the ties of trade, molecular manufacturing threatens to weaken the glue that holds nations together. We need that glue, though, to deal with the arms control issues raised by molecular manufacturing itself. This problem, caused by the potential for decentralisation, may loom large in the coming years. Lester Milbrath, professor of sociology and political science, observes, "Nanotechnologies will create the problem of how to meaningfully and sustainably occupy the time of people who need not perform much work in order to have a sufficiency of life's goods. Our society has never faced this problem before, and it is not clear what social restructuring will be required to have a good society in those circumstances. We face much deep social learning." The world has had little experience with what anthropologists call "abundance economies."

The native American tribes of the Pacific Northwest were one of those rarities. Ruth Benedict, in her classic book *Patterns of Culture*, wrote, "Their civilisation was built upon an ample supply of goods, inexhaustible, and obtained without excessive expenditure of labor." The Kwakiutls became famous for their "potlatches": contests in which they sought to shame their rivals by heaping more gifts upon them than they could ever return. The potlatches would often be a year in preparation, last for days, and occasionally involve destruction of entire buildings. It was certainly a colorful form of keeping up with the Joneses. What will motivate us, once we have achieved an abundance economy? What will we regard as worthwhile goals to pursue? Increased knowledge, new art, improved philosophy, eliminating human and planetary ills? Will we find ourselves creating a better, wiser world, or sunk in boredom and jaded now that we have all and want nothing?

If boredom gets out of hand, the lively spectacle of wealthy donors seeking to outdo each other to endow the arts, aid the poor, and do other good deeds for the sake of prestige would be welcome. What will happen as life spans continue to lengthen and the time needed to make a living decreases? Even today, there are people who, when confronted with the prospect of a significantly longer life span, exclaim that they couldn't imagine what they would do with all that time. This response can be hard to understand, when it would take a thousand years to walk all the world's roads, more thousands of years to read all the world's books, and another ten thousand years to have a dinner conversation with each of the world's people—but tastes differ, and even a few decades of bad television might make anyone long for the peace of the grave. A major concern, and certainly the single area of greatest upheaval, is employment (which may become hard to distinguish from leisure). Once, people had little choice of employment. To keep a full belly, most had to work at the only job available: peasant farming. Eventually, people will have a complete choice of employment: they will be able to keep a full belly and a wealthy lifestyle while doing whatever they please.

Recently, we are about halfway between those extremes. In advanced economies, many different jobs are deemed useful enough that other people will offer an adequate income in exchange for the result. Some people can make a living doing something they enjoy—is this work, or leisure? The impact of nanotechnology on patterns of employment will depend on when it arrives. Current demographics show a shrinking supply of young people entering the work force. Agriculture, the assembly line, and entry level service jobs are experiencing a labor shortage, and no relief is in sight. If these trends continue, nanotechnology may show up in the midst of a shortage of labor. If it arrives late enough, it may compete with industries that are already nearing full automation; "job displacement" may mean replacing an industrial robot with a nanomachine. Employment patterns have shifted radically in the past. One hundred and fifty years ago, the United States was an agricultural nation—69 per cent of all people worked the land and a growing percentage worked in industry doing things like building steam locomotives for Baldwin Locomotives Works or tanning leather for the giant Central Leather monopoly.

By the early twentieth century, agriculture was waning in numbers but increasing in productivity; most people worked in industry, and the tiny information and service sector was beginning to grow. Today, the picture has reversed: 69 per cent of employed Americans work in information or service jobs, only 28 per cent work in industrial production, and 3 per cent in agriculture. This tiny fraction feeds the other

97 per cent of Americans, exports hugely to other countries, and receives subsidies and price support payments to stop them from growing even more food. Manufacturing, even without nanotechnology, seems to be heading toward a similar condition. With an ever-declining percentage of our population working in manufacturing, we have as everyday products things that were once available only to kings and the high nobility. Yet owning multiple suits of clothes, having personal portraits of ourselves and family members, having music upon our command, having a personal bedroom, and having a coach awaiting our need—these are now regarded as being among the bare necessities of life.

It may be possible to adjust to even greater wealth with even less required labor, but the adjustment will surely cause problems. In a world in which nanotechnology reduces the need for workers in agriculture and manufacturing still further, the question will be asked, "What jobs are left for people to do once food, clothing, and shelter are very inexpensive?" Again, the twentieth century provides some guidelines. As technology has reduced costs by efficiently producing many units of an identical item, people have begun to demand customisation to meet individual needs or preferences. As a result, there are ever more jobs in producing custom goods. Today, semi-custom goods that try to help us meet our needs or express our taste abound: designer linens, ready-to-wear fashions, cosmetics, cars, trucks, recreational vehicles, furniture, carpeting, shoes, televisions, toys, sports equipment, washing machines, microwave ovens, food processors, bread bakers, pasta makers, home computers, telephones, answering machines—are all available in large and ever-changing variety. Just as varied is the fabulous wealth and diversity of information produced in the twentieth century. Information products are a large factor in the economy: Americans buy 2.5 billion books, 6 billion magazines, and 20 billion newspapers each year. In recent years, new magazines have been invented and launched at the rate of one every business day of the year.

A visit to a well-stocked magazine rack shows only a hint of the wealth of highly specialised publications, each one focused on a specialised interest or attitude: hotdog skiing, low-fat gourmet cooking, travel in Arizona, a magazine for people with a home office and a computer, and finely tuned magazines on health, leisure, psychology, science, politics, movie stars and rock stars, music, hunting, fishing, games, art, fashion, beauty, antiques, computers, cars, guns, wrestling. Motion pictures, which started as a flock of independent production companies and then consolidated into the great studios of the 1930s, have since followed the decentralisation and diversification trends of recent years.

Now an expanding range of film entertainment comes via network TV, cable channels, private networks, videotapes, music videos. Independent producers are aided by the technology innovations of cable, direct broadcast satellites, videotape technology, laser disks, videocameras. The arts have burgeoned, with the general public as the new patron of the arts. Any artist or art form that could find and satisfy a market boomed in the twentieth century. Not just the traditional arts of actors, writers, musicians, and painters, but all forms of "domestic" artistry have grown to unprecedented levels: landscape and interior design, fashion design, cosmetics, hairstyling, architecture, bridal consulting. Providing for these demands are some of the "service and information" jobs created in the late twentieth century.

"Service" jobs include many ways of helping other people: from nursing to computer repairs to sales. In "information" jobs, projected to have the fastest percentage growth over the next decade, people find, evaluate, analyze, and create information. A magazine columnist or TV news producer obviously has an "information" job. But so do programmers, paralegals, lawyers, accountants, financial analysts, credit counselors, psychologists, librarians, managers, engineers, biologists, travel agents, and teachers. "Increasingly," states *Forbes* magazine, "people are no longer laborers; they are educated professionals who carry their most important work tools in their heads. Dismissing them from their jobs, cutting them off from their places of employment may hurt them emotionally and financially. But it doesn't separate them from their vocation in the same way that pushing a farmer off his freshly seeded land does. For centuries workers were more dependent on a particular physical setting than they are now. Modern occupations generally give their practitioners more independence—and greater mobility—than did those of yesteryear." These human skills that people carry with them will continue to be valued: managing complexity, providing creativity, customising things for other people, helping people deal with problems, providing old services in new contexts, teaching, entertaining, and making decisions. A reasonable guess would be that many of the service and information industries of the twentieth century will continue to evolve and exist in a world with nanotechnology. What is harder to imagine would be what new industries will come into being once we have new capabilities and lower costs. Along with the old economic law of supply and demand is another governing factor: price elasticity effects.

People's desire for something is "elastic": it expands or contracts when the cost of something valuable goes down or up. If the price of a flight to Europe is five hundred dollars, more people will take a European vacation than if the price is five thousand dollars. When you had to hire a highly trained mathematician to do equations, calculation was slow and expensive. People didn't do much of it unless they absolutely had to. Today, computers make calculation cheap and automatic. So now businesses do sophisticated financial modeling, chemists design protein molecules, students calculate orbital trajectories for spaceships, children play video games, moviemakers do ever more amazing special effects, and the cartoon—virtually extinct because of high labor costs—has returned to movie theaters, all because computers permit cheap calculation. Nanotechnology will offer new, affordable capabilities to these and other people. Today, it's as hard to predict what new industries will be invented as it would have been for the creators of the ENIAC computer to have predicted cheap, handheld game computers for children. So rather than producing drastic unemployment, nanotechnology seems likely to continue the trend already seen today, away from jobs that can be automated and into jobs where the human perspective is vital. But the true possibilities are, as always in the modern world, beyond predicting. Major shifts in demographics always cause disruptions. Even when we know they are coming, we never prepare for them. Our plans are based on expectations of what will happen. If things don't go as expected, we find that we have "malinvested." Houston real estate was valuable and looked to become even more so when times were good for the oil business there; when the fortunes of the oil business changed, Houston real estate was found to have been overbuilt, overpriced, and many millions of dollars were lost. Lengthening life spans push people toward taking a longer-term perspective, but rapid rates of change force a shorter-term perspective in investments. Turbulence in technology and in governmental monetary policy have already shortened time horizons. Businesspeople once routinely built plants with a thirty-year useful life. Today, the rate of change is too fast, and uncertainty regarding inflation and potential changes in tax laws is too great for such investments to make sense. Faster change will shrink time horizons further. Governments have taken on themselves the burden of looking a lifetime ahead, and the Social Security Administration is in for some rough times. When Otto von Bismarck, Germany's Iron Chancellor, came up with the notion of a guaranteed old age pension, it was a cynically clever and low-cost way to gain popular goodwill. So few people lived to age sixty-five that the amounts paid out in pensions were a pittance. After watching the German experiment for a handful of years, other governments began following suit. None of them expected a world like ours where a baby girl born in the United States today has an average life expectancy of 78.4 years—double that of Bismarck's time—and even this estimate is based on the faulty assumption that her medical care will be no better than her great-grandmother's was. At present, the Social Security Administration has two models: one they call "positive" and one they call "negative."

In the "positive" model, people work like dogs until old age, retire, and promptly die—presumably before they've had a chance to collect substantial social security or medical benefits. In the "negative" model, people retire early, develop illnesses that require medical intervention, and then live a long time making doctor visits and hospital stays during those years. Plans based on these models deserve to be disrupted. A better, more realistic scenario would have people living and able to support themselves for a long time, with illnesses that can be handled easily and inexpensively. Present social security benefits are enough to provide a certain standard of living—food, housing, transportation, and so forth. In a future of great material wealth, these benefits will be easy to provide, and present projections of economic woe resulting from an aging population will seem quaint. Back in the seventies, author Alvin Toffler brought out a book called *Future Shock*, describing how disturbing rapid change is for people. The book was a best-seller, but how much actual future shock has been seen in the past decade? Most people seem to have come through the last two decades pretty much all right, not in a state of shock at all. Rather than being shocked by technology, they are instead annoyed about pollution and traffic. Does this mean Toffler was wrong in predicting future shock? It's true that technology has been advancing rapidly in many areas over the past twenty years. But consider the average person's home life: How much of this rapid technological advance has shown up there?

A great deal, yet most of it is hidden, unlike the earlier part of the century, where obvious change was the norm. Electric lights and appliances, automobiles, telephones, airplanes, radio, and television affected almost everyone's private life. One person's life could span the time from horse-and-buggy travel to watching the Moon landings on television. In contrast, the past twenty years have seen new technologies move more quietly into the home. The VCR and microwave oven don't seem nearly as revolutionary as earlier inventions. Telephone answering machines are useful but haven't caused major changes in lifestyles. Fax machines are handy, but they're much like having very fast mail, and as this is written, fax machines aren't yet in most homes. So it's not surprising that the average person has felt little future shock lately. New medicines taken as pills—which may be radically improved—look just like the earlier pills.

The computerised bills that come in the mail aren't any more exciting to pay than the old human-prepared bills. This situation is unlikely to last. How much longer can technology advance so rapidly in so many fields without major effects on our lifestyles? There's been a respite from future shock in the last three decades; people have had a chance to catch their breath. When nanotechnology arrives, will future shock arrive with it? Some segments of society today are already getting practice in dealing with rapid technological advance. Those getting the most vigorous workout are in the computer field, where a machine two years old is regarded as obsolete, and software must be updated every few months to keep abreast of the new developments. But has this terrific rate of progress been dizzying or overwhelming? Not for the consumer—on the contrary, computers have become easier to use. In the 1960s, the New Math that was introduced into American grade schools and junior high schools included extensive study of arithmetic using numbers written in something other than the familiar base 10.

This was to prepare the "Adults of Tomorrow" for "The Computer Age" in which we would all be writing assembly language computer programmes in binary (base 2) code. But customers now purchase software rather than write it themselves—they need never deal with computer languages at all, much less a primitive assembly language. The rapid increase of computer speed has helped make computers easier to use. This progression has occurred many times before: Cars started off with external hand cranks, then advanced to starters you could yank from the comfort of the driver's seat; now starters perform invisibly when you turn the key in the ignition. This pattern will surely continue. First, some people will adapt to the technology, but in the long run the technology will adapt to us.

The more flexible and powerful the technology, the more easily it will adapt. Seen from a distance, seemingly trivial patterns of adaptation form part of a larger process that has marked the last century: the Western world has begun to invent mechanisms to handle a world of persistent change. Our mechanisms are by no means perfect or painless, as any unemployed person can testify. Employment agencies and headhunters for job seekers; unemployment and severance packages to ease job transitions; on-the-job training, continuing education, retraining, specialised seminars to update professional skills, professional associations, networking, community resources centers, government training programmes, and volunteer agencies are just a few of the inventions dealing with change and transition. Consumer information services, regulatory agencies, and environmental organisations are others. The most effective will endure. More options will continue to be invented. intellectual exercise of Feynman's had its obvious seed in the then recent (1953) discovery of the DNA double helix by Watson and Crick. If the genome has such a small size, Feynman thought, given the awesome amount of information it encodes, is there anything preventing scientists and engineers from emulating this natural feat? Let us keep in mind such inheritance. Biology thus gave birth to the whole field of nanotechnologies. To go by the empirical criterion, Feynman's call was that of a pioneer, it went unheeded for a generation.

After Heinrich Rohrer and Gerd Binnig devised in 1981 a scanning tunneling microscope, a collective leap forward ensued. About that time, George B. Whitesides initiated a series of studies which showed the kind of contributions chemists might make to nanosystems. The nanotechnology bandwagon started rolling in the 1990s. It was pulled by two horses. There was Moore's Law. Engineers and scientists had for decades looked at the horizon of computers foregoing silicon chips in favour of molecular devices. And there was the demographic factor. The steady increase in the number of scientists, which continued well into the 1980s, was unmatched by a corresponding increase in public money. Accordingly, young scientists as a group partitioned themselves between industrial and pharmaceutical work. That biology gave rise to nanotechnology raises a counterclaim. Whenever a group of chemists discuss nanotechnologies, they remind one another "this is nothing but chemistry, after all". The declaration can be seen as narcissistic, self-centered, and narrow-minded. Nevertheless, it has a degree of truth. Moreover, like any such stereotypic assertion on the part of a group of scientists, it is worthy of in-depth scrutiny.

Let us then assume for the new sub-discipline to be yet another spin-off from chemistry, in like manner to biochemistry, solid-state physics, molecular biology, or materials science. Because of some unfortunate connotations of the word 'chemistry', its ancient roots on the one hand, its age-old associations with explosives, poisons, and pollutants on the other, people have refrained from using it as a label on a field holding both novelty and promise of lucrative applications. Hence, the renaming as nanotechnologies. Chemists are prompt to recognize in nanotechnologies lines of forces to them familiar and maybe even *passé*. These include colloids and Langmuir-Blodgett films, which have regained a new hold on both scientists' imagination and grantsmanship. At a more fundamental level, though, chemists have injected into nanotechnologies a key notion, that of specificity. Contrary to what is routinely the case in physics, with fractals an example, chemical phenomena fail to obey scale invariance. They are liable to show quite different features depending on the scale of the observation. Indeed, this serves as a dividing line between chemistry as the science and chemistry as the industry. The latter is interested only in bulk matter and in bulk phenomena, whether heat transfer or equations for overall transformations, with more concern for the bottom line in financial reports than for the specific and microscopic (or nanoscopic) mechanisms involved, in catalysis for instance. However, what makes nanotechnologies a unique and viable hybrid is that their other parent has been, unmistakably, biology. This is an important legacy, all the more so in view of many of the present applications being in areas distant from biology such as electronics, optoelectronics, and other highly applied applications, from cosmetics to self-cleaning windows. Biological phenomena are a continued inspiration and biological applications are around the corner. The former? While it was known that many organisms orient themselves from the earth's magnetic field, Blakemore's 1975 finding of bacteria from marine marsh muds heading rapidly north was seminal. Blakemore then used transmission electron microscopy to reveal iron-rich crystals, above 100 nm in diameter, in these bacteria.

Several crystals would align into chains along their major axis. Frankel, Blakemore, and Wolfe proceeded to report that an unclassified magnetotactic spirillum (MS-1), when cultured in solutions containing ferric salts, precipitated uniform single crystals that consisted of mostly spinel Fe_3O_4. Microbial magnetite and the ferrimagnetic greigite g-Fe_3S_4, present in magnetotactic bacteria grown in marine, sulfidic environments, are inorganic crystallisations impressive for their quality. The magnetic particles have the optimal single domain size for the specific mineral. Each magnetic particle contains only a single magnetic domain. Moreover, within the magnetosome, the crystals are aligned so that their assembly bears a permanent magnetic dipole, making an effective compass. Biogenic magnetite provides great inspiration to laboratories active in nanotechnology, as for instance the book by Günther Schmid testifies to. The latter? Innocuous quantum dots (QD) bioconjugates have been injected into *Xenopus* embryos to monitor the development to the tadpole stages. QD label *Xenopus* embryos at different stages and with specific QD intracellular localisations. But let us return to the attitude of chemists who discard nanotechnologies as a combination of old, well-known chemistry and a public relations ploy to raise money from industry and from granting agencies. Could it be true that the field of nanotechnologies amounts only to renaming a subset of well-established chemical knowledge? Does it lack originality and impetus? Are we dealing with old wine, which time may have started to sour and to make unpalatable and thus needed a new label?

Conversely, is this new wine whose novel taste suggests brand-new flasks? Consider colloidal gold as a test case. Gold nanoparticles have been the subject of considerable attention over the ages. A back-of-the-envelope calculation shows that a small nanocrystal of one nanometer diameter will have as many as 30% of its atoms on the surface while a larger nanocrystal of 10 nm (ca. 1000 atoms) will have about 15% of its atoms on the surface. Metal particles with dimensions in the nanometer scale thus display unusual properties. The mean free path of an electron in a metal at room temperature is in the range 10-100 nm. As a metallic particle shrinks to that dimension, unusual effects might be observed. Indeed, gold nanoparticles of diameters about 100 nm or less appear red, not gold, when suspended in transparent media. They enjoy an interesting history dating back to the pioneering work of Faraday on the synthesis of gold hydrosols, gold nanoparticles dissolved in water. As one may recall, early in 1856 Michael Faraday (1791-1867) became interested in the interaction of gold with light. Gold leafs show different colors in transmitted light, green, blue, and purple, than they do in reflected light.

This observation intrigued Faraday. Together with his friend, Warren de la Rue, he prepared gold films. He started preparation of gold precipitates from reducing with ferrous sulfate a solution of auric salts, made presumably by attack of the metal with *aqua regia*. In so doing, Faraday discovered colloidal gold. This was the subject of intense experimenting and thought. Between the onset of this work, February 2, and its completion at the end of the same year 1856, Faraday wrote in his diary no fewer than 300 manuscript pages (1,160 numbered entries) on this topic. He submitted an article to the Royal Society on November 15, it appeared on February 15, 1857. In general and since Faraday, metal sols possess fascinating colors and have long been used as dyes. Smaller gold nanoparticles of diameters about 3 nm are no longer 'noble' and unreactive, but can catalyze chemical reactions. Consider nanocatalysts comprised of gold atom clusters, Au_8, Au_4, Au_3Sr, adsorbed at an F center of a MgO (100) plane. The optimal adsorption geometries of the O_2 molecule to these nanocatalysts differ markedly. Another example is from the temperature-programmed reaction spectra of the different products of the polymerisation of acetylene on small supported, monodispersed palladium clusters. Striking atom-by-atom size-dependent reactivities and selectivities are observed.

Only the three reaction products C_6H_6 (benzene), C_4H_8 (butene), and C_4H_6 (butadiene) form. Remarkably no C_3H_n, C_5H_n, and C_8H_n are detected, indicating the absence of C-C bond scission, as already observed on palladium single crystals and on palladium particles. Up to Pd_3, only benzene formation is catalysed, reflecting a high selectivity for cyclotrimerisation of acetylene. Pd_n clusters ($4 < n < 6$) reveal a second reaction channel by catalysing in addition the formation of butadiene. The third reaction product, butene, desorbing at a rather low temperature of 200 K, is clearly observed for Pd_8. For this cluster size the abundance of the three reaction products is similar. For even larger clusters ($13 < n < 30$) benzene formation increases with cluster size, whereas the conversion of acetylene to butene reaches a maximum for Pd_{20}. Note that Pd_{30} selectively suppresses the formation of butadiene (see for instance the Nanocatalysis chapter in the book by Rao, Müller, and Cheetham). What this example of metallic nanoparticles shows, which many of us have always suspected to be true but which conventional wisdom and standard practice – why bother reading any scientific publication older than say five years? – have both condemned summarily, is that casks of old wine connect with containers of the precious liquid from the most recent harvest. New science can be at its most innovative when it grafts itself onto its historical roots. Prediction starts with retrodiction, a point to which I will return. Who is most responsible for inflating the nanobubble? Are scientists, journalists, or venture capitalists the guiltiest? Whatever the answer, a host of start-up companies have seen the light of day, and have issued shares. The Yellow Pages will have to make room for quite a few names with the nano- prefix: Nanometrics, Nanogen, Nanologic, Nanopierce, Nanoproprietary, Nanosys, Nanotechnology Development Corporation are just a few.

Some universities have invested heavily into setting-up industrial parks specialised in nanotechnology. SUNY Albany and the University of South Carolina are two examples. The promise is there, because some of the devices already made actually work. IBM was able to use a nanotechnique to increase by a factor 20 the amount of memory on a hard drive. John Wolfe, who started the first nanotech venture-capital firm, compares it to the Internet circa 1993, before Netscape went public. Is it to say that this new bubble is doomed from the start and that it will burst, sooner or later? Assuredly. Is it to say that it will sink without leaving a trace? Assuredly not. Some companies will survive and be lasting.

Already, one can differentiate between fly-by-night operations buoyed only by conning the gullible, and reputable firms such as Nanosys doing solid work, and worth investing into. Nanosys collaborates with well-established corporate giants such as Intel, Matsushita, and DuPont. Peugeot-Citroën and Ford inject sub-10 nm CeO_2 nanoparticles into diesel fuel before it is burnt in the engine. This results in an *in situ* catalyst that not only improves fuel efficiency but also greatly reduces the emission of particulates. More than a million automobiles will incorporate that system by 2005 already. The American Federal government is also among the investors in the Nano Rush. Washington is spending $ 3.7 billion on nanotech research over the next four years. To this writer, the key issue is representation. Social groups create self-images, they project somewhat different images to the public.

Nanotechnology, as mentioned at the beginning of this essay, arose from one of the icons of the 20th century, the DNA double helix. It was successful, early on, in giving itself a similar and logo-like icon. IBM scientists created it and the advertising arm of the multinational corporation was ready to pounce and was quick to hammer it irreversibly into public consciousness. This is the STM image of a corral of 48 iron atoms placed in a circle of 7.3 nm radius. This particular image has become emblematic of nanotechnologies, at least for the time being. With it, scientists exhibited to the public, not without intense pride, an atom-made fence. What a feat! But this construction can also be seen as an unselfconscious picture of the proverbial ivory tower. In yet another image, nanotechnology proudly depicted in 1990 on the cover of *Nature* its archetype with an STM image of DNA. Another graphic depiction of nanotechnologies brings up my next point, representation as re-presentation. The cover page in issue number 24 of *Angewandte Chemie International Edition in English,* volume 37 for 1998, was adorned with a schematic representation of the Mo_{132} cluster. Its similarity to Kepler's early model of the universe, featuring the ubiquity of the icosahedron, was emphasised in the illustration. Thus, nanotechnology in the time of the Millennium rediscovered the microcosm-macrocosm correspondence prevailing in medieval metaphysics. With some modification, needless to say. Humanism is out: a metallic cluster as the nanocosm replaces man as the microcosm. Just as for medieval theologians, the stress is on the unity of nature, on the uniformity of physical law throughout the universe. But physical science has replaced God in Western thought. One can interpret also the bow to Kepler as one of the numerous reincarnations of the Platonic archetype of the five regular solids, which has marked the history of chemistry in the braudelian *longue durée*: Kepler and the snowflake, the constituent particles of crystalline solids as proposed by René-Just Haüy, molecular shape according to André-Marie Ampère, Achille Le Bel's and Jacobus Henricus van't Hoff's tetrahedral carbon, Alfred Werner's complexes, Aaron Klug's icosahedral viruses and, closer yet to us, Richard Smalley's and Harold Kroto's buckminsterfullerene. Thus, nanochemists gave themselves a self-image of continuity within the geometric, Pythagorean tradition. Let me sum up with representation as re-presentation. I have been at pains in this review to point to all the aspects of nanotechnology that predated it.

Nanoscientists have moved existing pieces to create new patterns. There is nothing wrong with this efficient heuristics. Discovery is not the only path to knowledge, fame, and riches. But this raises an interesting question in philosophy and historiography, which will be my final point. Science rewrites its own history, not only as history but as science fodder too. Which should not be construed in the weak sense of Whig history-telling as indulged in by scientists. While it is very much true that scientists, in a Whig perspective, tend to see the past as heralding the present, their purpose is not the mundane one of setting up an artificial history, approximating to a linear genealogy. The nanotechnologies paradigm shows a considerably more interesting attitude. Science reconfigures past events and developments in order to refocus itself into new directions. Such a reshuffling of the past, which of course may be deeply upsetting to professional historians who see their work voided and having to be restarted anew, the undersigned included, is an essential part of the dynamics of science. And it shows historians of chemistry that there is life after Partington.

Nanotube Transistor, Rings and Theory

Transistors are the basic building blocks of integrated circuits. To use nanotubes in future circuits it is essential to be able to make transistors from them. We have successfully fabricated and tested nanotube transistors using individual multi-wall or single-wall nanotubes as the channel of a field-effect transistor (FET). We measure the electrical characteristics of our nanotube-FETs and find that we can change the amount of current (I_{SD}) flowing through the nanotube channel by a factor of 100,000 by changing the voltage applied to a gate (V_G), as can be seen in our data below. G is the low bias conductance of the tube. This work has been published in Applied Physics Letters, vol 73, p. 2447 (1998). As we cool the FET down from room temperature to 4 degree Kelvin (minus 460 degree Fahrenheit) we see the device behaviour change dramatically. While the device acts like a field-effect transistor at room temperature, at 4K it behaves like a single-electron transistor (SET). The three plots below show this change in device behaviour for different temperatures.

We have found a way to convert ropes of straight nanotubes into nanotube rings. The rings are composed of many layers of single-walled nanotubes, and have a radius of typically 0.7 micron. Coiling has been observed in proteins and other biomolecules, where hydrogen bonding is thought to provide the main force for coiling. Carbon nanotubes however present a novel behaviour where coiling involves only van der Waals forces. The rings, which we can position on metal electrodes, allow us to study novel electric transport phenomena. Shown below is an AFM micrograph of a one micron-diameter ring (the purple circle) placed over gold electrodes (the light blue objects). SEM (Scanning Electron Microscope) image of nanotube rings on a silicon substrate. The image is magnified 8000 times. The nanotubes used to form the rings are extremely small; their diameter is only 1.4 nm. They are 1-dimensional conductors and at low temperatures, quantum interference phenomena dominate electrical transport through the tubes. The ring geometry allows us to observe such quantum effects in these 1D conductors. We find clear evidence of *weak localisation*, which enables us to determine the electron's phase coherence length, which can be as large as 0.5 mm at 3 K. The plot below shows the resistance of a ring as a function of the magnetic field applied through it. The resistance is maximum at zero field because electrons propagating around the ring in opposite direction interfere with each other constructively at the point of injection, forming a standing wave. Applying a magnetic field breaks time-reversal invariance, and the contructive interference is lost. Effect of twisting on the properties of a metallic armchair (6.6) nanotube. Three twisted configurations are shown. Twisting is found to transform the metallic nanotube to a semiconducting one with a band-gap that varies with the twist angle as shown. Variation of the local density of states (LDOS) along the length of a (6,6) nanotube as a function of the bending angle.

Understanding Carbon Nanotubes and Its Potential Applications

Here, we focus on our recent work on *carbon nanotubes*; their structure, properties and uses in nano-electronic devices. Carbon nanotubes are extremely thin (their diameter is about 10,000 times smaller than a human hair), hollow cylinders made of carbon atoms. This study consists of the following sections (to access them you may also use the navigation bar on the left):

- *Nanotube manipulation:* We can manipulate the nanotube positions, change their shape, cut them and place them on electrodes.
- *Molecular mechanics:* We can simulate the mechanical behaviour of nanotubes by calculating the forces acting between nanotubes and other objects such as the substrate.
- *Nanotube Field-Effect Transistor:* We have successfully used semiconducting single and multi-walled nanotubes as channels of field-effect transistors.
- *Nanotube rings:* While normally nanotubes are straight, we have devised ways to prepare them in a ring form.
- Nanotube theory: Computation and theory of the electrical and mechanical properties

Nanotubes, depending on their structure, can be metals or semiconductors. They are also extremely strong materials and have good thermal conductivity. The above characteristics have generated strong interest in their possible use in nano-electronic and nano-mechanical devices. For example, they can be used as nano-wires or as active components in electronic devices such as the field-effect transistor shown in this study.

Understanding Fundamentals and Applications: Objectives, Schedules and Site Listings

The small scale and the one dimensional structure of carbon nanotubes are directly related to their unique properties which find more and more applications as the understanding and progress in synthesis advances. Carbon nanotubes represent an exemplary system where the bottoms-up approach to synthesis results in perfect structures with sizes less than 10nm, a range which remains inaccessible for advanced projection lithography techniques. Applications of carbon nanotubes range from reinforcement of composites or conductive plastics to electrodes for batteries or flat screens, field effect transistors, chemical or force sensors and electromechanical memory. These applications have been demonstrated in bulk, on surfaces and on individual tubes. This course is focused on the fundamental properties of carbon nanotubes and provides the basic knowledge needed to follow the important developments in this field.

We will show how the helical structure of the tube influences the electronic properties of the tube in a fundamental way. In the first part we will explain the structural peculiarities, give an overview of the main synthesis techniques and then show how, starting from the known electronic properties of graphite, how those of the nanotubes can be described. The second part of the course covers electronic transport, electron emission properties, optical properties, mechanical properties and composites. We will show to what extend analytical techniques can provide information about individual tubes and we will see the importance of the immediate environment of the nanotube surface. In summary the goal of the course is:

- To give an introduction to the elementary properties of carbon nanotubes.
- To give the context and current status of the scientific research on carbon nanotubes.
- Synthesis Methods
- Characterisation Methods
- CNT Based Materials
- Carbon Nano Industrial Applications

Listings of Sites

- *Nanotube Site, The*—facilitates the exchange of ideas among researchers by concentrating links to sites dedicated to nanotubes.
- *Nanoscale Science and Technology Group—IBM T. J. Watson Research Center*—investigating carbon nanotubes, nanolithography, and silicon nanoelectronics.

- *PhysicsWeb: Carbon Nanotubes*—1998 article looking at how experiments to test the theory that carbon nanotubes have a variety of useful properties are just becoming possible.
- *NASA JSC Carbon Nanotube Project*—focused on developing bulk nanotube production, purification, and application of single-wall carbon nanotubes.
- *Physical Properties of Carbon Nanotubes*—compendium of the currently accepted physical properties of carbon nanotubes.
- *Nanotubes & Buckyballs*—provides descriptions of the one-dimensional fullerene with a cylindrical shape and the most symmetrical large molecule known to man. Also includes links to other scientific nanotube articles and papers.
- *Biosensor for Fullerenes and Carbon Nanotubes*—paper reporting on binding experiments for anti-fullerene antibodies with several different fullerene derivatives as well as carbon nanotubes.
- *Nanotubes: Surprising Sensitivity to Oxygen Creates New Possibilities*—article examining how exposure to air can convert a semiconducting nanotube into a metallic conductor.

Meaning of Buckytubes: Towards Undestanding Its Dimensions

The beginning of this story has been told many times. In 1985, a confluence of events led to an unexpected and unplanned experiment with a new kind of microscope resulting in the discovery of a new molecule made purely of carbon – the very element chemists felt there was nothing more to learn about. Buckyballs – sixty carbon atoms arranged in a soccer ball shape – had been discovered and the chemical world, not to mention the physical and material worlds, would never be the same. The mechanical (stiffness, strength, toughness), thermal and electrical properties of pure buckytube materials enable a multitude of applications, from batteries and fuel cells to fibres and cables to pharmaceutics and biomedical materials. Scores of additional applications emerge when one thinks of blending nanotubes with other materials to improve existing properties or to provide new ones. Using nanotubes as fillers in thermoplastics and thermosets, for example, has been discussed for several years, and is only recently undergoing rapid investigation and development as sufficient quantities of high-quality buckytube material is becoming available to enable such investigations. One of the most important technology developments of the last half of the twentieth century was the substantial replacement of metals with plastics. Most of this replacement has been in structural applications, where plastics have been engineered to outperform steel and other structural metals by providing adequate strength or stiffness at lower weight and cost. A key property that metals will always have over plastics, however, is in electrical conductivity. Plastics are amasingly good electrical insulators; in fact, this property gives rise to many of the most widespread and important uses of plastics. Nevertheless, the applications for plastics would be broadened substantially if good solutions existed to make these materials conductive. These application areas include: antistatic, electrostatic dissipative, and electromagnetic shielding and absorbing materials. Electromagnetic interference and radiofrequency interference (EMI/RFI) shielding, for example, is essential in laptop computers, cell phones, pagers and other portable electronic devices to prevent interference with and from other electronic equipment. At present, there is no suitable plastic material for this purpose, and metal, in one form or another, is typically added to provide this function in electronic equipment cases, imposing substantial weight and manufacturing expense.

For some applications, plastics have been loaded with conductive materials for years to provide conductivity. The most common filler is carbon black, which is relatively inexpensive and works well in many applications. One drawback to carbon black as a conductive filler, however, is the high loading required to provide the desired level of conductivity. It is well known that conductivity of a filled insulator, such as a polymer resin, increases with filler loading in a classic S-curve. That is, up to a critical loading, the bulk conductivity changes little, but increases very rapidly upon adding just a bit more filler. This is because high bulk conductivity requires the presence of many long conductive pathways, which are obtained only when the loading is so high that when randomly distributed, the conductive particles (e.g., carbon black) are likely to form long chains. This critical loading threshold is actually many times higher than would be required if somehow these particles could be placed in the optimal positions to form long chains with the minimal loading. But, of course, being sub-micron in size, this can't be done. A great deal of carbon black is wasted in the redundancy required to build up above the threshold level where these long chains form.

Also well known is the fact that the critical loading threshold decreases dramatically as the aspect ratio (length to width) of the filler particles increases. This is because longer particles cover a greater distance of the conductive pathway, whereas carbon black, which is spheroidal, has to form a chain of touching particles to cover the distance that a fibre-shaped filler would cover by itself. Why is filler loading important? Aside from weight savings, when a plastic is loaded with carbon black at 30, 40, even 50% by volume, which is often the level needed to reach the desired bulk conductivity, the mechanical properties of the composite are severely degraded. Often it is not usable at all, and typically it is no longer mouldable, which is frequently the most critical property of plastic parts.

Buckytubes offer a solution. First, Buckytubes are terrific electrical conductors, as described above. No polymer is a better conductor and none better is likely to be found. (So-called conductive polymers, a class of long-chain molecules with a conjugated backbone, would be better described as molecular resistors; they are intrinsically semiconducting.) Second, Buckytubes have a phenomenally high aspect ratio. Individual tubes are about 1 nm in diameter (about half the diameter of DNA, and about 1/10,000th the diameter of graphitc fibres), and 100-1000 nm in length. Thus, the aspect ratio of buckytubes is around 100-1000, compared with about 1 for carbon black particles. This already changes the game entirely, by pushing the critical loading level downward, as described above. Finally, buckytubes naturally form, in fact are born with, a morphology that is probably ideal for conductive filler applications. Buckytubes self-assemble into "ropes" of tens to hundreds of aligned tubes, running side by side, branching and recombining. When examined by electron microscopy, it is exceedingly difficult to find the end of any of these ropes. Thus, ropes form naturally occurring very long conductive pathways that can be exploited in making electrically conductive filled composites. Initial indications are that dramatically lower loadings of buckytubes are required to reach a given level of conductivity than for any other conductive fillers.

The opportunities for conductive plastics, as well as thermosets, filled with buckytubes are abundant. Very low loadings (<0.1%) provide for antistatic and electrostatic dissipative applications. One example is in painting automobile body parts, which are increasingly made of plastics. Because they are insulators, plastic parts charge up, which cause them to repel electrostatically the paint droplets formed in spray-painting of the body parts. This results in a great deal of wasted paint, which is both an economic and an environmental problem. A conductive primer coat can be applied, but that extra processing step is also quite costly. the ideal situation is to make the part itself sufficiently conductive to drain away charge build up by connecting the part to ground during the painting process.

Another broad area of application for buckytube-filled plastics is in EMI/RFI shielding, which has uses, as described above, in portable electronics, and defence applications. If, as appears likely, good attenuation of electromagnetic radiation can be attained at buckytube loadings on the order of 1% or less, good mechanical stability should be maintained, allowing it to be moulded. This would represent a significant breakthrough in plastics and enable broadening of their uses. Other defence uses of buckytube composites are similarly significant such as radar-absorbing or modifying material for aircraft and missiles. The application areas just described make use of the electrical conductivity of buckytubes where they contribute the highest added value. However, there is great promise in the development of applications that exploit the thermal and mechanical properties of buckytubes as well. Like the electrical applications described here, the possibilities are mind boggling. Stay tuned! The special nature of carbon combines with the molecular perfection of buckytubes (single-wall carbon nanotubes) to endow them with exceptionally high material properties such as electrical and thermal conductivity, strength, stiffness, and toughness. No other element in the periodic table bonds to itself in an extended network with the strength of the carbon-carbon bond. The delocalised pi-electron donated by each atom is free to move about the entire structure, rather than stay home with its donor atom, giving rise to the first molecule with metallic-type electrical conductivity. The high-frequency carbon-carbon bond vibrations provide an intrinsic thermal conductivity higher than even diamond. In most materials, however, the actual observed material properties—strength, electrical conductivity, etc.—are degraded very substantially by the occurrence of defects in their structure. For example, high strength steel typically fails at about 1% of its theoretical breaking strength.

Buckytubes, however, achieve values very close to their theoretical limits because of their perfection of structure—their molecular perfection. This aspect is part of the unique story of buckytubes. Buckytubes are an example of true nanotechnology: only a nanometer in diameter, but molecules that can be manipulated chemically and physically. They open incredible applications in materials, electronics, chemical processing and energy management. Buckytubes have the following physical, chemical and mechanical properties that make them such an outstanding material:

- Electrical conductivity- the first polymer with truly metallic conductivity.
- Thermal conductivity—higher than diamond along the tube axis.
- Mechanical—the stiffest, strongest, and toughest fibre known.
- Chemistry of carbon—can be reacted and manipulated with the richness and flexibility of other carbon molecules. Carbon is the basis of most materials we use every day.
- Molecular perfection—essentially free of defects.
- Self-assembly – as noted above, strong van der Waals attraction leads to spontaneous roping of many nanotubes. Important in certain applications.

Electrical Conductivity & Thermal Conductivity

Depending on their precise structure, buckytubes can be either metallic conductors or semiconductors. The semiconducting tubes are of two types: one has a band gap of hundredths of an electron volt, while others have about a 1-eV band gap. Ropes have been measured with a resistivity of 10-4 ohm-cm at 300 K [A. Thess et al., Science 273, 483 (1996)], making them the most conductive fibres known. Individual tubes have been observed to conduct electrons ballistically, that is, with no scattering, with coherence lengths of several microns [Tans, et al., Nature 393, 49 (1998)]. In addition, they can carry the highest current density of any known material, measured [B.Q. Wei, et al., Appl. Phys. Lett. 79 1172 (2001)] as high as 109 A/cm^2, and Phaedon

Avouris, the Director of IBM's Center for Nanoscale Science and Technology, has suggested that it could be as high as 1013 A/cm^2.

Prior to buckytubes, diamond was the best thermal conductor. Buckytubes have now been shown to have a thermal conductivity at least twice that of diamond [Hone J, Carbon Nanotube Topics in Applied Physics, 273 (2001)]. Since this property, like electrical conductivity, is highly anisotropic in buckytubes, a macroscopic crystal of buckytubes, close-packed and aligned, would have the unique property of feeling cold to the touch, like metal, on the sides with the tube ends exposed, but similar to wood on the other sides.

Mechanical Properties, Chemistry and Manipulation

Buckytubes are the stiffest known fibre, with a measured Young's modulus of 1.4 TPa [M.-F. Yu et al., Phys. Rev. Lett. 84, 5552 (2000)]. They have an expected elongation to failure of 20-30%, which combined with the stiffness, projects to a tensile strength well above 100 GPa (possibly higher), by far the highest known. For comparison, the Young's modulus of high-strength steel is around 200 GPa, and its tensile strength is 1-2 GPa. Their toughness has been demonstrated in simulations and experiments to be remarkable for something so stiff; buckytubes can be doubled over, snapping back with no resulting defect [Yakobson]

Buckytubes are polymers of pure carbon, and thus possess all the richness of carbon chemistry. Carbon is by far the most chemically versatile of elements, permitting an infinite number of combinations and derivatives. Closed ends of buckytubes are like fullerenes, whose chemistry is well understood. Open ends have the chemistry of the edges of graphite, also well understood. The possibility of attaching to either end of a buckytube essentially any chemical, structure or surface creates an entirely new situation whereby these end-entities can communicate with one another via true metallic transport of electrons. This is unprecedented since never before have there existed molecules with metallic conductivity, and this opens up worlds of possibilities for technological applications. Technology also exists for chemically modifying the sidewalls of buckytubes, both covalently, by attaching chemical groups to the sides with chemical bonds, and non-covalently. Examples of the latter are the use of either simple or chemical surfactants, to modify the surface energy of the buckytubes. In addition, the ability to manipulate buckytubes with chemical precision enables, for example, their dispersion in solvents, making them compatible with plastic resins for composites, etc.

Knowing Major Applications in Conductive Plastics, Adhesives and Connectors

Much of the history of plastics over the last half century has been as a replacement for metal. For structural applications, plastics have made tremendous headway, but not where electrical conductivity is required, plastics being famously good electrical insulators. This deficiency is overcome by loading plastics up with conductive fillers, such as carbon black and graphite fibres (the larger ones used to make golf clubs and tennis racquets). The loading required to provide the necessary conductivity is typically high, however, resulting in heavy parts, and more importantly, plastic parts whose structural properties are highly degraded.

It is well established that the higher aspect ratio of a filler, the lower loading required to achieve a given level of conductivity. Buckytubes are ideal in this sense, since they have the highest aspect ratio of any carbon fibre. In addition, their natural tendency to form ropes provides inherently very long conductive pathways even at ultra-low loadings. Applications that exploit this behaviour of Buckytubes include EMI/RFI shielding composites and coatings for enclosures, gaskets, and other uses; electrostatic dissipation (ESD), and antistatic materials and (even transparent!) coatings; and radar-absorbing materials. The same issues that make Buckytubes attractive as conductive fillers for use in shielding, ESD materials, etc., make them attractive for electronics materials, such as adhesives and other connectors (e.g., solders). The special nature of carbon combines with the molecular perfection of buckytubes (single-wall carbon nanotubes) to endow them with exceptionally high material properties such as electrical and thermal conductivity, strength, stiffness, and toughness. No other element in the periodic table bonds to itself in an extended network with the strength of the carbon-carbon bond.

The delocalised pi-electron donated by each atom is free to move about the entire structure, rather than stay home with its donor atom, giving rise to the first molecule with metallic-type electrical conductivity. The high-frequency carbon-carbon bond vibrations provide an intrinsic thermal conductivity higher than even diamond. In most materials, however, the actual observed material properties—strength, electrical conductivity, etc.—are degraded very substantially by the occurrence of defects in their structure. For example, high strength steel typically fails at about 1% of its theoretical breaking strength. Buckytubes, however, achieve values very close to their theoretical limits because of their perfection of structure—their molecular perfection. This aspect is part of the unique story of buckytubes. Buckytubes are an example of true nanotechnology: only a nanometer in diameter, but molecules that can be manipulated chemically and physically. They open incredible applications in materials, electronics, chemical processing and energy management. Buckytubes have extraordinary electrical conductivity, heat conductivity and mechanical properties. They are probably the best electron field-emitter possible. They are polymers of pure carbon and can be reacted and manipulated using the tremendously rich chemistry of carbon. This provides opportunity to modify the structure and to optimise solubility and dispersion. Very significantly, buckytubes are molecularly perfect, which means that they are free of property-degrading flaws in the nanotube structure. Their material properties can therefore approach closely the very high levels intrinsic to them. These extraordinary characteristics give buckytubes potential in numerous applications.

Buckytubes are the best known field emitters of any material. This is understandable, given their high electrical conductivity, and the unbeatable sharpness of their tip (the sharper the tip, the more concentrated will be an electric field, leading to field emission; this is the same reason lightening rods are sharp). The sharpness of the tip also means that they emit at especially low voltage, an important fact for building electrical devices that utilize this feature. Buckytubes can carry an astonishingly high current density, possibly as high as 1013 A/cm^2. Furthermore, the current is extremely stable [B.Q. Wei, et al. Appl. Phys. Lett. 79 1172 (2001)]. An immediate application of this behaviour receiving considerable interest is in field-emission flat-panel displays. Instead of a single electron gun, as in a traditional cathode ray tube display, here there is a separate electron gun (or many) for each pixel in the display. The high current density, low turn-on and operating voltage, and steady, long-lived behaviour make buckytubes attract field emitters to enable this application. Other applications utilising the field-emission characteristics of buckytubes include: general cold-cathode lighting sources, lightning arrestors, and electron microscope sources.

Much of the history of plastics over the last half century has been as a replacement for metal. For structural applications, plastics have made tremendous headway, but not where electrical conductivity is required, plastics being famously good electrical insulators. This deficiency is overcome by loading plastics up with conductive fillers, such as carbon black and graphite fibres (the larger ones used to make golf clubs and tennis racquets). The loading required to provide the necessary conductivity is typically high, however, resulting in heavy parts, and more importantly, plastic parts whose structural properties are highly degraded. It is well established that the higher aspect ratio of filler, the lower loading required to achieve a given level of conductivity. Buckytubes are ideal in this sense, since they have the highest aspect ratio of any carbon fibre. In addition, their natural tendency to form ropes provides inherently very long conductive pathways even at ultra-low loadings. Applications that exploit this behaviour of buckytubes include EMI/RFI shielding composites and coatings for enclosures, gaskets, and other uses; electrostatic dissipation (ESD), and antistatic materials and (even transparent!) coatings; and radar-absorbing materials.

Buckytubes have the intrinsic characteristics desired in material used as electrodes in batteries and capacitors, two technologies of rapidly increasing importance. Buckytubes have a tremendously high surface area (~1000 m^2/g), good electrical conductivity, and very importantly, their linear geometry makes their surface highly accessible to the electrolyte. Research has shown that buckytubes have the highest reversible capacity of any carbon material for use in lithium-ion batteries [B. Gao, Chem. Phys. Lett. 327, 69 (2000)]. In addition, buckytubes are outstanding materials for supercapacitor electrodes [R.Z. Ma, et al., Science in China Series E-Technological Sciences 43 178 (2000)] and are now being marketed. Buckytubes also have applications in a variety of fuel cell components. They have a number of properties including high surface area and thermal conductivity that make them useful as electrode catalyst supports in PEM fuel cells. They may also be used in gas diffusion layers as well as current collectors because of their high electrical conductivity. Buckytubes' high strength and toughness to weight characteristics may also prove valuable as part of composite components in fuel cells that are deployed in transport applications where durability is extremely important.

The same issues that make buckytubes attractive as conductive fillers for use in shielding, ESD materials, etc., make them attractive for electronics materials, such as adhesives and other connectors (e.g., solders).

The idea of building electronic circuits out of the essential building blocks of materials—molecules—has seen a revival the past five years, and is a key component of nanotechnology. In any electronic circuit, but particularly as dimensions shrink to the nanoscale, the interconnections between switches and other active devices become increasingly important. Their geometry, electrical conductivity, and ability to be precisely derived, make buckytubes the ideal candidates for the connections in molecular electronics. In addition, they have been demonstrated as switches themselves. The record-setting anisotropic thermal conductivity of buckytubes is enabling applications where heat needs to move from one place to another. Such an application is electronics, particularly advanced computing, where uncooled chips now routinely reach over 100°C. CNI's technology for creating aligned structures and ribbons of buckytubes [D.Walters, et al., Chem. Phys. Lett. 338, 14 (2001)] is a step toward realising incredibly efficient heat conduits. In addition, composites with buckytubes have been shown to dramatically increase the bulk thermal conductivity at small loadings.

The world-record properties of buckytubes are not limited to electrical and thermal conductivities, but also include mechanical properties, such as stiffness, toughness, and strength. These properties lead to a wealth of applications exploiting them, including advanced composites requiring high values in one or more of these properties. Fibres spun of pure buckytubes have recently been demonstrated [R.H. Baughman, Science 290, 1310 (2000)] and are undergoing rapid development, along with buckytube composite fibres. Such super strong fibres will have applications including body and vehicle armour, transmission line cables, woven fabrics and textiles. Buckytubes have an intrinsically high surface area; in fact, every atom is not just on a surface—each atom is on two surfaces, the inside and outside! Combined with the ability to attach essentially any chemical species to their sidewalls provides an opportunity for unique catalyst supports. Their electrical conductivity may also be exploited in the search for new catalysts and catalytic behaviour. The exploration of buckytubes in biomedical applications is just underway, but has significant potential. Cells have been shown to grow on buckytubes, so they appear to have no toxic effect. The cells also do not adhere to the buckytubes, potentially giving rise to applications such as coatings for prosthetics and anti-fouling coatings for ships. The ability to chemically modify the sidewalls of buckytubes also leads to biomedical applications such as vascular stents, and neuron growth and regeneration.

REFERENCES

Braun, T.; Schubert, A. & Zsindely, S.: 1997, 'Nanoscience and nanotechnology on the balance', *Scientometrics*, 38, 321-325.

Brown, N.: 2004, 'Needle on the Real: Technoscience and Poetry at the Limits of Fabrication', in: N.K. Hayles (ed.), *Nanoculture: Implications of the New Technoscience*, Bristol, UK: Intellect Books, pp. 173-190.

Bueno, O.: 2004, 'The Drexler-Smalley Debated on Nanotechnology: Incommensurability at Work?', *Hyle: International Journal for Philosophy of Chemistry*, 10(2), 83-98.

Bueno, O.: 2004, 'Von Neumann, Self-Reproduction and the Constitution of Nanophenomena', in: D. Baird, A. Nordmann & J. Schummer (eds.), *Discovering the Nanoscale*, Amsterdam: IOS Press, pp. 101-115.

Einsiedel, E.F.; Goldenberg, L.: 2004, 'Dwarfing the Social? Nanotechnology Lessons from the Biotechnology Front', *Bulletin of Science, Technology & Society*, 24, 28-33.

F. Buot, "Mesoscopic Physics and Nanoelectronics: Nanoscience and Nanotechnology," *Physics Reports*, pp.73-174, 1993.

F. Capasso and S. Datta, "Quantum Electron Devices," *Physics Today*, pp.74-82, February 1990.

Feynman, R., "There's Plenty of Room at the Bottom: An invitation to Enter a New Field of Physics," Talk at the Annual Meeting of the American Physical Society, 29 December 1959. Reprinted in Appendix B of Crandall and Lewis. See citation above.

Fielder, F.A.; Reynolds, G.H.: 1994, 'Legal Problems of Nanotechnology: An Overview', *Southern California Interdisciplinary Law Journal*, 3, 593-629.

Fleischer, Th.; Decker, M. & Fiedeler, U. (eds.): 2004, *Große Aufmerksamkeit für kleine Welten ' Nanotechnologie und ihre Folgen*, special issue of *Technikfolgenabschätzung ' Theorie und Praxis*, 13 (2), 5-85.

Fogelberg, H. & Glimell, H.: 2003, 'Molecular Matters: In Search of the Real Stuff', in: H. Fogelberg & H. Glimell, *Bringing Visibility to the Invisible: Towards A Social Understanding of Nanotechnology*, Göteborg: Göteborg University, pp. 5-32.

Fogelberg, H.: 2003, 'The Grand Politics of Technoscience: Contextualizing Nanotechnology', in: H. Fogelberg & H. Glimell, *Bringing Visibility to the Invisible: Towards A Social Understanding of Nanotechnology*, Göteborg: Göteborg University, pp. 33-54.

Fogelberg, H.: 2003, 'The Material Culture of Nanotechnology', in: H. Fogelberg & H. Glimell,

Bringing Visibility to the Invisible: Towards A Social Understanding of Nanotechnology, Göteborg: Göteborg University, pp. 99-114.

Frazier, G., "An Ideology For Nanoelectronics," in *Concurrent Computations: Algorithms, Architecture, and Technology*, Plenum Press, New York, 1988.

Gorman, M.E.; Groves, J.F. & Shrager, J.: 2004, 'Societal Dimensions of Nanotechnology as a Trading Zone: Results from a Pilot Project', in: D. Baird, A. Nordmann & J. Schummer (eds.), *Discovering the Nanoscale*, Amsterdam: IOS Press, pp. 63-73.

Grinbaum, A. & Dupuy, J.-P.: 2004, 'Living with Uncertainty: Toward the Ongoing Normative Assessment of Nanotechnology', *Techne: Research in Philosophy and Technology*, 8(3).

Grinbaum, A.: 2004, 'La condition de l'homme moderne et les nanotechnologies', in: G. Nivat (ed.), *Les limites de l'humain. 39èmes Rencontres Internationales de Genève*, L'Age d'Homme, Genève, p. 141.

Grunwald, A.: 2004, 'Ethische Aspekte der Nanotechnologie. Eine Felderkundung', *Technikfolgenabschätzung ' Theorie und Praxis*, 13 (2), 71-78.

Gupta, V.K. & Pangannaya, N.B.: 2000, 'Carbon nanotubes: bibliometric analysis of patents', *World Patent Information*, 22, 185-189.

Hansson, S.O.: 2004, 'Great Uncertainty about Small Things', *Techne: Research in Philosophy and Technology*, 8(3) (forthcoming).

Hayles, N.K. (ed.): 2004, *Nanoculture: Implications of the New Technoscience*, Bristol, UK: Intellect Books.

Hayles, N.K.: 2004, 'Connecting the Quantum Dots: Nanotechscience and Culture', in: N.K. Hayles (ed.), *Nanoculture: Implications of the New Technoscience*, Bristol, UK: Intellect Books, pp. 11-26.

Hemerén, G.: 2004, 'Nano Ethics Primer', in: European Commission (Community Health and Consumer Protection): *Nanotechnologies: A Preliminary Risk Analysis on theBasis of a Workshop, Brussels,* 1-2 March 2004.

Hennig, J.: 2004, 'Changes in the Design of Scanning Tunneling Microscopic Images from 1980 to 1990', *Techne: Research in Philosophy and Technology*, 8(3).

Hessenbruch, A.: 2004, 'Nanotechnology and the Negotiation of Novelty', in: D. Baird, A. Nordmann & J. Schummer (eds.), *Discovering the Nanoscale*, Amsterdam: IOS Press, pp. 135-144.

Hullmann, A. & Meyer, M.: 2003, 'Publications and Patents in Nanotechnology. An overview of previous studies and the state of the art', *Scientometrics*, 58 (3) 507-527.

Huo, Q., D.I. Margolese, U. Ciesla, P. Feng, T.E. Gier, P. Sieger, R. Leon, P.M. Petroff, F. Schuth, and G.D. Stucky. 1994. *Nature* 368:317.

Jena, P., S.N. Khanna, and B.K. Rao. 1996. In *Science and technology of atomically engineered materials*, ed. P. Jena. River Edge, NJ: World Scientific.

Johansson, M.: 2003, 'Plenty of room at the bottom: Towards an anthropology of nanoscience', *Anthropology Today*, 19 (No. 6), 3-6.

Johnson, A.: 2004, 'The End of Pure Science: Science Policy from Bayh-Dole to the NNI', in: D. Baird, A. Nordmann & J. Schummer (eds.), *Discovering the Nanoscale*, Amsterdam: IOS Press, pp. 217-230.

Kresge, C.T., M.E. Leonowicz, W.J. Roth, J.C. Vartuli, and J.S. Beck. 1992. *Nature* 359:710.

Kupperman, A., S. Nadimi, S. Oliver, G. Ozin, J. Garcés, and M. Olken. 1993. *Nature* 365:239.

Kuusi, O.; Meyer, M.: 2002, 'Technological generalizations and leitbilder ' the anticipation of technological opportunities', *Technological Forecasting & Social Change*, 69, 625-639.

Landon, B.: 2004, 'Less is More: Much Less is Much More: The Insistent Allure of Nanotechnology Narratives in Science Fiction', in: N.K. Hayles (ed.), *Nanoculture: Implications of the New Technoscience*, Bristol, UK: Intellect Books, pp. 131-146.

Laszlo, P.: 2004, 'Is There Life After Partington?', *Hyle: International Journal for Philosophy of Chemistry*, 10(2), 169-178.

Lenhard, J.: 2004, 'Nanoscience and the Janus-Faced Character of Simulations', in: D. Baird, A. Nordmann & J. Schummer (eds.), *Discovering the Nanoscale*, Amsterdam: IOS Press, pp. 93-100.

Lent, C.S., Tougaw, P.D., Porod, W., Bernstein, G.H., "Quantum Cellular Automata," *Nanotechnology*, Vol. 4, p. 49, 1993.

Lewak, S.E.: 2004, 'What's the Buzz? Tell Me What's A-Happening: Wonder, Nanotechnology, and Alice's Adventures in Wonderland', in: N.K. Hayles (ed.), *Nanoculture: Implications of the New Technoscience*, Bristol, UK: Intellect Books, pp. 201-.

Lide, D.R., ed. 1993-1994. *CRC Handbook of Chemistry and Physics*, 74th ed.

Lin-Easton, P.C.: 2001, 'It's Time for Environmentalists to Think Small ' Real Small: A Call for the Involvement of Environmental Lawyers in Developing Precautionary Policies Molecular Nanotechnology', *Georgetown International Law Review*, 14, 106-134.

López, J.: 2004, 'Bridging the Gaps: Science Fiction in Nanotechnology', *Hyle: International Journal for Philosophy of Chemistry*, 10(2), 129-152.

Lösch, A.: 2004, 'Nanomedicine and Space: Discursive Orders of Mediating Innovations', in: D. Baird, A. Nordmann & J. Schummer (eds.), *Discovering the Nanoscale*, Amsterdam: IOS Press, pp. 193-202.

Paschen, H.; Coenen, C.; Fleischer, T.; Grünwald, R.; Oertel, D. & Revermann, C.: 2004, *Nanotechnologie: Forschung, Entwicklung, Anwendung*, Berlin: Springer (*Nanotechnologie, TAB-Arbeitsbericht 92*, Berlin: Büro für Technikfolgen-Abschätzung beim Deutschen Bundestag).

Roberts, J.A.: 2004, 'Deciding the Future of Nanotechnologies: Legal Perspectives on Issues of Democracy and Technology', in: D. Baird, A. Nordmann & J. Schummer (eds.), *Discovering the Nanoscale*, Amsterdam: IOS Press, pp. 247-255.

Robison, W.L.: 2004, 'Nano-Ethics', in: D. Baird, A. Nordmann & J. Schummer (eds.), *Discovering the Nanoscale*, Amsterdam: IOS Press, pp. 285-300.

Roco, M.C. & Bainbridge, W.S. (eds.): 2001, *Societal implications of nanoscience and nanotechnology*, (Proceedings of a workshop organized by the National Science Foundation, September 28-29, 2000), Kluwer: Dordrecht.

Roco, M.C. & Tomellini, R. (eds.): 2002, *Nanotechnology: Revolutionary Opportunities and Societal Implications, Workshop, Lecce (Italy), 31 January - 1 February 2002*, Luxemburg: Office for Official Publications of the European Communities, [200 pp.]

Roco, M.C.; Bainbridge, W.S. (eds.): 2002, *Converging Technologies for Improving Human Performance: Nanotechnology, Biotechnology, Information Technology and the Cognitive Science*, Arlington, VA: National Science Foundation.

Rohlfing, E.A., D.M. Cox, and A. Kaldor. 1984. *J. Chem. Phys*. 81:3846.

Rosen, A. 1998. A periodic table in three dimensions: A sightseeing tour in the nanometer world. In *Advances in quantum chemistry*. In press.

Ruthven, D.M., S. Farooq, K.S. Knaebel. 1994. *Pressure swing adsorption*. New York: VCH Publishers.

Sarewitz, D.; Woodhouse, E.: 2003, 'Small is Powerful', in: A. Lightman, D. Sarewitz & Chr. Desser, (eds.), *Living with the Genie: Essays on Technology and the Quest for Human Mastery*, Washington, DC: Island Press, pp. 63-83.

Schiemann, G.: 2004, 'Dissolution of the Nature-Technology Dichotomy? Perspectives on Nanotechnology from an Everyday Understanding of Nature', in: D. Baird, A. Nordmann & J. Schummer (eds.), *Discovering the Nanoscale*, Amsterdam: IOS Press, pp. 209-213.

Schmidt, J.C.: 2004, 'Unbounded Technologies: Working Through Technological Reductionism of Nanotechnology', in: D. Baird, A. Nordmann & J. Schummer (eds.), *Discovering the Nanoscale*, Amsterdam: IOS Press, pp. 35-50.

Schuler, E.: 2004, 'Perception of Risks and Nanotechnology', in: D. Baird, A. Nordmann & J. Schummer (eds.), *Discovering the Nanoscale*, Amsterdam: IOS Press, pp. 279-284.

Schummer, J.: 2003, 'The Notion of Nature in Chemistry', *Studies in History and Philosophy of Science*, 34 (2003), 705-736.

Schummer, J.: 2004, 'Naturverhältnisse in der modernen Wirkstoff-Forschung', in: K. Kornwachs (ed.), *Technik ' System ' Verantwortung*, Münster: LIT, pp. 629-638.

Schummer, J.: 2004, 'Why do Chemists Perform Experiments?', in: D. Sobczynska, P. Zeidler, E. Zielonacka-Lis (eds.), *Chemistry in the Philosophical Melting Pot*, Peter Lang (Frankfurt/M.), 2004, pp. 395-410.

Schummer, J.; Baird, D. (eds.): 2004-5, *Nanotech Challenges*, joint special issue of *Hyle: International Journal for Philosophy of Chemistry & Techne: Research in Philosophy and Technology* (forthcoming).

Schwarz, A.E.: 2004, 'Shrinking the 'Ecological Footprint' with NanoTechnoScience?', in: D. Baird, A. Nordmann & J. Schummer (eds.), *Discovering the Nanoscale*, Amsterdam: IOS Press, pp. 203-208.

International Encyclopaedia of
NANO TECHNOLOGY, SCIENCE AND PHYSICS

International Encyclopaedia of NANO TECHNOLOGY, SCIENCE AND PHYSICS

Volume 2

M.C. Narayan

ANMOL PUBLICATIONS PVT. LTD.
NEW DELHI - 110 002 (INDIA)

ANMOL PUBLICATIONS PVT. LTD.
H.O.: 4374/4B, Ansari Road, Daryaganj,
New Delhi-110 002 (India)
Ph.: 23278000, 23261597
B.O.: No. 1015, Ist Main Road, BSK IIIrd Stage
IIIrd Phase, IIIrd Block,
Bangalore - 560 085 (India)
Visit us at: www.anmolpublications.com

International Encyclopaedia of Nano Technology, Science & Physics

First Edition, 2008
ISBN 978-81-261-3680-3 (Set)

PRINTED IN INDIA

Printed at Mehra Offset Press, Delhi.

Contents

Preface

In general terms, the physics of nano-scale and molecular systems studied with scanning probes is called Nano Physics. The nanoscale physics uses various experimental techniques to examine the physical properties of objects in the nanoscale size range, that is, a little bit larger than the size of atoms. Some interesting physical properties at this range include conductivity of small numbers of atoms and molecules, forces arising between objects on this scale, and the transition between the quantum nature of a few atoms and a large number of atoms. In Physics, the prefix "Nano" is used to describe a billionth of a unit, so a nanometer is a billionth of a meter. This is the equivalent of approximately five to ten atoms laid end-to-end. These dimensions are also reflected in the English translation of the Greek word "nanos"—"dwarf". A single human hair is around ten thousand times thicker than a nanometer, and another impressive comparison – a nanoparticle is to a soccer ball as a soccer ball is to the Earth. But nanotechnology is not just microtechnology on a smaller scale; it is much more than a mere improvement on previous forms of technology. Nanotechnology is a completely new production process in a molecular world, opening up possibilities far beyond what we have known before. Nature provides the inspiration for this new technology – from time immemorial, nanomachines have been at work in all organic cells, building our world of plants, animals and people from a store of chemical elements. The discovery in recent years of the mechanisms for accessing individual building blocks of matter, and with it the increasing understanding of how these building blocks organize themselves has sparked off the industrial conquest of the nanosphere throughout the world. For instance, Nanotube is real. Strictly speaking, any tube with nanoscale dimensions, but generally used to refer to carbon nanotubes (a commonly mentioned non-carbon variety is made of boron nitride), which are sheets of graphite rolled up to make a tube. The dimensions are variable (down to 0.4 nm in diameter) and you can also get nanotubes within nanotubes, leading to a distinction between multi-walled and single-walled nanotubes. Apart from remarkable tensile strength, nanotubes exhibit varying electrical properties (depending on the way the graphite structure spirals around the tube and other factors), and can be insulating, semiconducting or conducting (metallic). Carbon nanotubes are considered by some to be fullerenes (strictly speaking, this is probably correct, but they are more commonly treated as distinct) and are sometimes referred to as buckytubes. Microelectromechanical systems (MEMS) are generally used to refer to systems that can respond to a stimulus or create physical forces (sensors and actuators) and that have dimensions on the micrometer scale. They are almost exclusively made using the same lithographic techniques that are used to make silicon chips for computing. Miniature accelerometers are the most successful product in this field and are used to trigger air bags in cars. When such systems can be made with nanoscale dimensions they can be classified as NEMS. The success story of the giant magnetoresistance effect (GMR) shows in what a short space of time nanotechnology can make it from research laboratory to complete market penetration. Some of the contemporary research areas in global Nano Technology, Science & Physics research include:

- Low temperature measurements on structured graphene
- Low-Temperature Microscopy
- Semiconductor nanowires
- Mesoscopic transport through multi-terminal nanostructures
- Time-resolved single electron transport in mesoscopic structures
- Hole transport in p-type GaAs nanostructures
- Two-dimensional electron gas systems
- Quantum dots and quantum point contacts
- Two-dimensional electron gas systems
- Nanoparticles and clusters
- Nanotubes
- Spin injection and magnetic semiconductors
- Superconductors
- Molecular Beam Epitaxy
- Scanning Tunneling Microscopy
- Spintronics
- III-V Materials
- X-STM
- Optical Transitions

This book titled "International Encyclopaedia of Nano Technology, Science & Physics" deals briefly with the subject areas of nano-optics, nanotube, nanofibre, nano-sensor, nanowire and nanoribbon. Keeping in mind the applications and utilities of Nano Physics for meeting the contemporary global energy requirements, it seems that this subject will become the core area of future nanoscience and nanotechnology research. Nanostructured materials do not represent a new phenomenon. However, the ability to probe, manipulate, understand and engineer matter at atomic scales has only recently come within our grasp. Developments such as the invention of the

Scanning Tunneling Microscope have since made nanoscale science a reality. Today, the term nanotechnology is used to describe specific construction and manipulation performed on an atomic and molecular scale. Generally, systems with a maximum size of less than a hundred nanometers are classified as nanotechnology. This book tends to cover the aforementioned subject areas in their broadest possible terms. All useful supplementary research and reference tools have been included for readers' further investigations in the subject area of Nano Physics. Subject matters of Nano Physics are covered in following chapters in this Encyclopaedia: Basics of Nano Physics and Nanotechnology; Integrating Physics, Computers and Nanotechnology; Historical Evolution and Development of Nanotubes; Applications of Nanotubes and Buckytubes in Conductive Plastics, Adhesives, Connectors and Other Areas; Recent Advances in Other Major Fields of Nanoscience, Nano Physics and Nanotechnology; Use of Nano Physics in Nano Labs and Nanotechnology Products; Major Applications & Utilities of Nano-optics; and Towards Explaining the Differences among Nanoscale Scanning, Nanosensors, Nanowires and Nanoribbons. This Encyclopaedia is research friendly and future oriented.

—EDITOR

5

Recent Advances in Other Major Fields of Nanoscience, Nano Physics and Nanotechnology

The ability to create and exploit devices on the nanoscale is beginning to have a major and practical impact upon biomedicine. For instance, the development of chemical and biological sensors (e.g., the IBM 'NOSE' project) will make real-time diagnostics possible and eventually enable personal therapeutic regimens to be developed. Likewise, the ability to examine biological processes at a scale below that of visible light will yield a new biology—'mesophysiology'—that operates between high-resolution molecular imaging and the bulk biochemical and physiological techniques of traditional biology. The programme of work in the nanobiology section of the Centre will build upon the innovation and expertise of the medical co-applicant and his collaborators in the field. A requirement for success will be the ability to interact with the existing skills of mainstream physical measurement methodologies and the new nanofabrication techniques which will become available as part of this proposal. An interdisciplinary approach is essential to capture the wealth of UCL physics and electrical engineering expertise for the benefit of medicine and nanobiology.

Development of AFM systems for biology. The results are images of surface and intracellular structures of living cells (at left, live cells on glass with prominent cytoskeletal fibres) and maps of physical properties—e.g. receptor binding forces—of cell materials. We have also integrated a confocal microscope with the AFM to accomplish simultaneous fluorescence analysis (at right, a 3D isosurface reconstruction of bone cells on a mineralised substrate (blue) labelled in two colours (red, green) to independently demonstrate the distribution of two membrane proteins). The future lies in developing the bio-AFM further and introducing a range of more sophisticated physical techniques, electrophysiological applications and micromanipulation methods. Each image measures 50 x 50 mm2.

Recently, we have developed several of the technologies that will enable nanoscale analysis of processes in living cells.96-100 We have designed and built a life science optical interface between a commercial AFM and inverted microscope and modified the z-piezo controller to allow imaging and analysis over the large heights required for use with living cells which have a large, irregular and moving surface landscape. Finally, we have combined confocal microscopy, time lapse image capture and patch clamping with the modified AFM to measure real time cell signalling responses by optical and electrophysiological methods. This work has been made possible by 'blue-skies' pharmaceutical industry support and collaboration with instrument manufacturers—we believe that this collaborative approach will be essential for long term exploitation of our findings.

The outcomes from this engineering/development programme have been: (i) the first estimates of single molecule ligand-receptor binding forces in cells using integrin cell adhesion receptors as a paradigm in cells; (ii) generalisation of these techniques to other receptors and binding molecules; (iii) development of mapping techniques to demonstrate the combined location of active receptors with 3-D topography in cells; (iv) micromanipulation and signalling experiments: introduction of a few ligand molecules to a cell using the AFM cantilever at defined positions and under known and controlled mechanical forces with simultaneous recording of signal processes such as intracellular calcium levels, kinase enzyme activation, and induction of apoptosis; and (v) use of an AFM to analyse the material properties of cells and of associated ion channels, some of which are activated by the tip itself.

Based on the above, the first research areas to be pursued in nanobiology in the new Centre will focus on the further development and utilisation of life sciences force microscopy. This will include work on: (i) AFM tips with special shapes and coatings, to be made using the clean room facilities of the Centre, and designed taking advantage of the Centre staff's combined expertise in fabrication and modelling of tip interactions with soft surfaces; (ii) the continuation of our work on the study of cellular signalling events ensuing from receptor activation via reactive ligands on AFM tips; (iii) the analysis of the biomechanics of cellular mechanotransduction; and (iv) the design and manufacture of an AFM-based 'cell scraper' which we will use to study total cellular interaction forces between cells and newly fabricated materials for use in tissue engineering (with members of the UCL Tissue Engineering Centre and the Eastman Dental Institute) and 'lab on a chip' applications within the Centre.

The activities just described will eventually be only a very small part of the nanobiology activities in the Centre. We will be working with a large number of collaborators to further the development of SPM for biology and medicine, as well as to fabricate special purpose chips and even micromachines for assays and eventually even disease treatments. To illustrate this commitment, specific collaborations that we are planning include:

Surface and intracellular properties of cells and their inhomogeneities (e.g., lipid rafts and polarised cell domains) by such methods as surface electrochemical properties (Prof. David Williams, Dept. of Chemistry UCL), ion conductance imaging (Prof. Chris Pitt, Dept. of Electronic & Electrical Engineering UCL), electrophysiology (Prof. Jonathan Ashmore FRS, Dept. of Physiology UCL), cell thermal imaging, and X-ray reflectometry.

Development of new probe techniques such as: single molecule FRET (fluorescence energy transfer) at the tip-cell/macromolecule interface;108 near-IR scanning near-field optical microscopy using microfabricated devices;109-112 magnetic property imaging; probing electron transport in cells (Prof. David Delpy FRS, Dept. of Medical Physics UCL).

Sequence analysis of receptors 'fished' from cells using specific ligand-derivatised AFM tips made in the Centre (Prof. Tony Segal FRS, Dept. of Medicine UCL in the Wellcome Trust proteomic facility).

AFM imaging and structure of 2D crystals and proteins in artificial lipid bilayers (Prof. Helen Saibil, Birkbeck Crystallography and BBSRC Biomolecular Structure Centre); and analysis of structure of process-engineered plasmids, proteins and other biomolecules (Prof. Peter Dunnill, ACBE UCL). The hope here is to learn about proteins occurring in or near membranes, where their conformations may be quite different than in the crystals required for traditional (X-ray) protein structure determination.

Large-scale sensor arrays for diagnostics of biological fluids. The arrays in question can consist of a wide variety of devices, ranging from micromechanical cantilevers, each with a slightly different chemical binding property, to microcalorimeters to measure energy released upon chemical reactions with different molecules resident within each calorimeter (Prof. David Williams).

Electronic properties and exploitation of biomolecules. Although the electrical characteristics of biomolecules may not generally be relevant to their biological function, they do represent a fascinating class of new materials for possible use in electronics. We plan to use scanning tunnelling microscopy (STM) as well as special-purpose substrates prepared in the nanofabrication facility to measure and eventually to make use of these characteristics (Prof. David Delpy FRS).

Our biological research programme is absolutely dependent upon unfettered access not only to the nano-fabrication facility, but also on the presence in a single location of scientists with expertise in theory, hard and soft condensed matter physics, and electrical engineering. All of these will be available in the Centre, and as can be seen from the previous sections of this proposal, are keen to pursue applications of their work in nanobiology. What the Centre will therefore provide is much more than the essential new physical infra-structure, but the unique interdisciplinary environment needed for meeting the challenges of nanobiology.

Rapid advances in nanotechnology are creating new opportunities for interfacing nanotechnology and biology in order to more effectively tackle the nation's biodefense agenda. The advances being made in nanobiology are contributing to new biodefense solutions in detection, diagnostics, therapeutics, and biodefense immunology. Although many of these advances are still in their infancy, the field has progressed with leaps and bounds just within the past year. What's more, the leaders who are pioneering the application of nanobiology to biodefense are enthusiastic about the results to date and the future potential of the technology.

Applications of Nanobiology to Biodefense will provide an excellent opportunity for nanobiologists, bioengineers, microbiologists, and immunologists to get the latest information about the current state of the art and obtain more insight into how they can apply nanobiology technology and nanosystems biology to immunological problems, targeted drug delivery and increased accuracy and sensitivity in analysing laboratory samples. The goal of the conference is to enable these groups to seize the potential of this powerful new technology. To date, nanobiologists, bioengineers, microbiologists, and immunologists have not engaged in a vigorous interaction, and, therefore, the technology and the bench science have not been sufficiently connected. The conference will bring together these groups so that they can better appreciate the opportunities for the successful application of nanobiology to biodefense and can better assess the opportunities for working together.

Towards Understanding the Nano Physics and Nano Biology Imperatives

All living things are nanofoundries. Nature has perfected the artform of biological nanotechnology for billions of years. Now, an emerging technology domain is poised to present a toolkit from which new lifeforms can be crafted, the inner molecular workings of living cells can be directly manipulated, even aging may be treated not as a disease, but as a reversable pathology. The very definition of life itself is perched at the edge of the next great revolution in medicine: nanobiology. Currently existing are technologies and applications in the arenas of biomolecular components and biocompatible surfaces integrated into microscale systems, implantable biochip devices, synthetically engineered quasi-viral components, modified DNA, structured proteomics, pseudoproteins, biomolecular "devices". What is coming are artificially engineered organelles and cells, technologies which combine organic and inorganic materials and substrates into integrated nanoscale systems, "biomolecular prosthetics", and intra-cellular modification strategies which will redefine the very essense of what is commonly referred to as life. 3D rendering of a theorectical quasi-viral component, by Charles Ostman, 1993. First patents actually issued for such "viral nano-machinery", a modified baccilovirus utilised for proteomic targeting and delivery, issued to Onyx Pharmceuticals in 1996 Dendrimer molecules, spherical branching strucutures which can be grown to very precise size and complexity configurations. This example has antibodies attached to its outer extremities, as a biomedical "device", but dendrimer components can be utilised for a very wide range of applicaitions, including self assembling molecular systems and structures.

Molecular protein delivery to targeted, specific cell types is currently being developed as a method for instigating physiological adjustment or modifications within living cells. In this context, protein can be viewed as being the functional equvilant of "software" which instructs the activities of biomolecular mechanisms and organelles within the cell, somewhat in the same way that machine code instructs computer processor chips to perform various functions. Among many such examples of this process being currently studied, is the utilisation of the P53 protein, which can act like a "switch" to literally shut down the metabolism of a living cell.

This has potential, for instance, as a possible cure for various forms of cancer, in that the cancer cells have a genetic identity different than the original cell types from which they emanated, and therefore can be "targeted" for delivery of this P53 protein. Photograph of a colony of live neurons living on the surface of a biochip, from Dr. Gunther Gross, University of N TexasBiolithography, the process of photonically instigating the binding of biomolecules to a variety of target surfaces, such as metals, glass, plastic, even teflon, was orginally patented by BSI (now Surmodics) over a decade ago. Now the procedure has become very robust, enabling the precise "attachment" of proteins, lipids, enzymes, even entire living cells, with molecular precision to almost any type of surface. Above and below: 3D rendered scences from the film Nanotechnology indicating a neural prosthetic biochip, and a "nanobot" repairing a damaged nerve axon.

Charles Ostman, and the production team from AAC, San Francisco. Self assembling organic nanotubes as a strucutural system, and (left) hexagonal nanophotonic structures composed from rhodopsins (retinal proteins). Phodopsins are extremely interesting becuase of their potential application in spatial light modulation, which facilitates optical computing and storage with biological materials—from the Theoretical Biophysics Group, U of Illinois, Urbana DNA, the software for all living things, can also be utilised as a structural proteomic system, and a "logical" component for molecular computing. The example configuration shown here is one of many developed by Ned Seeman at NYU, and others exploring similar concepts Self assembling / self organising molecular systems, emanating from the development of applied nano-biology, foster the creation of molecular computing platforms.

The biotech industry is extremely IT and computational resource intensive, which in turn benefits directly from the advent of evermore powerful and diverse forms of computing enabled by developments in nanocomputing systems. Biological metaphors in computing, such as genetic and evolutionary computation, currently implemented on reconfigurable computing platforms, further acclerates the pace biotech development via bioinformatics and in-silico biology processes. Various implementations of nanocomputing components will also be able to support related versions of biolgical metaphors in computing. Note the computer generated "virtual protein" model. This complex molecular structure was actually evolved, utilising genetic algorithms on a reconfigurable computing platform to derive this proteomic solution. This is an example of biological metaphors in computing being utilised to create the physical biological materials of our future evolution.

Dave Fogel, Natural Solutions Inc.3D procedural rendering of a theoretical pseudo-proteomic subunit Charles OstmanThe worlds of biotechnology and nanotechnology currently converge into nanobiology. The first patents for utilising a modified virus as a proteomic delivery vehicle, that being a molecular scale "device" which can seek out specific cells and deliver various materials to the inner parts of those cells, were issued to Onyx pharmaceuticals in 1996. In 1997, the Nanobiological Systems Group of Searle Labs (a division of Monsanto Chemical) was using dendrimer molecules, a sort of complex nanoscale branching structure which can be "grown" into very specific geometric forms, as another type of cellular targeting and delivery system. These same dendrimer molecules also are part of a growing collection of self-assembling and self organising "components", the leggo blocks of nanotechnology. Now several companies have emerged into the marketplace with their versions of these dendrimer molecular systems.

Biological materials systems provide a rich collection of options from which self-assembling and self-organising molecular systems can be "harvested", enabling access to materials to facilitate the construction of 2D structures and 3D manifolds. Many development programmes utilising proteomic components and other forms of biomolecular substrates which can either be used directly as functional components, or as nano-scale manipulation mechanisms for positioning other molecular subsystems, have been underway for a number of years.

Nano Physics, Nanotechnology and Nano Biology: Their Coming Together

The first examples of quasi-viral component molecular "devices" (synthetically engineered viral "entities") utilised for delivery of molecular protein components to specifically targeted cells are already in development. Synthetic organisms and biobots, quasi-viral components and engineered microbes as molecular scale medical "devices"; these are just a few of the technologies residing at the beginning edge of a new domain of development which redefines the very essense of what is traditionally referred to as "medicine".

Explaining the Neurochip

NeuroChipThis 3D rendering represents the latest development in a series of "biochip" devices, in this case, the precursory example technology which could have applications ranging from neural prosthetics to the next generation of biological computing and sensory system components. This example indicates molecular protein "snippets" induced into bonding with a receptor site, as a dynamic contact/ molecular conduction node embedded into a silicon micro device substrate. Potential applications—Neural prosthetics, sensory enhancement systems, direct neural interconnect to implantable biochip components and integrated systems. Potential risks—applications which extend beyond benign or "situationally neutral" engineered solutions for human sensory and behavioural capacities

The topic of nanobiology is, in fact, not a singular "science" or technical discipline at all, but rather a concept, as fundamental as life itself, and will eventually affect what can even be described as life, as it is currently defined. And yet, it is the very nature of such a complex and diverse arena of development which is precisely what the latest genre' of computer visualisation graphics techniques, incorporating the combination of complex molecular modelling, volumetric topographical scanning of biomolecular and subcellular objects, and 3D rendering and visuallisation techniques that enables such a rapidly developing technology domain.

The paradigm of the near future (and even today, in some arenas of development) will be biomedical "systems" which are reconstructed or "assembled" from biomolecular components, and mimic the physiology of living organisms in their operational and behavioural characteristics. As a solution to a highly diverse range of medical conditions, to which the only treatment has been the delivery of drugs or physiologically altering treatments of chemicals or radiation, this is the beginning edge of a radical new design concept. There is a state of change at hand, the rate and complexity of which is an acceleration vector never before experienced in human history. The relationships between rapidly evolving technologies, ever exapnding volumes and complexities of of information resources, and socio-economic system substrates are going to reshape almost every aspect of life as it is currently understood to be. To study the nuances and implications of this synergistic codependancy is to become an adaptable entity, capable of evolving and utilising the tools embedded in this interactive, bi-directional experiential knowledge conveyance system which will be referred to here as the "media matrix".

Examine the details of this conceptual paradigm as a learning experience at the Multimedia Sudies Programme, San Francisco State University, and as an ongoing media research and development programme manifest in a variety of broadcast and published formats, and also via theInstitute for Global Futures.

The Institute for Global Futures

The Institute for Global Futures is a strategic technologies research group which engages in trend analysis, technology development stream topology identification, and complex forecasting which examines synergistic thread linkages between various technologies and socio-economic systems. The activities of the institute include consulting services, strategic planning, reports, analysis, media programmes, conferences, and strategic alliance building for a variety of global clients.

Complex new technologies are rapidly transforming markets, economies, and industries. The Institute for Global Futures was founded to promote the productive, sustainable, and strategic use of advanced leading edge technologies.

The included imgages are computer generated 3D renderings of theoretical and applied nanocomponents, created by the author as a visualisation of various implementations of such molecular substructures and interactive symbiotic components. These are a few examples from a much larger collection of rendered, theoretical and applied nanosubcomponent systems, consisting of various "species" types with specific interactive and symbiotic properties.

In the current issue of Science (20 MAY 2005) Robert F. Service has reported that in a meeting (NSTI Nanotech 2005, Anaheim, California,) chemist Paul Alivisatos from University of California (UC), Berkeley,has described about a system developed by them, which can be used to measure distances out to 70 nanometers and can keep track of their targets indefinitely.This system uses pairs of gold nanoparticles & works theoratically similar to FRET (fluorescent resonance energy transfer).But neither it has the restriction for the dye molecules to be within the distance of 10 nanometer nor it has problem of photobleaching.

He reports that Alivisatos, Carsten Sönnichsen, Björn Reinhard, and Jan Liphardt—all colleagues at UC Berkeley and Lawrence Berkeley National Laboratory has first immoblised gold nanoparticles on a glass slide. When they shined these particles with white light they found that light scattering from particles changed steadily with distance.On the basis of these results they have attached a pair of nanoparticles at the either end of sigle stranded DNA. When they have added complementry DNA strand they have found a red shift in the wavelength of light scattered by the nanoparticles. Alivisatos hopes that his "longer-lived, longer-range nanoparticle-based FRET will eventually overtake its organic cousin for measurements in which background lightscattering is low".

Scientists Observe Nanosize Microtubules Treadmilling Across Plant Cells

A study in the journal Science is offering new insights into a long-standing mystery about plant growth. The scientists who conducted the experiment say their results could open new avenues of research for developing more effective herbicides and pharmaceuticals. Plant biologists from Stanford University and the Carnegie Institution of Washington report their new findings in the April 24 online edition of Science Express. The researchers are the first to witness the birth and growth of individual "microtubules"—nanosize tubes of protein that form inside living plant cells.

"We have a strong interest in cell growth, form and development, and microtubules play a prominent role in all three," said Stanford scientist Sidney L. Shaw, lead author of the Science study. Shaw is a research associate in the laboratory of Sharon Long, a professor of biological sciences and dean of Stanford's School of Humanities and Sciences.

"Imagine the plant cell as a cylinder with many thin rods, or microtubules, laying along parallel lines just under the surface," noted co-author David W. Ehrhardt, a staff scientist in the Carnegie Institution's Department of Plant Biology located on the Stanford campus. The microtubules form specific patterns that determine the eventual shape of the cell and ultimately the shape of the plant, he explained.

"If plant cells didn't have a particular shape, we wouldn't see trees and grass grow as they do. They'd be squat and lumpy, lacking much of the diversity of form that plays important roles in physiological and ecological function and also makes plants fun to garden with," Ehrhardt added.

A microtubule is only about 25 nanometers in diameter, roughly 2,000 times narrower than a human hair. Found in all plant and animal cells, these hollow rods of protein assemble spontaneously and carry out a number of important functions. Some act as miniature girders that support the cell, while others form tracks that transport vital cellular cargo or guide the separation of chromosomes during cell division.

In animal cells, microtubules lengthen and grow from a fixed site called the "centrosome." But plant cells lack centrosomes, so the origin and development of plant microtubules had been something of a mystery—until now.

"The novelty of this study is that, for the first time, we've been able to watch the birth of microtubules in plants and watch them go streaking across the cell," Shaw explained.

"In plants, microtubules start out in the cell wall in a random pattern as if you'd thrown down a bunch of pick-up sticks," Ehrhardt noted. "But over time, they organise in a really remarkable way to form a three-dimensional array at the periphery of the cell that resembles a helix."

The Science study revealed for the first time how individual microtubules come together to form these large arrays—a discovery that may have ramifications beyond the field of plant biology. That's because the arrays, which look like overgrown DNA helices, are believed to play an important role in the distribution of cellulose, the most abundant organic substance on Earth and a major component of the cell wall. Cellulose is the principal ingredient in paper, lumber, cotton, rayon and a wide range of other products we take for granted. This tough, water-resistant compound also is found in fruits, vegetables and grains. When doctors recommend a high-fiber diet, they mean a diet rich in cellulose.

"In most plant cells, cellulose is laid down, not randomly, but in a very organised way," Ehrhardt said. "It's thought that microtubules control and guide the machinerv that builds the cell wall. Being able to engineer how cellulose is laid down could provide bioengineering opportunities for generating cell walls with different properties for a variety of biomaterials."

In their Science study, the researchers used a confocal microscope to observe single microtubules in the cells of Arabadopsis plants, a member of the mustard family. By decorating the microtubules with green fluorescent protein, the scientists were able to watch them grow and develop.

"One nice thing about our study is that we were looking at the living, intact organism," Ehrhardt said. "Most animal studies are done with isolated cells, so the behaviour of their neighboring cells aren't observed, whereas we were looking at live cells in their completely native, multicellular context."

Using time-lapse imagery, the scientists discovered that new microtubules emerged near the outer wall of the plant cell, not in the cell interior where animal microtubules originate. Time-lapse imaging also revealed that plant microtubules appear to move in the cell by a process known as "treadmilling," which occurs when bits of protein (known as "subunits") are added to the leading end of the microtubule and simultaneously removed from the trailing end.

"With treadmilling, the microtubule looks like it's moving in one direction, but in reality, one end is growing while the other is shortening," Ehrhardt noted.

"Treadmilling has only been rarely seen in some animal cells," Shaw added. "What's remarkable about the plant system is that almost all the microtubules we can see are undergoing a treadmilling motility."

The way protein subunits are added and removed from a microtubule also proved interesting, Ehrhardt said: "It isn't the case that they are being smoothly added to the growing end and smoothly removed from the shrinking end. Instead, polymers are quickly added and removed in spurts to one end, while being less quickly removed in spurts on the other.

"The growing end dances back and forth—it grows, it shrinks, it grows, it shrinks. But there is a bias in this dance that results in growth over time. On the other end, the activity is smoother. It shortens slowly, it pauses, then some more gets removed and it pauses again. So the behaviour at the two ends is very different, and the combination of the two behaviours 'moves' the assembled microtubule from one location to another in the cell."

Time-lapse imagery revealed that an individual microtubule starts out in a random "pick-up stick" position, then treadmills across the cell at an average rate of 0.5 microns per minute (one micron equals one-millionth of a meter), encountering other microtubules along the way.

"Then sometimes it bumps into an organised bundle of microtubules, and like a train following a track, it turns direction and follows them, becoming a part of the helical array," Ehrhardt said.

"There's nothing actually physically moving anything, and that makes treadmilling a really unexpected mechanism for assembling higher-ordered structures," Shaw observed.

What guides the microtubule to its new orientation? The researchers hope that subsequent experiments will provide the answer.

"When a microtubule is crawling along the surface of the cell, how does it know it has encountered another microtubule?" Ehrhardt asked. "Are there other proteins that connect microtubules together and guide traveling microtubules into bundles? We suspect that might be the case, but we don't know."

The researchers hope to expand their observations to include other kinds of microtubule arrays, such as those involved in chromosome separation during cell division.

"The drugs that are responsible for slowing cancers, such as taxol, are actually plant molecules than bind to microtubules to prevent a cell from dividing," Shaw explained. "In fact, most of the major microtubule-acting drugs are derived from plants."

Many herbicides are also anti-microtubule, Ehrhardt added: "By understanding more about how microtubules behave, we'll probably get some insights into how these herbicides are actually acting, and that could open up opportunities for developing better herbicides."

The third co-author of the Science study is Roheena Kamyar, a former graduate student in Ehrhardt's laboratory now enrolled at the University of Michigan School of Medicine. Funding for the study was provided by the Carnegie Institution of Washington, the Howard Hughes Medical Institute and the U.S. Department of Energy.

Golden Ruler: A Powerful Prospective Tool of Nanobiology

- Syndicated content from Journal of Nanobiotechnology.
- Syndicated content from Yahoo News.
- Syndicated content from Feedster (different sources).
- Syndicated content from Blogdigger (Different Sources).
- Syndicated content from Topix NanoTechnology.
- Syndicated content from HubMed Nanobiology Alerts from PubMed.
- Syndicated content from NanoBiology on PubMed.
- Syndicated content from AzoNano.
- Syndicated content from NanoVIP.com.
- Submitted by Jameel Ahmad Khan on Mon, 2005-05-23 14:56.

DNA Scaffolding

The next frontier for information processing may lie at the interface of nanoelectronics and biotechnology.

An interdisciplinary team led by electrical and computer engineering professor Richard Kiehl is exploring the use of DNA as a programmable scaffolding for the self-assembly of nanoscale electronic components. As a model for fabricating and designing semiconductor devices and circuits, DNA offers two key advantages: size scale and programmability.

Most industry experts believe that within the next 10 to 15 years the ability to scale down conventional technologies will reach its limit. At that point, the operating principles of conventional devices—and the techniques used to fabricate them—will break down. The basic elements of the DNA molecule are at just the right scale, says Kiehl.

Self-assembly uses bio-recognition, a natural process in which one molecule is attracted to and binds with another to form small structures. In the case of DNA, the attraction can be programmed so that the molecules will spontaneously assemble in solution to achieve a desired result.

"It's possible to synthesize small versions of DNA molecules in the laboratory and programme in whatever code you want," says Kiehl. "And because the two strands of DNA have complementary codes that match up, you can design one strand of DNA in a certain way so it will match another strand and assemble a nanoscale structure this way."

The matched segments form a scaffolding on which nanoparticles are affixed at highly selective attachment points. It's an approach that offers the programmability and precision needed for assembling electronic circuitry on the nanoscale.

"We have to make a real paradigm shift," Kiehl says. "Not only do we have to keep improving performance, but we also must look at the kinds of devices we can make at those scales and how we want to use them to process information."

To that end, the researchers are turning to the human brain for inspiration. They envision devices whose electrical characteristics resemble those of neuron-like electrical waveforms in the brain. Like certain regions of the brain, the devices would process information based on pattern recognition rather than on individual bits of information. It's a more sophisticated level of information processing than can be achieved using conventional computers. Kiehl predicts there will be a wide range of applications for this technology, including signal processing, communications systems, and computer systems. "The higher end of this [work] will be things that computers can't do very well today because the operations they use are too restrictive. One is the ability to recognize a pattern, such as identifying a letter as being an 'A' or a 'B', or being able to identify a face.

"It won't be just making things faster and faster in the conventional way," he says. "It will really be opening up new ways to process information in machines."

Shrinking Sizes in Micro Manufacturing

The ubiquitous mobile 'phone has had a revolutionising impact on modern life and not just in the social and commercial spheres but also in manufacturing. The need for miniaturisation in the build has spawned developments in the areas of micro and nanotechnology (MNT) and these currently have the potential for exploitation in the manufacture of a wide range of products and not just in cellular communications. At least that is the view of the UK Manufacturing Technologies Association (MTA), an organisation with strong roots in the machine tool field.

Says the MTA, "The mould and die sector and its equipment suppliers will enjoy particular benefits from micro and nano manufacturing through the requirement for high volumes of micro components made from plastics and light alloys. Telecommunications are typical of this trend. Micro and nano manufacturing will be increasingly employed in a huge variety of consumer products and industrial components. Current mobile telephone technology incorporates such aspects: this will permeate not only other product areas but also the tools, machine tools and processes used in their manufacture".

MTA's fear is that the UK will "miss out" on these opportunities unless it increases its awareness and expertise in the micro and nanotechnologies. It was for this reason that it devoted its fifth seminar in the series "Sharing Tomorrow's Technology Today" to the topic of "Micro and Nano Manufacturing", which was held at the University of Birmingham in partnership with the Machining Research Group of its School of Engineering, a team with specialised knowledge and facilities in these technologies. The seminar was supported by demonstrations on these facilities that included high-speed micromachining and micro die sink and wire cut electro discharge machining (EDM).

Focussed energy

CVLM involves using pulsed (laser) light energy focussed onto a spot size of less than 5?ìm, to remove material by vapourisation and ablation. It is much faster than micro-EDM due to its high material removal rate (through a combination of high pulse repetition rate and an ability to remove material to depths of 10?ìm in each pulse). Another advantage is the large aspect ratios with up to 40:1 being reported for stainless steel. On the debit side is a variable surface finish, slight degree of taper in holes and lower resolution compared to the EDM process.

Micro-EDM was described in the presentation by Dipl-Ing Axel Demmer of the Fraunhofer Institute of Production Technology (IPT) in Aachen, Germany. He cited some of the advantages of the process, including the ability to machine hard materials and therefore well suited to mould manufacture, the low process forces involved and aspect ratios greater than in micromilling.

The "micro version" of EDM operates at frequencies in the kilohertz to megahertz spectrum in contrast to the kilohertz range in the conventional discharge process, with shorter pulse durations (less than 0.1?ìs vs 100?ìs) and smaller spark gaps (under 10?ìm vs 100?ìm). However, it suffers electrode wear rates over 30 per cent compared to under 5 per cent in conventional EDM.

In his talk, Hanspeter Gruber of Agie in Switzerland differentiated between die sink micro-EDM and the wire cut process. For die sinking, Mr Gruber quoted ribs

20?ìm thick with aspect rations of up to 25, and up to 100 for holes. Electrode sizes were stated to be 10?ìm diameter ×1.5?mm long. Other parameters he listed included, form errors of ±3?ìm, position errors of ±1?ìm and surface qualities of less than 0.1?ìm Ra.

Micro Definitions

The status and scope for micro and nanomanufacturing in the UK was explored by Professor David Allen of Cranfield University. He first established some definitions, including "Micromachine, a machine made from micrometer scale components", and "Nanotechnology, the study, development and processing of materials, devices and systems, in which structure definition on a dimension of less than 100?nm is essential to obtain the required functional performance". He also identified the commonly used terms of Microsystems Technology (MST), which is prevalent in Europe, and MicroElectroMechanical Systems (MEMS), widely used in the USA. Originally, MST or MEMS was a spin-off from the silicon-based microelectronics industry, which relied extensively on photolithography as the manufacturing process. Currently, miniaturised mechanical components are produced and the range of materials has been increased to include metals, polymers and ceramics. In addition, optical components, thermal components and fluidic components are being miniaturised and integrated.

Professor Allen illustrated that the market for MST products has shown a steady rise growing from approximately $25 million in 2000 to an estimated $45 million in 2004, more than half of which is in just two products: disc drive heads and ink-jet printing heads. The former are manufactured using photochemical machining (PCM) mainly in the USA. PCM is also a very active process in the UK with companies such as Precision Micro in Birmingham, which has one of the largest automated PCM facilities certainly in Europe and possibly in the world according to Professor Allen, and Weymouth-based Tecan, who offer "one stop" contract manufacturing of precision mechanical parts and tooling. In Professor Allen's view such facilities are under utilised by the UK MEMS industry.

The other major non-silicon MEMS product, the jet printing head is produced using either laser micromachining or micro-EDM. According to Professor Allen, the UK is very active in two laser micromachining processes, copper vapour laser machining (CVLM) by Oxford Lasers of Didcot, and excimer laser machining (ELM) by Exitech of Oxford. In the latter, the beam of an excimer excited laser is projected through a mask and focussed on the material to create complex fine detail features.

New Micro-EDM Machine

The latest Agie Vertex wire cut EDMs have a significantly improved process performance at speeds eight times that of the earlier models. They are capable of shaping 2D components using wires down to 20?ìm with the smallest thread hole being just 50?ìm larger than the wire diameter. They have all the attributes of modern EDM machines including twin wire cutting and auto threading of the fine wires. A further possibility is to combine the die sink and wire cut EDM processes, taking the benefits of each to produce intricate micro parts. Mr Gruber cited the example of a four-blade fan measuring 6?mm in diameter and a finished blade leading edge thickness of 15?ìm. The first step was to wire cut the 2D outline shape of the fan with 50?ìm wire and then shape the blade on both surfaces by die sink EDM using a precision milled convex-concave electrode. Mr Gruber went on to list a wide variety of microcomponent for which micro-EDM offers an economic solution. These were:

- mould inserts for electronic devices such as connectors;
- pins and holes in microvalves and nozzles;
- forms for extrusion nozzles;
- punching tools for microgears and lead frames; and
- precision parts for the watch industry.

Moving from the micro to the nanoworld is a three-places-of-decimal step, for which the normal "sliding and rolling" movement of machine tools is totally inadequate. Agie is working with the École Polytechnique Fédérale de Lausanne (EPFL) to solve this problem and has developed the Agie Nano, a prototype nano-EDM machine that uses flexures to create and support nanometre movements. A flexure is simple and robust and has the benefits of zero stick-slip, friction, backlash and wear, a high resonant frequency and a high rigidity. The prototype is a parallel kinematics machine with an 8?mm motion range and a 5?nm static resolution. First tests of this machine, reported by Mr Gruber, show it capable of precise movements with high repeatability and high precision 2D contouring.

Gantry Design

Rigidity of the machine structure is also crucial to high speed and micromilling. Matsuura has based the LX-1 on an overhead gantry design optimised using 3D finite element method (FEM) analysis. To overcome the yawing effect symptomatic of the gantry structure, two linear motors are fitted to drive the Y-axis, each synchronised with the other and mounted on the outside of the gantry side limbs in isolation from the machining area to eliminate any swarf "pollution" problems.

To further ensure its micromachining attributes, the Matsuura LX machine is fitted with 0.1?ìm resolution linear scales giving a positional feedback to 10?nm. The overall result is a machine able to machine the finest of microcavities with circularity errors of 3?ìm even at the highest feed rate on the LX-1 and under 1?ìm on the LX-0. For the future, Matsuura is planning an 80,000?rpm spindle machine, which is due in 2005, that will be able to precision grind as well as high speed micromill. Another hybrid machine will be the Photon M, which combines milling and laser machining for mould and die rapid prototyping work, also due in 2005.

A major consideration in micromachining is the CAD data available and its use in programming. As Roger Onions of Delcam emphasised in his presentation, the main priority in CAM is to protect the cutting tool from sudden changes in load, and with micro-sized cutters this is even more important than in conventional machining. Further, because of the fine tolerances, more detailed toolpaths with many more points are needed, which necessitate a sophisticated CAM system, able to complete calculations quickly, and consequently more computing power. He recommended looking out for errors during data exchange as any missing surfaces or gaps can be fatal. Extra time spent in CAD will be rewarded with successful CAM. Mr Onions suggested, "Model what you intend to machine and don't leave features like internal fillets to be formed implicitly by the cutter as this will cause excessive load and load changes in forming the radii". Other pieces of advice were that holes should be drilled (an all-round load) rather than milled (side load) and when programming, cap over holes or the cutter will try to plunge into material. And, ensure that trimming is accurate as poor trimming is more likely to cause gouging with microcutters.

Micromilling

Another rapidly developing microcomponent process discussed at the MTA seminar was micromachining, primarily microdrilling and micromilling. IPT Aachen's Mr Demmer itemised some of the recent advances in this process, including ultrafine carbide tools able to mill hardened tool steel up to 55?HRc, 100?ìm diameter ball and square end mills with and without coating, tolerances of ±5?ìm and 10:1 aspect ratios. With these developments, micromilling is able to compete with micro-EDM and offers superior metal removal rates.

Because of the small diameter cutters involved, micromilling requires high speed spindles on machine tools utilising high speed machining (HSM) technology. The design and performance of such machines was described by David Edwards of Matsuura Machinery. He highlighted two machines from the Matsuura range, the LX-1 and the LX-0 series. These are CNC HSM machining centres with large table capacities (700×500?mm on the LX-1) but with micromilling capability. This allows for the machining of microcavities in relatively large tool blocks, for instance.

One of the most important HSM features of the LX-1 for micromilling is the linear motor drive motors for the three major axes. In addition to the high fast traverse and feed rates capability of linear motors – up to 90?m/min for the LX-1 – compared to conventional ball screws, there is no dependency on friction, no backlash or motor reversal errors, no loss of accuracy through component wear and no accumulative errors from the ballscrew-coupling-motor configuration. Also, the same linear motor technology is used for the fourth axis table rotary motion instead of the conventional mechanical worm and gear arrangement.

The other piece of HSM technology on the LX-1 is the spindle, which runs on ceramic bearings lubricated by an oil/air system and is capable of 60,000?rpm. To exploit this potential a shrink fit tool holder is employed with a two-face contact and tool locking system. This ensures both a high precision and long tool life.

Government funding

These moves are an indication that attitudes towards microsystems are changing, a view further confirmed last year when the government announced a new funding initiative. Trevor Fraser of the DTI told the seminar that up to £90 million will be provided to support the development of MNT manufacturing in the UK, in which the emphasis will be on outputs—the products and processes – rather than the inputs – the science and technology.

The funding will be split into two parts; the first of £50 million will be for an "Applied Research" programme, managed by Ernst & Young. It is open to large and small companies and academic institutions and supported on a 50:50 basis. The funds will be released in three annual calls of around £15 million, the first of which closed earlier in 2004.

The second element of the initiative, amounting to £40 million, is to support an MNT network in two areas, the establishment of the "Virtual Hub" directed by Hugh Clare, formerly of Unilever, and the setting up or enhancement of Regional Centres to provide facilities in MNT.

However, the £90 million is not the only source of funds for any organisation wishing to develop MNT. Mr Fraser revealed the wider funding picture in the UK with grants available through other government agencies that could be applied to MNT, including the Research Councils (over £60 million) and the Development Agencies (over £200 million). In addition, £1 billion of EU "Framework 6" money is "up for grabs". All this Europe-wide finance does indicate that the politicians, at least, are taking very seriously the technology necessary to "shrink sizes and tighten tolerances".

Raceline Machining

Delcam has a number of powerful tools within its PowerSHAPE CAM software to minimise tool loads. These include "Raceline machining", a patented Delcam strategy for high-speed roughing that smoothes out tool paths as it moves away from the component form. It minimises sharp direction changes and so allows faster machining and longer cutter life. Another tool is "Trochoidal machining", which uses a "shaving" approach to machine slots and avoid full-width cuts to protect the cutter at higher speeds. However, it does mean employing a cutter that is smaller than the slot width.

One of the principle aims of the MTA seminar was to encourage "UK Limited" to participate and exploit micro and nanomachining. This is no easy task as in Cranfield Professor Allen's view there was no apparent no national focus less than 4 years ago. He said, "Whilst it was recognised that some centres of excellence existed within the UK, it was felt that the national MEMS effort was un-coordinated and that the links between academia and industry were weak". This became particularly apparent following DTI Trade Missions to study MEMS in France and Switzerland (1999) and Taiwan and Korea (2000). As a result members of these Missions set out to promote national coordination and subsequently formed the Microsystems Manufacturing Association (MMA) under the auspices of the Institution of Mechanical Engineers (IMechE). The MMA has been lobbying the Government to set-up a National Microsystems Manufacturing Centre (NM2C) and this is currently planned for the South West of England. Also, the Northwest Development Agency is setting up a National Microsystems Packaging Centre (NMPC) in Liverpool.

Scientists in Britain and USA are taking inspiration from the gecko to try and develop the ultimate in adhesive tape. The gecko has evolved with toe surfaces covered in the perfect adhesive that allows them to stick in the most difficult of circumstances—even upside down on wet glass. Geckos have micro and nano-scale hairs on their toes. Each main hair is a thousand times thinner than a human hair. On the end of each hair are more hairs—each a thousand times smaller again. Molecular attraction then occurs by way of Van der Waals forces that allow the gecko to cling on. Synthetic versions of these hairs have now been developed. Applications include incorporating the new adhesive material into lightweight rescue robots that can readily climb walls in search and rescue missions. NASA has plans to use the material on space flights as early as next year. The theory is that by being able to have astronauts stick to the exterior of a space station or the space shuttle, then spacewalks would become far less dangerous.

Researchers in the US are working on a Nano-velcro that could have a holding power tens of times greater than epoxy adhesives. The researchers at Michigan State University are working on the reusable material made from carbon nanotubes. Unlike conventional velcro that uses loops and hooks, nano-velcro only has hooks. It is believed that the material could become a strong, self-repairing, heat resistant micro-fastening system to hold together components in ultra-small robots and electrical devices.

Nanocapsules and Dendrimers—Properties and Future Applications

A nanocapsule is any nanoparticle that consists of a shell and a space, in which desired substances may be placed. Technologies for microencapsulating materials have been around for several years, primarily for applications involving minimisation of hygroscopy and chemical interactions, elimination of oxidation, and controlled release of nutraceuticals.

Man-made liposome's have been used in cosmetics for some years to control the release of substances or protect them from the environment. Recently many other materials, such as polymers, have been used to make nanocapsules.

Polymeric nanocapsules can be made in specific sizes, shapes, and in reasonable quantities. Nanocapsules can be made to function in various ways. They can be produced as monodisperse particles with exactly defined biochemical, electrical, optical, and magnetic properties. They can be tailored to suit the complexity of whatever application they are intended for, such causing the release of the contents in response to a particular bimolecular triggering mechanism in targeted drug-delivery systems.

Nanocapsules can be used as smart drugs that have specific chemical receptors and only bind to specific cells. It is this receptor that makes the drug 'smart,' allowing it to target cancer or disease. The advantages of nano-encapsulation technologies for pharmaceutical applications include:

- Higher dose loading with smaller dose volumes
- Longer site-specific dose retention
- More rapid absorption of active drug substances
- Increased bioavailability of the drug
- Higher safety and efficacy
- Improved patient compliance

Beyond the ability to deliver existing drugs to their target, nanocapsules would allow for as much as a 10,000-fold decrease in drug dosages, reducing the harmful side effects of drugs used in chemotherapy. Quite often, drugs don't make it to market is because they have too many unwanted side effects. However, placing the same drug inside a nanocapsule and delivering it directly to its intended target in a reduced dosage, eliminates some of those side effects, or at least reduces them to an acceptable level.

Nanocapsules also have potential applications in agrochemicals, cosmetics, genetic engineering, wastewater treatments, cleaning products, and adhesive component applications. They can be used to encapsulate enzymes, catalysts, oils, adhesives, polymers, inorganic micro- and nanoparticles, latex particles, or even biological cells.

A dendrimer is an artificially manufactured or synthesised large moleculc comprised of many smaller ones linked together—built up from branched units called monomers. Technically, dendrimers are a unique class of a polymer, about the size of an average protein, with a compact, tree-like molecular structure, which provides a high degree of surface functionality and versatility. Their shape gives them vast amounts of surface area, making them useful building blocks and carrier molecules at the nanoscale and they come in a variety of forms, with different physical (including optical, electrical and chemical) properties.

Dendrimers can act as biologically active carrier molecules in drug delivery to which can be attached therapeutic agents and as scavengers of metal ions, offering the potential for environmental clean-up operations because their size allows them to be filtered out with ultra-filtration techniques.

Benefits of Nanomaterials and Nanoparticles in Agriculture and Environment

A number of projects around the world are exploring the use of nanoparticles on the farm for purposes other than pesticides—from enhanced photosynthesis to better germination and soil management. Researchers at Kyoto University (Japan) have discovered a method of producing ammonia using buckyballs. Ammonia is a key component of fertiliser, but it is not clear if the resulting product for use in the fields would contain buckyballs. Scientists at the University of Korea have applied for a patent on a liquid after which time it dissolves in the groundwater and becomes indistinguishable from naturally occurring iron. The Russian Academy of Sciences reports that they have been able to improve the germination of tomato seeds by spraying a solution of iron nanoparticles on to fields.

In 2003, ETC Group reported on a nanotech-based soil binder called SoilSet developed by Sequoia Pacific Research of Utah (USA). SoilSet is a quick-setting mulch which relies on chemical reactions on the nanoscale to bind the soil together. It was sprayed over 1,400 acres of Encebado mountain in New Mexico to prevent erosion following forest fires, as well as on smaller areas of forest burns in Mendecino County, California.

A number of approaches are being developed to apply nanotechnology and particularly nanoparticles to cleaning up soils contaminated with heavy metals and PCBs. Dr. Wei-Xang Zhang has pioneered a nano clean-up method of injecting nano-scale iron into a contaminated site. The particles flow along with the groundwater and decontaminate en route, which is much less expensive than digging out the soil to treat it. Dr. Zhang's tests with nano-scale iron show significantly lower contaminant levels within a day or two. The tests also show that the nano-scale iron will remain active in the soil for six to eight weeks, after which time it dissolves in the groundwater and becomes indistinguishable from naturally occurring iron.

Uses in Agriculture, Fishing and Veterinary Medicine

Livestock tracking has been a problem for farmers since before Little Bo Peep lost her sheep. Nano-Bo-Peep, however, would have no such problems. Just as converging technologies in crop production will use nanosensor networks to continuously monitor the health of plants, so, too, will sensors monitor livestock. The USDA envisions the rise of 'smart herds'—cows, sheep and pigs fitted with sensors and locators relaying data about their health and geographical location to a central computer.

This is a vision of precision agriculture on the hoof. The long-term aim is not merely to monitor, but also to automatically and autonomously intervene with pharmaceuticals using small drug delivery devices that can be implanted into the animal in advance of illness. The notion of linking in-built sensors to in-built smart delivery systems has been called "the fuel injection principle" since it mimics the way modern cars use sensors to time fuel-delivery to the engine. The closest applications to market are implantable insulin-delivery devices or "drug chips" that will be linked with glucose sensors for (human) diabetics to automatically regulate blood sugar levels. Over time, this could become the model for all drug delivery, in both humans and animals. One of the current barriers to implantable medical devices is that their composite materials (e.g., metal or plastics) are often incompatible with living tissue. New materials, engineered at the nano-scale to be biocompatible, seek to address this problem.

Implanting tracking devices in animals is nothing new—either in pets, valuable farm animals or for wildlife conservation. Injectable microchips are already used in a variety of ways with the aim of improving animal welfare and safety—to study animal behaviour in the wild, to track meat products back to their source or to reunite strays with their human guardians. In the nanotech era, however, retrofitting farm animals with sensors, drug chips and nanocapsules will further extend the vision of animals as industrial production units. Animals also are likely to be used as the testing ground for less savoury or more risky applications that could later be extended to human beings. Using microfluidics for breeding is likely to accelerate genetic uniformity within livestock species and also opens the possibility of applying new nano-eugenic technologies to humans in the future. The ability to remotely regulate animals may have adverse affects as livestock go longer periods without direct human care. The same technologies transferred to humans raises profound concerns about quality of life and civil liberties. In October 2004, the US Food and Drug Administration approved the use of implantable microchips in humans to provide easy access to an individual's medical records—the first approval of microchips for medical uses in the United States. As healthcare is driven more and more by the bottom line, the future use of implantable chips for automated drug delivery may become economically preferable to nursing. When dealing with the elderly or those with different cognitive abilities or with any condition requiring regular treatment, ethical questions may arise about who decides to make an individual 'fuel injected.' Automated drug delivery could allow some people to live independently who would otherwise be institutionalised. However, the absence of human caretakers is also a factor.

The world's fastest growing area of animal production is the farming of fish, crustaceans and molluscs, particularly in Asia. According to the FAO there were 45.7 million tonnes of aquaculture production in 2000 and it is growing at a rate of more than 9% per year. With a strong history of adopting new technologies, the highly integrated fish farming industry may be among the first to incorporate and commercialise nanotech products. Nevada-based Altair Nanotechnologies makes a water cleaning product for swimming pools and fishponds called 'NanoCheck.' It uses 40 nm particles of a lanthanum-based compound which absorbs phosphates from the water and prevents algae growth. NanoCheck is currently undergoing large-scale testing in swimming pools and Altair is expected to launch a swimming pool cleaner in early 2005. Altair has its eye on a potentially large demand for NanoCheck for use in thousands of commercial fish farms worldwide where algae removal and prevention is costly at present. According to Altair, the company plans to expand its tests to confirm that its nanoparticles will not harm fish, but no mention is made of the tests that will be undertaken to examine the impacts of nanoparticle-laden run-off on human health or on the environment.

The USDA is completing trials on a system for mass vaccination of fish using ultrasound. Nanocapsules containing short strands of DNA are added to a fishpond where they are absorbed into the cells of the fish. Ultrasound is then used to rupture the capsules, releasing the DNA and eliciting an immune response from the fish. This technology has so far been tested on rainbow trout by Clear Springs Foods (Idaho, US)—a major aquaculture company that produces about one third of all US farmed trout. Scientists from the Russian Academy of Sciences have reported that young carp and sturgeon exhibited a faster rate of growth (30% and 24% respectively) when they were fed nanoparticles of iron.

Using Nanotechnologies in Food Production

By allowing nanotech products to come to market in the absence of public debate and regulatory oversight, governments, agribusiness and scientific institutions have already jeopardised the potential benefits of nano-scale technologies. First and foremost, society—including farmers, civil society organisations and social movements—must engage in a wide debate about nanotechnology and its multiple economic, health and environmental implications.

In keeping with the Precautionary Principle, all food, feed and beverage products (including nutritional supplements) that incorporate manufactured nanoparticles should be removed from the shelves and new ones prohibited from commercialisation until such time as laboratory protocols and regulatory regimes are in place that take into account the special characteristics of these materials, and until they are shown to be safe. Similarly, nano-scale formulations of agricultural input products such as pesticides, fertilisers and soil treatments should be prohibited from environmental release until a new regulatory regime specifically designed to examine these products finds them safe.

Governments must also move immediately to establish a moratorium on lab experimentation with—and the release of—"synthetic biology" materials until society can engage in a thorough analysis of the health, environmental and socio-economic implications. Any efforts by governments or industry to confine discussions to meetings of experts or to focus debate solely on the health and safety aspects of nano-scale technologies will be a mistake. The broader social and ethical issues must also be addressed.

At the intergovernmental level, the Food and Agriculture Organisation's (FAO) standing committees and commissions on agriculture, fisheries, forestry and genetic resources should be monitoring and debating the new technologies—with active input and feedback from peasant and small farmers' organisations. FAO's Committee on Commodity Problems should immediately begin to examine the socio-economic implications for farmers, food safety and national governments. The UN/FAO Committee on World Food Security should be discussing the implications for agro-terrorism as well as food sovereignty. Additionally, the UN Convention on Biological Diversity should review nanobiotech's potential impact, especially on bio-safety.

Other UN agencies such as the United Nations Conference on Trade and Development (UNCTAD) and International Labour Organisation (ILO) should join with FAO to examine the impact of nanotech on the ownership and control of the world's food supply, commodities and labour. The international community should establish a body dedicated to tracking, evaluating and monitoring new technologies and their products through an International Convention for the Evaluation of New Technologies (ICENT).

Thai physicists are looking to nanotechnology to develop a new rice strain and improve the quality of silk. They will alter the rice by making a nanosize hole through a rice cell. A nitrogen atom shot through the hole will rearrange rice DNA to produce a new rice strain. Their goal is a fragrant rice plant with short stems that isn't overly sensitive to sunlight. Other engineers are planning to use plasma to coat the surface of silk so it doesn't get wet easily and attracts less dust. The plasma causes a nanolevel change to the silk surface. The surface treatment causes water to bead on the surface and simply roll off.

Nanotechnology and Molecular Food Manufacturing

Some people claim that in the future, molecular engineering will enable us to "grow" unlimited quantities of food without soil, seed, farms or farmers—and that it will wipe out global hunger in the process. Consider the views outlined below. The Experts Saying About Molecular Food Manufacturing?

- "Nanomachines could create unlimited amounts of food by synthesis at the atomic level, which would eradicate hunger."—Carmen I. Moraru et al., Professor of Food Science, Cornell University (USA), on nanotech's potential impact on food science.
- "Molecular biosynthesis and robotic replenishment may allow quick replacement of production, so we wouldn't have to depend on centralised systems to grow and deliver our food. In the first, primitive stages of molecular assembly, we'd build packaged greenhouses, radically different from those today, that would allow local or individualised production by millions who know nothing about farming...At the next stage of molecular manufacturing, food synthesis could occur directly, without growing crops or livestock."—Douglas Mulhall, 'Our Molecular Future'.
- "Why can't human beings imitate nature's methodology? Instead of harvesting grain and cattle for carbohydrates and protein, nanomachines (nanobots) could assemble the desired steak or flour from carbon, hydrogen, and oxygen atoms present in the air as water and carbon dioxide. Nanobots present in foods could circulate through the blood system, cleaning out fat deposits and killing pathogens."—Dr. Marvin J. Rudolph, Director, DuPont Food Industry Solutions, in 'Food Technology', January 2004.

Producing food by molecular manufacturing is the most ambitious goal of nanotech—and the least likely to materialize any time soon. To those who have followed the biotech debate over the past two decades, enthusiastic claims that a new technology will feed hungry people is a tired and empty refrain. Nano-optimists see the future through the biotech industry's rose- (and green-) coloured glasses: now it's nanotech, they claim, that will eradicate hunger by increasing agricultural yields, enhancing the nutritional content of food and eliminating the risk of food allergens.

ETC Group concludes that present-day "nanotech for tummies" is following the same trajectory as other nano-scale R&D, with the earliest applications in the area of "smart" materials and sensors. More revolutionary applications, such as the atomic modification of food, are perhaps more distant. But it's worth noting that a few ambitious scientists are trying to create food in the lab.

Tissue engineers at Touro College (New York City) and at the Medical University of South Carolina (USA) are experimenting with growing meat by "marinating" fish myoblast (muscle) cells in liquid nutrients to encourage the cells to divide and multiply on their own. The first goal is to keep astronauts in space from going hungry.

Uses of Nanotechnology in Industries

Our nantechnology industry portal provides information about the science and business of nanotechnology as well as applications of nanotechnology. Find out what a nanotube is, how nanotechnology affects art, or what nanotechnology stocks are on the market. Review this definition on nanotechnology to begin developing an understanding of nanotechnology, a field of science which will have significant impacts on the future of Logistics and Supply Chain Management.

Molecular Logistics: Although it will be many years before commercial applications of these nanotechnology emerges, developments are now coming fast and furious. It is important for logisticians to stay abreast of nanotechnology breakthroughs and provide industry input into the development of nanotechnology and teleportation. To learn more about the latest in Nanotechnology, visit our nanotechnology article archive.

Teleportation & Molecular Logistics: Many of us remember the Transporter on Star Trek and the "beam me up Scotty" phrase that typified use of this particle beam transportation technology. Most all of you will also probably scoff at this as a topic for a serious nanotechnology article, but following our recent review of this topic and the related field of Molecular Nanotechnology, we thought an introduction and/or update about them for our members was in order.

Nanotechnology Industry Information

Review our subject pages to find more information on the Nanotechnology Industry including an Industry Overview, Nanotech Companies, Discussions, Events, News and Publications.

- Industry Overview
- Nanotechnology Companies

- Discussion
- Nanotech News
- Nanotech Publications

General Nanotechnology Resources

Our focused Nanotechnology background subject pages provide rapid access to our resources including nanotechnology history, methods, people involved in nanotechnology, education, nanotechnology research, nanotechnology terminology and more.

- Nanotechnology History
- Nanotechnology Methods
- People in Nanotechnology
- Nanotechnology Education
- Nanotechnology Research
- Nanotechnology Terminology
- Future Implications of Nanotechnology

Nanotechnology Applications

Whichever application or aspect of the Nanotechnology Industry interests you, we have developed menus to explore all, including Materials, Computing, Microtechnology and Nanolithography.

- Materials
- Microtechnology
- Computing
- Nanolithography
- Scanning Probe Microscopy
- Nanotechnology Images

Bluesign: The Bluesign technologies ag in Sevelen (Switzerland), established in October 2000, is a subsidiary company of the Albers Group.

Beijing-Introduction of Technical Products: This technology will be widely applied in textile industry, clothing industry, packaging industry, pipes for petroleum transportation etc. Presently the application tests in textile industry have been successfully done.

"Focuses on the Future" of the U.S. Textile Industry: Annual Meeting of the American Textile Manufacturers Institute (ATMI), President Doug Ellis, Southern Mills, Inc. talked about the challenges the industry has faced recently and makes some personal predictions about the future of the textile industry.

Future Tech: Wonder Wear: Can we interest you in a suit that banishes dirt, sweat, and germs, sir? by Wendy Marston. DISCOVER Vol. 21 No. 1 (January 2000)

Have you spoken to your jacket today?: Avantex—Innovation Forum for Garment Textiles and Technologies of the Future Special review. by JENS BLOCH.

India's Textles: Progress and Challenges: Ths paper gives good insight as to how suppliers, customers, and competitors are viewing the future of the textile industry.

MOD2 Inc.: A software firm located in Los Angeles and we have developed a software product that specializes in automation and management of the day to day activities of textile importers, converters, and wholesalers.

Nano-Tex, LLC.: A research company founded on the principles of nanotechnology creating new or improved textile properties through molecular engineering.

Nanocomposite Fibers: University of Massachusettes. Dr.Yong, K.Kim, Dr.Armand F.Lewis, Dr.Steven, B.Warner, Dr.Prabir K.Patra and Subbarao Kota;Textile Sciences Department

Opportunities at Nano-Tex: Designing innovative products for the textile industry.

Nanotechnology: A site of links to sources and news with nonotechnology.

Textiles of the Future: Dr Niall Finn. Textiles of the future will include 'smart' fibres that inform or assist the wearer. Potential applications include clothes that monitor the medical condition of the wearer, that warn of the presence of toxic chemicals, or that adjust to suit the wearer's environment.

Trends in Japanese Textile Technology: John E. Berkowitch for U.S. Department of Commerce,Office of Technology Policy Asia-Pacific Technology Programme. Phyllis Genther. Yoshida Project Director.

The Future Impact of Molecular Nanotechnology on Textile Tec: The Future Impact of Molecular Nanotechnology on Textile Technology and on the Textile Industry. Rev. 1.8 David R. Forrest.

The Impact of Nanotech IP in the Global South

Over the past two decades the role of intellectual property in all areas of science and technology has exploded globally—primarily due to rules prescribed by the World Trade Organisation's Trade-Related Aspects of Intellectual Property (TRIPs) and by bilateral/regional trade agreements. The TRIPs agreement obligates all WTO member countries to adopt and enforce minimum standards of intellectual property. WTO has 150 members, and claims that it accounts for over 97% of all world trade.50 In 1996 the World Intellectual Property Organisation (WIPO) and WTO established a collaborative relationship in order to implement the TRIPs Agreement. (WIPO has 182 member states, which represents over 90% of the world's countries.) Among other IP rules, WTO members must allow patenting in *all* fields of technology.

The initial grace periods and flexibility allowed by TRIPs for developing country members have nearly expired. By 2006, the so-called "least developed countries" are required to adopt the WTO/TRIPs standards. Over the past decade, civil society, social movements, the UN Human Rights Commission and some governments have warned of the inequities of IP for the global South. Recently, even at WIPO – the UN body whose mission is to promote and protect intellectual property – the uneven IP playing field and the negative impacts of TRIPs have become undeniable and untenable for many developing nations. In September 2004 the "Geneva Declaration on the Future of the World Intellectual Property Organisation" warned that current IP regimes are having negative impacts in the developing world, resulting in lack of access to essential medicines, anticompetitive practices that hinder innovation and the misappropriation of social and public goods.

At WIPO's General Assembly meeting (September 27-October 5, 2004), Brazil and Argentina, supported by 14 developing country co-sponsors, proposed that WIPO adopt a "development agenda," stating that Intellectual property protection cannot be seen as an end in itself, nor can the harmonisation of intellectual property laws leading to higher protection standards in all countries, irrespective of their levels of development. The role of intellectual property and its impact on development must be carefully assessed on a case-by-case basis. IP protection is a policy instrument the operation of which may, in actual practice, produce benefits as well as costs, which may vary in accordance with a country's level of development.

Action is therefore needed to ensure, in all countries, that the costs do not outweigh the benefits of IP protection. WIPO's General Assembly adopted the decision to welcome a development agenda. But the US, UK and other industrialised nations are balking at the decision to give development concerns a higher profile within WIPO, acknowledging only that WIPO should give greater technical assistance to developing countries.52 Reports will be prepared and the issue will be considered at WIPO's General Assembly meeting in September 2005.53 Meanwhile, developing nations are now facing a new technology wave – and the requirement to accommodate nanotechnology-related inventions – even while still grappling with unresolved controversies over biotechnology and information technologies. By next year, ready or not, most of the world's developing nations will be obligated to evaluate and enforce nanotech patents.

ETC Group is not suggesting that nanotech, unencumbered by patents, will provide solutions for the South's most pressing needs. On the contrary, ETC Group believes that a technological fix can never right social wrongs. However, much ink has been spilled of late on the benefits nanotech will bring to developing nations while ignoring the realities of technology transfer and intellectual property.54 Multinational corporations, universities and nanotech start-ups (primarily in the OECD countries) have already secured numerous patents on essential nanotech tools, materials and processes. To the extent that these are "foundational" patents – that is, seminal breakthrough inventions upon which later innovations are built, researchers in the developing world could be shut-out. Nanotech patent thickets are already causing concern in the US and Europe. Researchers in the global South are likely to find that participation in the proprietary "nanotech revolution" is highly restricted by patent tollbooths, obliging them to pay royalties and licensing fees to gain access. Ultimately, nanotech will profoundly impact the South's economy, regardless of its level of direct participation or its handling of intellectual property. Nano-scale manufacturing platforms could make geography, raw materials, as well as labour, irrelevant. By employing nanotech to build from the bottom-up rather than processing down, the quantity of raw materials required could be sharply reduced. In short, nano-scale technologies are poised to become the strategic platform for global control of materials, food, agriculture and health in the immediate years ahead. Intellectual property monopoly is a crucial and powerful tool for realising that strategy. As a starting place for further debate, ETC Group offers the following recommendations related to nanotech IP.

Six "Pro-South Science" Policy Proposals:

1. *Poor applause:* When the G8 meets in July to discuss its dubious "Pro-Poor Science" strategy, leaders of rich nations shouldn't ignore the restrictions implicit in intellectual property that make it difficult or impossible for the South to develop its own independent technology solutions and to have access to the useful technologies of others. A truly "Pro-South" science policy would establish a global ten-year sunset clause on all monopoly patents.

2. *Patent pause:* WIPO should initiate a global suspension of patent approvals related to any applications that meet the US PTO's Class 977 standard (the criteria for nanotechnology patents) until further social review, including wide public debate, is undertaken on the impacts of nanotech IP.

3. *Technology flaws:* In close cooperation with social movements, including trade

unions, the Food and Agriculture Organisation (FAO) and the United Nations Conference on Trade and Development (UNCTAD) should cooperate with WIPO in producing a study on the impact of nanotech-related intellectual property on monopoly practices, technology transfer and trade. The UNCTAD Commission on Science and Technology for Development, also in conjunction with social movements, should examine the implications for technology transfer and the needs and interests of developing countries.

4. TRIPs clause: South governments and countries-in-transition should suspend any Class 977 equivalent patent grants or applications pending a full evaluation of their impacts. In particular, governments should determine whether or not such patents compromise access to the basic elements of nature or contravene national legislation or international agreements, such as WTO TRIPs, concerning intellectual property over living material.

5. Ordre public laws: National governments and relevant international organisations such as WIPO and UNCTAD should examine the social and *ordre public* significance of Class 977 equivalent patents that could compromise access to the fundamental components of nature.

6. Diversity laws: With input from indigenous peoples and peasant farmers' organisations, the Convention on Biological Diversity (CBD), the Commission on Sustainable Development (CSD) and FAO should consider the impact of intellectual property on fundamental elements of nature with respect to biodiversity and national sovereignty over genetic resources (especially in such fields as synthetic biology or nanobiotechnology). In a wider context, either the CBD or the CSD should adopt a permanent agenda item to monitor developments in nano-scale technologies.

REFERENCES

B. Aksak, M. Sitti, A. Casell, J. Li, M. Meyyappan, and P. Callen, "Friction of partially embedded vertically aligned carbon nanofibers inside elastomers," Applied Physics Letters, vol. 91, no. 061906, 2007.

B. Bhushan and R. A. Sayer, "Surface characterization and friction of a bio-inspired reversible adhesive tape," Microsystem Technologies, pp. 71-78, Vol. 13, no. 1, January, 2007.

B. Bhushan R. Sayer, "Gecko Feet: Natural Attachment Systems for Smart Adhesion," Applied Scanning Probe Methods Vol. VII - Biomimetics and Industrial Applications, Springer-Verlag, Heidelberg, Germany, 2007, pp. 41-76.

B. Bhushan, "Adhesion of multi-level hierarchical attachment systems in gecko feet," J. Adhesion Sci. Technol., Vol. 21, No. 12-13, pp. 1213-1258 (2007).

B. Bhushan, A.G. Peressadko, T.-W. Kim, " Adhesion analysis of two-level hierarchical morphology in natural attachment systems for 'smart adhesion' ", Journal of Adhesion Science and Technology, Volume 20, Number 13, 2006, pp. 1475-1491

B. Schubert, C. Majidi, R. E. Groff, S. Baek, B. Bush, R. Maboudian, R.S. Fearing, "Towards friction and adhesion from high modulus microfiber arrays," Journal of Adhesion Science and Technology, 21(12-13), pp. 1297-1315 (2007).

B. Schubert, J. Lee, C. Majidi, R.S. Fearing, "Sliding induced adhesion of stiff polymer microfiber arrays: 2. Microscale behaviour," Journal Royal Society, Interface, Jan. 22, 2008. 10.1098/rsif.2007.1309

B. Yurdumakan, N.R. Raravikar, P.M. Ajayan, and A. Dhinojwala, "Synthetic gecko foot-hairs from multiwalled carbon nanotubes," Chemical Communications, vol. 2005, pp. 3799-3801.

B. Zhao, N. Pesika, K. Rosenberg, Y. Tian, H. Zeng, P. McGuiggan, K. Autumn, and J. Israelachvili, "Adhesion and Friction Force Coupling of Gecko Setal Arrays: Implications for Structured Adhesive Surfaces," Langmuir, Dec. 2007, 10.1021/la702126k.

B.N.J. Persson and S. Gorb, "The effect of surface roughness on the adhesion of elastic plates with applications to biological systems," Journal of Chemical Physics, vol. 119, no. 21, pp. 11437-11444, 2003.

B.N.J. Persson, "Biological adhesion for locomotion: basic principles" Journal of Adhesion Science and Technology, Volume 21, Numbers 12-13, pp. pp. 1145-1173, 2007.

B.N.J. Persson, "On the mechanism of adhesion in biological systems," Journal of Chemical Physics, vol. 118, no. 16, 22 April 2003.

C. Greiner, A.D. Campo, E. Arzt, "Adhesion of bioinspired micropatterned surfaces: effects of pillar radius, aspect ratio, and preload," Langmuir, Vol. 23, no. 7, pp. 3495-502, Mar. 2007.

C. Majidi, R. Groff, and R. Fearing, "Clumping and Packing of Nanohairs Manufactured by Nanocasting" 2004 ASME Int. Mechanical Engineering Congress and Exposition. IMECE2004-62142, Anaheim, CA Nov. 13-19, 2004.

C. Menon, M. Murphy, and M. Sitti, "Gecko Inspired Surface Climbing Robots," IEEE International Conference on Robotics and Biomimetics (ROBIO), Shenyang, China, Aug 2004.

C. S. Majidi, R.E. Groff, and R.S. Fearing, "Attachment of Fiber Array Adhesive through Side Contact," Journal of Applied Physics, vol. 98, 103521 (2005).

C.-Y. Hui, N.J.Glassmaker, T.Tang and A. Jagota, "Design of biomimetic fibrillar interfaces: 2. Mechanics of enhanced adhesion," Jnl. Royal Society London Interface, 7 July 2004.

D. Campolo, S.D. Jones and R. S. Fearing, "Fabrication of Gecko foot-hair like nano structures and adhesion to random rough surfaces," IEEE Nano 2003 Aug 12-14, San Francisco.

D.S. Kim, H.S. Lee, J. Lee, S. Kim, K-H. Lee, W. Moon, and T.H. Kwon, "Replication of high-aspect-ratio nanopillar array for biomimetic gecko foot-hair prototype by UV nano embossing with anodoc aluminum oxide mold," Microsystem Technology, DOI 10.1007/s00542-006-0220-1, 2006.

Daltorio K.A., Gorb S., Peressadko A., Horchler A.D., Ritzmann R.E. and Quinn R.D. (2005) " A Robot that Climbs Walls using Micro-structured Polymer Feet," International Conference on Climbing and Walking Robots (CLAWAR), London, U.K., Sept. 13-15, 2005.

E. Arzt, S. Gorb, R. Spolenak, "From micro to nano contacts in biological attachment devices," Proc. Nat. Acad. Sciences, vol. 100, no. 19, pp. 10603-10606, 2003.

E. Cheung, M.E. Karagozler, S. Park, B. Kim, and M. Sitti, "A New Endoscopic Microcapsule Robot using Beetle Inspired Microfibrillar Adhesives," Proc. 2005 IEEE/ASME Int. Conf. on Advanced Intelligent Mechatronics, pp. 551-5557, Monterey, CA 24-28 July 2005.

E.P. Chan, C. Greiner, E. Arzt, and A.J. Crosby, "Designing model systems for enhanced adhesion," MRS Bulletin, vol. 32, pp. 496-503, June 2007.

E-S. Yoon, R.A. Singh, H. Kong, B. Kim, D-H. Kim, H.E. Jong, and K.Y. Suh, "Tribological properties of bio-mimetic nano-patterned polymeric surfaces on silicon wafer," Tribology Letters pp. 31-37, vol. 21, no. 1, Jan. 2006.

G. Huber, H. Mantz, R. Spolenak, K. Mecke, K. Jacobs, S.N. Gorb, and E. Arzt, "Evidence for capillarity contributions to gecko adhesion from single spatula nanomechanical measurements," Proc. National Academy of Sciences, vol. 102, no. 45, pp. 16293-16296, 2005.

G. Huber, S.N. Gorb, N. Hosoda, R. Spolenak, E. Arzt, "Influence of surface roughness on gecko adhesion," Acta Biomaterialia, vol. 3, pp. 607-610, 2007.

G. La Spina, C. Stefanini, A. Menciassi, and P. Dario, "A novel technological process for fabricating micro-tips for biomimetic adhesion," Journal of Micromechanics and Microengineering, vol. 15, pp 1576-1587, 2005.

G. Shah, M. Sitti, "Modeling and Design of Biomimetic Adhesives Inspired by Gecko Foot-Hairs" IEEE International Conference on Robotics and Biomimetics (ROBIO), Shenyang, China, Aug 2004.

G.C. Hill, D.R. Soto, S.J. Lue, A.M. Peattie, R.J. Full, and T.W. Kenny, "Investigating the role of orientation angle on gecko setae adhesion using a dual-axis MEMS force sensor," 14th Int. Conf. on Solid State Sensors, Actuators, and Microsystems, Transducers and Eurosensors 2007, pp. 2263-2266, Lyon France, June 10-14, 2007.

Gorb, S.N.; Varenberg, M. "Mushroom-shaped geometry of contact elements in biological adhesive systems" Journal of Adhesion Science and Technology, Volume 21, Numbers 12-13, pp. pp. 1175-1183, 2007. pp. 1175-1183(9)

Gravish, N., M. Wilkinson, and K. Autumn. "Frictional and elastic energy in gecko adhesive detachment," Journal of The Royal Society Interface, 2007. DOI:10.1098/rsif.2007.1077

H. E. Jeong, S. H. Lee, P. Kim, and K. Y. Suh, "Stretched Polymer Nanohairs by Nanodrawing," Nano Letters, vol. 6, no. 7, pp. 1508-1513, 2006.

H. Gao, Wang X, Yao H, Gorb S and Arzt E "Mechanics of hierarchical adhesion structures of geckos," Mechanics of Materials, vol. 37, pp. 275-285, 2005.

H. Wang, T. Mei, and X. Wang, " Modelling and Rough Surface Compliance Analysis of Biomimetic Micro/Nano Adhesive Array," Int. Conf. on Information Acquisition ICIA '07, 8-11 July 2007, pp. 168-172.

H. Yao and H. Gao, "Mechanics of robust and releasable adhesion in biology: bottom up designed hierarchical structures of gecko," Journal of the Mechanics and Physics of Solids, vol. 54, pp. 1120-1146, 2006.

H. Yao, G. Della Rocca, P.R. Guduru, and H. Gao, "Adhesion and sliding response of a biologically inspired fibrillar surface: experimental observations," Journal of the Royal Society Interface, doi:10.1098/rsif2007.1225, 2007.

Hui, C-Y.; Jagota, A.; Shen, L.; Rajan, A.; Glassmaker, N.; Tang, T., "Design of bio-inspired fibrillar interfaces for contact and adhesion theory and experiments," Journal of Adhesion Science and Technology, Volume 21, Numbers 12-13, pp. pp. 1259-1280, 2007.

J. Berengueres, M. Urago, S. Saito, K. Tadakuma, H. Meguro, "Gecko inspired electro-static chuck," Robio 2006, Kunming China, Dec. 17-20, 2006.

J. Berengueres, S. Saito and K. Tadakuma, "Structural properties of a scaled gecko foot-hair," Bioinspiration and Biomimetics, p. 1-8, vol. 2, 2007.

J. Berengues, K. Takahashi, S. Saito, and K. Tadakuma, "Magnetic Hair for Wall Mobility," IEEE/RSJ Int.Conf. on Intelligent Robots and Systems, pp. 3934-3939, Oct 9-15, 2006, Beijing, China.

J. Kwon, E. Cheung, S. Park, and M. Sitti, "Friction Enhancement via Micro-Patterned Wet Elastomer Adhesives on Small Intestinal Surfaces," Biomedical Materials, vol. 1, pp. 216-220, 2006.

J. Lee, C. Majidi, B. Schubert, R.S. Fearing, "Sliding induced adhesion of stiff polymer microfiber arrays: 1. Macroscale behaviour," Journal Royal Society, Interface, Jan. 22, 2008. 10.1098/rsif.2007.1308

J. Shan, T. Mei, L. Ni, S. Chen, and J. Chu, "Fabrication and adhesive force analysis of biomimetic gecko foot-hair array," Proc. of the 1st IEEE Int. Conf. on Nano/Micro Engineered and Molecular systems, Zhuhai, China, Jan. 18-21, 2006, pp. 1546-1549.

J.Y. Chung and M. K. Chaudhury, "Roles of discontinuities in bio-inspired adhesive pads," Jnl. Royal Society Interface, vol. 2, pp. 55-61, 2005.

K. Autumn and N. Gravish, "Gecko adhesion: evolutionary nanotechnology," Phil. Trans. R. Soc. A doi:10.1098/rsta.2007.2173

K. Autumn and W. Hansen, "Ultrahydrophobicity indicates a non-adhesive default state in gecko setae," J. Comp. Physiol. A, vol. 192, pp. 1205-1212, 2006.

K. Autumn, "Gecko adhesion: structure, function, and applications," MRS Bulletin, vol. 32, pp. 473-478, June 2007.

K. Autumn, A. Dittmore, D. Santos, M. Spenko, and M. Cutkosky, "Frictional Adhesion: a new angle on gecko attachment," Journal of Experimental Biology, vol. 209, pp. 3569-3579, 2006

K. Autumn, C. Majidi, R.E. Groff, A. Dittmore, and R. Fearing, "Effective elastic modulus of isolated gecko setal arrays," Journal of Experimental Biology, vol. 209, pp. 3558-3568, 2006.

K. Autumn, M. Sitti, Y.A. Liang, A.M. Peattie, W.R. Hansen, S. Sponberg, T. Kenny, R. Fearing, J.N. Israelachvili, and R.J. Full, "Evidence for van der Waals adhesion in gecko setae," *Proceedings National Academy of Sciences*, vol. 99, no. 19, pp. 12252-12256, September 17, 2002.

K. Autumn, Y. Liang, T. Hsieh, W. Zesch, W.-P. Chan, T. Kenny, R. Fearing, and R.J. Full, "Adhesive force of a single gecko foot-hair", *Nature*, vol. 405, pp. 681-685, June 8, 2000.

K. Gjerde, J. Kjelstrup-Hansen, C.H. Clausen, K.B.K. Teo, W.I. Millne, H-G. Rubahn, and P. Boggild, "Carbon nanotube forests: a non-stick workbench for nanomanipulation," Nanotechnology, pp. 4917-4922, 2006.

K. Takahashi, J.O.L. Berengueres, K. J. Obata, and S. Saito, "Gecko's foot hair structure and their ability to hang from rough surfaces and move quickly," Int. Journal of Adhesion and Adhesives, vol. 26, pp. 639-643, 2006.

L. Ge, S. Sethi, L. Ci, P.M. Ajayan, and A. Dhinojwala, "Carbon nanotube-based synthetic gecko tapes," Proc. Nat. Acad. Sciences, June 19, 2007, 10.1073/pnas.0703505104.

L. Qu and L. Dai, "Gecko-Foot-Mimetic aligned single-walled carbon nanotube dry adhesives with unique electrical and thermal properties," Advanced Materials, vol. 19, pp. 3844-3849, 2007.

M. Jin, X. Feng, L. Feng, T. Sun, J. Zhai, T. Li, L. Jiang, "Superhydrophobic Aligned Polystyrene Nanotube Films with High Adhesive Force," Advanced Materials, pp. 1977-1981, vol. 17, 2005.

M. Sitti and R. S. Fearing, "Nanomolding based Fabrication of Synthetic Gecko Foot-Hair Micro/ Nanostructures," *Proc. of the IEEE Nanotechnology Conference*, pp. 137-140, Washington, DC, USA, August 2002.

M. Sitti and R.S. Fearing, "Synthetic Gecko Foot-Hair Micro/Nano Structures for Future Wall-Climbing Robots," Int. Conf. Robotics and Automation 2003, pp. 1164-1170, Taipei, Taiwan, Sept. 2003.

M. Sitti and R.S. Fearing, "Synthetic Gecko Foot-Hair Micro/Nano-Structures as Dry Adhesives," Journal of Adhesion Science and Technology, vol. 17, no.8, pp. 1055-1074, 2003.

M. T. Northen, K.L. Turner, C. Greiner and E. Arzt, "A hierarchical gecko-inspired switchable adhesive," Hilton Head, 2006.

M. Varenberg and S. Gorb, "Close-up of mushroom-shaped fibrillar adhesive microstructure: contact element behaviour," Journal of the Royal Society Interface, doi:10.1098/rsif.2007.1201

M. Varenberg and S. Gorb, "Shearing of fibrillar adhesive microstructure: friction and shear-related changes in pull-off force," Journal of the Royal Society Interface, vol. 4, pp. 721-725, 2007.

M. Varenberg, A. Peressadko, S. Gorb, and E. Arzt, "Effect of real contact geometry on adhesion," Appl. Phys. Lett., vol. 89, 121905 (2006)

M.T. Northen and K.L. Turner, "A batch fabricated biomimetic dry adhesive," Nanotechnology, vol. 16, pp. 1159-1166, 2005.

M.T. Northen and K.L. Turner, "Batch fabrication and characterization of nanostructures for enhanced adhesion," Current Applied Physics, vol. 6, pp. 379-383, 2006.

M.T. Northen and K.L. Turner, "Meso-scale adhesion testing of integrated micro- and nano-scale structures," Sensors and Actuators A, vol. 130-131, pp. 583-587, 2006.

Murphy, M. P.; Aksak, B.; Sitti, Metin "Adhesion and anisotropic friction enhancements of angled heterogeneous micro-fiber arrays with spherical and spatula tips," Journal of Adhesion Science and Technology, Volume 21, Numbers 12-13, pp. pp. 1281-1296, 2007.

N. J. Glassmaker, A.Jagota, C.-Y.Hui, and J. Kim, "Design of biomimetic fibrillar interfaces: 1. Making contact" Jnl. Royal Society London Interface, 7 July 2004.

N. Pesika, Y. Tian, B. Zhao, K. Rosenberg, H. Zeng, P. McGuiggan, K. Autumn, and J. Israelachvili. 2007. Geometry-based model of tape peeling applied to the gecko adhesive system. Proceedings of the 30th annual meeting of the Adhesion Society:99-101.

N. Ren, X. Wang, T. Mei, and H. Wang, "Detachment Control Analysis for Nanofabricated Adhesive Mimicking Gecko Foot-Hairs," IEEE International Conference on Mechatronics and Automation, 25-28 June 2006, pp. 267-271.

N.J. Glassmaker. A. Jagota, C.Y. Hui, W.L. Noderer, and M.K. Chaudhury, "Biologically inspired crack trapping for enhanced adhesion," Proc. National Academy of Sciences, vol. 104, no. 26, pp. 10786-10791, 26 June 2007.

N.W. Rizzo, K.H. Garner, D.J. Walls, N.M. Keiper-Hrynko, T.S. Ganzke, and D.L. Hallahan, "Characterization of the structure and composition of gecko adhesive setae," J. Royal Soc. Interface, vol. 3, pp. 441-451, 2006.

P. Choi, J.Kim, L.J. Guo, "Fabrication of synthetic gecko tape and characterization of adhesion forces," 50th Int. Conf. on Electron, Ion, and Photon Beam Technology and Nanofabrication, Baltimore, May 30-June 2, 2006.

P.K. Porwal and C. Y. Hui, "Strength statistics of adhesive contact between a fibrillar structure and a rough substrate," J. Royal Society Interface, 10.1098/rsif.2007.1133.

P.L. Dickrell, S.B. Sinnott, D.W. Hahn, N.R. Raravikar, L.S. Schadler, P.M. Ajayan, and W.G. Sawyer, "Frictional anisotropy of oriented carbon nanotube surfaces," Tribology Letters, vol. 18, no. 1, pp. 59-62, Jan. 2005.

R. Spolenak, S. Gorb, and E. Arzt, "Adhesion design maps for bio-inspired attachment syystems," Acta Biomateriala, in press. Sept. 2004.

S. Chen and A. K. Soh, "Tuning the geometrical parameters of biomimetic fibrillar structures to enhance adhesion," Journal of the Royal Society Interface, doi: 10.1098/rsif.2007.1121.

S. Gorb and M. Scherge, "Biological microtribology: anisotropy in frictional forces of orthopteran attachment pads reflects the ultrastructure of a highly deformable material," *Proc. Royal Soc. London, B.*, vol. 267, pp. 1239-1244, 2000.

S. Gorb, M. Varenberg, A. Peressadko, and J. Tuma, "Biomimetic mushroom-shaped fibrillar adhesive microstructure," Journal of The Royal Society Interface, doi:10.1098/rsif.2006.0164, 2006.

S. Kim and M. Sitti, "Biologically inspired polymer microfibers with spatulate tips as repeatable fibrillar adhesives," Applied Physics Letters, vol. 89, no. 261911, 2006.

S. Kim and M. Sitti, C.Y. Hui, and R. Long, A. Jagota, "Effect of backing layer thickness on adhesion of single-level elastomer fiber arrays," Applied Physics Letters, vol. 91, no. 161905, 2007.

S. Kim, B. Aksak, and M. Sitti, "Enhanced friction of elastomer microfiber adhesives with spatulate tips," Appl. Phys. Lett. 91, 221913 (2007); DOI:10.1063/1.2820755

S. Reddy, E. Arzt, and A. del Campo, "Bioinspired surfaces with switchable adhesion," Advanced Materials, vol. 19, pp. 3833-3837, 2007.

S. Sethi, L. Ge, L. Ci, P.M. Ajayan, and A. Dhinojwala, "Gecko-inspired carbon nanotube-based self-cleaning adhesives," Nanoletters, 10.1021/nl0727765, 12 Feb. 2008.

S.N. Gorb, M. Sinha, A. Peressadko, K.A. Daltorio and R.D. Quinn, "Insects did it first: a micropatterned adhesive tape for robotic applications," Bioinspiration and Biomimetics, S117-S225, 2007.

Santos, D.; Spenko, M.; Parness, A.; Kim, S.; Cutkosky, M. "Directional adhesion for climbing: theoretical and practical considerations" Journal of Adhesion Science and Technology, Volume 21, Numbers 12-13, pp. pp. 1317-1341, 2007.

T. S. Kustandi, V. D. Samper, W. S. Ng, A. S. Chong and H. Gao, "Fabrication of a gecko-like hierarchical fibril array using a bonded porous alumina template," J. Micromech. Microeng. vol. 17 pp. N75-N81, 2007.

T. Tang, C-Y. Hui, and N.J. Glassmaker, "Can a fibrillar interface be stronger and tougher than a non-fibrillar one?" Jnl. Royal Society Interface, vol. 2, pp. 505-516, 2005.

T. Tong, C. Majidi, Y. Zhao, L. Delzeit, R.E. Groff, P. Reddy, A. Majumdar, A. Kashani, and M. Meyyappan, "Compressive properties of dense vertically aligned multi-walled carbon nanotube arrays," Proc. of NANO2005, ASME Integrated Nanosystems: Design, Synthesis and Applications, Sept. 14-16, 2005 Berkeley CA.

T. Tong, Y. Zhao, L. Delzeit, A. Kashani, and A. Majumdar, "Multiwalled carbon nanotube/ nanofiber arrays as conductive and dry adhesive interface materials," Proc. of ASME Nano2004, Nano2004-46013, Sept. 22-24, 2004, Pasadena, CA.

T.W. Kim and and B. Bhushan, "Effect of Stiffness of Multi-Level Hierarchical Attachment System on Adhesion Enhancement," Ultramicroscopy doi:10.1016/j.ultramic.2006.11.008, 2007.

T.W. Kim, and B. Bhushan, "Optimization of Biomimetic Attachment System Contacting with a Rough Surface," Journal of Vacuum Science and Technology A, pp. 1003-1012, 2 July 2007.

W. Sun, P. Neuzil, T. S. Kustandi, "The nature of the gecko lizard adhesive force," Biophysical Journal: Biophysical Letters," L14-L17, 2005.

W.L. Noderer, L. Shen, S. Vajpayee, N.J. Glassmaker, A. Jahgota, and C.-Y. Hui, "Enhanced adhesion and compliance of film-terminated fibrillar surfaces," Proc. Royal Soc. A, vol. 463, pp. 2631-2654, 2007.

W.R. Hansen and K. Autumn, "Evidence of self-cleaning in gecko setae," Proc. National Academy of Science, vol. 102, no.2, pp. 385-389, January 11, 2005.

Wang. Mei, Wang, Lin "Interaction and Simulation Analysis between the Biomimetic Gecko" Adhesion Array and Rough Surface IEEE Int Conf Mechatronics and Automation Jul 2005

Y. Tian, N. Pesika, H. Zeng, K. Rosenberg, B. Zhao, P. McGuiggan, K. Autumn, and J. Israelachvili, "Adhesion and friction in gecko toe attachment and detachment," Proc. Nat. Acad. Sciences, vol. 103, no. 51, Dec. 19, 2006.

Y. Zhao, T. Tong, L. Dezeit, A. Kashani, M. Meyyapan, and A. Majumdar, "Interfacial energy and strength of multiwalled-carbon-nanotube-based dry adhesive," Journal of Vacuum Science B, pp. 331-335, vol. 24, no. 1, Jan. 2006.

6

Use of Nano Physics in Nano Labs and Nanotechnology Products: An Integrated Approach

Bottom-up Methods for Making Nanotechnology Products

There are two general ways available to produce nanomaterials, as shown in the following figure. The first way is to start with a bulk material and then break it into smaller pieces using mechanical, chemical or other form of energy (top-down). An opposite approach is to synthesise the material from atomic or molecular species via chemical reactions, allowing for the precursor particles to grow in size (bottom-up). Both approaches can be done in either gas, liquid, supercritical fluids, solid states, or in vacuum. Most of the manufacturers are interested in the ability to control: a) particle size b) particle shape c) size distribution d) particle composition e) degree of particle agglomeration.

Methods to produce nanoparticles from atoms are chemical processes based on transformations in solution e.g. sol-gel processing, chemical vapour deposition (CVD), plasma or flame spraying synthesis, laser pyrolysis, atomic or molecular condensation. These chemical processes rely on the availability of appropriate "metal-organic" molecules as precursors. Sol-gel processing differs from other chemical processes due to its relatively low processing temperature. This makes the sol-gel process cost-effective and versatile. In spraying processes the flow of reactants (gas, liquid in form of aerosols or mixtures of both) is introduced to high-energy flame produced for example by plasma spraying equipment or carbon dioxide laser. The reactants decompose and particles are formed in a flame by homogeneous nucleation and growth. Rapid cooling results in formation of nanoscale particles.

These are chemical processes to materials based on transformations in solution such as sol-gel processing, hydro or solvo thermal syntheses, Metal Organic Decomposition (MOD), or in the vapour phase chemical vapour deposition (CVD).

Most chemical routes rely on the availability of appropriate "metal-organic" molecules as precursors. Among the various precursors of metal oxides, namely metal b-diketonates and metal carboxylates, metal alkoxides are the most versatile. They are available for nearly all elements and cost-effective synthesis from cheap feedstock have been developed for some.

Controlling the Construction and Growth of the Nanoparticles

Two general ways are available to control the formation and growth of the nanoparticles. One is called arrested precipitation and depends either on exhaustion of one of the reactants or on the introduction of the chemical that would block the reaction. Another method relies on a physical restriction of the volume available for the growth of the individual nanoparticles by using templates.

Sol-gel and Aerosol-Based Processes

The sol gel technique is a long-established industrial process for the generation of colloidal nanoparticles from liquid phase, that has been further developed in last years for the production of advanced nanomaterials and coatings. Sol-gel-processes are well adapted for oxide nanoparticles and composites nanopowders synthesis. The main advantages of sol-gel techniques for the preparation of materials are low temperature of processing, versatility, and flexible rheology allowing easy shaping and embedding. They offer unique opportunities for access to organic-inorganic materials. The most commonly used precursors of oxides are alkoxides, due to their commercial availability and to the high liability of the M-OR bond allowing *facile tailoring in situ* during processing. Aerosol-based processes are a common method for the industrial production of nanoparticles. Aerosols can be defined as solid or liquid particles in a gas phase, where the particles can range from molecules up to 100 μm in size. Aerosols were used in industrial manufacturing long before the basic science and engineering of the aerosols were understood. For example, carbon black particles used in pigments and reinforced car tires are produced by hydrocarbon combustion; titania pigment for use in paints and plastics is made by oxidation of titanium tetrachloride; fumed silica and titania formed from respective tetrachlorides by flame pyrolysis; optical fibres are manufactured by similar process.

Traditionally, spraying is used either to dry wet materials or to deposit coatings. Spraying of the precursor chemicals onto a heated surface or into the hot atmosphere results in precursor pyrolysis and formation of the particles. For example, a room temperature electro-spraying process was developed at Oxford University to produce nanoparticles of compound semiconductors and some metals. In particular, CdS nanoparticles were produced by generating aerosol micro-droplets containing Cd salt in the atmosphere containing hydrogen sulphide.

Chemical Vapour Deposition and Atomic or Molecular Condensation

CVD consists in activating a chemical reaction between the substrate surface and a gaseous precursor. Activation can be achieved either with temperature (Thermal CVD) or with a plasma (PECVD: Plasma Enhanced Chemical Vapour Deposition). The main advantage is the nondirective aspect of this technology. Plasma allows to decrease significantly the process temperature compared to the thermal CVD process. CVD is widely used to produce carbon nanotubes.

This method is used mainly for metal containing nanoparticles. A bulk material is heated in vacuum to produce a stream of vaporised and atomised matter, which is directed to a chamber containing either inert or reactive gas atmosphere. Rapid cooling of the metal atoms due to their collision with the gas molecules results in the condensation and formation of nanoparticles. If a reactive gas like oxygen is used then metal oxide nanoparticles are produced.

Using Gas-Phase Condensation to Produce Metal Nanopowders

The theory of gas-phase condensation for the production of metal nanopowders is well known, having been first reported in 1930. Gas-phase condensation uses a vacuum chamber that consists of a heating element, the metal to be made into nanopowder, powder collection equipment and vacuum hardware.

Ways by which Gas-Phase Condensation Process Works

The process utilises a gas, which is typically inert, at pressures high enough to promote particle formation, but low enough to allow the production of spherical particles. Metal is introduced onto a heated element and is rapidly melted. The metal is quickly taken to temperatures far above the melting point, but less than the boiling point, so that an adequate vapour pressure is achieved. Gas is continuously introduced into the chamber and removed by the pumps, so the gas flow moves the evaporated metal away from the hot element. As the gas cools the metal vapour, nanometer-sized particles form. These particles are liquid since they are still too hot to be solid. The liquid particles collide and coalesce in a controlled environment so that the particles grow to specification, remaining spherical and with smooth surfaces. As the liquid particles are further cooled under control, they become solid and grow no longer. At this point the nanoparticles are very reactive, so they are coated with a material that prevents further interaction with other particles (agglomeration) or with other materials.

Supercritical Fluid Synthesis and Spinning to Make Thin Polymer Fibers

Methods using supercritical fluids are also powerful for the synthesis of nanoparticles. For these methods, the properties of a supercritical fluid (fluid forced into supercritical state by regulating its temperature and its pressure) are used to form nanoparticles by a rapid expansion of a supercritical solution. Supercritical fluid method is currently developed at the pilot scale in a continuous process.

An emerging technology for the manufacture of thin polymer fibers is based on the principle of spinning dilute polymer solutions in a high voltage electric field. Electro spinning is a process by which a suspended drop of polymer is charged with thousands of volts. At a characteristic voltage the droplet forms a Taylor cone, and a fine jet of polymer releases from the surface in response to the tensile forces generated by interaction of an applied electric field, with the electrical charge carried by the jet. This produces a bundle of polymer fibers. The jet can be directed to a grounded surface and collected as a continuous web of fibers ranging in size from a few μm's to less than 100 nm.

Using Templates to Form Nanoparticles and Self-Assembly of Nanoparticles

Any material containing regular nano-sized pores or voids can be used as a template to form nanoparticles. Examples of such templates include porous alumina, zeolites, di-block co-polymers, dendrimers, proteins and other molecules. The template does not have to be a 3D object. Artificial templates can be created on a plane surface or a gas-liquid interface by forming self-assembled monolayers. Nanoparticles of a wide range of materials—including a variety of organic and biological compounds, but also inorganic oxides, metals, and semiconductors—can be processed using chemical self-assembly techniques. These techniques exploit selective attachment of molecules to specific surfaces, biomolecular recognition and self-ordering principles (e.g. the preferential docking of DNA strands with complementary base pairs) as well as well-developed chemistry for attaching molecules onto clusters and substrates (e.g. thiol (-SH) end groups) and other techniques like reverse micelle, sonochemical, and photochemical synthesis to realise 1-D, 2-D and 3-D self-assembled nanostructures. The molecular building blocks act as parts of a jigsaw puzzle that join together in a perfect order without an obvious driving force present.

Molecular Nanotechnology Offers Visions for the Future

Long-term and visionary nanotechnological conceptions, however, go far beyond these first approaches. This applies in particular to the development of biomimetic materials with the ability of self-organisation, self-healing and self-replication by means of molecular nanotechnology. One objective here is the combination of synthetic and biological materials, architectures and systems, respectively, the imitation of biological processes for technological applications. This field of nanobiotechnology is at present still in the state of basic research, but is regarded as one of the most promising research fields for the future.

Nanotech's "Second Nature" Patents

On the 25th anniversary of Diamond vs. Chakrabarty,* the US Supreme Court's landmark decision (June 16, 1980) that opened the floodgates to the patenting of living organisms, ETC Group releases a new report, "Nanotech's 'Second Nature' Patents." Since Chakrabarty, the biotech industry has worked hand-in-hand with governments to allow for the patenting of all biological products – the first monopoly grab over life. Chakrabarty set the stage for today's nanotechnology patents, where the reach of exclusive monopoly is not just on life – but the building blocks of life – nanotech's 'second nature' patents," explains Hope Shand, Research Director of ETC Group.

ETC Group's new report examines current trends in intellectual property and nanotechnology and the implications for the developing world. Nanotechnology refers to the manipulation of matter at the scale of atoms and molecules, where size is measured in billionths of meters. The world's largest transnationals, leading academic labs and nanotech start-ups are all racing to win monopoly control of tiny tech's colossal market. "Control and ownership of nanotech is a vital issue for all governments and civil society because nanomaterials and processes can be applied to virtually any manufactured good across all industry sectors," said Kathy Jo Wetter of ETC Group. "Patents are being granted that cut across multiple industry sectors – a single nano-scale innovation may span pharma, food, electronics and materials alike," continues Wetter. The US National Science Foundation predicts that nanotechnology will capture a $1 trillion dollar market within six or seven years.

ETC Group finds that breathtakingly broad nanotech patents have been granted that cut across multiple industry sectors and include sweeping claims on entire areas of the Periodic Table. Although industry analysts assert that nanotechnology is in its infancy, "patent thickets" on fundamental nano-scale materials, tools and processes are already creating thorny barriers for would-be innovators. Claims are often broad, overlapping and conflicting—a scenario ripe for massive patent litigation battles in the future. ETC Group's report provides case studies of patent activity involving four of nanotech's hottest and potentially most lucrative nanomaterials and one essential tool: carbon nanotubes; inorganic nanostructures; quantum dots; dendrimers; scanning probe microscopes.

G8: Downsizing Development? When the G8 Summit meets in Scotland next month, the leaders of the world's most powerful countries will unveil a "Pro-Poor Science" strategy to turn new technologies like nanotech into a silver bullet for social injustice. "Despite rosy predictions that nanotech will provide a technical fix for hunger, disease and the environment, the extraordinary pace of nanotech patenting suggests that developing nations will participate primarily via royalty payments," said Pat Mooney, Executive Director of ETC Group. "In a world dominated by proprietary science, researchers in the global South are likely to find that participation in the nanotech revolution is highly restricted by patent tollbooths, obliging them to pay royalties and licensing fees to gain access," said Mooney. "Ultimately, nanotech will profoundly affect the South's economy, regardless of its handling of intellectual property," explains Silvia Ribeiro from ETC Group's Mexico City office. "Nano-scale technologies will revolutionize the way that new materials are designed and manufactured – changes that could turn commodity markets upside-down and make geography, raw materials, even labour, irrelevant. Nanotech underpins a new strategic platform for global control of materials, food, agriculture and health, and patent monopoly is a powerful tool for realising that strategy," said Ribeiro.

Many South nations are still grappling with unresolved controversies over biotechnology, but by the end of this year, ready or not, even the world's "least developed" nations who are members of the World Trade Organisation will be obligated by its Trade-Related Aspects of Intellectual Property (WTO-TRIPs) to evaluate and enforce nanotech patents.

Lessons learned from Diamond v. Chakrabarty: Despite all the hype about Mr. Chakrabarty's oileating microbe and how it would gobble up oil spills, the patented microorganism never worked. Instead of curing environmental ills, the biotech industry has introduced its own contamination problems – unwanted gene flow from genetically modified crops, a particularly serious problem for centres of genetic diversity in the developing world. Unlike 25 years ago, today's nanotech-related patents have not required major rule changes. As a result, many governments are unaware of the nanotech patent rush. ETC Group recommends that the World Intellectual Property Organisation (WIPO) initiate a global suspension of patent approvals related to nanotechnology until South governments and countries-in-transition can undertake a full evaluation of their impacts, and until social movements can cooperate with WIPO, the Food and Agriculture Organisation (FAO) and the United Nations Conference on Trade and Development (UNCTAD) to examine the impact of nanotech-related intellectual property on monopoly practices, technology transfer and trade. The full text of the 32-page report is available for downloading, free-of-charge, on the ETC Group.

The Action Group on Erosion, Technology and Concentration, formerly RAFI, is an international civil society organisation headquartered in Canada. The ETC Group is dedicated to the advancement of cultural and ecological diversity and human rights. www.etcgroup.org. The ETC Group is also a member of the Community Biodiversity Development and Conservation Programme (CBDC). The CBDC is a collaborative experimental initiative involving civil society organisations and public research institutions in 14 countries. The CBDC is dedicated to the exploration of community-directed programmes to strengthen the conservation and enhancement of agricultural biodiversity. Implications for the Global South is Nanotech "Pro-Poor Science" or is the G8 Downsizing Development? On the 25th anniversary of the US Supreme Court decision that opened the floodgates to the patenting of all life forms, ETC Group reports on current trends in intellectual property relating to nanoscale technologies. With nanotechnology, the reach of exclusive monopoly patents is not just on life, but all of nature. Accordingly, ETC Group refers to nanotech's "second nature" patents. Nanotech patent claims offer the South (and society) an advance look at who's likely to own nanotechnology and dominate 21st century commodity markets. Breathtakingly broad nanotech patents are being granted that span multiple industry sectors and include sweeping claims on entire classes of the Periodic Table. Yet, when the G8 nations meet this July they will be unveiling a "Pro-Poor Science" strategy to turn nanotechnology into a silver bullet antidote to social injustice. Is nanotech the solution – or just another big downer for development?

The race is on to win monopoly control of tiny tech's colossal market. The US National Science Foundation predicts that the immensely broad power and scope of nano-scale technologies will revolutionize manufacturing across all industry sectors – capturing a $1 trillion market within six or seven years. Although industry analysts assert that nanotech is in its infancy, "patent thickets" on fundamental nano-scale materials, building blocks and tools are already creating thorny barriers for would-be innovators. Industry analysts warn that, "IP roadblocks could severely retard the development of nanotechnology." Some insist that nano-scale technologies will address the most pressing needs of the South's marginalised peoples. But in a world dominated by proprietary science, it is the patent owners and those who can pay license fees who will determine access and price. Even as South governments are grappling with confusion and controversies over biotechnology, the World Trade Organisation's Trade- Related Aspects of Intellectual Property (TRIPs) obligates even "least developed" countries to enforce nanotech patents by 2006.

The world's largest transnationals, leading academic labs and nanotech start-ups are all racing in the patent gold rush. Increasingly, universities are licensing on an exclusive basis. Nanotech's "second nature patents" are positioning multinational matter moguls to own and control novel materials, devices and their manufacturing processes.

While WIPO ponders a "development agenda" in Geneva, patent offices in Washington, Munich and Tokyo are deciding who will gain exclusive monopoly over a technology that will bring profound changes in demand for raw materials and manufacturing around the globe. Despite rosy predictions that nanotech will provide a technical fix for hunger, disease and environmental security in the South, the extraordinary pace of nanotech patenting suggests that developing nations will participate via royalty payments. Accordingly, national governments must protect their sovereignty and intergovernmental agencies must move quickly to prevent multi-sector monopolies, technology barriers and the formation of a powerful new oligopoly that seeks control over the economy's 'second nature.'

Control and ownership of nanotechnology is a vital issue for all governments and civil society because nanomaterials and processes can be applied to virtually any manufactured good across *all* industry sectors. Nanotechnology refers to the manipulation of matter at the scale of atoms and molecules, where size is measured in billionths of meters (one nanometer = one-billionth of a meter). Nanotech isn't a single technology – but a range of technologies converging at the nano-scale – including biotechnology, genomics, neurosciences, robotics and information technologies. For a brief introduction to nanotechnology, please see, *A Tiny Primer on Nano-Scale Technologies* [http://www.etcgroup.org] Worldwide, industry and governments invested more than $10 billion in nanotech R&D last year, with two-thirds coming from corporate and private funds.3 There are an estimated 1200 nanotech start-up companies, half of which are US-based.4 Virtually all Fortune 500 companies invest in nanotech R&D. A recent survey conducted by MIT's *Technology Review* predicts that products involving nanotech will account for over $100 billion by 2008; the US government predicts that the nanotech market will explode to $1 trillion by 2012. Intellectual property (IP) will play a major role in deciding who will capture nanotech's trillion dollar market, who will have access to nano-scale technologies and at what price. At stake is control over innovations that span multiple industry sectors – from electronics, energy, mining and defense to new materials, pharmaceuticals and agriculture. As the *Wall St. Journal* put it, "companies that hold pioneering patents could potentially put up tolls on entire industries." Even industry insiders admit that current intellectual property trends related to nanotech are chaotic.

Many broad patents on nanotech-related materials, tools and processes have been granted too early and too often. In 2002, the US-based industry trade group, Nanotechnology Business Alliance, was already warning in testimony before the US Congress, "...several early nanotech patents are given such broad coverage, the industry is potentially in real danger of experiencing unnecessary legal slowdowns."

More recently, nanotech industry analysts observe that the "euphoria for patenting" in the US combined with the US Patent & Trademark Office's inability to handle a flood of patent applications has resulted in "the rejection of valid claims, the issuance of broad and over-lapping claims, and a fragmented and somewhat chaotic IP landscape." The writers warn, "These IP roadblocks could severely retard development of nanotechnology."

This report aims to provide information to civil society and South policymakers about current trends in nanotechnology and intellectual property. What is the current IP landscape? What is being patented and by whom? What are the implications for innovation and development in the South? How will intellectual property affect the South's participation in the nanotech revolution? "Like biotechnology in the early 1980s, the IP landgrab mentality pays off for early pioneers. Don't bet

Patents on nano-scale materials, tools and processes are a fair measure of the tsunami-like strength of this latest industrial revolution. Estimates vary on the number of nanotech patents issued since the early 1990s, but all agree that both companies and public sector entities are "rushing to the patent office in record numbers to patent nanotechnology inventions."

The lack of uniform definitions for nanotechnology means that identifying the number of nanotechrelated patents granted over the past decade is a very imprecise science. When the US PTO announced it had created a new classification for nanotechnology patents in October 2004 it defined nanotechnology patents narrowly: Classification 977 includes only those patents 1) whose subject matter is in the scale of approximately 1-100 nanometers in at least one dimension; and 2) that involve materials, structures, devices or systems that have novel properties and functions because of their nano-scale size. Nanotech patent searches often use broad search terms (for example, the prefix "nano"), which can result in exaggerated counts. There is wide consensus, however, that major patent offices worldwide are granting nanotech patents at an extraordinary pace. The chart below illustrates the overall trend, which is similar at the World Intellectual Property Organisation's Patent Cooperation Treaty and at the US PTO. In response to the demand for nanotech patents, more than a dozen major law firms in the US have recently established nanotech patent law specialties. Researchers from the University of Arizona and the US National Science Foundation examined nanoscale science and engineering patents at the US Patent & Trademark Office from 1976-2003. 12 They found that 8,630 nanotech-related patents were issued by the US PTO in 2003 alone, an increase of 50% over the previous three years. The top 5 countries represented were: US (5,228 patents), Japan (926), Germany (684), Canada (244) and France (183). The top 5 entities winning nanotech-related patents included four multinational electronic firms and one university: IBM (198 patents), Micron Technologies (129), Advanced Micro Devices (128), Intel (90) and University of California (89). A new report by Lux Research, Inc. identifies far fewer nanotech patents granted by the US PTO. In April 2005 Lux announced that it had identified 3,818 nanotech-related patents issued between 1985-March 2005, with an additional 1,777 patent applications.

ETC Group takes a Closer Look at Nanotech Patents Granted by the US PTO: In October 2004, the US PTO announced it had created a new classification for nanotechnology patents – Class 977 – which would serve as a cross-reference to help examiners, among others, search prior art. Before Class 977 existed, examiners relied on keyword searches to find relevant information and related patents.15 As defined by the US PTO, nanotechnology patents in Class 977 must meet the following criteria:

- relate to research and technology development in the length scale of approximately 1-100 nm in at least one dimension
- provide a fundamental understanding of phenomena and materials at the nanoscale and create and use structures, devices, and systems that have size-dependent novel properties and functions.16 The US PTO is now reviewing patents issued before the classification was created and re-classifying those that meet the 977 criteria. According to Bruce Kisliuk, a patent examiner group director, the Patent Office hopes to be caught up reviewing already-issued patents within a year.17 Until the review is complete, the PTO is referring to Class 977 as a "digest." Eventually, Class 977 will be broken down further into sub-classes. As of May 25, 2005, Class 977 included 726 patents issued by the US PTO, dating from May 24, 2005 back to 1980. In its incomplete state, Class 977 should be seen as a sample rather than a comprehensive picture of nano-patenting in the US. (At present, Class 977 contains only six nano-patents dating from the second half of 2004, for example.) Even though the classification is incomplete, the sample is worth looking at because it offers an advantage over more random patent searches (e.g., searching for keywords such as "nano" or "quantum"): we can be reasonably confident that the patents in Class 977 are bona fide nano-patents because they have been reviewed and each was determined to meet the classification criteria – relating to the nanometer scale (1-100 nm) in at least one dimension and exploiting size-dependent properties. A search through Class 977 reveals:
- The USPTO has issued nano-patents in the fields of electronics (including cameras, computers and other devices), instruments and tools (including scanning probe microscopes and sensors), pharmaceuticals (including drug delivery), food, agrochemicals, devices, materials (including fullerenes, nanotubes and quantum dots) and processes for creating and engineering nano-sized particles.
- More than 290 different primary patent examiners were assigned to evaluate these Class 977 patents. The extraordinarily large number of examiners raises suspicions that over-lapping and conflicting patents were granted by different examiners at about the same time; it also illustrates the multi-disciplinary and cross-sectoral nature of nano-scale science.
- Out of the 235 Class 977 patents issued in 2003 (the most complete year):

- 112 (48%) are assigned to US-based companies
- 29 (over 12%) are assigned to US universities
- 70 (30%) are assigned to companies based outside the US (of these, 26 are assigned to companies based in Japan, 10 in Germany, 9 in Australia – 8 of which are assigned to one company, Silverbrook Research. Companies from Canada, Venezuela, France, Korea, Belgium, Denmark, Italy, Ireland, Netherlands, Taiwan and Antilles make up the other non-US assignees.)
- 15 (6.5%) are assigned to non-US government entities or national research agencies
- The US government has rights to 31 patents (either because it provided funding for research or through outright ownership), which represents 22% of the patents assigned to US entities

Class 977 patents issued by the USPTO range from mundane to alarming to other-worldly. One would expect only great things from the Delphi Oracle Corporation and it does not disappoint: Delphi owns patent 6,493,910, "a shoelace with enhanced knot retention and method of manufacture." (It's a nanopatent because the shoelace makes use of nano-scale fumed silica.)

On the disturbing end of the spectrum, the US government won patent 5,805,657 back in 1998 for "nuclear fuel elements made from nanophase materials," which makes one wonder in the event of infringement – will there be anyone around to sue? Zyvex Corporation owns patent 6,510,359, a "method and system for self-replicating manufacturing stations." In case anyone devises one of these in the next 20 years, it looks like Zyvex will own it. They claim a "non-biological self replicating manufacturing system" comprising:

- a translating machine capable of translating at least a first surface in relation to at least a second surface, wherein said at least a first surface comprises at least one assembly station thereon and wherein said at least a second surface comprises parts arranged thereon for use in constructing at least one other assembly station; and
- a control system for controlling the operation of said translating machine and for controlling said at least one assembly station to cause said at least one assembly station to construct at least one other assembly station from said parts on said second surface.

Patent Paradox: For most nanotech start-up companies (especially those that operate without products or profits), intellectual property is the most essential asset. In the words of one industry executive, the strength of a nanotech company's patent portfolio is what separates the winners from the losers.19 Consider, for example:

Nanogen's stock jumped over 50 percent on the day it announced it had won a patent on a technology to detect genetic variants; NVE Corporation's stock surged 41 percent when it announced its new patent on magnetic random access memory.

But the transaction costs of filing for and defending patents are enormous. Competing in the high-stakes patent game requires not only winning and defending patents, but also using them as bargaining chips to cross-license other proprietary technologies. Intellectual property is one of the biggest expenses for fledgling start-ups, and nanotech industry insiders predict that long and costly patent litigation battles could inflict mortal wounds. Consider, for example:

- In the US, depending on the complexity, claims, and length of the application, filing fees can go as high as $1000.21
- For nanotech start-up companies, patent fees are often a major cost of doing business – sometimes second only to payroll.22 Legal fees for winning a patent in the US can cost $25,000- $30,000, and up to $250,000 for international patent claims.23
- Outside the US, countries charge around $5,000 per year on each pending patent. The translation fees alone to win a patent in Japan are between $12,000 and $20,000.24
- The number of patent lawsuits filed in the US more than doubled over the decade of the 1990s. Based on a survey of intellectual property lawyers in 2000, the cost of defending a large (more than $25 million at risk) patent infringement suit ranges from $2 million to $4.5 million. For cases with less than $1 million at risk, the cost was $300,000 to $750,000, or about half the amount in dispute.
- Most shocking of all, it is estimated that only about two percent of all issued patents end up generating more revenue than the cost of obtaining the patent. Among this small group of "profitable" patents, only some will end up being worth the price of litigation.

Many intellectual property experts in the US are predicting that large-scale nanotech patent litigation is inevitable, and it's likely to be ugly. Because of the large number of over-lapping and conflicting patents being granted, nanotech companies must be prepared to vigorously defend their patents in court. It's conventional wisdom that, in most patent battles, it's the largest enterprises – not the most innovative – that will prevail. According to authors Josh Lerner and Adam Jaffe, "the firm with the best lawyers or the greatest capacity to withstand the risk of litigation wins the innovation wars – rather than the company with the brightest scientists or most original, valuable ideas."

"Nobody's making any money right now, so there's no particular reason to fight the battles yet. But a lot of people are coming out with patents, and a lot of claims are certainly overlapping, if not certainly *conflicting. The fights are going to be brutal."* – Matthew Nordan, vice-president of research, nanotech analyst firm, Lux Research Firms can use patents as "strategic weapons to strangle competition." If a larger firm believes that its dominance in a nanotechnology market is threatened, it "can slowly starve their start-up competitors into extinction by waging a protracted battle on the IP front."30 As with biotech in the 1980s and 1990s, nanotech start-ups holding key patents may also become attractive takeover targets because it's less costly for a multinational firm to acquire the company than to litigate in court. In part two of this report ETC Group provides case studies of nanotech and intellectual property, focusing on some of the key nanomaterials or tools that are the subject of multiple (and likely) over-lapping patent claims. The case studies examine: carbon nanotubes, Nanosys Inc.'s patents on inorganic nanomaterials, quantum dots, dendrimers and scanning probe microscopes. What are the major trends in Nanotech-related IP?

Patents on nanotech's fundamental building blocks and tools. Stanford University law professor Mark Lemley asserts that nanotechnology "is the first new field in a century in which people started patenting the basic ideas at the outset." In contrast to most other major enabling technologies of the 20th century (such as computer hardware, software, the Internet, and even biotechnology), writes Lemley, the most basic ideas and fundamental building blocks in nanotechnology "are either already patented or may well end up being patented." ETC Group does not share Lemley's view that biotech's fundamental enabling tools were not patented early on – but we do share his view that nanotechnology's most basic ideas and building blocks are being patented early and often.

In the nanotech arena, it's not just the opportunity to patent the most basic enabling tools, but the ability to patent the nanomaterials themselves, the products they are used in and the methods of making them. At the US Patent & Trademark Office there are three primary types of patent claims:

1) composition of matter claims (that is, nanomaterials such as nanotubes, nanowires and nanoparticles),
2) device, apparatus or system claims (including, for example, tools used to characterize and control nanomaterials—or devices incorporating nanomaterials),
3) method claims (processes for synthesising nanomaterials or constructing nano-scale devices).

The CEO of Nanosys, Inc., Larry Bock, explains the once-in-a-lifetime attraction of being able to get in on the ground floor by consolidating fundamental nanotech patents: "One of the key things that got me excited about this whole area three years ago is that from an IP standpoint you could control IP from things like composition of matter to the simplest functional devices to the end application all under one IP portfolio. That doesn't come around very often in one's career."

Nanomaterials are chemical elements or compounds less than 100 nm in size. Taking advantage of quantum physics, nanotech companies are engineering novel materials that may have entirely new properties never before identified in nature. The "raw materials" for creating nanomaterials and devices are the chemical elements of the Periodic Table – the building blocks of *everything* – both living and nonliving. Whereas biotechnology patents make claims on biological products and processes – nanotechnology patents may literally stake claim to chemical elements, as well as the compounds and the devices that incorporate them. With nano-scale technologies the issue is not just patents on life – but on all of nature. In short, atomic-level manufacturing provides new opportunities for sweeping monopoly control over both animate and inanimate matter. In essence, patenting at the nano-scale could mean monopolising the basic elements that make life possible.

Proprietary Periodic Table: Exclusive monopoly patents on chemical elements are not new. Glenn Seaborg, the 1951 Nobel Prize-winning physicist, won US patent #3,156,523 for the chemical element *Americium* (element no. 95 on the periodic table) on November 10, 1964. Seaborg's patent is recognised for having the shortest patent claim on record: "What is claimed is Element 95." Seaborg's second patented element was Curium #96 – US patent # 3,161,462 granted on December 15, 1964. "It is true that one cannot patent an element found in its natural form; however, if you create a *purified form of it that has industrial uses—say, neon—you can certainly secure a patent."* – Lila Feisee, Biotechnology Industry Organisation's Director for Government Relations and Intellectual Property

It's Elemental: When Harvard University's Charles Lieber obtained a key patent (US patent 5,897,945) on nano-scale metal oxide nanorods, he didn't claim nanorods composed of a single type of metal—but instead claimed a metal oxide selected from up to 33 chemical elements. Harvard's claims on nanorods include those comprised of titanium, zirconium, hafnium, vanadium, niobium, tantalum, chromium, molybdenum, tungsten, manganese, technetium, rhenium, iron, osmium, cobalt, nickel, copper, zinc, cadmium, scandium, yttrium, lanthanum, a lanthanide series element, boron, gallium, indium, thallium, germanium, tin, lead, magnesium, calcium, strontium, and barium. In a single patent, Lieber's claims extend to nearly one-third of the chemical elements in the Periodic Table – spanning 11 of the 18 Groups. Patent lawyers have identified Harvard's patent (licensed to Nanosys, Inc.) as one of the top 10 patents that could influence the development of nanotechnology.35 Similarly, a key patent on semiconductor nanocrystals (quantum dots) held by the

University of California (licensed to Nanosys, Inc. and Quantum Dot Corp.) claims semiconductor nanoparticles from elements in Groups III-V of the Periodic Table. The claims in US patent number 5,505,928 extend to boron, aluminum, gallium, indium, nitrogen, phosphorus, arsenic, antimony as well as those compound semiconductors that result from combining elements in Groups III-V (such as gallium arsenide).

Cross-industry Patent Claims: Nanotechnology is not only cross-disciplinary, a single nano-scale innovation may have diverse applications that span multiple industry sectors. Mark Lemley of Stanford Law School observes, "a significant number of nanotechnology patentees will own rights not just in the industry in which they participate, but in other industries as well."36 When ETC Group examined the 700+ patents that the US Patent & Trademark Office had identified as nanotechnology patents as of May 25, 2005, it was not surprising to find that the patents had originally been assigned to *all* of the major patent classes – including electricity; human necessities; chemistry/metallurgy; performing operations and transporting; mechanical engineering (lighting, heating, weapons, blasting); physics; fixed construction; textiles and paper.

Masters of the Industrial Universe? The crucial aspect to understand about nano IP is not simply that the patents span a broad range of fields, but that a single invention can be relevant for widely divergent applications. It doesn't come as a big surprise to find an invention that is applicable to related fields (e.g., pharmaceuticals and foods), but consider the following examples from US PTO's Class 977:

- US5,874,029 – University of Kansas, 23 February 1999: Methods for particle micronisation and nanonisation by recrystallisation from organic solutions sprayed into a compressed antisolvent: The invention can be used in the pharmaceutical, food, chemical, electronics, catalyst, polymer, pesticide, explosives, and coating industries, all of which have a need for smalldiameter particles.
- US6,667,099 – Creavis Gesellschaft für Technologie und Innovation mbH, 23 December 2003: Meso-and nanotubes: The invention relates to mesotubes and nanotubes (hollow fibers) having an inner diameter of 10 nm-50 μm and to a method for the production thereof ...*The hollow fibers* are used in separation technology, catalysis, micro-electronics, medical technology, material technology or in the clothing industry.
- US6,641,773 – The USA as represented by the Secretary of the Army, 11 November, 2004: Electro spinning of submicron diameter polymer filaments: An electro spinning process yields uniform, nanometer diameter polymer filaments...*The filament is particularly useful for weaving body* armor, for chemical/biological protective clothing, as a biomedical tissue growth

support, for fabricating micro sieves and for microelectronics fabrication. The reason that the same invention can be used inside the human body, in clothing and in computers, as in the third example above, is that at the molecular level biological and non-biological material can be integrated – whether this is a seamless integration is a matter yet to be determined by toxicological research.

Second Nature Patents: While biotech's raw materials are biological, nano-scale technologies involve the manipulation of both living and non-living materials, sometimes in combination. When this is the case, the discipline is known as *nanobiotechnology*. A nanostructured material used inside the body as a bone replacement is one example of nanobiotechnology, but so is a hybrid organism created from living and non-living materials, such as the nano-scale silicon and muscle-tissue hybrid announced by researchers in early 2005.37 Closely related to and sometimes overlapping nanobiotech is the new field of "synthetic biology" in which living systems are built to order and then programmed to perform specific tasks. These, too, often combine biological and non-biological parts. Patents on the products of nanobiotechnology provide the opportunity to monopolize the basic elements that are the building blocks of the entire natural world, bringing a whole new dimension to the notion of "life patenting." The table below provides examples of the possible range of nanobiotechnology and synthetic biology patents recently issued by the US PTO. It includes, for example: hybrid devices combining a nanomaterial and muscle tissue, which generate electrical power and which the inventor has described as "absolutely alive"38 (Montemagno); membranes made from biological and non-biological materials to be used in electricity production or water purification (Montemagno); a method for controlling the properties of semiconductor nanoparticles by creating them with the help of biological material (Belcher); synthetic DNA base pairs that do not occur in nature (Benner); a method for genetically modifying cells by pricking them with carbon nanotube "needles" and injecting foreign DNA (McKnight); a gene switch that uses "switching agents" to control gene expression by turning them on or off.

Role of Public Sector Universities in Nanotech IP: One of the unique features of nanotechnology, according to Stanford Law School professor, Mark Lemley, is that universities and public research foundations hold "a grossly disproportionate share of nanotech patents" that he believes are critically important to downstream nanotech products. In 2004 a patent attorney specialising in nanotechnology identified 10 key patents that he believed could have the greatest impact on the development of nanotechnology. Seven of the 10 patents are owned by universities.

Because they conduct basic research, it's not surprising that universities are the early-stage engines for nanotechnology. But unlike early-stage researchers 25 years ago, the new generation of US public researchers has become "extremely aggressive patenters" largely because of the Bayh-Dole Act of 1980 – US legislation designed to encourage technology transfer by permitting universities to patent their federally funded research projects. Before 1980, universities worldwide were granted about 250 US patents per year. By 2003, the number of university-owned patents increased almost 16-fold, to 3,933.41 Because university labs aren't in the business of commercialising products, they try to re-coup their research costs by patenting their employees' early scientific innovations – in the hope of earning royalties or licensing fees. Exclusive licensing is generally the more lucrative deal – and therefore the most appealing to technology transfer offices. In general, universities are acting more and more like businesses. Not only are universities patenting nanotech early and often, they are more frequently licensing their inventions on an exclusive basis. US policymakers who favour Bayh-Dole would argue that universities are benefiting society by transferring science and technology to the private sector for commercialisation. But in many cases, consumers end up paying twice – once by paying taxes to support governmentfinanced research, and again when they purchase a new, proprietary technology developed with taxpayer funds. With the emphasis on winning exclusive monopoly patents, the traditional academic culture of open communication and exchange is also undermined and eroded.

The largest single holder of nanotech patents in the world is a Chinese researcher, Yang Mengjun, who is taking ancient Chinese medicinal herbs, reducing them to nano-scale formulations, and claiming exclusive monopoly over the herbs or the process used to nano-size them. He holds over 900 patents on nanoscale versions of traditional Chinese medicinal plants.43 Similar patents are being granted in the US and Europe. For example, the Pacific Corporation (Korea) has won a European patent on nanoscale ginseng for use in cosmetic products.44 The Pacific Corporation claims that an emulsion of ginseng at the nano-scale (reduced to small particles between 50-500nm) allows it to penetrate the skin, exerting an antiaging effect.

Patent claims on nano-scale formulations of traditional herbal plants are providing insidious pathways to monopolize traditional resources and knowledge – one more reason why the Convention on Biological Diversity and FAO should address the implications of nanotechnology. Diamond vs. Chakrabarty Remembered June 16, 2005 marks the 25th anniversary of Diamond vs. Chakrabarty, the landmark US Supreme Court decision that opened the floodgates to the patenting of all life forms. The anniversary offers a timely opportunity to examine current trends in intellectual property relating to nano-scale technologies – the world's newest technological wave.

In 1971, Ananda Chakrabarty, an employee of General Electric, applied for a patent on a genetically modified, oil-eating microbe. His patent application was rejected by the US Patent & Trademark Office (US PTO) on the grounds that animate life forms were not patentable. When Chakrabarty won his case on appeal, the PTO Commissioner, Sidney Diamond, took the case to the US Supreme Court. On June 16, 1980 by a narrow 5-4 margin, the US Supreme Court ruled that Chakrabarty's oil-eating microbe was not a product of nature; living organisms could be seen as human made inventions and are therefore patentable subject matter. An ironic footnote to the saga is that the "invention" didn't work. The monumental importance of the Chakrabarty decision did not register with the Court – or the public – at the time. (Some environmentalists were eager to embrace life patenting if it meant microbes could devour oil spills.) In 1980 the Supreme Court specifically noted that the Chakrabarty decision was a narrow case that would not affect the "future of scientific research."45 The Court got it wrong. According to lawyer and activist Andrew Kimbrell, "The complete failure by the Court to correctly assess the impacts of the Chakrabarty decision may go down as among the biggest judicial miscalculations in the Court's long history."

As a result of Chakrabarty, the slippery slope of IP on living organisms became a patent landslide, and a bonanza for the biotech industry. Over the course of a single decade, the US government re-interpreted intellectual property laws to allow for exclusive monopoly control over *all* biological products and processes. After Chakrabarty, the once unthinkable patenting of genes, plants, animals, microorganisms and human genetic material would become common practice in the US – positioning industry and the US government to set the precedent for IP regimes worldwide via the World Trade Organisation and through bilateral and regional trade agreements.

Timeline

1980 – Diamond v. Chakrabarty – US Supreme Court case establishes precedent for the patenting of living organisms.

1980 – US Bayh-Dole Act allows businesses, universities and non-profit organisations to retain title to patents that result from federally-funded research, and to exclusively license them.

1984 – University of California wins US patent on a cell line developed from cancerous tissue of John Moore, a leukemia patient whose cancerous spleen cells were patented and commercialised without his knowledge.

1985 – US PTO rules that genetically engineered plants, seeds and plant tissue are patentable subject matter.

1986 – Uruguay Round of General Agreement on Tariffs and Trade (GATT) begins with Trade-Related Intellectual Property (TRIPs) on the table.

1987 – US PTO rules that genetically modified animals are patentable subject matter.

1988 – US PTO issues first patent on a living animal – a transgenic mouse. Harvard licenses the "OncoMouse" to Dupont.

1993 – Patent claim by the US Secretary of Commerce on the cell line of a 26-year old Guaymi indigenous woman from Panama. Following worldwide controversy, the US government abandoned its patent claim on the Guaymi cell line in November 1993.

1994 – EPO issues species-wide patent on all genetically modified soybeans to Agracetus (later acquired by Monsanto).

1995 – WTO comes into force at the conclusion of the Uruguay Round of GATT.

1995 – Civil society organisations and social movements in Europe defeat European Patent Directive that aims to harmonize patenting of genetic material within the EU.

1995 – US Supreme Court rules in Asgrow vs. Winterboer that farmers no longer have the right to harvest and re-sell proprietary seed for reproductive purposes (that is, proprietary seed protected by Breeders' Rights – plant variety protection – may be saved only for the purpose of re-planting the farmers' own acreage).

1995 – US PTO issues patent to US National Institutes of Health for an unmodified human cell line from an indigenous person from Papua New Guinea. Due to the international controversy, the US government is forced to disclaim the patent in December 1996.

1998 – European Parliament gives final approval to a controversial biotechnology "patent directive" that aims to harmonize national legislation on the patenting of genetic material within the EU. The Directive creates, for the first time in European history, an explicit legal right to patent higher organisms such as plants and animals.

2002 – Canadian Supreme Court rules against patenting of higher life forms– rejects patenting of genetically modified mouse.

2004 – Canadian Supreme Court (Monsanto v. Schmeiser) affirmed Monsanto's right to prosecute farmers who are found to have proprietary GM crops growing on their land – whether they wanted them or not.

2004 – US Patent & Trademark Office establishes nanotechnology patent class.

Lessons learned from Chakrabarty:

- Historic decisions allowing exclusive monopoly control of all biological products and processes involved no public input or wider societal debate; these decisions were made by a handful of individuals in the courts and patent offices – not by the US Congress. In essence, it was the courts and not citizens who gave biotech the green light in the US. Similarly, at the international level, intellectual property rules have been crafted by and for a narrow group of corporate interests.

- Following Chakrabarty, the US government's aggressive life patenting policies set the bar for the rest of the world – especially at the World Trade Organisation.
- The slope is slippery indeed. The history of patent monopoly demonstrates that patent holders typically seek wider patentability, more expansive scope of patent claims, longer patent terms and greater harmonisation of patent rules worldwide.
- For many developing nations the rationale for accepting stronger IP regimes has been the argument that their economies would prosper from increased technology transfers and foreign direct investment. In the case of biotechnology, however, the vast majority of key enabling technologies are proprietary products and processes, tightly concentrated in the hands of multinational gene giants. Under these conditions, stronger levels of IP obligate developing countries to make a massive transfer of resources to the North, in order to acquire licenses for proprietary technologies. A new study by the World Bank concludes that the effects of stronger intellectual property regimes in creating greater trade flows to developing countries are "theoretically ambiguous." The authors conclude, however, that stronger levels of intellectual property in developing countries are not a factor in spurring high-technology trade flows.

From Dust to Dust: A concise history of patent monopoly

The rallying cry "no patents on life" has become a line in a technological and legal sandstorm. Although the notion of intellectual monopolies can be traced back to early Greece, patents did not come into their own until Britain's Industrial Revolution when the inventors of textile machinery demanded "protection." Recognising that patents would make technology accessible only to well-heeled manufacturers, smaller enterprises protested. The response: "Don't worry. We only seek to patent the machines we invented."

In the 1920s and 30s, when rose and chrysanthemum breeders demanded intellectual property for their flowers, they argued that it was unfair to grant patents to machine inventors but to deny equal rights to ornamental inventors. Although some were repelled by the idea that living things could be patented, the flower companies replied, "Don't worry. These patents protect only decorative plants – not food crops."

In the 1960s, when plant breeders called upon governments to grant them intellectual property over food crops, they said it was unfair to recognize the minor contributions of ornamental breeders without recognising the contributions of the breeders of crop varieties. The companies chided their critics by saying, "Don't be alarmed. We just want breeders' rights to protect plant varieties; we're not patenting plants, animals or human genetic material, and we would never stop farmers from saving seed."

In 1980, the Gene Giants won patents on genetically modified microbes. A few years later they applied for patents on plants and animals. When civil society protested, industry responded, "Why all the fuss? If you allow the patenting of micro-organisms, why not plants and lab rats?"

In the 1990s, corporations and governments began to patent genes, snippets of DNA, and entire human cell lines. When indigenous peoples protested, patent offices responded, "Don't worry. Human cell lines are just microorganisms."

Meanwhile, patents made it illegal for farmers to save and re-use proprietary seed. The seed/biotech industry denounced the 12,000-year old right of farmers to save harvested seed as patent infringement.

With the advent of nano-scale technologies, corporations are patenting essential building blocks of all living and non-living things. Industry is redefining life to create hybrid organisms that will take on machine functions. When we tell them they have gone too far, they will reply, "Don't worry. We're all just machines."

Nanotechnology and the Transportation Industry

The Benefits of Advanced Materials

Advanced materials that will allow for longer service life and lower failure rates. Among the key applications are nanocoating of metallic surfaces to achieve super-hardening, low friction, and enhanced corrosion protection; 'tailored' materials for infrastructure and vehicles; and "smart" materials that monitor and assess their own status and repair any defects resulting from fatigue, fire, etc. New materials designed to have superior properties (stronger, lighter) using materials with nanoscale dimensions will lead to faster, cheaper, and safer transportation

Advances in Transportation Due to Nanotechnology

Nanotechnology will impact in all areas of transportation, providing a variety of tools to build the system for the twenty-first century. For example, advances in nanotechnology may make possible carbon-based fibres which are 100x stronger than steel, at only one-sixth the weight.

Major Related Improvement in Public Safety

Such applications may include smart sensors for collision avoidance in automobiles;

nano-reinforced, glare-resistant car windows and windshields; chemical and biological weapon detectors at airports; and sensors that monitor rail track structural integrity This includes:

- Improved catalysts could reduce or eliminate the emission of pollutants from engines
- New photonic nanodevices could replace the heavy and costly radio frequency transmission equipment on board aircraft, ships, or satellites or be exploited for inexpensive remote control of vehicles

Nanotechnology Underpinning Developments in The Automotive Industry

Nanotechnology is underpinning developments in the automotive industry in a number of areas. The automotive industry is a major user of sensors and components for new integrated miniaturised systems, and MEMS have been a key driver in many of today's advanced safety systems. MEMS are currently used in two standard automotive applications: the air bag accelerometer, which first went small tech in the early 1990's, and the manifold air pressure sensor, or MAP sensor, first used in the late 1970's. Total automotive sales of silicon accelerometers are expected to increase from £192m in 2000 to £364m in 2005 (www.strategyanalytics.com). The worldwide MST market (including MEMS) for automotive applications in 2000 was £725m and is projected to grow to £800m by 2004. (NEXUS/Roger Gracc Associates).

The Benefits of Advanced Communications

Advanced communications that maximise the benefits of intelligent transportation systems and obviate the need for some travel altogether; sensors that continually monitor the condition and performance of roads, bridges, and other infrastructure This includes:

- New materials that will permit ultra-miniaturisation of space systems and equipment, including the development of 'smart' sensors and probes. Applications include economical supersonic aircraft; low-power, radiation-hardened computing systems for autonomous space vehicles; and advanced aircraft avionics
- Reduction in transportation energy use and its impacts on the environment. Applications include nanosensors to monitor vehicle emissions and trigger traps for pollutants; nanoparticle-reinforced materials that replace metallic components in cars; replacement of carbon black in tyres with nanoparticles of inorganic clays and polymers, leading to tyres that are environmentally friendly and wear-resistant; and carbon-based nanostructures that serve as 'hydrogen supersponges' in vehicle fuel cells.

It will often be useful in nanomedicine to distinguish molecules containing different isotopes of the same chemical elements, for example to count the number of organic molecules in a sample that contain radioactive C^{14} atoms in place of stable C^{12} atoms or to make direct assays of heavy radiometal-contaminated tissues. Nanodevices may also sort isotopically heteronuclear molecules into homonuclear fractions, for example to separate deuterochemicals from hydrochemicals or to acquire isotopically pure materials with which to construct:

(a) Components with extremely fine size differences.

(b) High-speed rotors with precisely balanced mass distributions.

(c) Maximum thermal conductivity, or mixed isotopes to reduce thermal conductivity.

(d) Long-lived isotopically pure structures that generate no internal radioactivity, thus promoting longer device lifetimes.

The most reliable sensor for discriminating among isotopes within target molecules is the single-proton massometer. Isotopes of any chemical element differ in mass by at least ~1 neutron (~protonic) increments. Thus a massometer with single-proton resolution can resolve all isotopes starting from chemically pure samples. Isobars (atoms having the same mass number) and isotones (atoms having the same number of neutrons) may be finally resolved chemically; nuclear isomers differing in mass by ~0.001 amu cannot be resolved. Teragravity centrifugation and time-of-flight particle-beam deflection techniques may also prove useful. Isotopes vary in many subtle ways due to their mass differences, although it is not yet clear how a potential isotope nanosensor might exploit some or all of these distinctions:

(1) *Line Shifts*—the electronic energy states of atoms or molecules depend on the reduced nuclear mass, causing a spectral line shift of ~1.0005 between hydrogen and deuterium, less among heavier isotopes, detectable spectroscopically.

(2) Hyperfine Structure—isotopes differ in electrical quadrupole moment, nuclear spin and magnetic moment, giving rise to hyperfine structure in the optical spectra, especially in the heavier elements. For example, the magnetic moments of hydrogen and deuterium (2.79268 and 0.857387 nuclear magnetons, respectively) are distinguishable via NMR. The NMR frequency in a 10 kilogauss magnetic field is 42.5759 MHz for H, 6.53566 MHz for D. Of course, not all isotopes are NMR sensitive.

(3) *Vibrations and Rotations*—molecular spectra are especially sensitive to isotopic changes, since the quanta of vibrational and rotational energies are directly dependent upon the masses of the atoms involved, and there

are also some subtle effects from differences in nuclear spin. Separations are commonly achieved by laser excitation.

(4) *Cross-Sections*—isotopes have different thermal neutron capture cross sections; for instance, the cross-section of C^{12} is 3.7×10^3 barns (1 barn = 10^{-24} cm^2) compared to 9×10^{-4} barns for C^{13} and $<10^{-6}$ barns for C^{14}. (Electron scattering might be better, since the electron is less likely to cause the nucleus to fission and it is generally difficult to use neutrons to probe samples on the molecular scale.) Similarly, the neutron cross-section for H_2O is 0.6 barns compared to 9.2×10^{-4} barns for D_2O.

(5) *Other Properties*—isotopes differ slightly in various bulk properties, many of which could be measured in submicron-scale aliquots by nanosensors. The ability of both large mammals and micron-size photosynthetic plant cells to preferentially concentrate C^{12} or C^{13} in organic matter, and the predicted ability of carbon nanotubes to selectively extract tritium from low-concentration H_2-T_2 mixtures, constitute proofs of principle that at least some of the bulk properties.

This section describes the ability of medical nanodevices to detect and measure fluid pressure variations while functioning inside the human body. Generation of acoustic signals for power transmission, communication, and navigation are considered elsewhere.

A change in the ambient pressure in a fluid environment (DP) may be determined by measuring the change in volume (DV) of a fixed quantity of gas at constant temperature, relative to a known sensor volume at a known reference pressure. (Thermal sensors elsewhere in the nanorobot provide readings by which corrections can be made for temperature variations.) If the sensor volume consists of a cylinder of fixed cross-sectional area, the change in volume is converted into a linear displacement of a piston located at the interface between gas and fluid environment. At SNR = 2, a (22 nm)^3 = 10^4 nm^3 sensor ($\sim 5 \times 10^5$ atoms or $\sim 10^{-20}$ kg for a cube with 1 nm thick walls) can detect a minimum pressure change $DP_{min} \sim 0.03$ atm; a (680 nm)^3 = 0.3 micron^3 sensor ($\sim 3 \times 10^9$ atoms, $\sim 6 \times 10^{-17}$ kg) detects $DP_{min} \sim 10^{-6}$ atm, probably near the practical limit for mobile in vivo medical nanodevices. Note that DP_{min} scales approximately as the inverse cube of mean sensor dimension, and immunity to thermal noise scales exponentially with sensor volume. Current silicon micromachined pressure sensors are ~ 1 mm^3 with 10^{-3} atm sensitivity, a factor of $\sim 10^{12}$ poorer sensitivity than the theoretical minimum for their size. (A field-effect chemical sensor with an active sensing volume of 0.1 mm^3 that measures pressure indirectly has detected a 10^{-11} atm step in hydrogen gas with $t_{meas} \sim 3$ sec, only a factor of 1000 above the theoretical minimum for a sensor of that size.) The human ear can detect pressure waves as small as 2×10^{-10} atm. Two 0.3-micron^3 pressure sensors located at either end of a 1 micron nanorobot could detect pressure changes as small as 10^{-6} atm per micron, a ~1 atm/m spatial gradient. The maximum detectable gradient would involve adjacent measurements of minimum (10^{-6} atm) and maximum (~10^4 atm) absolute pressure, a theoretical peak spatial gradient of ~10^{16} atm/m, though a practical device could not survive exposure to such extreme shear force.

In biosystems and nanosystems alike, the detection of an optical photon is a high-energy event. A single photon of red light at l = 700 nm carries an energy of 280 zJ, a blue photon at 400 nm conveys 500 zJ, and an ultraviolet (UV) photon at 200 nm transmits 1000 zJ to its receptor. These energies, ranging from 65-234 kT, exceed thermal noise by a wide margin, hence are all easily detectable reliably by suitable nanosensors. Note that optical photon detection does not require a nanosensor the size of one wavelength or larger—for example, 1-nm chlorophyll molecules have no difficulty absorbing 660-nm photons.

Biological systems provide numerous examples of single-photon transducers, including photopigment molecules such as chlorophyll and rhodopsin which mediate electron transfers in <~1 picosecond by undergoing conformational changes, triggering a series of biochemical events. An estimated 100-500 cell membrane channels are affected by the detection of a single optical photon. The first documented case of a physiological role for chlorophyll in animals has been found in the fish *Malacosteus niger*, which uses a chlorophyll derivative to see far-infrared light. Mammalian cells also act as indirect UV sensors as for instance when energy absorption induces perturbations of membrane structure or a conformational change in membrane proteins, activating the JNK cascade below 300 pJ/micron^2, and centrosomes may serve as cellular directional IR sensors in single fibroblast cells. Integration times for photon sensors are briefly treated.

Artificial "bioelectronic" cyanine-quinone chromophore molecules ~1 nm in size can detect optical photons of wavelengths l = 580-630 nm, and can trigger and reset in 1-3 picosec (governed by the tunneling times associated with reformation of the zwitterionic states of the inputs), in theory permitting ~100 GHz modulations to be received. Exciton-generating poly(p-phenylene vinylene) polymer switches have the potential to achieve ~THz switching speed; a birefringent Kerr medium also switches in a few picoseconds. Chalcogenide glass has viscosity $>10^{13}$ kg/m-sec in darkness, but $\sim 5 \times 10^{11}$ kg/m-sec under illumination. Proposals for specific implementations of photoergic transducers and optical cables are elsewhere High energy photons including X-rays (l ~ 2 nm, ~10^5 zJ or 0.0001 picojoule) and gamma rays (l ~ 0.002 nm, ~0.1 picojoule), and ejecta from the decay of radioactive nuclei such as a-particles (helium nuclei, ~1 picojoule) and b-particles (free electrons, 0.01-0.1 picojoule) may be extremely destructive to nanomechanical systems and are difficult to reliably measure quantitatively. For example, a charged particle with 0.01 pJ of kinetic energy (K.E.) traversing a decelerative electrostatic field of 10^9 volts/m requires K.E./qE ~ 60 microns to slow to a halt.

However, particle-induced visual sensations have been observed by dark-adapted astronauts in space, and Ruderfer has suggested that even solar neutrinos may be detectable by the human brain (though most physicists would find this implausible). The goal of cellular bioscanning is the noninvasive and non-destructive in vivo examination of interior biological structures. One of the most common nanomedical sensor tasks is the scanning of cellular and subcellular structures. Such tasks may include localisation and examination of cytoplasmic and nuclear membranes, as well as the identification and diagnostic measurement of cellular contents including organelles and other natural molecular devices, cytoskeletal structures, biochemical composition and the kinetics of the cytoplasm. The precision and speed of medical nanodevices is so great that they can provide a surfeit of detailed diagnostic information well beyond that which is normally needed in classical medicine for a complete analysis of somatic status. Except in the most subtle cases a malfunction in any one component of the cellular machinery normally provokes a cascade of pathological observables in many other subsystems.

Detection of any one of these cascade observables, if sufficiently unique and well-defined, may provide adequate diagnostic information (see) to plan a proper reparative procedure. DNA is one of the few cellular components that is regularly inspected and repaired. Most homeostatic systems adopt a more simple philosophy of periodic replacement of components regardless of functionality. Nanomedicine allows the philosophy of inspection and repair to be extended to all cellular components. The following discussion briefly describes a few of the sensory techniques that may be useful in achieving this objective. Tactile topographic scanning provides the most direct means for examining cellular structures in vivo. For example, ex vivo live cell scanning in air by commercially available atomic force microscopy using a 20-40 nm radius tip allows nondestructive feature resolutions of ~50 nm across the top 10 nm of the cell.

Investigators have used the scanning tip to punch a hole in the cell, then pull back and scan the breach, observing the membrane heal itself via self-assembly in real time. Up to ~KHz scanning frequencies are possible, with up to 1024 data points per scan line. The nanomechanical manipulator arm achieves comparable tip velocities and positional accuracies. Assuming a scan rate of ~10^6 pixels/sec, a micron-scale nanodevice, once securely anchored to the surface of a (~20 micron)^3 human cell, could employ a tactile scanning probe to image the 0.1 per cent of plasma membrane lying within its (1.4 micron)^2 vicinity in ~2 sec to ~1 nm^2 resolution (~1 mm/sec tip velocity), or ~50 sec to ~0.2 nm (e.g., atomic) resolution (~0.2 mm/sec tip velocity).

Inside the cell, and again post-anchoring, an entire 6 micron^2 mitochondrial surface could be imaged to atomic resolution in ~100 sec; the surface of a 100 nm length of 25 nm diameter microtubule could be atomically resolved in 0.2 sec—though of course in a living cell these structures may be changing dynamically during the scanning process. The continuous power dissipation of a (0.1 micron)^2 scan head moving at 1 mm/sec through water is ~0.002 pW (~kT/pixel at ~10^6 pixels/sec), though of course the energy cost of sensing, recording, and processing each pixel must be at least ~10 kT/pixel (ignoring the possibility of pre-computational image compression), so the total scanner power draw could be as high as 0.02-0.1 pW in continuous operation. Special scanning tips and techniques should allow topographic, roughness, elastic, adhesive, chemical, electrostatic (charge density), conductance, capacitance, magnetic, or thermal surface properties to be measured.

For large (~0.1-1 micron) cellular components, identification and preliminary diagnosis of improper structure may be possible using measured surface characteristics which may be matched to entries in an extensive onboard library, perhaps combined with dynamic monitoring of anomalies in cross-membrane molecular traffic using chemical nanosensors. Most membranes are self-sealing, so it should also be possible to gently insert a telescoping member into the target organelle, which member then slowly reticulates and extrudes smaller probes with sensory tips in a fixed pattern and step size, allowing the acquisition of detailed internal structural and compositional information. Small (~1-10 nm) protein-based components of the cell such as enzymes, MHC carriers, and ribosomes are self-assembling or require only modest assistance (e.g., molecular chaperones) to self-assemble. Reversible mechanical denaturation of these smaller proteins, using diamondoid probe structure to displace water molecules (thus reducing hydrophobic forces) and by using specialised handling tools akin to functionalised AFM tips and molecular clamps, may be followed by precise nondestructive amino acid sequencing to allow identification and diagnostic compositional analysis. Afterward, the protein molecule may be refolded back into its original minimum-energy conformation, possibly with the assistance of chaperone-like structures in some cases. Power constraints are somewhat less severe in the case of acoustic transmission microscopy (acoustic tomography), a technique which requires a minimum of two physically separated components (e.g., a transmitter and a receiver). Consider an acoustic emitter radiating a series of omnidirectional pressure pulses of frequency n which travel across a tissue cell of width X_{path}, to be detected by a receiver on the other side after a signal transit time t_{cell} ~ X_{path}/v_{cell}, where v_{cell} is the speed of sound in the cytosol. In the simplest case, assume that the cytosol is clear except for one target of interest along the path (e.g., an organelle) in which the speed of sound is v_{target} # v_{cell}. The minimum detectable difference in travel times between a signal that passes through the target and one that does not is Dt (~10^{-9} sec for diamondoid systems), so the minimum resolvable target size is r_{min} ~

v_{target} v_{cell} Dt/2 abs(v_{target} - v_{cell}) when r_{min} >~ v_{cell}/n. For soft tissue targets in water at 1.5 GHz, r_{min} ~ 30 microns; for tooth enamel in water, r_{min} ~ 1 micron. However, the use of sampling gates will permit the detection of phase shifts that are much less than the characteristic time constant of the system in incoming waves, hence the minimum thickness of objects in theory detectable by acoustic tomography can be far smaller than the values for r_{min} estimated above.

In either mode of operation, data gathering is accelerated by positioning more than one receiver around the target. Sensitivity is boosted by increasing emitter size, taking multiple measurements, or by illuminating the target with some large external source located elsewhere in the tissue or even outside of the organ. However, cells and body tissues are mesoscopic "junkyards"—highly heterogeneous media which may produce large numbers of nontarget scattering events, thus increasing the difficulty of extracting signal from noise. Electrostatic scanning is largely ineffective because of Debye-Huckel shielding. Magnetic stray field probes allow resolution of 10 nm physical features, but only for materials with substantial magnetic domains—which most biological substances lack. But nuclear magnetic resonance (NMR) imaging may allow cellular tomography by creating 3-D proton (hydrogen atom) density maps. Atomic density maps of other biologically important elements with nonzero nuclear magnetic moments (including D, Li^6, B^{10}, B^{11}, C^{13}, N^{14}, N^{15}, O^{17}, F^{19}, Na^{23}, Mg^{25}, P^{31}, Cl^{35}, K^{39}, Fe^{57}, Cu^{63} and Cu^{65}) may also be compiled. For example, sodium imaging is already used clinically to assess brain damage in patients with strokes, epilepsy, and tumors. In a hypothetical NMR cytotomographic nanoinstrument, a large permanent magnet is positioned near the surface of the cell or organelle to be examined. This creates a large static background magnetic field that polarizes the protons.

The spatial gradient of this field establishes a unique resonant frequency, called the Larmor frequency, within each isomagnetic surface throughout the test volume. A second time-varying magnetic driver field (e.g., vibrating permanent magnet or weak rf field) is then scanned through the full range of resonant frequencies, exciting into resonance the protons (nuclear spins) in each isomagnetic surface, in turn. Depending on the sensor implementation chosen, each resonance detected may cause absorption of driver field energy, an increase in measured impedance, or even a return echo of magnetic energy (if the driver field is operated in pulse echo mode) as the excited protons relax to equilibrium in ~1 sec. The large polarising magnet is then rotated to a new orientation, moving the isomagnetic surfaces to new positions within the test volume, and the scan is repeated. A 3D map of the spatial proton distribution may be computed after several scan cycles.

Submicron laser emitters have been available since the late 1980s. It should be possible to use NSOM-like nanoprobes to optically scan the surfaces of cells or organelles to <~1 nm resolution, mapping their topography and spectroscopic characteristics to depths of tens of nanometers without penetrating the surface. However, a proper membrane-sealing invasive light funnel ~20 nm in diameter might not seriously disrupt some cellular or organelle membranes and thus could be inserted into the interiors of these bodies or through the cytoskeletal interstices to permit deeper volumetric scanning. The thermal conductivity of water at 310 K is 0.623 watts/m-K and the energy per 500 nm photon is 400 zJ, so for 1 micron^3 of watery tissue the maximum scan rate is ~35 micron^3/sec-K, if e^{SNR} photons are used to image each 1 nm^3 voxel with SNR = 2. Thus a 1-sec volumetric optical scan of an aqueous ~1-micron^3 sample volume to 1 nm^3 resolution requires a ~1 nanowatt scanner running at ~GHz bit rates, raising sample volume temperature by ~0.03 K. NSOM permits the determination of five of the six degrees of freedom for each molecule, lacking only the optically inactive rotation around the dipole axis.

In principle, it should be possible to produce <~1 nm resolution near-field optical scans of in situ protein molecules, since with atomically precise fabrication and single lines of atoms as conductors a minimum light guide (metal-dielectric-metal) is 3 atoms wide. Given a molecular laser and adequate collimation, photons passing through folded proteins will scatter according to the molecular structure. Detection of sufficient photons comprising these scattering patterns should allow the noninvasive determination of protein structure; polarised photons provide information on chirality. Absorption and fluorescent signals will be visible from phenyl rings, tryptophan, and bound cofactors such as ATP and adenine (which is fluorescent). Positions of monoclonal antibodies on virus surfaces are now identifiable experimentally using NSOM; it should be possible to map binding sites on virus and cell surfaces using fluorescently labelled antibodies. Single molecule detection has been proposed as a tool for rapid base-sequencing of DNA. Other optically-based cellular imaging techniques must be distinguished. Optical Coherence Tomography (OCT) uses Michelson interferometry to achieve ~10 micron spatial resolutions over tissue depths of 2-3 mm in nontransparent tissue in the near infrared. However, OCT requires numerous physical components not easily implemented on a micron-size detector (e.g., beam splitter, lens-grating pair, galvanometer mirror, optical prism), femtosecond pulse shaping, and illumination power levels of ~10^7 watts/m^2 >> 100 watts/m^2 "safe" continuous limit in tissue.

Bioluminescence techniques in which the light source is placed inside the tissue (e.g., a transgenic mouse with a luciferase gene in every cell of its body) has a spatial resolution limited to ~10 per cent of the depth, or ~10 microns resolution at a ~100 micron depth. Coherent anti-Stokes Raman scattering (CARS) already permits organelle imaging in living cells and can in theory create a point-by-point chemical map of a cell using two intersecting lasers. Three-dimensional observation of microscopic biological nonliving structures by means of X-ray holography requires a high degree of spatial coherence and good contrast between target and surroundings. Good contrast may be achieved in the wavelength range between the K absorption edges of carbon (l = 4.37 nm) and oxygen (l =2.33 nm), the "water window" where carbon-containing biological objects absorb radiation efficiently but water is relatively transparent.

A 50-micron diameter emitter has been tested that uses near-IR 5-femtosecond laser pulses impinging upon a helium gas target to create a well-collimated (<1 milliradian) beam of coherent soft X-rays at a 1 KHz repetition rate producing a brightness of 5×10^8 photons/mm^2-milliradian^2-sec, a peak X-ray intensity of $>10^{10}$ watts/m^2 on the propagation axis behind the He target. Many cellular parameters need not be measured directly in order to be detected by a medical nanodevice. Cell volume sensing is a case in point. An intracellular nanodevice can indirectly monitor changes in the volume of the cell in which it resides by one of two methods. First, measurements of the mechanical deformation of the cellular membrane, stretch-activated channels, or cytoskeletal strains and structural changes are quite sensitive to alterations in total cell volume. Second, concentration or dilution of the cytoplasmic environment through cell shrinkage or swelling leads to the activation of various volume-regulation responses which may be detected by the nanorobot. Changes in the concentration of soluble cytosolic proteins may nonspecifically affect enzyme activity via "macro- molecular crowding".

Minor changes in cell volume can cause severalfold changes in ion transport. Cellular signalling entities that have been linked to the transduction and amplification of the primary volume signal include Ca^{++} transients, phosphoinositide turnover, eicosanoid metabolism, kinase/phosphatase systems such as JNK and p38, cAMP, and G-proteins. These signal amplification pathways may be monitored using chemical concentration sensors aboard the medical nanodevice, allowing the nanorobot to eavesdrop on the natural sensory channel traffic of the cell. In many ways the neuron is the most nanomedically important class of cell in the human body. Nanomedical applications regarding neurons and the brain are addressed in. The following is a brief summary of noninvasive (i.e., no axonal membrane penetration) nanotechnological methods for monitoring the electrical traffic of individual neurons. Noninvasive measurement of axonal traffic within nerve bundles will require multiple sensors and greater sensitivity to compensate for shielding by the perineurium, a tight resistive sheath enclosing the bundle ~20 microns thick with resistivity ~4000 ohm-cm.

The "typical" ~20 micron human neuron discharges 5-100 times per second, moving from -60 mV potential to +30 mV potential in ~10^{-3} sec. Thus the variation in electric field at the axonal surface is ± 4500 volts/m. Since electric field sensors can detect 100 volt/m fields up to GHz frequencies, an electric sensor attached by circumaxonal cuff or pressed against the axonal surface (possibly at the node of

Ranvier) should readily detect each action potential discharge. By 1998 silicon-to-neuron extracellular junctions already permitted direct stimulation of individual nerve cells in vitro without killing the cells, and extracellular electrodes were commonly used to detect neuronal electrical activity noninvasively, both in vivo and in vitro. Artificial electric fields may also be employed to trigger or moderate neural signals. Cell membrane capacitance is typically ~0.01 picofarads/micron^2, and varies with the state of the health of the cell. The magnetic flux density caused by a single action potential discharge is ~0.1 microtesla at the axonal surface, which may be detected by a "compass oscillator" type magnetosensor with a ~KHz maximum sampling rate. It might be possible for artificial magnetic fields to directly influence neural transmissions. Even a static 65 millitesla field has been shown to reduce frog skin Na^+ transport by 10-30 per cent. Each neuronal discharge develops an electrical energy of ~20 picojoule (~10^{10} kT), far smaller than the magnetic energy stored in a B = 1.4 tesla field of a permanent micromagnet traversing an L^3 = (20 micron)^3 volume which is $B^2 L^3/2 m_0$ ~ 6000 pJ. If properly manipulated, such a field may be sufficient to enhance, modulate, or extinguish a passing neural signal. While neurons come in many shapes and sizes our "exemplar" ~14,000-micron^3 neuron discharging ~90 mV into an input impedance of ~500 Kohms produces ~0.2 microampere current per pulse and generates a continuous (average) 100-300 pW of waste heat as measured experimentally. The discharge rate of 5-100 Hz can produce brief surges up to ~2000 pW during a high-frequency train, but the duty cycle of such trains is far less than 100 per cent, reducing time-averaged dissipation to the observed 100-300 pW range. Single impulses are measured experimentally to produce 2-7 microkelvin temperature spikes in cold or room-temperature mammalian non-myelinated nerve fibers and ~23 microkelvins in non-myelinated garfish olfactory nerve fibers at an energy density ranging from 270-1670 joules/m^3-impulse from 0-20°C. In non-myelinated fibers the initial heat occurs in two temperature-dependent phases: a burst of positive heat, followed by rapid heat reabsorption (called the negative heat).

The positive heat derives from the dissipation of free energy stored in the membrane capacity, and from the decrease in entropy of the membrane dielectric with depolarisation. An L ~ 20-micron neuron in good thermal contact with an aqueous heat sink at 310 K has thermal conductance $L K_t$ ~ 10^{-5} watts/K, so trains of 5-100 Hz impulses lasting 1 second should raise cellular temperature by 10-30 microkelvins; up to ~200 microkelvin thermal spikes from such trains have been observed experimentally. These events are easily detectable by nanoscale thermal sensors capable of ~1 microkelvin sensitivity up to ~1 KHz. The ~microkelvin heat signature of individual impulses or very short pulse trains can probably be temporally resolved because the minimum pulse repetition time is ~10 millisec (at 100 Hz) which is much longer than the thermal time constant for an L ~ 20-micron neuron (thermal conductivity K_t = 0.528 watts/m-K and heat capacity C_V = 3.86 × 10^6 joules/m^3-K for brain tissue; which is t_{EQ} ~ $L^2 C_V/K_t$ ~ 3 millisec. The synaptic cleft between the axonal presynaptic terminal and the dendritic postsynaptic membrane is 10-20 nm in most synapses, although in the vertebrate myoneural junction it may be as large as 100 nm. Contact area per bouton is ~1 micron^2, giving a total gap volume of ~10^7-10^8 nm^3. The density of acetylcholine receptors is highest in muscles along the crests and upper thirds of the junctional folds (~10,000/micron^2), and is lowest in the extrasynaptic regions (~5/micron^2). Each action potential discharge triggers the release of ~10^4-10^5 molecules of acetylcholine into the gap volume of an active neuromuscular junction (diffusion time ~1 microsec), raising c_{ligand} from near zero to ~3 × 10^{-4} molecules/nm^3 (~0.0005 M) in ~1 millisec, followed by near-complete hydrolysation by acetylcholinesterase during the 1-2 millisec refractory period.

Into the gap volume may easily be inserted a ~10^5 nm^3 neurotransmitter concentration sensor able to measure ~100 acetylcholine molecules in ~1 millisec, thus detecting pulses at the fastest discharge rate. A similar device could be used to precisely regulate neuro-transmitter concentration at the junction, and hence the neural signal itself, under nanodevice control. Simple electrochemical and mechanochemical artificial synapses have been demonstrated. Many other noninvasive neurosensing techniques are readily conceivable. For instance, counting rotors or sodium magnetic resonance imaging could detect changes in the ionic composition (e.g., Na^+, K^+) of the periaxonal fluid before, during, and after discharge. Starting in 1968, H. Frohlich, observing that millivolt electrical potentials maintained across cell membranes ~10 nm thick give rise to huge fields ~10^7 volts/m possibly producing an electret state, theorised that membrane molecules must be highly electrically polarised and thus could interact to produce coherent surface acoustic vibrational modes in the 10-100 GHz (microwave) frequency range; the longest wavelength is about twice the membrane thickness. Interestingly, this frequency span is very close to the maximum trigger/reset frequency for bioelectronic molecules).

However, the direct detection of 10-100 GHz millimeter radiation by non-nanotechnological means is experimentally difficult and controversial because the tests must be performed in vivo in close proximity to an actively metabolising cell in water—and water strongly absorbs microwaves over macroscale ranges (e.g., ~99 per cent absorption in 3 mm at 100 GHz). Nevertheless, active cells have shown enhanced Raman anti-Stokes scattering, an effect ascribed to the converse of the Frohlich oscillations. In one study, the normalised growth rate of yeast cultures was enhanced or inhibited when irradiated by CW microwave fields of ~30 watts/m^2 of various frequencies; growth rate data spanning 62 separate runs revealed a repeatable frequency-dependent spectral fine structure with six distinct peaks of width ~10 MHz near 42 GHz. Investigations of related phenomena are ongoing and voluminous; the interested reader should peruse *Bioelectromagnetics*, the archival journal of this field. 100 GHz waves attenuate only ~1 per cent after passing through ~3 microns of soft tissue. A single electron injected into an integral membrane protein could act as an oscillating dipole, making a 300 volt/m signal 10 nm from the protein antenna with an energy transfer of ~0.004 kT per cycle; ~1000 oscillating electrons could produce a measurable field. A 20-micron diameter cell modeled as a nonuniform spherical dipole layer with transmembrane dipoles located 10 nm apart and embedded in a dissipative medium could produce 10^2-10^5 volt/m microwave fields 1-10 microns from the cell surface.

Thus, a variety of rf and microwave electromagnetic emanations may in theory be detectable both within and nearby living cells which could prove diagnostic of numerous internal states. Such states may include cytoskeletal dynamics, metabolic rates, plasmon-type excitations due to the collective motion of ions freed in chemical reactions, positional, rotational or conformational changes in biological macromolecules and membranes, internal movements of organelles and nerve traffic conduction, cellular pinocytosis, cellular reproduction events, cell membrane identity (e.g., distinguishing erythrocyte, Gram-positive and Gram-negative cell coat conductivities at 10 KHz), and cell-cell interactions. Macrosensing is the detection of global somatic states (inside the human body) and extrasomatic states (sensory data originating outside the human body). While the treatment here is necessarily incomplete, the discussion nevertheless gives a good feel for the kinds of environmental variables that internally-situated nanodevices could sense. Not all capabilities outlined here need be available on every nanorobot, since injection of a cocktail of numerous distinct but mutually cooperative machine species allows designers to take full advantage of the benefits of functional specialisation.

In many cases, a given environmental variable can be measured by several different classes of sensor device. However, since these devices are microscopic it is in theory possible to operationalize almost all of the macrosensing capabilities described below in one patient using just a billion devices (~10 mm^{-3} whole-body deployment density), a total volume of ~1 mm^3 of nanorobots or ~0.1 per cent of the typical ~1 cm^3 therapeutic dose. Can sounds generated by a single cell be detected, and thus be useful for diagnosis? Probably not, given that low frequency acoustic radiators are notoriously inefficient. For example, mitochondrion organelles of the giant amoeba *Reticulomyxa* are shuttled back and forth by 1-4 cytoplasmic dynein motors while riding on the outside of a bundle of 1-6 microtubules. Each dynein motor generates 2.6 pN of force and drives the mitochondria at up to ~10 micron/sec, developing 0.3-1.0 × 10^{-16} watts of mechanical power within each 320-nm diameter organelle. Taking each organelle as a cylindrical acoustic radiator with a mechanical input power of P_{in} ~ 10^{-16} watts at n ~ 1000 Hz, the output acoustic pressure at the organelle surface is only ~10^{-9} atm an acoustic power intensity I ~ 10^{-15} watts/m^2 which is not detectable by micron-sized nanorobots. Nevertheless, cells and intracellular elements are capable of vibrating in a dynamic manner with complex harmonics that can be altered by growth factors and by the process of carcinogenesis, so the possibility cannot be completely ruled out. Blood pressure ranges from 0.1-0.2 atm in the arteries to as low as 0.005 atm in the veins. The systolic/diastolic differential ranges from 0.05-0.07 atm in the aorta and 0.01-0.02 atm in the pulmonary artery, falling to 0.001-0.003 atm in the microvessels, or 0.003-0.005 atm if the precapillary sphincter is dilated. In venous vessels, pulse fluctuations are 0.002-0.010 atm in the superior vena cava, 0.004-0.006 atm in the subclavian vein, ~0.004 atm in venules generally, and ~0.0005 atm in the brachial vein. There is also a ~0.05 Hz random fluctuation in the microvessels with amplitude on the order of 0.004-0.007 atm. Both blood pressure and pulse rate can be reliably monitored by a medical nanodevice virtually anywhere in the vascular system using a $(68\ nm)^3$ pressure sensor with ~0.001 atm sensitivity. Pulse propagation through body tissue is somewhat muted due to absorption in compressible fatty membranes, but most cells lie within 1-3 cell-widths of a capillary so the cardiac acoustic signal should still be measurable using more sensitive detectors. The time-averaged interstitial pressure in subcutaneous tissue is 0.001-0.004 atm.

Arterial pulse waves (vascular oscillations) carry subtle messages about the health of internal organs and the arterial tree. The idea of using pulse waves for diagnosis has a long history dating back 2000 years in China. For example, in the *Book on Pulse Waves* by Wang Shu-He (201-285 AD), waves detected by manual probing are classified using such subjective and qualitative descriptors as floating, deep, hidden, rapid, slow, moderate, feeble, replete, full, thready, faint, weak, soft, slippery, hesitant, hollow, firm, long, short, swift, running, intermittent, uneven, taut, string-tight, gigantic, or tremulous. Abnormal waves were empirically related to disease states. Wave data gathered by nanodevices could make possible a theoretically sound, quantitative system of noninvasive observation, classification, and diagnosis as a supplement to other nanomedical tools. The variation of mechanical pressure over a complete respiratory cycle is ~0.003 atm in the pleura, ~0.002 atm at the alveoli, detectable by nanomedical pressure sensors positioned in the vicinity of the respiratory organs. Holding a deep breath further stretches the pulmonary elastic tissue, up to 0.02 atm. However, turbulent flows at Reynolds number $N_R > 2300$ in the trachea, main bronchus and lobar bronchus produce a whooshing noise that may be the loudest noncardiac sound in the human torso during conventional auscultation.

Many other mechanical body noises should be globally audible to properly instrumented medical nanodevices. If normal chewing motions (of hard foods) release 1-10 milliwatts in a ~100 cm^3 oral volume with a ~1 sec jawstroke, power density is ~10 $watts/m^3$ or ~10^{-4} atm of tooth-crunching noise. A stomach growl registering 45 dB (vs. 30 dB whisper, 60 dB normal conversation) at 2 meters has a source power of 160 milliwatts; released from a 10 cm^3 gastric sphincter volume gives a ~2×10^{-6} atm acoustic wave, detectable throughout the body. Walking and running releases 20-100 joules/footfall for a 70 kg man; assuming the energy is absorbed within a ~1 cm thickness or within ~1 second by the sole of the foot, an upward-moving planar compression wave of 0.4-2.0 atm, easily detectable by acoustically instrumented nanodevices body-wide. (Shoe insoles dissipate energy and alter the shock wave pulse shape.) Handclapping generates 0.02-0.2 atm pulses, also easily detectable. Lesser noises including ~30 millisec hiccups at (4-60)/min, intestinal and ureteral peristalsis, sloshing of liquid stomach contents, heart murmurs, a tap on the shoulder by a friend, nasal sniffling and swallowing, clicks from picking or drumming fingernails, crepitations, manustuprations and ejaculations, the rustling noise of clothing against the skin, flapping eyelids, anal towelling, bruits (including murmurs and thrills) due to vascular lesions, dermal impact of water while showering, copulatory noises, urethral flow turbulence during urination, transmitted vibrations from musical instruments, creaking joints, and squeaking muscles can be detected locally if not globally. Implantation of significant interconnected in vivo diamondoid structures may produce increased sensitivity to internal noises, due to the extremely low acoustic absorption coefficient of diamond. Average source power for conversational speech in air is ~10 microwatts at the vocal cords (60 dB), up to ~1000 microwatts for shouting (90 dB) and as little as 0.1 microwatts (30 dB) for whispering. Vocal cord surface area ~1 cm^2, giving an acoustic intensity I ~ 0.001-10 watts/m^2. (Using the decibel notation, dB = 10 $\log_{10}(I/I_0)$, where $I_0 \sim 5 \times 10^{-13}$ watts/m^2 in air, $I_0 \sim 1 \times 10^{-16}$ watts/m^2 in water.) For water at 310 K, r = 993.4 kg/m^3 and v_{sound} = 1500 m/sec, therefore A_p = 0.0005-0.05 atm for speech, detectable by nanodevices throughout the body due to minimal attenuation at audible frequencies.

Other easily detectable vocalisations include whistling, humming, coughing, sneezing, rales, wheezing, expectorating, eructations, flatus, vomiting, hawking and noseblowing. In the case of spherical waves diverging from an omnidirectional transmitter of radius r at a distance X from the transmitter, average intensity declines inversely as the square of the distance and so. Of course, an ex vivo acoustic nanosensor may receive sound that has passed through no interface, hence may detect pressure waves ~3 orders of magnitude lower in amplitude. Assuming $A_{transmit} = A_{incident}$, P_R ~ 600 microwatts, so ex vivo nanorobots with a 0.3 micron^3 sensor could hear people shouting at x_R = 2 meters. To hear talking (~10 microwatt source) requires a 2.4 micron^3 ex vivo pressure sensor (limit ~10^{-7} atm). Optimally positioned and calibrated nanomedical pressure sensors could directly measure changes in the ambient barometric pressure to within ± 10^{-6} atm. Normal atmospheric variation due to weather ranges from 0.94-1.05 atm; such slow moving changes are readily monitored. Very near the Earth's surface, the air pressure P at altitude h above sea level is approximated by $P = e^{-k_p h}$ (atm), where $k_p = 1.16 \times 10^{-4}$ m^{-1} at 20°C; at sea level, a 10^{-6} atm change in pressure reflects a change in altitude of only ~1 cm.

However, the opening or closing of a door inside a (~5 m)^3 room that displaces >125 cm^3 of air also causes a minimally detectable >10^{-6} atm pressure pulse. Other sources of environmental pressure variation such as infrasonic (~0.2 Hz) microbaroms from offshore ocean storms, wind entering through open windows, forced-air currents from central heating or A/C systems, or even the movements of nearby people and pets may be detectable and thus, may further confuse the measurement, reducing absolute accuracy unless suitable corrections are made. Using a navigational transponder network, a population of ~10^{11} nanodevices each spaced an average ~100 microns apart throughout the body tissues can determine relative location to a positional accuracy of ~3 microns and an angular accuracy of ~2 milliradian, with the data updatable once every millisecond. Thus, dispersed, the network can continuously monitor and record the relative positions of all limbs with worst-case (cumulative error) ~0.8 mm accuracy over a 2-meter span. These devices can prepare high-resolution dynamic maps of body position, velocity, acceleration and rotation, allowing precise real-time digital kinesthesia during sports or artistic activities such as gymnastics, pole vaulting, or ballet; during transportative activities such as driving cars around hairpin turns, roller-coaster rides, military aircraft maneuvers and space launches; during precision tool-using such as needle-threading, antique watch repair, or while using the fingers as measurement calipers; during self-defense activities requiring complex motions such as karate or judo; and during emergency situations such as automobile crashes and tumbling motions during falls from great heights. All such sensory data is readily outmessaged to the patient or user in real time.

Nanorobots could assist in path integration or the reconstruction of experienced limb trajectories. Given the ability to determine rotation rates and to measure applied forces and torques, the network should also be able to forecast the anticipated future positions of limbs. By comparing these projections to actual results, the network can then infer the viscosity of the medium in which the activity is taking place (e.g., air or water), whether the environment is stable or is translating or rotating in some direction, and whether the human user is physically supported or in free-fall. Internal nanodevices can directly measure if a body is sitting, standing, laying prone, inverted, falling, floating or diving, and then communicate that information directly to the human user. Detection of patient activity states—e.g., the patient is sitting, standing, or walking—can be used to control nanorobot behaviours, activate or deactivate outmessaging displays, and so forth. A simple macroscale wearable tactile "compass belt" that would convey desired geographical directions has already been proposed.

Since nanopendular sensors can reliably determine the direction of the local gravity field vector to within ~2 milliradians in ~10^{-4} sec, the navigational network can poll its members and arrive at an accurate consensus on which direction is up. Nanodevices affixed to relatively stable hard body parts will exhibit more consistent orientational readings. Measurement of the gravity vector allows the vertical orientation of the rest of the body to be precisely fixed in space, especially useful for gymnasts, trapeze artists, underwater divers in murky lakes, and vestibular-impaired individuals, to whom this information may be outmessaged directly.

The gravitational vector may also be indirectly measured, albeit more slowly and less accurately, by monitoring the body's natural reactions to changes in the axis of gravitational loading. For instance, the variation in head-to-toe hydrostatic pressure for a 1.7 m tall, 70 kg human standing in a 1-g gravity field is ~0.17 atm; the ability to measure a systematic cross-body differential of 10^{-6} atm allows the detection of a ~10^{-5} g change along the lengthwise aspect, or ~10^{-4} g change along the transverse aspect. Thus a continuously updated whole-body barostatic map allows the human body to serve as a three-dimensional gravity/orientation sensor. A network of nanodevices can inventory the volume and density of each of the body's ~10^{8} $(1\ mm)^{3}$ voxels using a combination of acoustic ranging, somatic mapping, and flowmetry. Each voxel is identified as fat, muscle tissue, bone mass, interstitial fluid, and so forth. These measurements allow the body's weight to be precisely computed as volume times density, rather than the usual method of determining weight using gravitational force data.

The existence of a natural physiological "ponderostat," a crudely analogous humoral body-mass detector, has been proposed; indeed, the insulin-leptin system has been found to serve a related purpose, and is easily eavesdropped by medical nanorobots. To achieve such phenomenal positional accuracies, the nanodevice must be able to computationally resolve several complicating factors. First, localised mass concentrations representing nonuniformities in crustal density produce residuals of up to ±0.0006 m/sec^2, which may be removed from the data using a standard map of known terrestrial isostatic variations and anomalies. Indeed, matching observations to such a map could provide useful longitudinal information as well. Another complication is the variation in gravity due to tidal forces amounting to ~3×10^{-7} g's, twice daily, which lies at the limits of detectability for a 20-micron gravimeter. Other minor geodesic and terrain-related corrections, too complicated for discussion here, may also need to be applied in certain circumstances. Note that the presence of nearby heavy objects does not influence measurement accuracy: a 100-ton building 10 meters away adds a lateral acceleration of only 7×10^{-9} g's to a human body. One final complication is that patient movements create kinematic accelerations that must be distinguished from the gravitational accelerations. Gravity readings can be corrected by taking derivatives of the signals from kinesthetic monitoring to give gravity in the reference frame of the patient's room, and many individual measurements may be averaged to improve accuracy since the gravity vector normally changes only very slowly. In vivo studies of the electric properties of blood vessels shows that the walls of veins and arteries present a specific electric resistance ~200 times greater than the conductive medium of blood (plasma)—roughly 200 ohm-m vs. 0.7 ohm-m.

Thus blood vessels are properly regarded as relatively insulated conducting cables which can electrically connect an injured tissue with surrounding noninjured tissue. Capillaries form an electric junction with the interstitial fluid, so the electric gradient can be canceled by ionic transport. Cell membranes are insulating dielectrics with resistance and capacitance, penetrated by ionic channels or gates for ionic transport. Additionally, in 1941 Szent-Gyorgyi suggested the possibility of semi-conduction in proteins, a theory which has since been elaborated by many investigators. B. Nordenstrom has proposed that the above schema represents a system for selective electrogenous mass transport of material between blood and tissue which he calls Vascular-Interstitial Closed (electric) Circuits (VICCs)—in effect, an additional electro-circulatory system operating in parallel with the well-known diffusive, osmotic, and hydrodynamic mechanisms of regular bloodstream materials transport. As long as ions can leak through open pores and migrate through ion channels, long distance transport cannot take place and the VICC system remains primarily a local (e.g., within tissues), not global, electrical circuit.

Connections also exist with conductive media including cerebrospinal fluid, bile, and urine. Nanorobots capable of monitoring the status of local VICCs may rapidly and efficiently acquire a wealth of systemic information without the need for direct inspection of the affected tissues. For example, when a working muscle produces catabolic products such as lactic acid, an electrochemical potential gradient develops between the muscle and surrounding tissue in a known manner. Injured tissues become polarised in relation to surrounding tissue, as initial catabolic degradation products acidify the tissue. Malignant neoplasms, benign neoplasms, and internally necrotic granulomas all may polarize electrically in relation to surrounding tissue; a vascular thrombus also contains ionised material. Large deflections in the diffusion potential of blood occur if blood is deoxygenated or is infected with Gram-negative bacteria. Blood changes its electric potential from +500 mV to +1000 mV during spontaneous coagulation, and the spontaneous electric potential at the site of a crush injury of the iliac crest in rats oscillated several times between +190 mV and -60 mV over a four day experimental trial.

Nordenstrom proposes that in vivo electrophoresis via the VICC system may also play a role in mediating leukotaxis. When tissue is artificially polarised by electrodes, accumulation of granulocytes with margination and development of pseudopods is the result. In one experiment, exposure of a mesenterial membrane to a 1 microampere 1 volt field for 30 minutes stimulated diapedetic bleeding near the anode.

At higher power, the arterio-capillaries contracted, narrowed, and emptied of blood cells, while the veno-capillaries and venules widened and filled with granulocytes. Monitoring such actions of local VICCs experiencing natural voltage fluctuations could alert medical nanorobots to the presence of local injuries, tissue changes, or pathologies that otherwise might go unnoticed for a considerable time. Detection of electric fields to ~100 volts/m, the normal vertical atmospheric gradient, in theory might allow determination of aboveground altitude to ~1 m accuracy if the time between recalibrations is very brief. However, a human body (which is a good conductor), standing on level ground, acquires a slight negative surface charge and becomes "part of the ground" electrically. This distorts upward the equipotentials that usually run parallel to the ground, thoroughly corrupting an altitude measurement. Nearby lightning hits and storms are detectable, which is useful. However, medical nanodevices with appropriate magnetometers can measure variations in magnetic field to ~0.1 microtesla using a $(660 \text{ nm})^3$ permanent magnet sensor, in a measurement time t_{meas} ~ 0.6 millisec. Geomagnetic field maps of the Earth's surface are highly nonisotropic in both latitudinal and longitudinal directions; knowledge of the field plus the absolute direction of true north (e.g., using a nanogyroscope) gives longitude information from the separation of the planetary magnetic and spin poles.

This implies that properly equipped in vivo nanodevices can establish their latitude and longitude on Earth's surface ~1000 times every second by this means. In particular, the horizontal component of the geomagnetic field vector ranges from 0-41 microtesla from magnetic pole to magnetic equator; the independent vertical field component ranges from 0-70 microtesla. Thus a 0.1-microtesla sensor should resolve each component to an accuracy of ~11 arcmin or ~20 km in both latitude or longitude. Terrestrial surface magnetic anomalies (e.g., iron deposits) produce local variations of 0.03-30 microtesla and are thus detectable. Additional complications to these measurements include artificial magnetic field sources, the 0.01-0.1 microtesla solar daily variation, occasional magnetic storms producing erratic geomagnetic fluctuations of 0.01-5 microtesla, sudden-commencement ionospheric electrojets up to 0.3 microtesla, and various long-term secular variations in the geomagnetic field. Radio, television, and direct broadcast satellite signals probably are not detectable by individual nanodevices, but such detection may be indirectly achieved in vivo using nanorobot-accessible dedicated macroscale antennas implanted in the body.

The piezoelectric effect is "the production of electrical polarisation in a material by the application of mechanical stress". Many materials in the human body are piezoelectric, including tendon and elastin, dentin and bone. This polarisation, or surface charge, varies as the physical stress imposed on the material changes over time. Thus, measurement of the piezoelectric effect in bone or tendon can provide information on the physical loads that are being carried by these materials. For example, a shear stress applied along the long axis of a bone alters the polarisation voltage that appears on a surface at right angles to the axis. In another experiment, piezoelectric surface charges on a loaded human femur were measured to range from -131 to +207 picocoulombs/cm^2 (-8 to +13 charges/$micron^2$) from one end of the bone to the other end, depending upon position along the shaft. Real-time monitoring of this data would permit specific inferences as to the amount of load the bone was carrying and from what direction, including bending, shearing, and twisting forces, from which whole-body activity states could subsequently be inferred. Knowledge of these surface charge variations might also be exploited in osteographic or functional navigation.

In vivo medical nanodevices can gather little useful optical information from the external environment. Here's why. In biological soft tissues, scattering dominates absorption except in the pigmented layers of the epidermis and stratum corneum. Thus the propagation of light in tissues may be regarded as occurring in two steps. Over the optical band from 400-700 nm, incident intensity I_0 = 100-400 watts/m^2 when standing in direct sunlight; artificial lighting in homes and offices is typically 0.1-10 watts/m^2; moonlight provides only $I_0 \sim 10^{-4}$ watts/m^2; and the absolute threshold for human vision is ~10^{-8} watts/m^2. For $s_d \sim 900$ m^{-1}, the intensity ratio of transmitted/incident visible light falls to $I_d/I_0 = 0.1$ at z = 2.5 mm depth (~eyelids), ~10^{-4} at z = 1 cm depth, and ~10^{-11} at z = 2.8 cm—a depth at which the tissue would be completely dark to the human eye even under direct sunlight illumination at the outermost skin surface. Let us assume that a patient is standing in $I_0 \sim 400$ watt/m^2 direct unfiltered sunlight, and that t_{meas} = 1 sec, SNR = 2, N_{sensor} = 25,000 elements giving a nanorobot eyespot of $A_e N_{sensor}$ = 0.025 micron^2, and that s_d = 900 m^{-1}. The maximum tissue depth for optical photon detection is $z_{max} \sim 17$ mm ($I_d \sim 10^{-4}$ watt/m^2) which includes a volume of tissue comprising up to ~30 per cent of total body volume.

Changes in illumination at the minimum indoor artificial level of ~0.1 watts/m^2 are visible to $z_{max} \sim 7$ mm depth. This eyespot, if exposed to air on the outermost surface of the skin, is just sensitive enough to detect full moonlight. Our general conclusion is that variations in normal indoor lighting may be directly measurable by nanorobots stationed within the outermost ~1 cm of body tissues. Imaging, as opposed to the simple illumination detection system described here, is a much more difficult design challenge The ability to detect individual neural cell electrical discharges noninvasively in many different ways, coupled with the abilities (A) to recognize and identify specific desired target nerve cells and (B) to pool data gathered independently by spatially separated nanodevices in real time, offers the possibility of indirect neural macrosensing of complex environmental stimuli by eavesdropping on the body's own regular sensory signal traffic. For example, facultatively mobile nanodevices may swim into the spiral artery of the ear and down through its bifurcations to reach the cochlear canal, then position themselves as neural monitors in the vicinity of the spiral nerve fibers and the nerves entering the epithelium of the organ of Corti (cochlear or auditory nerves) within the spiral ganglion.

These monitors can detect, record, or rebroadcast to other nanodevices in the communications network all auditory neural traffic perceived by the human ear.

Advanced speech recognition systems may permit recovery of spoken words and identification of individual speakers (including vocalisations by the user), recognition of background noises, reception and validation of cues or commands spoken directly to in vivo nanosystems by authorised medical personnel, and so forth. Since properly configured monitors can also modulate or stimulate nerve impulses, these devices may add audible signals to the audio traffic, thus may be employed as hearing aids (using feedback loops), real-time language translation mechanisms, continuous vocalisation/audition recorders, voice-stress analyzers, or nanodevice-user communications links. Nanomonitors positioned at the afferent nerve endings emanating from hair cells located in the otolithic membrane of the utricle and saccule and in the cristae ampullaris of the semicircular canals allow medical nanodevices to directly record, amplify, attenuate, or modulate the body's own sensations of gravity, rotation, and acceleration, although kinesthetic sensory management might also be required for complete control.

Motor neurons likewise can be monitored to keep track of limb motions and positions, or specific muscle activities, and even to exert control. Neuron-resident nanorobots may detect auditory effects induced in the brain by pulsed microwaves from external sources. Feline cochlear neurons sensitive to audible frequencies of >300 Hz respond to single microwave pulses at a threshold-specific absorption rate of 6-11 watts/kg-pulse, while in humans the threshold for effect, which depends on energy per pulse, may be as low as ~0.02 joule/m^2-pulse for people with low hearing threshold. Olfactory and gustatory sensory neural traffic similarly may be eavesdropped by nanosensory instruments.

Nerve taps in the medulla oblongata or at the phrenic nerve that drives the diaphragmatic muscles allow direct monitoring of respiratory activity. Pain signals may be recorded or modified as required, as can mechanical and temperature nerve impulses from other receptors located in the skin. Even psychological variables such as emotionality, vigilance, and mental workload may be directly monitored by measuring ANS activity in sympathetic efferent fibers outside of the brain. The most complex and difficult challenge in neural macrosensing will be optic nerve taps. The retina is thoroughly vascularised, permitting ready access to both photoreceptor (rod, cone, bipolar and ganglion) and integrator (horizontal, amacrine, and centrifugal bipolar) neurons. However, the optic nerve bundle itself has ~10^6 tightly bunched individual nerve fibers, a 10-100 MHz signal bandwidth, and significant natural data compression techniques which all must be untangled in real time.

Developing algorithms capable of interpreting raw optical nerve traffic, say, to recognize a specific human face or a specific scene in the vision field, would prove a significant research challenge. (Direct monitoring of photo-receptors or retinal membrane potentials and other techniques may simplify untangling of the signal compression.) Rapid visual field identification using artificial neural nets is also a subject of much current research interest. Eyeball rotations (e.g., via resident intradevice nanogyroscopes, eyelid position, pupil aperture, and lens accommodation under ciliary muscle control must be monitored to supplement the vision field analysis. We have only scratched the surface of the potential for nanorobot macrosensing.

Various chemical substances cyclically increase and decrease in serum concentration with time of day, permitting crude temporal macrosensing. Other substances significantly vary in serum concentration depending upon whether the patient is supine or standing, thus permitting limited biochemical postural macrosensing. If nanorobots can leave and re-enter the body, then macrosensing may include direct sampling of the external environment upon demand and the possibilities for remote data acquisition become virtually limitless.

Sensors pervade many aspects of modern living. They are built into many consumer electronic devices, cars, medical devices, security and safety devices, and systems for monitoring pollution and environmental conditions. Many applications demand miniaturisation to reduce power consumption for integration into portable devices. Affordable mass production is also a prerequisite for sensors for consumer products, and for disposable devices such as sensors for medical diagnostics and pollution monitoring. Sensors support applications across the economy—industrial processes, and those in construction, extractive industries, agriculture, health care and so on—and can be incorporated into new or existing products.

Sensors can model various parameters: physical parameters such as temperature, displacement, acceleration, flow and so on; and chemical and biochemical parameters, such as concentrations of gases, ions or molecules, and molecular interactions)

The application of nanotechnology to sensors should allow improvements in functionality. In particular, new biosensor technology combined with micro and nanofabrication technology can deliver a huge range of applications. They should also lead to much decreased size, enabling the integration of 'nanosensors' into many other devices.

We can also expect to see actuators that control movement on the nanoscale. Sensor/actuator combinations will deliver 'smart' and precise functions in products and processes. For example, nanofabrication and inspection tools require sensors and actuator systems that can position objects with nanometre accuracy. In this way, sensors and actuators constitute another enabling technology.

MS Modeling's DMol^3 has been used to investigate the nitrogen dioxide sensing mechanism of tin dioxide nanoribbons. Understanding the sensing mechanism will enable the efficient design and manufacture of nanoscale chemical sensors—an important commercial application of nanotechnology. Practical challenges with carbon nanotubes involving cost of synthesis, control of chirality and diameter, separation from bundles, and attachment of functional groups have prompted researchers to explore other types of one-dimensional nanostructures. Two of the most promising candidates in this regard are nanowires and nanoribbons. As the name suggests, nanoribbons are solid objects (unlike nanotubes, which are hollow) with a near-uniform rectangular cross-section.

So far, nanoribbons have primarily been synthesised from the oxides of metals and semiconductors. In particular, SnO_2 and ZnO nanoribbons have been materials systems of great current interest because of potential applications as catalysts, in optoelectronic devices, and as chemical sensors for pollutant gas species and biomolecules. Although they grow to tens of microns long, the nanoribbons are remarkably single-crystalline and essentially free of dislocations. Thus they provide an ideal model for the systematic study of electrical, thermal, optical, and transport processes in one-dimensional semiconducting nanostructures, and their response to various external process conditions.

Recent experiments with SnO_2 nanoribbons indicate that these are highly effective in detecting even very small amounts of harmful gases like NO_2. Upon adsorption of these gases, the electrical conductance of the sample decreases by more than an order of magnitude. More interestingly, it is possible to get rid of the adsorbates by shining UV light, and the electrical conductance is completely restored to its original value. Such single-crystalline sensing elements have several advantages over conventional thin-film oxide sensors: low operating temperatures, no ill-defined coarse grain boundaries, and high active surface-to-volume ratio. To be able to fully commercialize their potential, it is important to better understand the sensing mechanism of such systems. With this goal in mind, researchers at the Brookhaven National Laboratory, Lawrence Berkeley National Laboratory, and Accelrys used MS Modeling's $DMol^3$ to investigate the nitrogen dioxide sensing mechanism of SnO_2 nanoribbons. Reporting in *Nano Letters*, the researchers examined the NO_2-sensing mechanism of SnO_2 nanoribbons with exposed (1 0'1) and (0 1 0) surfaces. The density functional theory (DFT) calculations revealed that:

- The most stable adsorbed species involved an unexpected NO_3 group doubly bonded to Sn centers
- An orders-of-magnitude drop in electrical conductance can be explained by significant electron transfer to the adatoms
- Computed binding energies were consistent with adsorbate stability up to 700 K, with X-ray absorption spectroscopy indicating predominantly NO_3 species on the nanoribbon surface.

The ability of the nanoribbons to sense O_2 and CO was also investigated. In the case of O_2, the response of the nanoribbon was highly sensitive to the concentration of O-vacancies on the surface. Thus, in the absence of any surface vacancies, the calculation predicted negligible charge transfer. However, when surface vacancies were present, an O_2 molecule can adsorb as a peroxide bridge, and withdraw a significant amount of electronic charge from the nanoribbon surface, thereby decreasing its electrical conductance. In the case of CO adsorption, there was a net electron transfer from the CO to the nanoribbon surface. Thus the calculation predicted an increase in nanoribbon electrical conductance upon CO-adsorption, in agreement with experimental results.

Uses in Drug Discovery

The pharmaceutical industry has to discover and develop innovative medicines for a wide range of diseases in a marketplace that is (1) likely to experience growing regulatory and pricing pressures and (2) that is increasingly targeted towards diseases that have been traditionally resistant to pharmacotherapy (e.g., solid tumors). This chapter profiles a number of tools that are emerging from nanotechnology that are believed will be important aids in the process of drug discovery in this environment. Today, nearly all pharmaceutical companies follow common technology processes for discovering drugs. These include cloning and expressing human receptors and enzymes in formats that allow high-throughput, automated screening and the application of combinatorial chemistry. Thus, random screening can now be achieved with libraries sufficiently large and diverse to have a relatively high probability of finding a novel molecule. These libraries are possible because they can be generated by the techniques of combinatorial chemistry (combichem). Backing this are important developments in the application of genetics and genomics to understand associations between diseases and gene products. Importantly, bioinformatics are beginning to identify putative targets for a number of diseases.

An important consequence of today's approach is that more lead molecules are being discovered for diverse targets, giving the medicinal chemist more scope to find a candidate molecule. Yet the probability of success of launching a candidate molecule into the market remains basically unchanged at around 8 percent, and the perception is that the time to market is not reducing as might be expected with greater automation. Numerous problem areas remain:

- Current methodologies for compound synthesis.
- Storage and retrieval are inefficient and inadequate.
- Better information is needed to improve target selection and validation.
- More predictive screens are essential.

- More emphasis must be placed on acquiring good ADMET data earlier in the discovery process.
- Novel screening methodologies are needed that dramatically increase throughputs and reduce reagent consumption.

It is becoming clear that many of the important solutions will be nanotechnological in nature. This is because the size range that holds the most interest in drug discovery is more or less the same as that which defines nanotechnology from 100 nm down to the atomic level (approximately 0.2 nm). Nanotechnology cannot provide the ultimate solutions to all these problems. However, it is becoming clear that many of the tools developed to pursue nanotechnology and nanoscience will have important roles in drug discovery.

While AFM microscopes are invaluable for imaging objects at the nanoscale, until recently they haven't been able to see how components of a cell react in biological processes, such as their response to a specific chemical or compound. New imaging techniques involve attaching antibodies specific to individual proteins to the tip of an atomic force microscopes' probe. When an antibody reacts with the protein to which it is targeted, it creates a variance in the microscope's reading compared to a reading with a bare tip, showing the presence of a protein or other specific material in the region being scanned. This technique leads to a greater understanding of the chemical dynamics involved in how cells react to stimuli, and could prove particularly significant for drug discovery. However, although NanoMarkets believes that AFM is a useful tool to improve the understanding of molecular interactions, our research also indicates that the widespread use of this technology in bioanalytical applications has been hindered by the limited throughput of these techniques and the high experimental burden imposed by complicated and expensive instrumentation. Highly parallel micrometer and submicrometer cantilever arrays, which are currently being developed, might increase the throughput of AFM-based force spectroscopy.

NSOMs allow the study of optical properties of the sample surface with a resolution better than the wavelength of the light. By scanning the optical probe at very small distances from the sample (a few nanometers) "evanescent waves" from the surface are detected by the probe. The use of evanescent waves allows bypassing of the wavelength limitation of traditional optical techniques (the traditional limit of resolution being half the wavelength in use). NSOM can be used to image biological samples and could ultimately be a significant contributor to drug discovery. However, there is an obstacle to such deployment at the present time manufacturing the probes reliably, the requirement for a very small hole through which the light passes being problematic.

SPR is a phenomenon that occurs when light is reflected off thin metal films and a small amount interacts with electrons in the film, reducing the light intensity. The refractive index of the materials sandwiching the film dictates the angle at which the light reduction (essentially a shadow) occurs. Using an SPR-based approach, the interaction of biomolecules can be detected in real time, offering applications largely for observing biological systems in action on a very small scale, but also with potential for biosensors. Surface plasmons were the subject of some interesting general research in late 2002 and early 2003. There is still a lot to be learned in this field, but NanoMarkets believes that there will be important improvements for imaging systems (and thus for drug discovery) as this learning process proceeds.

Current technologies used for proteomic studies are based on a variety of separation techniques followed by identification of the separated proteins and proteolytic peptides using mass spectrometry (MS). One popular technique is high-resolution two-dimensional (2D) gel electrophoresis, which is capable of resolving 2500 or more distinct protein spots from complex samples. The in-gel protein spots are then identified using sensitive MS and sequence database searching. An alternative approach to 2D gel electrophoresis is chromatographic separation of peptides with electrospray ionisation (ESI)-MS or tandem MS (MS/MS) detection. The need for relatively large sample volumes is a major challenge for conventional ESI-MS analysis. There are now systems that address this challenge by applying nanotechnology to the electrospray nozzles. Thus, Advion's ESI Chip contains an array of nanoelectrospray nozzles, each one-fifth the diameter of a human hair, etched in a silicon wafer.

Nanolithography is the art and science of etching, writing, or printing at the nanoscopic level. DPN is a scanning probe nanopatterning technique in which an AFM tip is used to deliver molecules to a surface via a solvent meniscus, which naturally forms in the ambient atmosphere. This direct-write technique offers high-resolution patterning capabilities for a number of molecular and biomolecular "inks" on a variety of substrates,.such as metals, semiconductors, and monolayer functionalised surfaces. DPN is becoming a significant tool for the scientist interested in fabricating and studying soft- and hard-matter on the nanoscale. DPN enables precise multiple patterns with near-perfect registration. It's both a fabrication and imaging tool, as the patterned areas can be imaged with clean or ink-coated tips. The ability to achieve precise alignment of multiple patterns is an additional advantage earned by using an AFM tip to write, as well as read, nanoscopic features on a surface. NanoMarkets believes that together, all these attributes make DPN a valuable tool for studying fundamental issues on colloid chemistry, surface science, and nanobiotechnology.

The central paradigm in proteomics studies has been to identify differential protein levels in healthy and diseased cells, characterize these proteins and determine the protein's role in biochemical pathways. These proteins can then serve as diagnostic markers and potential drug targets. Traditionally, 2D gel electrophoresis has been the workhorse in proteomics on the front end of mass spectroscopy. This technology, however, has been described as a bottleneck in high-throughput proteomic studies. As a result there has been a significant push to develop alternative technologies that serve to alleviate this rate failing step in protein separation prior to further analysis. At the forefront of these emerging and growing technologies are protein and DNA microarrays that allow the highly specific capture and analysis of a large number of proteins expressed in various cell types exposed to given perturbations in a high throughput manner.

Microarrays are generally defined as a substrate and surface chemistry onto which a biomolecule or capture agent has been immobilised for the purpose of expression analysis and functional characteristic. The potential to decrease the time required for the critical steps or protein separation and characterisation are serving as a considerable driver in the protein microarray market. Microarray technologies can be considered platforms for nanoscale bioanalysis, and these products have already proven their value in the marketplace. However, currently available microarray technologies suffer from certain limitations that prohibit the exploitation of the full range of drug discovery applications. These limitations are already being addressed at the nano level with "nanoarrays," which are ultra-miniaturised versions of the traditional microarray that can measure interactions between individual molecules down to resolutions of as little as one nanometer. Nanoarrays are being touted as the next evolutionary step in the miniaturisation of bioaffinity tests for proteins, nucleic acids, and receptor-ligand pairs. It is believed that nanometer-scale resolution capabilities of nanoarray technology offer many advantages in the emerging field of functional proteomics. Unlike nucleic acids, which can easily be multiplied by amplification methods, individual proteins cannot be easily increased in quantity. Using nanoarray technology, very small quantities of individual proteins can be effectively screened against a large set of drug targets. In addition, nanoarrays can be incorporated as sensors in ways that would be impossible with larger microarrays. In addition, unlike traditional microarrays, a nanoarray can be used "in solution." Microarrays, once exposed to the target protein, must be removed from solution and rinsed before they are scanned to see whether or not there have been any "hits." Not only does this take time, but the rinsing process can also easily result in the degradation of the array itself.

QDots are nanometer-sized semiconductor crystals, or electrostatically confined electrons. While other bio-imaging tools such as fluorescent organic dyes have been available for many years. QDots hold the promise of a true revolution in bio-imaging quality and sensitivity. Since 1914, researchers studying everything from blood tests to the effects of cancer drugs on the workings of a cell have used fluorescent dyes to tag cells. The dyes, however, can be problematic. Each dye molecule requires a laser of the same wavelength to cause it to illuminate. The dyes are also imprecise and have a tendency to bleed together. They can only be lit up for very short periods of time, typically just a few seconds after a light source is applied.

QDots take advantage of the quantum confinement effect predicted by quantum mechanical theory to fluoresce extremely brightly when excited by a light source such as a laser. And they don't have any of the disadvantages of conventional technologies. By varying the size of the crystals one can cause a rainbow of colors to fluoresce. In addition, QDots stay lit for much longer periods of time than conventional dyes often hours or days. QDots enable the tagging of a variety of different biological components like proteins or different strands of DNA with specific colors. QDots are designed to bond with and illuminate individual biological targets of choice whether genes, nucleic acids, proteins, small molecules, cancer cells, or even entire blood vessels. It may be noted that some of the world's largest pharmaceutical and biotech companies, such as GlaxoSmithKline, Pfizer, and AstraZeneca, and leading biotech firm Genentech are applying QDots in high-content drug screening and have completed initial drug screens using QDots as the biological read-out. Genentech is applying QDots to the detection of breast cancer. This chapter demonstrates how nanotechnology offers several improvements in drug discovery technology. These include:

- Improved imaging (including real-time imaging) that enables researchers tc better understand how biological processes are affected by a new compound.
- The use of dip pen technology to improve fundamental research techniques for drug discovery applications.
- Major improvements in today's microarray and fluorescent dye technologies.

As the drug industry strives to meet the increasingly difficult task of producing new drugs and especially new blockbuster drugs, it is increasingly seeking new drug discovery technologies that can improve R&D success rates and time to market. It is therefore believed that these improvements are likely to lead to substantial new business revenues over the next few years.

Redesigning Platinum at the Nanoscale

The method has the potential of changing the metal's properties and benefiting emerging technologies. While we are in the early stages of research, we see the possibility of manipulating the nanoscale structure of platinum so that we can have control over the size, porosity, composition, surface species, solubility, stability and other functional properties of these metal nanostructures," says John Shelnutt, the Sandia scientist leading the research effort. "Such control means that the redesigned platinum could be used in many new applications, including catalysis, sensors and optoelectronic and magnetic devices." He adds that while research groups have reported a few platinum nanostructures — including nanoparticles, nanowires, nanosheets and others — the addition of new types of nanostructures is "highly desirable and potentially technologically important." Working with Shelnutt in the research are Frank van Swol from Sandia, UNM graduate student Yujiang Song and Eulalia Pereira from the University of Porto in Portugal. The new method of manipulating platinum was detailed in a paper in the *Journal of American Chemical Society* published in December. The idea for the technique is similar to photosynthesis, in which plants use the energy from sunlight to produce sugar. But instead of manufacturing sugar, the new method changes a platinum ion to the neutral metal atoms. The photosynthetic protein mimicks this repeatedly, allowing metal to be deposited as desired at the nanoscale. The method involves putting porphyrins — the active part of photosynthetic proteins — along with the platinum salt in an aqueous solution of ascorbic acid at room temperature. The porphyrins are placed in specific locations in the solution where it is intended that metal should be deposited. For example, the porphyrins may be confined to micelles, or liposomes. Micelles are spherical assemblies of detergent molecules in which the heads are exposed to the water and the tails stick together in the interior.

Liposomes are similar structures, but they are larger and have water on the inside and outside separated by a closed membrane — sort of like a cell. The membrane is composed of two layers of detergent molecules, with the heads on the inner and outer surface facing the water and the tails forming the interior of the membrane. When light is shined on the porphyrins located in these detergent structures, the porphyrins excite, becoming catalysts for platinum reduction and deposition. As this occurs, the metal grows onto the surfaces of the surfactant structures as a thin sheet or in other ways. In the case of micelles, the platinum grows into balls that look like the common toy "Koosh" ball. The ball size can be controlled by the amount of porphyrins and platinum in the solution, the amount of light illuminating the solution, and the amount of time the light is on. For the metals platinum and palladium that form these nanostructures, it is enough for the porphyrin molecule to grow only a small metal "seed" particle composed of about 500 atoms. When it reaches this size, the seed starts to catalyze its own rapid growth (by oxidation of ascorbic acid), budding off arms in all directions and creating the Koosh-ball-like nanostructures. The porphyrin provides a convenient method of making these seeds at the location and time desired, leading to a uniform and selectable nanostructure size The platinum nanostructures take on a different form when they are prepared under different conditions. When the porphyrin is in a micelle, the platinum nanostructures produced look like Koosh balls.

When the porphyrin is in the bilayer membrane of a liposome, the platinum grows in 2-nanometer thick sheet or platinum lace on the outer surface of the membrane, giving circular sheets — sort of like two-dimensional Koosh balls. Under solution conditions for which the liposomes aggregate, growth can occur along the interfaces of the liposomes to give platinum foamlike materials and foam nanoballs. The type of nanostructure is mainly determined by the type of surfactant assembly upon which the platinum grows and the extent of growth from the individual seed nanoparticles. Since the porphyrin remains attached to the platinum nanostructure and active in the presence of light, it can also perform other functions besides growing itself. For example, when illuminated with light, the platinum nanostructure evolves hydrogen from water. This reaction is similar to one of interest to car manufacturers looking for new ways to build automobiles powered by hydrogen fuel cells. Shelnutt says that in addition to structuring the platinum, the process also happens very fast. A few minutes in light will create many seeds, which then grow into the mature nanostructures in tens of minutes. And the process is easy to do. Today we encounter a strong trend towards nanoscience and nanotechnology. This trend is motivated by the fact that as we move to smaller and smaller scales the underlying physical laws change from macroscopic to microscopic. The exploitation of quantum effects for technological applications is the most obvious driving force behind the current miniaturisation.

The recent rapid advances are due in large part to our newly acquired ability to measure and manipulate individual structures on the nanoscale (scanning probe techniques, optical tweezers, and high-resolution electron microscopes). The increasing trend towards nanoscience and nanotechnology makes it necessary to address the key issues of optics on the nanometer scale. Since the diffraction limit does not allow us to focus light to dimensions smaller than roughly half a wavelength, traditionally it was not possible to interact selectively with nanoscale features. In recent years several new approaches have been put forth to 'shrink' the diffraction limit (confocal microscopy) or to even overcome it (near-field microscopy).

For example, with our tip-enhancement technique we are able to do Raman spectrocopy and multiphoton fluorescence imaging with a spatial resolution of less than 20nm. To date, this is the highest optical resolution of a spectroscopic measurement. The reason for the hard effort to advance the field of optics to the nanometer scale is the fact that the energy of light lies in the range of electronic and vibrational transitions in matter. Therefore, the interaction of light with matter renders unique information about the structural and dynamical properties of matter. These unique spectroscopic capabilities are of great importance for the study of biological and solid-state nanostructures.

We are applying near-field optical techniques to probe complex semiconductor nanostructures as well as individual protein molecules. We plan to explore the possibility to optically interact with semiconductor nanostructures on length scales smaller than the extent of their quantum wavefunctions. Probing and manipulating these wavefunctions might open up various exciting applications such as data storage and optical switching based on quantum logic. Nano-optics addresses the broad spectrum of optics on the nanometer scale covering technology and basic science. On the technological side, we find topics like nanolithography and high density optical data storage. On the basic sciences end, we might mention atom-photon interactions in the optical near-field and their potential applications for atom trapping and manipulation experiments. Compared with free propagating light the optical near-field is enriched by so-called virtual photons. These are the same sort of particles responsible for molecular binding (van der Waals/Casimir forces) and are therefore promising for selective probing of atomic structures. The consideration of virtual photons in the field of quantum optics will enlarge the range of fundamental experiments and applications.

Nano-optics deals with the controlled manipulation of evanescent optical fields and their exploitation in investigations of nanometer scale structures. The basic idea is to detect the evanescent field distribution, which is significant in the near-field region that extends to only a few tens of nanometers from the sample surface. The detection of the evanescent waves is typically done with a nanoprobe. The probe acts as a local scattering center transforming some of the evanescent field components into homogeneous (propagating) waves that can be detected in the far zone. By scanning the nanoprobe over the sample surface, the evanescent field distribution can be detected with a resolution limited mainly by the size of the nanoprobe and the scanning height. In practice, the resolution achieved in scanning near-field optical microscopy (SNOM) is on the order of a few tens of nanometers, i.e., the technique offers an order of magnitude improvement to the best conventional far-zone measurements. Our research has concentrated mainly on three specific areas.

The main efforts have been focused on developing the shear-force feedback system of our SNOM apparatus, on the near-field investigation of a broad-area semiconductor laser, and on the theoretical modeling of the near-field-matter interaction. Insight into the origin of the high spatial resolution in a SNOM measurement may be gained by investigating the contributions of the homogeneous and evanescent electromagnetic waves separately in some selected geometries. The simplest model for the sample-probe system in SNOM can be obtained by considering the probe tip and the sample structure as being composed of a collection of point-dipoles. Of fundamental importance in this approach is to understand how the electromagnetic field of a single dipole is decomposed into its evanescent and homogeneous parts. Knowing this one can then proceed to generalize these results to systems containing several point-dipoles by including the couplings between the dipoles in a self-consistent calculation. The point-dipole model has been widely employed as an aid in discussing the properties of near-field microscopy. Caution should, however, be exercised when interpreting the results. The point dipole does not offer an adequate model for describing mesoscopic structures in near-field investigations, since strong evanescent field gradients exist close to the sources. By applying the point-dipole model, however, we can investigate how the nanoprobe affects the near field and how the resolution depends on the scanning height of the probe over the sample in the different operation modes of SNOM. Furthermore, we can study the effects of different polarisation directions of the nanoprobe and the external polarising field on the resolution of a SNOM measurement. In particular, we can see the central role of the evanescent field components in breaking the classical diffraction barrier. The homogeneous, evanescent and total field distribution at the probe site when a z-polarised nanoprobe is scanned at the height of 40 nm over the sample.

The probe dipole locally illuminates the sample and the same probe is also used to collect the scattered light. As a result of the local interaction between the probe and the sample dipoles the scanned homogeneous field distribution now reveals the same information about the sub-wavelength scale structures of the sample as the evanescent field, although in lower magnitude. Nanooptics, i.e. optical spectroscopy with a spatial resolution of <100 nm, opens up fascinating insights into the physical processes occuring on a nanometer scale. In our group, we apply this technique to investigate *single* semiconductor quantum dots, sometimes refered to as 'artificial atoms' because of their d-like density of states. "Embossing" is an ancient technical term in English traceable to the period 1350 to 1400, and used by Chaucer in its present sense. The technical meaning, creating a raised design or pattern by pressing a die having the raised negative image of the design against the material to be patterned, has not changed during the ensuing 650 years. What has changed are the applications, materials used, and the scale of the design. Today's embossed patterns may have features measured in nanometers, bringing new capabilities to integrated optical systems. Nanoembossing, often referred to as nano-imprint lithography (NIL), can rapidly pattern relatively large area substrates with feature sizes below 100 nm.

Creating patterns at this scale previously required far slower and more costly techniques such as electron-beam writing. NIL, used in combination with related technologies, can create an entire wafer of integrated optical structures with nanometer features in minutes. An array of replicated lenses formed on the surface of a 4-in. silicon wafer. Wafer-level NIL promises reductions in the size, cost and complexity of integrated optics, while improving performance. Integrating embossed components on a wafer offers major size reductions. Producing truly "monolithic" optical systems could reduce centimeter-sized systems to millimeter-sized die. Optical performance is improved by locating components physically closer together, reducing the optical path length and the attendant losses. For example, optical elements such as lenses can be fabricated directly onto optical sources, such as lasers. Surprising to the non-expert, the small physical size of the nano-structures also creates optical elements that differ significantly in some characteristics from normal macro-optics. Optical elements with feature sizes smaller than the wavelength of the incident light, sometimes referred to as sub-wavelength elements, interact differently with the light than optical elements with larger feature sizes. More specifically, light interacts in different ways with periodic structures having spacing smaller than its wavelength. The altered interactions create a variety of devices with characteristics that differ from similar but larger structures. The sub-wavelength optical elements offer some new, desirable characteristics. Sub-wavelength grating structures, for example, may have a wider acceptance angle than conventional gratings. Wider acceptance angles relieve manufacturing tolerances and may permit automated assembly, further reducing costs. Sub-wavelength gratings also perform more consistently over a broader range of wavelengths than conventional gratings. NIL has demonstrated a growing variety of nano-optical components, including gratings, filters, anti-reflective coatings, refractive and diffractive components, fresnel lenses, waveguides, and lens arrays. With all of these potential products and proven benefits, why has nano-optics languished so long in the laboratories? Because formidable difficulties must be surmounted to replicate these tiny dimensions in volume manufacturing processes. To understand the manufacturing challenges that are now being overcome, look at the basic processing techniques that have been developed. NIL in its various forms has five processing steps:

(i) Design and fabricate a master stamp with features patterned by high-resolution techniques.

(ii) Impress the master stamp on a layer of suitably prepared material.

(iii) Maintain the master stamp pressure while the material first flows to duplicate the stamp, and then solidifies to preserve the duplicate pattern.

(iv) Separate the patterned replica from the master stamp, without damaging nanoscale features.

(v) Post-process the replica further for its desired optical function.

The starting point for NIL is the creation of a master stamp or mold. The ultimate resolution capability of the embossing depends solely on the master stamp. The design and fabrication of the master stamp is critical to minimum feature size, uniformity, fluid flow, and ease of separation. Designing a stamp to allow quick and complete fluid flow of the softened material is challenging. Once the design is satisfactory, it must be transferred to the master. Fine-line approaches, such as electron beam, are used to create nanoscale features on the master stamp.

Electron beams have fine-line capability, but direct e-beam writing has a working area only a few millimeters square, and is relatively slow. Stamps used for hot embossing or for injection molding are commonly made from nickel or steel. Stamps for UV embossing must be of UV-transparent materials. While the completed master stamp could be directly used to produce many replicas, wear or deterioration through repeated use would require creating a new master stamp. In practice, to preserve its pristine quality, the completed master stamp generally is cloned, using the replication techniques described here. Cloned sub-masters are used to produce the nanoscale features in volume production, either simultaneously over a whole substrate, or repetitively, at chip size, onto a substrate. The substrates may be silicon, quartz, glass, or other materials. Replication layers are generally coatings of heat-curable or UV curable polymers, with carefully controlled coating uniformity and thickness. While polymers are common materials for embossing, they may not offer the dimensional precision or the long-term stability under challenging environmental conditions demanded of some system components. This need has been filled by the development of special materials, such as of sol-gel organic/inorganic composite materials. These composites can be UV cured or thermally cured like polymers, and after curing display optical and mechanical properties close to those of glass. Embossing and curing the pattern on the substrate varies with different processes. Once the coating is cured, the stamp must be separated from the cured material, without damaging the replicated pattern. As feature size becomes smaller, clean separations are more difficult. A special coating may be applied to the stamp before molding, to facilitate separation. The embossed wafer may be post-processed by etching, deposition, or by repeating further lithographic processes. The cured patterned material can be the etch resist, for reactive ion etching of the underlying substrate. Alternatively, the patterned material may be used as a resist for metal deposition and lift-off. Post-processing might also include additional lithography, further depositions, and the addition of antireflective or protective coatings. Hot embossing of wafers uses the cloned master die as the wafer stamp. Wafers are first coated with a thermoplastic polymer, that is, a polymer that softens when heated. The wafer is positioned in a controlled-atmosphere stamping chamber.

After heating the wafer, the stamp is brought down with pre-determined pressure and temperature to soften the polymer and simultaneously form the image on the whole wafer. After allowing time for polymer flow during heating, the stamp and wafer are cooled to harden the image. To maintain the pattern, pressure must continue to be applied until both stamp and substrate are cooled below the thermoplastic softening point. The hot-embossed wafer may be used for its embossed features, or it may be post-processed as a resist for etching, material deposition, and lift-off patterning. For resist use, curing may be followed by a light etch to remove the remaining traces of materials from the clear areas. An alternative hot embossing approach is step-and-repeat stamping, using a smaller than wafer size master stamp Step-and-stamp imitates the action of step-and-repeat optical steppers used in semiconductor fabrication. The smaller master stamp is simpler to design, produce, and control. It is also less costly than a full wafer stamp. The step-and-repeat approach mitigates the dimensional, planarity, flow, and separation anxieties inherent in wafer embossing, by literally reducing the problem. The tradeoffs include the added time required for multiple repetitions of the stamping process, the need to confine heating to the stamp area, and the cost and complexity of a precision step-and-repeat apparatus. Cold embossing uses UV light to polymerize and solidify materials after forming by the stamp. The starting materials are thin films of UV-curable materials. Typical materials may be monomers, or inorganic-organic hybrid polymers, which are polymerised by UV to polymers with desirable optical properties. The films may be deposited onto silicon, gallium arsenide, or similar wafers. The UV-transmissive patterned master template is pressed against the film, imprinting the pattern, which is UV cured before pressure is released. UV embossing allows accurately aligning and overlaying multiple levels of features. UV embossing has a clear advantage over hot embossing in allowing multi-level applications, since the first level is polymerised after exposure, making it irrevocably formed and dimensionally stable, regardless of subsequent layer applications and exposures.

UV embossing is a low pressure, room temperature process, with little shrinkage of the pattern. It also allows partial UV casting, by limiting the illumination to maintain clear areas on the wafer. Cold embossing may be done on one or both sides of the wafer. Double-sided embossing allows micromolding. The laser lens array was formed by cold embossing. As with hot embossing, UV embossing may be used with a single wafer size stamp, or with a smaller stamp in a "step and flash" procedure similar to the "step and stamp" described above. Injection molding of optical components offers throughput and cost advantages for high-volume production. Injection molding can produce 3-D objects and integrated multicomponent systems. To move beyond conventional injection molding and capture nanofeatures, electroformed metal inserts are nanopatterned and included as part of the master mold. Double-sided injection molding at wafer scale has been used to create optical structures on both sides of a supporting substrate. These structures may be the same on both sides of the substrate, or they may be different; for example, combining a diffractive lens on one side with a refractive lens on the other.

The two sides may be processed separately, or simultaneously. In either case, precise alignment must be made between the features through the intervening substrate. Injection molding has been used to produce optical components in hundreds of thousands. However, the high tooling cost can only be justified for these large quantities. Injection molding draws upon a wider range of materials than direct embossing. Acrylics or polycarbonates may be used for all-plastic systems in low-cost, high-volume injection molding, producing inexpensive commercial products in the hundreds of thousands. A summary comparison of the three embossing processes. Hot embossing, at wafer level or by step-and-stamp, lends itself to single side patterning on thermoplastic materials. Process temperatures are typically 90°C above the glassivation temperature. Throughput is limited by the time required to heat, hold, and cool the die and substrate. Step and stamp avoids the high cost and tolerances of a wafer-size stamp, at a penalty in multiplying the time delays. Hot embossing has some well established and perhaps more profitable commercial applications. Music CDs are commonly replicated in volume by embossing. Roll-to-roll hot embossing has been used to add interesting color effects to wrapping paper, and to produce material with difficult to counterfeit optical patterns for security documents, such as identification cards.

Cold (UV) embossing may be preferred for critically dimensioned materials. It can be a room temperature process, eliminating the heating and cooling times and thermal expansions/contractions of hot embossing. It allows single or repeated multiple patterns to be layered on one or both sides of a substrate with accurate alignments. Both sides of the substrate may be embossed with identical or different patterns. For applications with more critical planarity, dimensional tolerances, or environmental requirements, cold embossing of photosensitive organic-inorganic sol-gel materials gives results close to those of glass. Injection molding produces 3-D molding, allowing entire systems to be molded as a unit. It is not a room temperature process; processing time and temperature depend on the materials used. These may be inexpensive commercial plastics, or highly accurate specialty materials. Patterning can be single- or double-sided, in 3-D, although not generally multilayered as in cold embossing. The high tooling cost makes injection molding practical for high-volume (e.g. 100,000 units and up) products. A mixture of techniques and equipment from the semiconductor and micropackaging industries has been adapted to nano-optics fabrication. Much of the development work reported above used one-of-a-kind laboratory equipment, built for that purpose, or modified equipment, originally developed for other purposes. One report referred to a modified semiconductor mask aligner. Another recent paper reported demonstrating hot embossing using a standard flip chip aligner-bonder. One nano-optics manufacturer has developed customised robotic embossing systems specifically for volume production. Only recently have major equipment manufacturers announced production equipment intended to meet the challenges of nanometer optics. Dedicated aligner-steppers with sub-20-nm embossing resolution are now available. Product applications based upon NIL include wavelength monitors, beam splitters, pattern generators, polarisation filters, switches, and many others.

Also commercially available now are replicated components that are high performance building blocks for fabricating optical assemblies. These offer manufacturing advantages in comprising a common family of similar products, allowing standardisation of manufacturing operations and equipment. A typical component size. Embossing has come a long way since Chaucer. After decades of laboratory development, nanoembossing promises to open new optical frontiers with truly monolithic production systems. The growing commercialisation of nano-optic systems will subject them to the scrutiny and judgment of the marketplace. Success in surmounting the manufacturing challenges will determine whether nanoembossed optical systems remain a technology niche, or become a broad new technology opportunity.

Optical Spectroscopy and Microscopy

Single walled carbon nanotubes (SWNTs) are tubular graphitic molecules that can be microns long and less than one nanometer wide. The combination of the unique properties of the carbon–carbon bond and their interesting molecular structure have given SWNTs unique physical properties and provided the potential for revolutionary applications. SWNT structure is characterised by two integers (n,m) that can be easily related to both its diameter and chirality. SWNTs have an unusual electronic configuration: each SWNT with a distinct (n,m), where n–m divisible by three is metallic, while the rest are semiconducting Further, SWNT electronic properties exhibit a strong structural dependence which helps drive potential applications. Simple one electron models indicate that the density of states (DOS) of a nanotube is characterised by singularities, with each singularity corresponding to a quantum subband. Recent studies of individual SWNT electronic structure using scanning tunneling spectroscopy (STS) agree with these general predictions. Simple one electron models also imply that the dipole-allowed optical transitions occur between mirror image peaks in the DOS. Absorption spectra of SWNT bundles contain three broad absorption regions corresponding to the first and second allowed transitions for semiconductor SWNTs, and the first allowed transition for metallic SWNTs, respectively. However, these spectra do not exhibit any features identified with a particular nanotube structure, since strong electronic coupling mixes the energy states from the different SWNTs in the bundle.

This strong intertube coupling, combined with the large inhomogeneity in nanotube structures present in an ensemble, completely obscures any fine structure in the absorption spectra. Isolation of individual nanotubes is critical for decoupling the optical response of one nanotube from another, and was recently demonstrated by O'Connell et al. for nanotubes manufactured by the high-pressure CO chemical vapor deposition (HiPCo) process (H-SWNTs). Absorption spectra of an ensemble of isolated SWNTs (Mic-SWNTs) show many more highly resolved features than the corresponding corresponding absorption spectrum from nanotube bundles. These distinct features correspond to dipole-allowed optical transitions from SWNTs with different (n,m) values and/or from higher interband transitions for a nanotube with a given (n,m) value. The general isolation procedure can be applied using numerous surfactants, as well as nanotubes produced by various methods such as the laser oven (L-SWNTs), and arc discharge (A-SWNTs) processes. Bundled SWNTs do not fluoresce since any metallic tubes present in the bundle provide an efficient nonradiative pathway for photoexcited electrons.

However, for HiPCo SWNTs, Mic-SWNT samples show strong, dipole allowed fluorescence at the bandedge. The measured Stokes shift between the absorbing and emitting lines is quite small (_40 cm_1). SWNTs are poor emitters; the fluorescence quantum yield (QY) of SWNTs is on the order of QY _10_4. The dependence of the SWNT electronic structure on diameter and chirality [(n,m) values] is fundamentally important and is also of technical interest. Simple one electron models predict that the energy of the dipole-allowed optical transitions depends linearly on the nanotube diameter. However, the observation of structurally dependent photoluminescence from SWNTs represented a major breakthrough, as the dependence of electronic transition energies was shown to depend on (n,m) values explicitly and not simply on the SWNT diameter.

In particular, the magnitude and sign of deviations from the predicted linear dependence were observed to depend on the value of (n-m) mod 3, suggesting that trigonal warping effects were playing a significant role. Empirical fitting parameters have been introduced modifying the simple one-electron picture to compensate for these effects, thus allowing for a precise prediction of the transition energies for different tube (n,m) structures. However, despite the observation of fluorescence spectral features attributable to specific nanotube (n,m) structures, it is not fully understood why the transition frequencies differ significantly from simple theoretical predictions, although many-particle interactions and/or trigonal warping effects likely play a role in this discrepancy.

Notwithstanding these recent advances in SWNT separation and isolation, the optical spectra clearly suffer from severe inhomogeneous broadening due to the distribution of different nanotube structures in the ensemble. Thus, early optical studies could only determine an averaged overall behaviour that obscured crucial electronic characteristics of the individual nanotube. For example, since spectra from SWNTs with different (n,m) values will overlap, it is difficult to make definite conclusions regarding important emission features such as the spectral line width or details of the band shape, or to observe evidence of coupling to vibrations. Therefore, new approaches are needed to further advance the understanding of SWNT optical properties and accelerate the de velopment of novel SWNT photonic applications, such as nanometer scale integrated electroluminescent devices.

Spin-casting the Mic-SWNT suspension onto a cover slip allows for isolation of individual nanotubes by standard confocal microscopy. A typical fluorescence image from the spun Mic-SWNT sample is shown, with a spatial resolution of 300 nm. The image was obtained by collecting an optical spectrum at every pixel and then binning regions of the individual spectra where strong fluorescence signals are expected according to the ensemble spectrum. Since atomic force microscopy (AFM) measurements reveal predominantly short SWNTs (length 200–300 nm) lying on top of residual surfactant patches, distinct tubular features in the Mic-SWNT fluorescence images are typically unresolved. Alternatively, by dispersing SWNTs directly in organic solvents such as dichloroethylene, followed by spin coating, occasionally individual nanotubes that fluoresce can be isolated, both the fluorescence and Raman images clearly show the characteristic one-dimensional features of the SWNT thus proving that the fluorescence in fact results from the single nanotube. Fluorescence images from the isolated Mic-SWNTs show distinct, bright spots at different positions for different emission wavelengths.

Displays representative spectra detected from one of these bright spots. Each spectrum now exhibits a single emission band with a smooth line shape; all traces of contributions to the fluorescence spectra from the different SWNT (n,m) structures have been eliminated. The fluorescence maximum for all three spectra matches three transitions observed in the ensemble spectrum. In so much as the ensemble fluorescence spectrum is a weighted superposition of spectra from individual SWNTs. Single nanotube spectroscopy clearly eliminates the overlapping spectral features associated with ensemble averaging and thus allows for the potential discovery of crucial SWNT optical and electronic properties. For example, SWNT fluorescence spectra are described by a single, featureless, symmetrical band corresponding to a single Lorentzian line shape function .

The fluorescence from individual semiconductor quantum dots and from small molecules exhibit an emission intermittency or on-off "blinking" behaviour for all excitation intensities, all temperatures, and on time-scales spanning many orders of magnitude. In fact, fluorescence intermittency is commonly used as the conclusive indicator of single chromophore optical emission. Surprisingly, the fluorescence intensity from a single SWNT at 300 K recorded using moderate excitation intensities (<70 kWcm_2) is constant over many seconds with no indication of fluctuations on the 40-ms to 100-second time scale. At extremely high excitation powers (>100 kWcm_2), fluorescence intensity fluctuations can be observed, likely due to laser-induced sample heating. The use of single-photon sources is extremely desirable for applications in quantum optics, such as quantum cryptography, since the presence of even two photons in a signal pulse is an opportunity for an eavesdropper to steal or manipulate information. Thus, single molecules or single nanoparticles are attractive as single-photon "on demand" sources in quantum optics. However, their unavoidable fluorescence intermittency drastically hampers their potential utility. For example, in quantum cryptography it is difficult for the receiver to determine whether an "off" period is related to the information being transferred, or just the result of a random blinking event. Some stable sources of single photons do exist, such as nitrogen vacancies in diamond nanoparticles. However, these sources have a very broad linewidth typically over 100 nm, thus limiting the bandwidth of any potential devices.

SWNTs have many advantages for possible use as single-photon sources for quantum optics applications. On the millisecond to minutes time scale, and at room temperature, SWNT fluorescence is stable showing no signs of intensity or spectral blinking. SWNTs also have an extremely narrow linewidth, and can be pumped electrically which allows for the exciting possibility of future nanometer scale integrated single-photon photonic devices. SWNTs also are resistant to photobleaching, as their fluorescence intensity does not change despite hours of intense photoillumination. Finally, unlike dye molecules or impurities in diamond, SWNTs have an easily size-tunable emission that is in the infrared at technologically relevant wavelengths (1.3–1.5 mm). Single molecule sources are also highly desirable for applications in biology and biotechnology. For example, the use of single fluorophores to track individual molecules inside a cell is a compelling method for studying intracellular processes. However, photobleaching of organic dye labels severely restricts the single molecule processes that can be studied in cells. Detection of a single analyte as reported through changes in the optical emission of a single fluorophore is also a promising approach to achieving the ultimate in detection sensitivity. On the other hand the fluorescence intermittency of single molecules makes the fluorometric detection of single analytes extremely difficult, since the modulation of fluorescence upon binding of the analyte cannot be distinguished from a random blinking event. SWNTs hold great promise for possible use as single-molecule fluorophores.

First, SWNT fluorescence does not photobleach at 300 K even after hours of continuous laser excitation. Thus, extremely long-lived biological processes on the singlemolecule level can be followed much more easily. Second, the fact that SWNTs do not exhibit fluorescence blinking allows them to be used as fluorometric reporters of a single biological agent. SWNT photonic devices show strong near-infrared (NIR) electroluminescence, allowing for easy "on-chip" integration of single SWNT sensors with silicon technology. Finally, SWNTs are optically active at wavelengths in the range of 800–1100 nm, which allows for NIR excitation and detection. This NIR optical activity has the tremendous advantage that background fluorescence from the sample can be dramatically reduced allowing for much improved sensitivity on the single fluorophore level. For some nanoparticles the processes that contribute to fluorescence blinking events slow down as the temperature is lowered.

For example, in CdSe semiconductor nanocrystals a higher fraction of nanoparticles are "on" (or bright) at low temperatures versus at room temperature. Interestingly, at low temperatures (T=1.8 K) a significant fraction (about 2/3) of SWNTs show an intermittency in their fluorescence intensity, and also a spectral wandering of their fluorescence energy as shown. Spectral diffusion cannot be observed at room temperature, since the fluorescence linewidth is much larger than the change in energy during a spectral jump. However, it is currently unclear whether the processes that cause blinking "turn on" at low temperatures, or whether SWNTs blink inherently on timescales much faster than 10 ms. With the time resolution of past experiments, fluctuations in the sub-millisecond regime were not observable.

Combined fluorescence and Raman scattering measurements on the single molecule level open up new ways to characterize SWNTs. Single nanotube studies enable the connection between fluorescence energy, (n,m) structural parameters, and RBM frequency of a particular SWNT with a single measurement, and without any complicated fitting procedures necessitated by working with macroscopic samples. Single nanotube studies also reveal new insights into nanotube optical properties, such as a variation of the fluorescence energies of nanotubes having the same structure. Unlike for single molecules, for SWNTs the photoluminescence unexpectedly does not show any intensity or spectral fluctuations at room temperature. The fact that SWNTs show no emission intensity blinking or bleaching demonstrates that SWNTs have the po tential to provide a stable, single molecule infrared photon source, with an extremely narrow linewidth, which would have a significant impact for future nanometer scale integrated photonic devices and which shows promise for applications in quantum optics and biological sensing.

Traditional NSOM utilizes a tapered, metal-coated optical fiber with a small aperture as either an excitation source or collection device. The spatial resolution attainable with this technique is roughly defined by the size of the aperture itself. Unfortunately, rough metallic coatings on glass fibers, coupled with the finite skin depth of metals at optical frequencies, place a limit on the spatial resolution attainable with fiber probes. Typical spatial resolution for the fiber approach to NSOM is ~50-100 nm, though careful probe and sample preparation have yielded ~30 nm resolution. Research groups have long been interested in mapping out the spatial distribution of the proteins in photosynthetic membranes. These proteins are approximately 10 nm in diameter, and electron microscopy indicates that they are closley packed in their native configuration. The density of this arrangement precludes optical mapping with either far-field measurement or traditional aperture NSOM. In the mid-1990's, a number of groups began working on what would prove to be the solution to the fundamental resolution limitations in aperture NSOM. The idea was to spatially map an optical signal from the surface of a sample by scattering that signal with a solid probe. The choice of probe ranged from tungsten STM probes to metallised AFM tips to silicon probes. While optical image interpretation was complicated by topographic coupling, the spatial resolution in the apertureless technique was improved to ~10's of nm. In 1999 the Xie group began work on a type of apertureless NSOM that takes advantage of the phenomenon of field enhancement by noble metal probes.

With the appropriate combination of excitation polarisation and probe geometry, a gold or silver probe is capable of strongly enhancing the magnitude of the incident illumination. This strong, localised field can be used as an excitation source for a spectroscopic transition of interest, such as fluorescence or Raman scattering. By locally exciting a spectroscopic transition in the sample and detecting the Stokes shifted light in the far field, the interpretation of apertureless NSOM images is greatly simplified. They utilize the nonlinear response of the sample by exciting two-photon fluorescence in the sample. This tip enhanced nonlinear optical microscopy (TENOM) amplifies the contrast. The enhancements achievable with specifically fabricated gold probes are sufficient to yield near-field two photon fluorescence signal above the background generated in the diffraction-limited far-field region. Furthermore, TENOM has improved spatial resolution to the sub-20 nm regime. This high resolution and strong signal generation should be capable of, for example, mapping the spatial distribution of PSI and PSII in the thylakoid membrane. While TENOM has presented a solution to issues of high resolution optical imaging, it does have several intrinsic drawbacks. These drawbacks include the quenching of non-energy-transferred fluorescence, the necessity to carefully tailor the probe geometry to yield high field enhancement, and the practical requirement of a nonlinear signal source.

Currently exploration is on to get a variety of approaches to eliminate the few drawbacks of TENOM and to expand the capabilities of apertureless NSOM in general. The first concern has been to address the problem of fluorescence quenching by the metal probe. To this end, an ion- and electron-beam assited deposition (IBAD and EBAD, respectively) of dielectrics has been applied to the apertureless probes. The growth of glass with these techniques allows us to build a spacer layer between the probe and the sample, reducing quenching while still maintaining high field enhancement at the sample. The theoretical treatment of fluorescence quenching by probes of complex geometries is currently being explored through the use of finite difference time domain (FDTD) simulation and Poynting vector integration. Due to the complex nature of light-metal interactions, electromagnetic simulation is vital in the TENOM and apertureless NSOM research.

The FDTD technique is a flexible, powerful method for fully three dimensional simulation of the response of metal probes to incident radiation. The power of the FDTD technique lies largely in its conceptual simplicity. A geometry of interest is defined by a number of cells, called Yee cells. The complex permittivities of each cell are assigned. After a geometry has been defined, an incident optical field is simulated, and the Maxwell's equations are integrated in a leapfrog fashion. The result is a method that can calculate the resultant fields for any arbitrary geometry illuminated with optical sources. FDTD has been applied to the study of NSOM in the past. Unfortunately, the large computational burden associated with this technique made three dimensional simulations quite difficult. Two dimensional calculations have generated insight into the aperture method of NSOM.

In this research a commercial FDTD package (RemCom's XFDTD) was utilised to simulate apertureless probes in three dimensions, working to adapt this microwave-focused package to the complex permittivities intrinsic to the optical regime behaviour of real metals. This commercial software, coupled with the recent tremendous improvement in computational power, has provided a general method for simulating complex geometries, such as groups of spheres. In order to expand apertureless NSOM to linear signal sources, intensity enhancements on the order of 10,000 were required in these experiments.

In order to develop a physical intuition for the nature of strong field enhancement, and to rationally design probes for high field enhancement, FDTD was used to model apertureless NSOM. The simulations indicated that probes with finite size on the scale of the excitation light were preferable to the "quasi-infinite" tips used in the past. Furthermore, by properly tuning the size of a right trigonal pyramid to the excitation light, field enhancements sufficient for exciting linear signal sources and yielding signal above the far field background were predicted. The simulated high field enhancement has been experimentally verified. A focused ion beam instrument was used to make apertureless NSOM probes. This instrument (FEI Strata DB-235) allowed to mill materials with ~10 nm resolution, image with ~5 nm resolution, and deposit glass or platinum with ~20 nm resolution. This powerful tool for nano-optics and near field microscopy applications, including surface enhanced Raman scattering (SERS) and chemical sensors was utilised.

Electromagnetic coherence theory for optical near fields refers to the electromagnetic field at the immediate vicinity of the sample where the evanescent or non-propagating waves are strong. The near field is a primary example of a highly non-paraxial field and, in general, it is composed of three orthogonal electric field components. The conventional tools of electromagnetic coherence theory are, however, valid mainly for beam-like fields. We have put forward the definitions for the degree of polarisation of general three-dimensional fields and for the electromagnetic degree of coherence. We have shown that in the limit of complete coherence, the electromagnetic coherence tensors, both in the space-time and space-frequency domains, behave analogously to the corresponding scalar functions. Furthermore, we have rigorously formulated the electromagnetic theory of coherent modes, and constructed the mode expansion for a class of statistically homogeneous and isotropic electromagnetic fields. Currently, we are investigating the optical coherence in a near-field detection system, and tightly focused electromagnetic fields.

We study the state of partial polarisation of tightly focused optical fields. In the vicinity of such a focus the electromagnetic field displays a complicated three dimensional structure. For example, the light is fully polarised on concentric rings around the optical axis even when the incident light is unpolarised in the 2D sense. The distribution of the degree of polarisation in the image plane of an imaging system of high numerical aperture. The rings of full polarisation are clearly resolved in the image.

We have modeled the electromagnetic field in various nanostructures using a boundary-integral method in two dimensions. In the boundary-integral method the electromagnetic scattering problem is casted into a form of an integral equation involving integrations only over the boundaries of the system. We have applied the method to analyze changes in the degree of polarisation in light transmission through a tip aperture of a scanning near-field optical microscope. Furthermore, we have carried out spectral analysis of light transmission through sub-wavelength slits in metal films. We study experimentally the detection and spectroscopy of very small gold nanoparticles. Such particles are an interesting model system to study the transition from individual atoms to the optical properties of bulk matter. Nanoparticles also have important applications in chemical sensing and as labels in biology. In our first experiments we used supercontinuum light produced in a photonic crystal fibre to detect sub 10-nm gold particles. A confocal microscope image of 15 nm gold particles. The cross section shows that the supercontinuum light can be focused to almost a diffraction limited spot. The spectrum of the light scattered by an individual particle. The scattering spectrum displays a clear resonance which can be used to distinguish gold nanoparticles from other scatterers in a biological medium. This work was performed in the nano-optics group of professor Vahid Sandoghdar at ETH Zürich. We are currently continuing this work at HUT with experiments on the absorption spectroscopy of individual metallic nanoparticles.

Optical Processing and Spectroscopy

The commercial introduction of wafer-based, nanofabrication techniques now makes it possible to create a new class of optical components called subwavelength optical elements (SOEs). With physical structures far smaller than the wavelength of light, the physics of the interaction of these fine-scale surface structures with light yields new arrangements of optical-processing functions. These arrangements have greater density, more robust performance and greater levels of integration when compared with many existing technologies and could fundamentally alter approaches to optical system design. However, as with any new technology, the designer must clearly understand the general concepts underlying the technology as well as its potential impact (in this case, on optical component design), as well as implications for the evolution of the technology.

Integrated optical component and module design has achieved a high degree of sophistication in the combination of discrete optical elements. The properties of optical materials and their combinations, clever optics design, and advances in alignment and assembly are applied to increase component and module density, reduce costs and increase reliability. But there are limits. The natural properties of many discrete optical elements are relatively fixed, limiting the flexibility of the optical component designer and reducing transferability from design to design. This is where SOEs come in. SOEs are the realisation of nanotechnology as applied to optical elements. As such, they offer excellent optical properties while being easy to integrate with other optical materials in a broad range of configurations. In addition, they self-integrate, allowing the combination of optical functions to reduce parts count and increase reliability, and to allow increased flexibility in optical component design. A broad range of optical effects can be achieved by the manipulation of nanostructures in the optical beam path.

The interaction of light with subwavelength grating structures-gratings with one or more dimensions one or more orders of magnitude smaller than the wavelength of the incident light-produces a range of controllable effects. To explain those effects, a simple subwavelength grating structure. When light is normally incident on the structure, the transmitted portion of the light-that which passes through the grating structure-is affected. The effect on the passed light can be varied by adjusting the size, shape and spacing of the grating structure. In particular, for the one-dimensional grating structure, appropriate selection of dimensions will result in behaviour as a polarizer, waveplate or polarisation-dependent filter. More sophisticated effects, such as polarisation-independent filters, can be achieved by two-dimensional gratings. The general class of these structures is referred to as SOEs. Because the diameter of the incident light beam is often much larger than the structures of the grating, the effect on the transmitted light is really the statistical sum of numerous local interactions between the light and the grating.

For example, if a light beam measuring 300 microns in diameter is incident on a two-dimensional grating such as that shown, where the structure size is roughly 100 x 100 nanometers, more than a million nanostructures will be illuminated. Therefore, varying the grating dimensions spatially across the incident beam front allows additional control over optical processing effects. The physics of the SOE depends on rigorous application of the boundary conditions of Maxwell's equations to describe the interaction of light with the structures. At the wavelengths of light used in telecommunications-980 through 1,800 nm-the structures required to achieve those effects have some dimensions on the order of tens to a few hundred nanometers. At the lower end of the scale, single electron or quantum effects may also be observed. Whereas the principles of reflection, refraction, diffraction and interference describe the behaviour of traditional optical elements, with SOEs the equations describing conventional optical behaviour do not fully cover the resultant phenomena because quantum mechanical effects now come into play.

In many applications, subwavelength structures act as a nanoscale diffraction grating whose interaction with incident light can be modeled by rigorous application of diffraction-grating theory and the abovementioned boundary conditions of Maxwell's equations. Consider refraction, a key property exploited in optical components. Traditionally, different materials must be used to obtain different refractive indices. In SOEs, however, different refractive indices can be obtained with the same material by adjusting the physical structure. As an example, SOE structures can be used to create an "artificial" birefringence effect. If a is the grating period and t is the grating width.

With that sort of flexibility, why are SOEs just now becoming available for telecommunications applications? The primary reason has been manufacturability. While these optical effects have been researched for at least 20 years, cost-effective manufacturing of the optical elements has not been available. Building subwavelength grating structures in a research environment has generally required high-energy techniques, such as electron-beam lithography, or extremely precise process control, for instance, to grow nanostructures via "self-alignment." Second, there must be a consistent process for creating a broad range of nanostructure patterns. Many techniques for creating nanostructure patterns are limited in the range of possible patterns. Nanoimprint lithography overcomes those limits and has four key steps: making a mold inscribed with the complement of the desired nanostructure, impressing this mold into a resist-coated wafer, separating the mold and selectively removing the resist with reactive ion etching to transfer the nanopattern to the target material. After that, postimprint processing may be used to add metal layers to enhance performance and to provide a protective overcoat for handling in a standard manufacturing environment. Testing and dicing follow. Because nanoimprint lithography forms the nanostructures of SOEs by a direct, physical process rather than by beams of energy, the wave diffraction, scattering and interference in the resist do not limit fabrication resolution. Silicon dioxide is commonly used to make a mold patterned with the desired nanostructures. Various techniques, including electron beam lithography, can be used to create the desired negative image of a nanostructure. Because the mold can be replicated and reused, complex multistep and multiprocess methods can be used to create the desired nanostructure. Since those initial tooling steps do not need to be repeated for each production wafer, they are amortised over the full production run of a particular SOE. Using different molds with different nanostructure patterns allows the same manufacturing process to create the full range of SOEs. Unpackaged for use in free-space applications, the final SOE is an optical chip with a subwavelength grating on one side of a substrate. The overall thickness of the element is dependent on the thickness of the substrate. The (x,y) dimensions depend on application requirements.

The practical consequences of using SOEs can be explored by looking at a specific realisation—an SOE polarisation beam splitter/combiner, or PBS/C. By appropriate selection of the dimensions of a one-dimensional subwavelength grating structure, one polarisation will be transmitted, while the orthogonal polarisation will be reflected, as illustrated. At that scale, the subwavelength-diffraction grating exhibits only zero-order diffraction, resulting in a number of useful element characteristics. These include consistent performance across a broad range of wavelengths, the grating structure performs the same for wavelengths from 980 nm to 1,800 nm, and wide variations in the angle of incidence of the light beam that are up to 20 degrees of deviation from normal. Typical performance parameters for an SOE PBS reveal insertion loss of less than 0.13 dB on both the reflected and the transmitted beams. Extinction ratios are greater than 40 dB and 20 dB for the transmitted and reflected beams, respectively. The flexibility of SOEs allows trade-offs around those parameters. The benefits of a PBS/C SOE can be realised in two areas. First, there are architectural benefits in optical component design. Because of their small form factor, they allow more compact component designs. Reduced insertion losses are also realised because SOEs can be abutted to other elements with minimum free-space separation. There are differences in the way that the elements handle light, for instance, the polarisation beam splitter is a reflective device with 180 degrees of effective separation obtained in less than 1 micron. Because of those differences, the beam path can be laid out to simplify device design and possibly eliminate other optical elements. Further, the reflective nature of the SOE PBS lets it support high-energy applications associated with laser transmitters and fiber amplifiers. Second, there are advantages for manufacturing. The wide acceptance angle simplifies the alignment process, reducing manufacturing time and cost. It is also possible to use automated "pick-and-place" manufacturing technologies. The SOEs are robust: with proper selection of materials they can be subjected to temperature ranges of –200°C to 400°C. Consequently, they can be subjected to a broad range of manufacturing processes. Finally, reduced size may both simplify packaging and reduce packaging costs. Applications of the SOE PBS/C have been explored for fiber amplifiers, circulators and isolators, interleavers, optical switches and variable optical attenuators. Here, their size and low power consumption are attractive relative to traditional technologies.

Current devices are the initial applications of SOEs, but the possibilities are expanding in several directions. SOE-based building-block functionality will be introduced both as chips and as packaged devices. Demonstrated SOE functionality spans polarizers, polarisation beam splitters/combiners, filters, photodetectors and photonic bandgap devices. Dynamic control for switching, attenuation and tuning is also possible. Monolithic integration of SOEs can be achieved by stacking SOE layers to create aggregated optical effects. Nanoimprint lithography permits direct layering on of SOEs without resorting to lamination techniques. Combining SOEs with optically active layers allows optical control circuits to be built, resulting in complex optical components "on-a-chip." Multilayer SOE integration has already been demonstrated by combining filters with photodetector arrays to create dynamic optical feedback loops. Because SOEs are are self-compatible and fabricated via wafer-scale manufacturing with relatively minor differences from device to device, the means exist to implement them.

Single walled carbon nanotubes (SWNTs) are tubular graphitic molecules that can be microns long and less than one nanometer wide. The combination of the unique properties of the carbon–carbon bond and their interesting molecular structure have given SWNTs unique physical properties and provided the potential for revolutionary applications. SWNT structure is characterised by two integers (n,m) that can be easily related to both its diameter and chirality. SWNTs have an unusual electronic configuration: each SWNT with a distinct (n,m), where n–m divisible by three is metallic, while the rest are semiconducting Further, SWNT electronic properties exhibit a strong structural dependence which helps drive potential applications. Simple one electron models indicate that the density of states (DOS) of a nanotube is characterised by singularities, with each singularity corresponding to a quantum subband.

Recent studies of individual SWNT electronic structure using scanning tunneling spectroscopy (STS) agree with these general predictions. Simple one electron models also imply that the dipole-allowed optical transitions occur between mirror image peaks in the DOS. Absorption spectra of SWNT bundles contain three broad absorption regions corresponding to the first and second allowed transitions for semiconductor SWNTs, and the first allowed transition for metallic SWNTs, respectively. However, these spectra do not exhibit any features identified with a particular nanotube structure, since strong electronic coupling mixes the energy states from the different SWNTs in the bundle. This strong intertube coupling, combined with the large inhomogeneity in nanotube structures present in an ensemble, completely obscures any fine structure in the absorption spectra. Isolation of individual nanotubes is critical for decoupling the optical response of one nanotube from another, and was recently demonstrated by O'Connell et al. for nanotubes manufactured by the high-pressure CO chemical vapor deposition (HiPCo) process (H-SWNTs). Absorption spectra of an ensemble of isolated SWNTs (Mic-SWNTs) show many more highly resolved features than the corresponding corresponding absorption spectrum from nanotube bundles. These distinct features correspond to dipole-allowed optical transitions from SWNTs with different (n,m) values and/or from higher interband transitions for a nanotube with a given (n,m) value.

The general isolation procedure can be applied using numerous surfactants, as well as nanotubes produced by various methods such as the laser oven (L-SWNTs), and arc discharge (A-SWNTs) processes. Bundled SWNTs do not fluoresce since any metallic tubes present in the bundle provide an efficient nonradiative pathway for photoexcited electrons. However, for HiPCo SWNTs, Mic-SWNT samples show strong, dipole allowed fluorescence at the bandedge. The measured Stokes shift between the absorbing and emitting lines is quite small (_40 cm_1). SWNTs are poor emitters; the fluorescence quantum yield (QY) of SWNTs is on the order of QY _10_4. The dependence of the SWNT electronic structure on diameter and chirality [(n,m) values] is fundamentally important and is also of technical interest.

Simple one electron models predict that the energy of the dipole-allowed optical transitions depends linearly on the nanotube diameter. However, the observation of structurally dependent photoluminescence from SWNTs represented a major breakthrough, as the dependence of electronic transition energies was shown to depend on (n,m) values explicitly and not simply on the SWNT diameter. In particular, the magnitude and sign of deviations from the predicted linear dependence were observed to depend on the value of (n-m) mod 3, suggesting that trigonal warping effects were playing a significant role. Empirical fitting parameters have been introduced modifying the simple one-electron picture to compensate for these effects, thus allowing for a precise prediction of the transition energies for different tube (n,m) structures. However, despite the observation of fluorescence spectral features attributable to specific nanotube (n,m) structures, it is not fully understood why the transition frequencies differ significantly from simple theoretical predictions, although many-particle interactions and/or trigonal warping effects likely play a role in this discrepancy.

Notwithstanding these recent advances in SWNT separation and isolation, the optical spectra clearly suffer from severe inhomogeneous broadening due to the distribution of different nanotube structures in the ensemble. Thus, early optical studies could only determine an averaged overall behaviour that obscured crucial electronic characteristics of the individual nanotube. For example, since spectra from SWNTs with different (n,m) values will overlap, it is difficult to make definite conclusions regarding important emission features such as the spectral line width or details of the band shape, or to observe evidence of coupling to vibrations. Therefore, new approaches are needed to further advance the understanding of SWNT optical properties and accelerate the de velopment of novel SWNT photonic applications, such as nanometer scale integrated electroluminescent devices.

Spin-casting the Mic-SWNT suspension onto a cover slip allows for isolation of individual nanotubes by standard confocal microscopy. A typical fluorescence image from the spun Mic-SWNT sample is shown, with a spatial resolution of 300 nm. The image was obtained by collecting an optical spectrum at every pixel and then binning regions of the individual spectra where strong fluorescence signals are expected according to the ensemble spectrum. Since atomic force microscopy (AFM) measurements reveal predominantly short SWNTs (length 200–300 nm) lying on top of residual surfactant patches, distinct tubular features in the Mic-SWNT fluorescence images are typically unresolved.

Alternatively, by dispersing SWNTs directly in organic solvents such as dichloroethylene, followed by spin coating, occasionally individual nanotubes that fluoresce can be isolated, both the fluorescence and Raman images clearly show the characteristic one-dimensional features of the SWNT thus proving that the fluorescence in fact results from the single nanotube. Fluorescence images from the isolated Mic-SWNTs show distinct, bright spots at different positions for different emission wavelengths. Displays representative spectra detected from one of these bright spots. Each spectrum now exhibits a single emission band with a smooth line shape; all traces of contributions to the fluorescence spectra from the different SWNT (n,m) structures have been eliminated. The fluorescence maximum for all three spectra matches three transitions observed in the ensemble spectrum. In so much as the ensemble fluorescence spectrum is a weighted superposition of spectra from individual SWNTs. Single nanotube spectroscopy clearly eliminates the overlapping spectral features associated with ensemble averaging and thus allows for the potential discovery of crucial SWNT optical and electronic properties. For example, SWNT fluorescence spectra are described by a single, featureless, symmetrical band corresponding to a single Lorentzian line shape function.

The fluorescence from individual semiconductor quantum dots and from small molecules exhibit an emission intermittency or on-off "blinking" behaviour for all excitation intensities, all temperatures, and on time-scales spanning many orders of magnitude. In fact, fluorescence intermittency is commonly used as the conclusive indicator of single chromophore optical emission. Surprisingly, the fluorescence intensity from a single SWNT at 300 K recorded using moderate excitation intensities (<70 kWcm_2) is constant over many seconds with no indication of fluctuations on the 40-ms to 100-second time scale. At extremely high excitation powers (>100 kWcm_2), fluorescence intensity fluctuations can be observed, likely due to laser-induced sample heating. The use of single-photon sources is extremely desirable for applications in quantum optics, such as quantum cryptography, since the presence of even two photons in a signal pulse is an opportunity for an eavesdropper to steal or manipulate information. Thus, single molecules or single nanoparticles are attractive as single-photon "on demand" sources in quantum optics. However, their unavoidable fluorescence intermittency drastically hampers their potential utility. For example, in quantum cryptography it is difficult for the receiver to determine whether an "off" period is related to the information being transferred, or just the result of a random blinking event. Some stable sources of single photons do exist, such as nitrogen vacancies in diamond nanoparticles. However, these sources have a very broad linewidth typically over 100 nm, thus limiting the bandwidth of any potential devices.

SWNTs have many advantages for possible use as single-photon sources for quantum optics applications. On the millisecond to minutes time scale, and at room temperature, SWNT fluorescence is stable showing no signs of intensity or spectral blinking. SWNTs also have an extremely narrow linewidth, and can be pumped electrically which allows for the exciting possibility of future nanometer scale integrated single-photon photonic devices. SWNTs also are resistant to photobleaching, as their fluorescence intensity does not change despite hours of intense photoillumination. Finally, unlike dye molecules or impurities in diamond, SWNTs have an easily size-tunable emission that is in the infrared at technologically relevant wavelengths (1.3–1.5 mm). Single molecule sources are also highly desirable for applications in biology and biotechnology. For example, the use of single fluorophores to track individual molecules inside a cell is a compelling method for studying intracellular processes. However, photobleaching of organic dye labels severely restricts the single molecule processes that can be studied in cells. Detection of a single analyte as reported through changes in the optical emission of a single fluorophore is also a promising approach to achieving the ultimate in detection sensitivity. On the other hand the fluorescence intermittency of single molecules makes the fluorometric detection of single analytes extremely difficult, since the modulation of fluorescence upon binding of the analyte cannot be distinguished from a random blinking event. SWNTs hold great promise for possible use as single-molecule fluorophores. First, SWNT fluorescence does not photobleach at 300 K even after hours of continuous laser excitation. Thus, extremely long-lived biological processes on the singlemolecule level can be followed much more easily. Second, the fact that SWNTs do not exhibit fluorescence blinking allows them to be used as fluorometric reporters of a single biological agent. SWNT photonic devices show strong near-infrared (NIR) electroluminescence, allowing for easy "on-chip" integration of single SWNT sensors with silicon technology. Finally, SWNTs are optically active at wavelengths in the range of 800–1100 nm, which allows for NIR excitation and detection.

This NIR optical activity has the tremendous advantage that background fluorescence from the sample can be dramatically reduced allowing for much improved sensitivity on the single fluorophore level. For some nanoparticles the processes that contribute to fluorescence blinking events slow down as the temperature is lowered.

For example, in CdSe semiconductor nanocrystals a higher fraction of nanoparticles are "on" (or bright) at low temperatures versus at room temperature. Interestingly, at low temperatures (T=1.8 K) a significant fraction (about 2/3) of SWNTs show an intermittency in their fluorescence intensity, and also a spectral wandering of their fluorescence energy as shown. Spectral diffusion cannot be observed at room temperature, since the fluorescence linewidth is much larger than the change in energy during a spectral jump. However, it is currently unclear whether the processes that cause blinking "turn on" at low temperatures, or whether SWNTs blink inherently on timescales much faster than 10 ms. With the time resolution of past experiments, fluctuations in the sub-millisecond regime were not observable.

Combined fluorescence and Raman scattering measurements on the single molecule level open up new ways to characterize SWNTs. Single nanotube studies enable the connection between fluorescence energy, (n,m) structural parameters, and RBM frequency of a particular SWNT with a single measurement, and without any complicated fitting procedures necessitated by working with macroscopic samples. Single nanotube studies also reveal new insights into nanotube optical properties, such as a variation of the fluorescence energies of nanotubes having the same structure. Unlike for single molecules, for SWNTs the photoluminescence unexpectedly does not show any intensity or spectral fluctuations at room temperature. The fact that SWNTs show no emission intensity blinking or bleaching demonstrates that SWNTs have the po tential to provide a stable, single molecule infrared photon source, with an extremely narrow linewidth, which would have a significant impact for future nanometer scale integrated photonic devices and which shows promise for applications in quantum optics and biological sensing.

Dimensions of Nanotechnology

In 1985 the carbon structure family was completed by a new modification of pure carbon. With the discovery of the fullerenes by Curl, Smalley and Kroto and additionally with the possibility for the production of fullerenes in large amouts shown by Krätschmer and Huffman in 1990 the scientific competition in experimental and theoretical research was started. Fullerenes are polyeders build up by n three times coordinated carbonatoms with 12 pentagons and (n) hexagons, were the minimum for n equal 20 is. Fullerenes fullfill the EULER's theorem, were a polyeder build up from pentagons and hexagons has to contain exact 12 pentagons, to build a closed structure. Following this rule, the dodekaeder with 20 carbonatoms is the smallest possible fullerene. Actual the smallest fullerene is the C_{60}, because important for the stability of the structure is, that no pentagons are side by side. This is discribed by the Isolated Pentagon Rule (IPR). If two pentagons are annalled, the tension of the binding is increasing and the structure is not anymore energetic stable. Fullerenes are also found in the natural environment, but only in the ppm-range. Some places are Shunga/Russia, New Zealand) and Sudbury/Canada. The carbon atoms in fullerenes are sp^m (m=2...3). The free valence elctrons on the cage building a strong localised p-electron system. This p-electron system influences chemical reactions of the fullerenes. In chemical reactions fullerenes are not reacting aromatic ("superbenzene"), they show aliphatic behaviour. Good solvents for fullerenes are CS_2, o-dichlorobenzene, toluene and xylene. Fullerenes are unsoluble in water and stabil at air. Thin layers of fullerenes are coloured from yellow to yellow-green. Because of pp-electron transitions solutions of fullerenes are coloured:

- C_{60} purple/violetred
- C_{70} brick-red
- C_{76} light yellow-green
- C_{2v}-C_{78} maroon, D_3-C_{78} golden
- C_{84} brown and
- C_{86} olive-green

By heating fullerenes without air up to 1500°C they transform to graphite.

Following species of fullerenes are known:

- fullerenes
- fullerites
- fullerides
- endohedral fullerenes
- exohedral fullerenes
- heterofullerenes
- metcars

Among all elements, C is the basis of entire life. The whole branch of chemistry—the organic chemistry—is devoted to the study of C-C bonds and different molecules originating from them. Carbon is only 4-valent element able to produce long homoatomic stable chains or different 4-regular nets. The other 4-valent candidate for this could be only Si, with its reach chemistry beginning to develop. After diamond and graphite—the hexagonal plane hollow shell, in 1985 was first synthesised by H.W. Kroto, R.F. Curl and R.E. Smalley the spherical closed pentagonal/hexagonal homoatomic shell: the fullerene C_{60}. Except from this, it possesses some another remarkable properties: the rotational symmetry of order 5, from the geometrical reasons (according to Barlow "crystallographic restriction theorem") forbidden in crystallographic space or plane symmetry groups, and highest possible icosahedral point-group symmetry. After C_{60}, different fullerenes (e.g. C_{70}. C_{76}. C_{78}. C_{82}. C_{84} etc.)

are synthesised, opening also a new field for research of different potentially possible fullerene structures from the geometry, graph theory or topology point of view.

From the tetravalence of C result four possible vertex situations, that could be denoted as 31, 22, 211 and 1111. The situation 31 could be obtained by adding two C atoms between any two others connected by a single bond, and situation 22 by adding a C atom between any two others connected by a double bond. Therefore, we could restrict our consideration to the remaining two non-trivial cases: 211 and 1111. Working in opposite sense, we could always delete 31 or 22 vertices, and obtain a reduced 4-regular graph, where in each vertex occurs at most one double bond (digon), that could be denoted by colored (bold) edge. First, we could consider all 4-regular graphs on a sphere, from which non-trivial in the sense of derivation are only reduced ones. In the knot theory, 4-regular graphs on a sphere with all vertices of the type 1111 are known as "basic polyhedra", and that with at least one vertex with a digon as "generating knots or links". From the chemical reasons, the vertices of the type 1111 are only theoretically acceptable. If all the vertices of such 4-regular graph are of the type 211, such graph we will be called a general fullerene. Every general fullerene could be derived from a basic polyhedron by "vertex bifurcation", this means, by replacing its vertices by digons, where for their position we have always two possibilities. To every general fullerene corresponds (up to isomorphism) an edge-colored 3-regular graph (with bold edges denoting digons). This way, we have two complementary ways for the derivation of general fullerenes: vertex bifurcation method applied to basic polyhedra, or edge-coloring method applied to 3-regular graphs, where in each vertex there is exactly one colored edge. For every general fullerene we could define its geometrical structure (i.e., the positions of C atoms) described by a non-colored 3-regular graph, and its chemical structure (i.e. positions of C atoms and their double bonds) described by the corresponding edge-colored 3-regular graph. In the same sense, for every general fullerene we could distinguish two possible symmetry groups: a symmetry group G corresponding to the geometrical structure and its subgroup G' corresponding to the chemical structure. In the same sense, we will distinguish geometrical and chemical isomers. For example, for C_{60}, $G=G'=I_h=S_5$ of order 120, but for C_{80} with the same G, G' is always a proper subgroup of G, and its chemical symmetry is lower than the geometrical. Hence, after C_{60}, the first fullerene with $G=G'=I_h=S_5$ will be C_{180}, then C_{240}, etc. Working with general fullerenes without any restriction for the number of edges of their faces, the first basic polyhedron from which we could derive them (after the trivial 1*) will be the regular octahedron or 6*, from which we obtain 7 general fullerenes. From the basic polyhedron 8* with v=8 we derive 30, and from the basic polyhedron 9* we obtain 4 general fullerenes. All the basic polyhedra with v<13 and their Schlegel.

Among general fullerenes we could distinguish the class consisting of 5/6 fullerenes with pentagonal or hexagonal faces. If n_5 is the number of pentagons, and n_6 the number of hexagons, from the relationship $3v=2e$ and Euler theorem directly follows that $n_5=12$, so the first 5/6 fullerene will be C_{20} with $n_6=0$ – the regular dodecahedron, giving possibility for two non-isomorphic edge-colorings, resulting in two chemically different isomers of the same geometrical dodecahedral form. The first basic polyhedra generating 5/6 fullerenes will be that with $v=10$ vertices. For $v=10$, there are three basic polyhedrons, but only 10^* and 10^{**} could generate 5/6 fullerenes, each only one of them. On the other hand, they generate, respectively, 78 and 288 general fullerenes. This way, we have two mutually dual methods for the derivation of fullerenes: (a) edge-coloring of a 3-regular graph, with one colored edge in each vertex; (b) introduction of a digon in every vertex of 4-regular graph, giving possibility for a double check of the results obtained. Their duality is illustrated by the example of two C_{20} chemical isomers derived, both of the same geometrical dodecahedral form with $G=[3,5]=I_h=S_5$ of the order 120, but the first with $G'=D_{5d}=[2^+,10]=D_5 x C_2$ of the order 20, and the other with $G'=[2,2]^+=D_2$ of the order 4. In this case, the symmetry of chemical isomers derived by the vertex bifurcation is preserved from their generating basic polyhedra.

For the enumeration of general fullerenes we used Polya enumeration theorem, applied to basic polyhedra knowing their automorphism groups, but with the restriction to 5/6 fullerenes its application is not possible. With the same restriction, the other derivation method: edge-coloring of 3-regular graphs is also not suitable for the application of Polya enumeration theorem, because of the condition that in every vertex only one edge must be colored. The basic polyhedra with $n<13$ vertices are derived by T.P. Kirkman, and used in the works by J. Conway (only for $n<12$), A. Caudron and S.V. Jablan (for $n<13$). The 3-connected 4-regular planar graphs (corresponding to basic polyhedra) are enumerated by H.J.Broersma, A.J.W.Duijvestijn and F. Göbel ($n<16$) and by B.M.Dillencourt ($n<13$), but given only as numerical results without any data about individual graphs. The 3-regular graphs with $n<13$ vertices and their edge-colorings producing 4-regular graphs are discussed by A.Yu.Vesnin. Proceeding in the same way, it is possible to prove that 5/6 fullerenes with 22 atoms not exist at all, and that they are seven 5/6 fullerenes C_{24} of the same geometrical form with $G=D_{6d}=[2^+,12]=D_{12}$. To distinguish different chemical isomers, sometimes even knowing their chemical symmetry group G' will be not sufficient. For their exact recognition we could use some results from the knot theory: the polynomial invariant of knot and link projections. Every 4-regular graph could be transformed into the projection of an alternating knot or link (and *vice versa*), and the correspondence between such alternating knot or link diagrams and 4-regular graphs is 1-1 (up to enantiomorphism).

Let be given an oriented knot or link diagram D with generators $g_1,..., g_n$. If the generators g_i, g_j, g_k are related, then $a_{ii}=-t$, $a_{ij}=1$, $a_{ik}=-1$; if they are related, then $a_{ii}=t$, $a_{ij}=1$, $a_{ik}=-1$; in all the other cases $a_{ij}=0$. The determinant $d(t)=|a_{ij}|$ is the polynomial invariant of D. In the case of links, in this polynomial invariant we use for generators of different components different variables (up to a permutation of variables). For example, let us show that two isomers of C_{20} are different. After converting their chemical Schlegel diagrams into the alternating knot diagrams, denoting their generators, and calculating the corresponding determinants we obtain, respectively, $D_1=t^{20}-10t^{18}+45t^{16}-120t^{14}+200t^{12}-197t^{10}+105t^8-40t^6+25t^4-10t^2$ and $D_2=t^{20}-10t^{18}+45t^{16}-120t^{14}+208t^{12}-250t^{10}+217t^8-130t^6+49t^4-10t^2$, proving their difference. On the other hand, for three-component alternating links corresponding to the diagrams, using the same multivariable invariant for link projections, we conclude that they both represent the isomorphic Schlegel diagrams of the same fullerene C_{24}.

Using the mentioned connection between alternating knot or link diagrams and 4-regular (chemical) Schlegel diagrams of fullerenes, it is interesting to consider all of them after such conversion. For example, two chemical isomers of C_{20} will result in knots, and from 7 isomers of C_{24} we obtain four knots, one 3-component, one 4-component and one 5-component link. Among the links obtained, two of them (3-component and 5-component one) contain a minimal possible component: hexagonal carbon ring (or simply, a circle). It is interesting that C_{60} consists only of such regularly arranged carbon rings, so maybe this could be another additional reason for its stability. Therefore, it will be interesting to consider the infinite class of 5/6 fullerenes with that property, that will be called "perfect". Some of "perfect" fullerenes are modeled with hexastrips by P. Gerdes, and similar structures: buckling patterns of shells and spherical honeycomb structures are considered by different authors (e.g. T. Tarnai). To obtain them, we will start from some 5/6 fullerene given in geometrical form (i.e. by a 3-regular graph). Than we could use "mid-edge-truncation" and vertex bifurcation in all vertices of the triangular faces obtained that way, transforming them into hexagons with alternating digonal edges. Let is given some fullerene (e.g. C_{20}) in its geometrical form (i.e. as 3-regular graph). By connecting the midpoints of all adjacent edges we obtain from it the 3/5 fullerene covered by connected triangular net and pentagonal faces preserved from C_{20}. After that, in all the vertices of the truncated polyhedron we introduce digons, to transform all triangles into hexagonal faces.

This way, from C_{20} we derived C_{60} (in its chemical form). The mid-edge-truncation we could apply to any 5/6 (geometrical) fullerene, to obtain new "perfect" (chemical) fullerene, formed by carbon rings. This way, from a 5/6 fullerene with v vertices we always may derive new "perfect" 5/6 fullerene with $3v$ vertices. Moreover, the symmetry of new fullerene is preserved from its generating fullerene. According to the theorem by Grünbaum & Motzkin, for every non-negative n_6 unequal to 1, there exists 3-valent convex polyhedron having n_6 hexagonal faces. Hence, from the infinite class of 3-regular 5/6 polyhedra with $v=20+n_6$ vertices, we obtain the infinite class of "perfect" fullerenes with $v=60+3n_6$ vertices. The "perfect" fullerenes satisfy two important chemical conditions: (a) the isolated pentagon rule (IPR); (b) hollow pentagon rule (HPR). The IPR rule means that there are no adjacent pentagons, and HPR means that all the pentagons are "holes", i.e. that every pentagon could have only external double bonds. The first 5/6 fullerene satisfying IPR is C_{60}, and it also satisfies HPR. The IPR is well known as the stability criterion: all fullerenes of lower order (less than 60) are unstable, because they don't satisfy IPR. On the other hand, C_{70} satisfies IPR, but cannot satisfy HPR. The same situation is with C_{80}, possessing the same icosahedral geometrical symmetry as C_{60}, but not able to preserve it after edge-coloring, because HPR cannot be satisfied. This is the reason that only "perfect" fullerenes, with $G=G'=[3,5]=I_h=S_5$, satisfying both IPR and HPR will be C_{60}, C_{180}, C_{240}, etc. We need also to notice that for $n_6= 0,2,3$ we have always one 3-regular 5/6 polyhedron (i.e. geometrical form of C_{20}, C_{24}, C_{26}), but for some larger values (e.g. $n_6=$ 4,5,7,9) there are several geometrical isomers of the generating fullerene, and consequently, the same number of "perfect" fullerenes derived from them. Hence, considering the fullerene isomers, we could distinguish "geometrical isomers", this means, different geometrical forms of some fullerene treated as 3-regular 5/6 polyhedron, and "chemical isomers" – different arrangements of double bonds, obtained from the same 3-regular graph by its edge-coloring.

For denoting different categories of symmetry groups, we will use Bohm symbols. In a symbol G_{nst}..., the first subscript n represents the maximal dimension of space in which the transformations of the symmetry group act, while the following subscripts st... represent the maximal dimensions of subspaces remaining invariant under the action of transformations of the symmetry group, that are properly included in each other. With regard to their symmetry, general fullerenes belong to the category of point groups G_{30}. The category G_{30} consists of seven polyhedral symmetry groups without invariant planes or lines: [3,3] or T_d, $[3,3]^+$ or T, [3,4] or O_h, $[3,4]^+$ or O, $[3^+,4]$ or T_h, [3,5] or I_h, $[3,5]^+$ or I, and from seven infinite classes of point symmetry groups with the invariant plane (and the line perpendicular to it in the invariant point): $[q]$ or C_{qv}, $[q]^+$ or C_q, $[2^+,2q^+]$ or S_{2q}, $[2,q^+]$ or C_{qh}, $[2,q]^+$ or D_q, $[2^+,2q]$ or D_{qd}, $[2,q]$ or D_{qh}, belonging to the subcategory G_{320}. For the groups of the subcategory G_{320}, in the case of rotations of order $q>2$, the invariant line (i.e. the rotation axis) may contain 0,1 or 2 vertices of a general fullerene. According to this, among all general fullerenes with a geometrical symmetry group G belonging to G_{320}, from the topological point of view we could distinguish, respectively, cylindrical fullerenes (nanotubes), conical and biconical ones. We could simply conclude that for polyhedral 5/6 fullerenes G could be only [3,3] (T_d), $[3,3]^+$ (T), [3,5] (I_h), $[3,5]^+$ (I), because of their topological structure ($n_5=12$), incompatible with the octahedral symmetry group [3,4] (O_h) or its polyhedral subgroups. In the case of nanotubes (or cylindrical fullerenes) we have infinite classes of 5/6 fullerenes with the geometrical symmetry group $[2,q]$ (D_{qh}) and $[2^+,2q]$ (D_{qd}), and the same chemical symmetry. The first infinite first class of cylindrical nanotubes with $G=G'=D_{5h}$ we obtain from a cylindrical 3/4/5 4-regular graph with two pentagonal bases, 10 triangular and $5(2k+1)$ quadrilateral faces ($k=0,1,2,...$) and with the same symmetry group. By the vertex bifurcation preserving its symmetry, we obtain the infinite class of nanotubes C_{30}, C_{50}, C_{70},..., with C_{70} as the first of them satisfying IPR. Certainly, the geometrical structure of C_{70} admits different edge colorings (i.e. chemical isomers). Starting from any two of them by "collapsing" (the inverse of "vertex bifurcation", i.e. by deleting digons) we could obtain different generating 4-regular graphs. This example of two different C_{70} isomers, with the same geometrical structure, and with the same G and G', shows that for the exact recognition of fullerene isomers we need to know more than their geometrical and chemical symmetry.

In the same way, from 4-regular graphs with two hexagonal bases, 12 triangular and $6(2k+1)$ quadrilateral faces ($k=0,1,2,...$) we obtain the infinite class of fullerenes C_{36}, C_{60} C_{84},... with the symmetry group $G=G'=D_{6h}$. The next symmetry groups $[2^+,2q]$ (D_{qd}) with $q=5,6$ we obtain in the same way, from 4-regular graphs with q-gonal bases, $2q$ triangular and $2kq$ quadrilateral faces ($k=1,2,...$ for $q=5$; $k=0,1,2,...$ for $q=6$). As the limiting case, for $q=5$ and $k=0$, we obtain C_{20} with the icosahedral symmetry group G, but with $G'=D_{5d}$, that could be used as the "brick" for the complete class of nanotubes C_{40}, C_{60},C_{80},... with $G= D_{5d}$, where all of them could be obtained from C_{20} by "gluing" the pentagonal bases. In the same way, fullerene C_{24} obtained for $q=6$ and $k=0$ could be used as the building block for the nanotubes C_{48}, C_{72},C_{96},... The geometrical structure of nanotube class with $G=D_{qd}$ ($q=5,6$) permits the edge coloring preserving the symmetry, so there always exist their isomers with $G=G'$. If the 3-rotation axis contains the opposite vertices of a fullerene, we have biconical fullerenes (e.g. C_{26}, C_{56}) with $G=D_{3h}$, $G=D_{3d}$, respectively. Certainly, after the edge coloring, their symmetry must be disturbed, and for them G' is always a proper subgroup of G. For example, for C_{26}, $G=D_{3h}$, $G'=C_{2v}$. Proceeding in the same way, it is possible to find or construct fullerene representatives of other symmetry groups from the category G_{320}: biconical C_{32} with $G=D_3$, biconical C_{38} or conical C_{34} with $G=C_{3v}$, conical C_{46} with $G=C_3$, or the infinite class of cylindrical fullerenes C_{42}, C_{48}, C_{54}... with $G=D_3$. In general, after edge coloring of their 3-regular graphs, symmetry could not be preserved in all conical or biconical fullerenes mentioned, so their geometrical symmetry is always higher than the chemical.

Different regular homoatomic C plane nets are discussed by T.Balaban. They could be derived in the same way as the general fullerenes: by introducing digons in the vertices of 4-regular graphs or by an edge-coloring of a 3-regular graph, resulting in 4-regular one. For example, we could start from the regular tessellation {4,4}, Archimedean tiling (3,6,3,6) or 2-uniform tiling ($3,4^2,6;3,6,3,6$) that all are 4-regular, and introduce digons in their vertices, or from regular tiling {6,3} that is 3-regular and color its edges. From 3-regular tilings we could also derive "perfect" plane nets in the same way as before. From such finite {6,3} we could simply derive the corresponding "perfect" hexagonal fullerene on torus. The proposed approach could be extended also to the double, triple, etc. torus with g=2,3,... Similar transformations of C nets from one surface to the other (e.g. from plane to cylinder, and then to torus) maybe could explain the formation of certain fullerenes and their growing process. Accepting that the faces could be also heptagons or octagons, from the relationship $2e=3v$ and Euler formula, follows that $n_5-n_7-2n_8= 12(1-g)$. For a sphere without octagons, $n_5-n_7=12$, and for a torus without octagons $n_5=n_7$. To obtain such general fullerenes with a higher degree of symmetry, we could start from different vertex-transitive structures (e.g. uniform polyhedra, stellated regular and semi-regular polyhedra or infinite polyhedra). For example, different uniform 4-valent polyhedra of the type $(3,q,3,q)$ (q=7,8,9,10,12,18) could be used for the derivation of the corresponding "perfect" fullerenes with q-gonal holes on a double torus (g=2). For this, we use the regular vertex-bifurcation of triangular faces, transforming all of them into hexagons. In the same way, the uniform tessellations of the type $(4,q,4,q)$ (q=5,6,8,12) or (5,10,5,10) of a double torus may result in different finite general fullerenes. The interesting classes of infinite general fullerenes with non-euclidean plane symmetry groups could be derived from the tessellations of hyperbolic plane H^2. For example, from the uniform tessellation (3,7,3,7) we derive the infinite perfect 6/7 fullerene in H^2 with heptagonal holes.

Scientific Application Semiconductor Superlattices in Nanotechnology

A brief review of semiconductor superlattices is presented—their definition, basics of theory, types of materials and methods used for their preparation, the most important and unusual electrical and optical properties, which have already found their practical applications.

When alternating layers of two semiconductors with different band gaps are deposited and thickness of these layers lies between 1-10 nm (which corresponds approximately to 10–100 atomic layers), what we obtain is semiconductor superlattice. The mostly known example of such a structure is GaAs/AlAs system depicted schematically together with its corresponding potential profile. Differences in band gaps of materials lead to discontinuities in valence and conduction band and a set of square potential wells (low band gap semiconductor) separated by potential barriers (high band gap semiconductor) forms. Let us consider one of these wells from the quantum mechanical point of view. If we assume that an electron mean free path within a well is substantially longer than well width and the barrier width is infinite, what we obtain is a classical problem of a particle confined in a square well of finite depth. Unfortunately, in the real-life systems we cannot obtain perfect separation of adjacent wells and this results in coupling of their wave functions. Thus, instead of discrete levels, this yields us subbands. An example of theoretical analysis of the GaAs/AlAs system shown, where an equal width, a, of well and barrier layer is assumed. Two main features common for other superlattice structures are easily seen when a is decreased. First, an energy distance between ground state and subsequent bound states increases, which happens because the well width is getting smaller. Second, the discrete levels split into the subbands as a result of thinner barrier layers and poorer separation of neighbouring wells. At this point, it is also important to notice, that the energy discontinuity, i.e. well depth, cannot be simply derived from such material constants as an energy gap, the conduction and valence band structure etc. in each semiconductor because of the lack of knowledge of the common energy reference. This parameter must be either calculated or measured.

Historically, the crystalline superlattices where the first to produce, starting from pioneering work of Esaki and Tsu (1970). What is crucial in fabrication of such structures is nearly perfect match in their lattice constants. Otherwise the density of defects becomes so large that it results in strong electron scattering and obscures quantum phenomena. This is why a molecular beam epitaxy technique is utilised in their deposition and the number of materials, which can be combined to create superlattices, is restricted. The best known and studied are GaAs/AlAs and GaAs/AsxGa1-xAs, InAs/GaSb are also worth mentioning. If interface defects cause such difficulties, why not to approach the problem from the opposite side—let us build the superlattice from the materials, where lattice constants no longer exist, i.e. amorphous semiconductors. This idea, apparently irrational, proved absolutely sensible in 1983, when Abeles and Tiedje reported in their paper fabrication of a-Si1-xNx:H/a-Si:H and a-Ge:H/a-Si:H systems. They stated that the strict requirements concerning lattice matching in crystalline superlattices can be relaxed due to amorphous structure by its very nature and a hydrogen content which plays an important role by passivating coordination defects. Since then, many other materials were combined to produce superlattices : a-Si:H/a-C:H, a-Si:H/a-SiN:H, a-Si:H/a-SiNx, a-Si:H/a-SiOx, a-Si:H/a-Ge:H, CdSe/SeTe, CdSe/Se, a-Si3N4:H/a-Si:H, a-C:H superlattices where the band gap is tailored by changing deposition conditions. An impressive, comparing to the crystalline case, number of types of amorphous superlattices results from the fact that in this case we have much more freedom in choosing components of our systems as we are not limited by the lattice matching condition.

It is also important to emphasise that the techniques utilised to produce amorphous films -plasma deposition, thermal evaporation, sputtering—are relatively cheaper than molecular beam epitaxy and easier to implement in industrial applications. On the other hand, in case of amorphous multilayers it is more difficult to obtain abrupt interfaces between subsequent layers. Different methods are employed : pulsed application of microwave plasma during continuos r.f. glow discharge in Si2H6/N2, by which N2 is decomposed and incorporated into the growing film, nitridation of a-Si:H to attain a-Si3N4:H, lowering the growth rate in order to avoid fabrication of transition layers when the deposition conditions change, quick exchange of source gases. However, sometimes smooth transition from one layer to another is intentionally introduced, which may lead to sinusoidal potential profile, instead of a step-like. For the sake of convenience, electrical properties, particularly conduction phenomena, may be considered in terms of what is happening when we study currents flowing in-plane (parallel to the layers) or perpendicular to them. As it has been stated above, quantum confinement causes shift of the first bound state allowed for charge carriers. This must of course result in the increase of an activation energy of conductivity, E_a. In general, the smaller well width, the higher E_a, thus in-plane electrical conductivity decreases as compared to the unlayered material. However, the opposite situation may occur as well. This is because in amorphous semiconductors, unlike crystalline ones, there is a finite density of states within a mobility gap. When two such materials are put into contact, charge carriers may be transferred through an interface, building up accumulation and depletion layers, and as a result, the conductivity in one of the materials increases to such a degree that obscures quantum effects. A completely different picture appears when conductivity in the direction perpendicular to the layers is considered. If there is no electric field, the current does not flow of course. When the small field is applied, a tunneling current can flow through the barriers. As the field is increased, an alignment between the energy subbands in adjacent wells is getting worse so observed maximum is a compromise between a driving force (electric field) and subbands mismatch. When the field is increased further, the subbands misalignment is so high, that charge transport becomes very inefficient and the current—field characteristics exhibits minimum.

If the field becomes higher, we can have the situation, that the lower subband in one quantum well is aligned with the upper one in the adjacent well. Transport is efficient again. The most important feature of this characteristics is a region of negative differential resistance where an increase of the electric field leads to a decrease of the electric current. More detailed theoretical considerations of this matter can be found elsewhere. This kind of behaviour has been observed in both crystalline and amorphous multilayers; however, in the crystalline case it is more distinct and resembling model behaviour. An example of characteristics for DLC superlattice is shown, where current-voltage profile together with its first and second differentials is presented. Bumps observed on the current-voltage curve are typical of resonant tunneling. The start of the negative differential resistance region at 3.1 V corresponds to the loss of subbands alignment. It should be noted that the conductivity in the direction perpendicular to the superlattice layers is by its nature very similar to what Esaki found in *p-n* germanium junctions and what is now applied in construction of tunnel diodes.

We have discussed electrical and optical phenomena which characterise superlattice structures which found their explanation on the basis of quantum size effect. However, alternative interpretations of experimental data do exist as well. Arsenault et al. claim that bumps observed in the current-voltage characteristics for perpendicular conductivity in amorphous multilayers, usually attributed to resonant tunneling can be explained in a different way. They suggest that at least one of these bumps is simply a transition between two transport mechanisms: space charge limited current (low electric field) and multiple hopping (high field). The most convincing justification of their hypothesis is that they observed the same bumps for both quantum well structures and single barriers. Quantum mechanical explanation of the *blue shift* has been criticised in a number of research papers as well. In our discussion it has been assumed that described multilayers were periodic system, characterised, from the geometrical point of view, only by one well and one barrier width. However it does not have to be the only approach. Another and very original view is presented in some researches. Superlattices of these researches consist of the layers of a-Ge and a-Si deposited using dual-target magnetron sputtering. The most interesting thing is their topology generated by Cantor sets. The modulation algorithm according to the so called "Cantor middle—third" set together with corresponding multilayer structure is presented.

Optoelectronic Space Applications

Diffractive optical elements, optoelectronic transducers and photonic components, which play an important role in optical data communication, can be substantially improved by lateral nanostructures. With the development of lateral optoelectronic nanostructures the way to controllable diffractive optics is paved. For this, elements with specific interference structures are necessary, which act as specific and possibly controllable transmission or reflection filters. Nanostructured optoelectronic components (e.g. quantum well or quantum dot lasers, photonic crystals) offer large market potentials in the future, e.g. for optical data communication or in the range of consumer electronics (for example laser television).

Nanostructured optoelectronic components offer promising application in space in the fields of optical satellite telecommunications or sensor technology (infrared sensors, high resolution charge-coupled devices, etc). With optical wireless data links (OWL) for intrasatellite communication as well as optical intersatellite links, significantly smaller and lighter devices a higher bandwidth could be realised in comparison to conventional microwave communications. Optical intersatellite links were demonstrated in the frame of the ARTEMIS mission of the ESA. For the data transmission, extreme frequency-stable solid state lasers (Nd:YAG lasers) are used, which are pumped with diode lasers.

The German company, Tesat, is a leading manufacturer of laser terminals for optical intersatellite communication. Such laser terminals are also interesting for scientific applications; for example, as an injection seeder for a satellite-based Doppler-Lidar (ALADDIN), as a satellite-based measuring device for gravitation wave detectors (LISA, with SMART as demonstration mission), or as a frequency normal for a satellite-based FT-spectrometer (POISON).

Semiconductor quantum dots, which have been manufactured in high quality by means of self-organisation for about the last 5 years, offer a new degree of freedom in selecting the working wavelength of photonic elements. They allow manufacturers to cover almost completely the entire spectral region from the ultraviolet to the far infrared, with a small number of substrate materials. Further advantages of quantum dot lasers are a small energy consumption through low threshold current densities, a high modulation range for high-speed applications as well as an improved temperature stability. Beyond that, an improved radiation hardness has already been proven by quantum dot lasers compared with quantum well lasers. First commercial uses of quantum dot lasers are expected in 2003.

Due to their radiation hardness and the low energy consumption, quantum dot lasers in principle are relevant for space applications, e.g. as pump lasers for solid state lasers, which are needed for different applications. In order to realise the potential of quantum dot lasers in space applications, appropriate measures have to be accomplished by the space industry for the specification, system integration and space qualification.

Photonic crystals are a further example of nano-optoelectronic components with application potential in optical data communication. Photonic crystals exhibit a periodic refractive index and possess an analogy to semiconductors in electronics, a "photonic band gap" for certain frequency in the visible and IR wavelength ranges. The lattice constant of photonic crystals lies in the range of half the wavelength of the light in the medium. For visible light this means that for the production of photonic crystals, a precision within the range of 10 nm is necessary. Two-dimensional structures today can be routinely manufactured with high precision.

At present, intensified efforts are made for the development of three-dimensional photonic crystals, e.g. with utilisation of lithography and self-organisation procedures, in which nanoscale colloids (e.g. from polymers or silicates) arrange spontaneously to a cubic lattice. These lattices are used as templates for lattices from more interesting materials such as metals and metal oxides. Three-dimensional photonic crystals would open up new possibilities in optical data communication (light could be guided and branched to arbitrary directions) and offer in principle the potential for the realisation of purely optical circuits (optical computing). Such photonic transistors are however at present still very far from realisation. In the long run, photonic crystals will find applications in optical satellite communications.

IR sensors offer a multiplicity of application potential in space, e.g. for the satellite-based earth observation and atmosphere research, for astronomy, as navigation aid for space systems or for optical data communication. Approaches for the miniaturisation and further improvement of infrared sensors are based among other things on the application of two- (quantum well), one- (quantum wire) or zero-dimensional (quantum dot) nanostructures. With the help of quantum well or quantum dot structures the detection characteristics of IR sensors can be adjusted selectively to the relevant spectral region (band gap engineering). Quantum well IR sensors, based on GaAs are developed, for example, by the Center for Space Microelectronics Technology of NASA for special space applications. This quantum well infrared photodetector (QWIP) consists of a GaAs-layer, which is embedded, sandwich-like, in two $Al_xGa_{1-x}As$ layers. The characteristics of the quantum wells can be adjusted by varying the thickness of the GaAs layer and the composition of the barrier layer. By means of molecular beam epitaxy, nm-thick layers can be produced on large areas with atomic precision. Ga-As QWIP can also be used for long-wave IR radiation > 6 μm.

The Technical University of Munich pursues a different approach in the frame of a BMBF (Bundesministerium für Bildung und Forschung—Federal Ministry of Education and Research) joint project on the self organisation of Si/Ge islands on silicon. The research activities are focused on controlling the characteristics of epitactical, defect free Si/Ge islands on silicon substrates, which can be produced with the self-organising, parallel Stranski-Krastanow procedure in the material system Si/SiGe. The objective here is to develop coupled systems with several layers of quantum dots in a homogeneous layer system, which exhibit new functions by charge transfer and electrostatic coupling and can be used as optical detectors particularly in the mid IR (MIR) range. With combined QD/QW structures a 50-fold increased photoresponse can be obtained compared with sole quantum dot structures. Expected advantages of QD MIR (mid-infrared) detectors to be mentioned are the extended spectral range, a high durability and reproducibility, radiation hardness and a low dark current. For the year 2003, the realisation of a prototype 2-colour IR detector in co-operation with DaimlerChrysler was done.

Nano-optics is a valuably novel class of technology that takes advantage of light's unique interaction with subwavelength, nanoscale patterned materials, and nanotechnology enabled fabrication methods to create a broadly applicable optical device and manufacturing platform. In optical circuits used in consumer electronics, communications, and security applications, nano-optic devices offer significant benefits by virtue of their novel behaviour, small size, ease of integration, and lower overall costs. Through these virtues, nano-optics delivers compelling alternatives to conventional bulk optical solutions, and offers functionality and form factors that conventional optics cannot match. The following is a basic review of nano-optics theory and applications.

The unique optical properties of nanooptic devices derive from appropriately patterning material on the nanoscale with structures whose critical dimensions are sized to be several times less than the wavelength of light at which they operate, thereby creating optical materials with highly useful modified optical functions. Through a proper combination of materials and structures, nano-optic devices can perform any passive optical function including polarisation filtering, phase retardation, spectral filtering, and propagation management (e.g., lenses and beam splitters). Various functions can be designed for both free-space- and waveguide- based applications. The former is far and away the most common in commercial applications. Nano-optic devices also can be designed to operate over any wavelength range. The basic methodology is applicable to UV, visible, and IR wavelengths with appropriate variations in structural dimensions and materials. Additionally, they can form the core of an optical system. Practical nano-optic applications require a complete optical system comprised of the nano-patterned material and adjacent materials including the optical substrate and thin-film coatings. The nano-optic structures are defined as a combination of material and physical attributes, as illustrated. Attributes can be divided in to three types:

- *Structural:* pattern (e.g., linear, mesh, and circular), dimensions (e.g., period, thickness, and duty cycle), and spatial variations (e.g., "chirping", arrays, and multiple layers).
- *Material:* nano-structure material (e.g., dielectrics, metals, and polymers), interstructure fill material, adjacent thin film materials, and substrate material (e.g., glass, plastics, dielectrics, and crystalline material).
- *System:* relationship of the nano-optic structures with the optical beam path and any actuator materials and structures (e.g., liquid crystal, solid state materials, electro-optic polymers, and MEMS).

Subwavelength structure size is a key to nano-optics' optical and physical advantages. Nano-optic elements employ microstructures one or more orders of magnitude smaller than the wavelengths of the incident light—with dimensions typically on the order of 10s to a few 100s of nanometers. At such dimensions, these structures may interact with light both according to the principles of classical optics and also in a realm of comparatively small-scale local interactions; rigorous application of the boundary conditions of Maxwell's equation describes the interactions of light with nano-optic structures. For very small structures, single electron or quantum effects may be observed. As a consequence, many nano-optic devices exhibit unique optical behaviours—such as high polarisation discrimination in a micron-thin space, achromatic phase retardance, and single layer combinations of optical functions like polarisation and spectral filtering—either by coincidence or intent. These nano-optics provide useful alternatives for optical circuit design; others can simply be designed as drop-in replacements for classical bulk optics. In addition to unique behaviours, the devices also achieve their optical effects over relatively short distances—nano-optic devices are generally very thin, usually one micron or less in thickness. This has strong positive repercussions for integrated optics since nano-optic structures can often be applied, similar to thin films, as a "coating layer" during a complex manufacturing process. Single electron and quantum effects also enable optical performance to be readily customised. Modifying material properties by adding nano-structures results in a continuous spectrum of optical functionality—essentially this combination creates hybrid, "artificial" optical materials—that allows performance specifications and optical wavelengths to be incrementally and accurately varied to meet specific application requirements.

Nano-optic structures can be realised through a variety of manufacturing methods. For commercial usefulness, an optical device is not optimally practical unless it can be manufactured in high volumes, in multiple application specific variations, and readily integrated with other relevant technologies. The first criterion is necessary to achieve economies of scale; the second, to ensure that the relevant market opportunity is broad enough to be interesting; and the third, to facilitate easy integration of nano-optic elements into more complex systems. The available fabrication approaches (self-assembly and nano-lithography) achieve these objectives to varying degrees. Some materials can be caused to form regular, nanoscale structures under appropriate, controlled conditions—a.k.a. self-assembly. The difficulty with this approach is lack of flexibility in achievable structures and usable materials, which limits the functions that can be realised. Nanolithography, which is akin to standard semi-conductor manufacturing, generally uses some method to create an image in a polymer resist layer; this image is then used as an etching mask to transfer the nanoscale pattern into the target material. Nanolithography methods include interference lithography, e-beam (electron beam) lithography, and nanopattern replication. Overall, each of these methods is optimal for some set of nanostructure patterns, materials, and volume requirements. As a manufacturing method, nanolithography supports a range of integration modes for optical circuits:

- Optimisation of traditional multi-element, multi-technology optical circuit architectures is possible by taking advantage of the unique optical behaviour of nano-optics.
- Monolithic "self-integration" can be achieved by layering one atop the other to create aggregate optical effects.
- Spatial integration can be accomplished by organising varying optical functions into an array structure via nanopattern replication.
- Hybrid integration is achieved by adding a nano-optic layer(s) to functional optical materials.

- Integration of nano-optics with optically active layers, such as liquid crystal or MEMS structures, can create electronically controllable optical devices.

Integrated optical devices and subassemblies based on nano-optic elements realise a range of benefits. Compared to conventional bulk optics, their size and optical properties make alignment in assembly more forgiving, less labor-intensive, and less costly. When physically combined with other nano-optics or other materials in manufacturing to create monolithic integrated optics, this eliminates the complication and cost of multidevice lamination, while improving reliability. Also, arrayed nano-optic devices can be applied in multi-beam or multi-path optical circuits, eliminating the need to individually align discrete optics. In general, the process by which nano-optic elements are fabricated is flexible and robust, allowing for cost-saving manufacturing process integration in creating hybrid and monolithic optics; again, the resultant optics are often smaller, more robust, more functional, and easier to assemble. Because nano-optic devices are manufactured utilising wafer-based processes proven in semi-conductor fabrication, this enables flexible sharing of manufacturing capacity and supports high-volume manufacturing throughput.

Nano-optics draws on a growing "design vocabulary" that is functionally equivalent to the design libraries used in semiconductor applications. Multiple design elements can be used with complex, multi-step, and multi-process methods to create particularly sophisticated nano-patterns and multi-layer devices. The broad range of different nano-structure patterns, materials, and modes of integration means that there are no inherent limits in the functionality and operational wavelength of nanooptic devices. These attributes are solely driven by the needs of the application and the optical circuit. To date, the commercialisation of nano-optics has demonstrated the portability of the manufacturing platform and its products across multiple markets including optical data storage, digital imaging, projection display, and telecommunications. As nano-optic technology continues to propagate, it promises to further expand both in applications and in benefits realised by users of nano-optic based devices.

Interference Lithography: Readily creates interference patterns with useful dimensions using UV light sources. The benefit of this method is simplicity; the difficulty is in creating complex shapes and arrays.

E-Beam Lithography: Writes arbitrary, complex patterns using a focused beam of electrons. The benefit is that almost any pattern can be created; a drawback for commercial production is that individual wafer processing times can be fairly long (tens to thousands of hours).

Nano-Pattern Replication: Relies on other methods, e.g., the two above—to create a master plate capturing the inverse of the desired pattern. A printing-like transfer method then patterns the polymer resist. The benefit is high-fidelity reproduction allowing extremely small feature size, complex patterns, effective material independence during repetitive production, and for different structures in the same production line. It is not applicable for limited runs or when only a fixed interference pattern is required.

A nano-optic polarisation beam splitter/combiner (PBS/C) provides 180 degrees of effective separation—meaning that one polarisation is transmitted and the other is reflected—via a submicron-thick nanoustructural layer. This device's beam separation and thin form allow adjacent components to be closer coupled compared with conventional optics, reducing existing optical path lengths by up to one-half. In turn, shorter optical path lengths reduce beam divergence, improving coupling efficiency and reducing power loss. Many optical circuit designs also incorporate a wave plate adjacent to the PBS/C.

Digital Imaging, Communications and Sensors: Physically, a nano-optic enabled variable optical attenuator (VOA) consists of a crossed pair of polarising nano-structures sandwiching a liquid crystal cell; applying an electrical field across the liquid crystal will rotate the liquid crystal molecules and control the fraction of light that is passed through the structure. This tunable optic addresses the frequent need to control and block the output power of polarised light sources, such as optical transceivers, in order to dynamically optimize their signals and to facilitate maintenance. Because the nano-optic polarizer is directly integrated into the liquid crystal cell construction, this allows the entire component to be less than 1 mm thick.

Nano-Enabled Drug Delivery Systems

Nanomedicine is widely expected to become a multi-billion dollar business over the next decade. Although some of the applications for nanomedicine are quite futuristic, others are on the verge of generating significant revenues. One of these is nano-enabled drug delivery systems which promise to increase the bioavailability of drugs and reduce the likelihood of toxicity. Researches pinpoint several areas where nano-engineering of drug delivery systems is leading to greater control over active compounds, resulting in the physician's ability to send the smallest dose possible in order to achieve the desired result in the patient. This chapter reviews the main opportunities that are emerging for nano-enabled drug delivery systems. In most cases, these are based on delivering drugs in nanoparticulate form. Most drugs perform better as nanoparticles, because they can be targeted better and there are fewer side effects. In addition, by using smaller amounts of drugs, treatment costs may be reduced. For example, according to researches, patients show a strong preference for nanoparticle inhalers as an alternative to the widely used injectable methods, which may permit far lower doses of expensive protein-based drugs like insulin.

It is expected that the dosing benefits of nano-enabled drug delivery systems will be extended to compounds used in treating both infectious disease and cancer, and has identified six types of drug delivery systems in which nanotechnology is likely to have an impact.

Although injectable delivery systems have some inherent discomfort for the patient and other types of delivery may be preferred by both the patient and the physician, nanotechnology may be applied to produce improvements and additional revenue streams for the drug company. A case in point is Johnson & Johnson, which, at the beginning of 2005, announced that its Élan NanoCrystal technology would be used in a Phase III clinical trial for an injectable formulation of paliperidone palmitate, an anti-schizophrenia drug. This is a new "nano formulation" of an older drug which overcomes the original's insolubility, by reducing the particle size to under 200 nm. This new formulation enables the product life cycle to be extended by a significant amount.

Implantable drug delivery systems are often preferable to the use of injectable drugs, because the latter frequently display first-order kinetics (e.g. initial concentration is elevated, which then drops exponentially over time). This may cause difficulties with toxicity, if and when the peak concentration remains above the therapeutic range. Conversely, drug efficacy often diminishes as the drug concentration falls below the targeted range. By contrast, implantable "time release" systems may help minimize peak plasma levels and reduce the risk of adverse reactions; allow for more predictable and extended duration of action; reduce the frequency of re-dosing; and improve patient acceptance and compliance. To these advantages nanotechnology can add the benefits listed at the beginning of this White Paper. Implantable drug delivery systems are expected to be used for the delivery of proteins, hormones, pain medications, and other pharmaceutical compounds. An example of how nanotechnology is already making its impact felt in this part of the drug delivery business is provided by pSivida's BioSilicon product. BioSilicon is a novel nanostructured material that effectively stores an active compound in nanosized pockets that controllably release minute amounts of drug as the silicon dissolves. pSivida is currently exploring biodegradable implantable methods for tissue engineering and ophthalmic delivery. Nano-implants will also be used in the not-too-distant future for treating cancer. Among the first nanoscale devices to show promise in anti-cancer therapeutics and drug delivery are structures called "nanoshells," which it is believed may afford a degree of control never before seen in implantable drug delivery products. Nanoshells typically have a silicon core that is sealed in an outer metallic core. By manipulating the ratio of wall to core, the shells can be precisely tuned to scatter or absorb very specific wavelengths of light. For example, Gold encased nanoshells have been used to convert light into heat, enabling the destruction of tumors by selective binding to malignant cells. A physician can use infrared rays to pass harmlessly through soft tissue, while initiating a lethal application of heat when the nanoshells are excited. Some researchers are experimenting with temperature-sensitive drug delivery control methods, using nanoshells that release their payload only when illuminated with the proper infrared wavelength.

The vast majority of consumers prefer an oral drug delivery system to implantables or injectables. With this in mind, various development groups are working to enhance traditional oral delivery systems with nanoengineered improvements. There are some areas where nano-enhanced drugs could make a big difference in increasing oral bioavailability and reducing undesirable side effects. By increasing bioavailability, nanoparticles can increase the yield in drug development and more importantly may help treat previously untreatable conditions. Because of the blood brain barrier (BBB) many new chemical entities aimed at treating brain disorders have proved not to be clinically useful. By contrast, nanoparticles have been demonstrated to cross the BBB with little difficulty. Companies such as Germany's NanoPharm have developed systems capable of reaching the brain for anesthesia (Dalargin, an analgesic), cancer drugs, and various therapeutics. The company claims several advantages over existing systems, including:

(1) no requirement to open the BBB,
(2) the ability to use potentially any drug, whether hydrophilic or hydrophobic, and
(3) the drug does not need to be modified.

Meanwhile, researchers at the University of Texas at Austin have described a means of using nanospheres for oral drug delivery. These nanosphere carriers are derived from hydrogels, which are highly stable organic compounds that swell when their environment becomes more acidic. They have been successfully formulated into controlled-release tablets and capsules, which release active compounds when the hydrogel body swells.

NanoMarkets' research indicates that nanomaterials provide a unique opportunity for rapid topical delivery of active compounds. Given their very small size, nanoparticles are able to enter human tissues and cells quickly. Companies such as Novavax have already developed regulated topical systems that take advantage of the unique properties of micellar nanoparticles. It has two hormone replacement therapies dubbed Estrasorb (which received FDA approval in October 2003) and Androsorb (which successfully completed Phase I human trials in 2003). The number of FDA-approved polymers available for use on-skin is increasing rapidly. With this in mind the industry will be presented with opportunities to create new transdermal platform designs that improve on-skin properties and diffusion of active molecules from patches. This interesting trend is expected to result in smaller and less invasive patches that increase the universe of available drug candidates. In some cases, electronics are even being integrated into patch-like platforms involving wound care, monitoring, and diagnostic methods. As such, patch-like platforms will play an important role in healthcare to treat, measure, diagnose, and generally improve quality of life.

Colloidal dispersions have been demonstrated to remove potentially lethal compounds from the bloodstream, including high concentrations of lipophilic therapeutics, illegal drugs, and chemical and biological agents. A team of scientists at the University of Florida and Clarkson University in Potsdam, N.Y., has demonstrated favourable results to this end, using biocompatible microemulsions. These oil-in-water systems have a rapid and efficient absorption capacity for many target molecules that are frequently overdosed, whether this be intentional or accidental. The microemulsions use a polymeric surfactant, in combination with an ionic co-surfactant. Other oil compositions have been made with varying absorption characteristics for different toxins, in all cases avoiding hemolysis and unwanted blood coagulation. The mechanism by which nanoparticles trap toxins involves aromatic or cyclodextrin receptors with electron deficiencies, which form complexes with selected toxins, sometimes referred to as "benzenoid moieties. The analysis above establishes that the impact of nanotechnology on the drug delivery business is likely to be very broad in three senses:

- Nanotechnological improvements promise to impact almost every common mode of drug delivery oral, implantable, injectable, etc.
- Such improvements will themselves cover a lot of ground, including lower drug toxicity, lowering cost of treatments, improved bioavailability and extension of the economic life of proprietary drugs.
- Nano enabled drug delivery will provide improvements in the treatment of some of the most widespread diseases, such as cancer and diabetes.

This is an impressive list. Also impressive is the fact that many of the categories of nano-enabled drug delivery systems are already close to or at the point of marketing. It is believed that not only will the nano-enabled drug delivery market be one of the first true nanomedicine markets to evolve, but as it does so, the revenues from nano-enabled drug delivery systems will be quite large.

7

Major Applications and Utilities of Nano-optics: An Overview of Global Trends

An Introduction to Nano-Optics and Near-Field Optical Microscopy

Today we encounter a strong trend towards nanoscience and nanotechnology. This trend is motivated by the fact that as we move to smaller and smaller scales the underlying physical laws change from macroscopic to microscopic. The exploitation of quantum effects for technological applications is the most obvious driving force behind the current miniaturisation. The recent rapid advances are due in large part to our newly acquired ability to measure and manipulate individual structures on the nanoscale (scanning probe techniques, optical tweezers, and high-resolution electron microscopes). The increasing trend towards nanoscience and nanotechnology makes it necessary to address the key issues of optics on the nanometer scale. Since the diffraction limit does not allow us to focus light to dimensions smaller than roughly half a wavelength, traditionally it was not possible to interact selectively with nanoscale features.

In recent years several new approaches have been put forth to 'shrink' the diffraction limit (confocal microscopy) or to even overcome it (near-field microscopy). For example, with our tip-enhancement technique we are able to do Raman spectrocopy and multiphoton fluorescence imaging with a spatial resolution of less than 20nm. To date, this is the highest optical resolution of a spectroscopic measurement. The reason for the hard effort to advance the field of optics to the nanometer scale is the fact that the energy of light lies in the range of electronic and vibrational transitions in matter. Therefore, the interaction of light with matter renders unique information about the structural and dynamical properties of matter. These unique spectroscopic capabilities are of great importance for the study of biological and solid-state nanostructures. We are applying near-field optical techniques to probe complex semiconductor nanostructures as well as individual protein molecules. We plan to explore the possibility to optically interact with semiconductor nanostructures on length scales smaller than the extent of their quantum wavefunctions. Probing and manipulating these wavefunctions might open up various exciting applications such as data storage and optical switching based on quantum logic.

Nano-optics addresses the broad spectrum of optics on the nanometer scale covering technology and basic science. On the technological side, we find topics like nanolithography and high density optical data storage. On the basic sciences end, we might mention atom-photon interactions in the optical near-field and their potential applications for atom trapping and manipulation experiments. Compared with free propagating light the optical near-field is enriched by so-called virtual photons. These are the same sort of particles responsible for molecular binding (van der Waals/Casimir forces) and are therefore promising for selective probing of atomic structures. The consideration of virtual photons in the field of quantum optics will enlarge the range of fundamental experiments and applications. Nano-optics deals with the controlled manipulation of evanescent optical fields and their exploitation in investigations of nanometer scale structures. The basic idea is to detect the evanescent field distribution, which is significant in the near-field region that extends to only a few tens of nanometers from the sample surface. The detection of the evanescent waves is typically done with a nanoprobe. The probe acts as a local scattering center transforming some of the evanescent field components into homogeneous (propagating) waves that can be detected in the far zone. By scanning the nanoprobe over the sample surface, the evanescent field distribution can be detected with a resolution limited mainly by the size of the nanoprobe and the scanning height. In practice, the resolution achieved in scanning near-field optical microscopy (SNOM) is on the order of a few tens of nanometers, i.e., the technique offers an order of magnitude improvement to the best conventional far-zone measurements. Our research has concentrated mainly on three specific areas.

The main efforts have been focused on developing the shear-force feedback system of our SNOM apparatus, on the near-field investigation of a broad-area semiconductor laser, and on the theoretical modeling of the near-field-matter interaction. Insight into the origin of the high spatial resolution in a SNOM measurement may be gained by investigating the contributions of the homogeneous and evanescent electromagnetic waves separately in some selected geometries. The simplest model for the sample-probe system in SNOM can be obtained by considering the probe tip and the sample structure as being composed of a collection of point-dipoles. Of fundamental importance in this approach is to understand how the electromagnetic field of a single dipole is decomposed into its evanescent and homogeneous parts. Knowing this one can then proceed to generalize these results to systems containing several point-dipoles by including the couplings between the dipoles in a self-consistent calculation. The point-dipole model has been widely employed as an aid in discussing the properties of near-field microscopy. Caution should, however, be exercised when interpreting the results. The point dipole does not offer an adequate model for describing mesoscopic structures in near-field investigations, since strong evanescent field gradients exist close to the sources. By applying the point-dipole model, however, we can investigate how the nanoprobe affects the near field and how the resolution depends on the scanning height of the probe over the sample in the different operation modes of SNOM. Furthermore, we can study the effects of different polarisation directions of the nanoprobe and the external polarising field on the resolution of a SNOM measurement. In particular, we can see the central role of the evanescent field components in breaking the classical diffraction barrier. The homogeneous, evanescent and total field distribution at the probe site when a z-polarised nanoprobe is scanned at the height of 40 nm over the sample. The probe dipole locally illuminates the sample and the same probe is also used to collect the scattered light. As a result of the local interaction between the probe and the sample dipoles the scanned homogeneous field distribution now reveals the same information about the sub-wavelength scale structures of the sample as the evanescent field, although in lower magnitude. Nanooptics, i.e. optical spectroscopy with a spatial resolution of <100 nm, opens up fascinating insights into the physical processes occuring on a nanometer scale. In our group, we apply this technique to investigate *single* semiconductor quantum dots, sometimes refered to as 'artificial atoms' because of their d-like density of states.

"Embossing" is an ancient technical term in English traceable to the period 1350 to 1400, and used by Chaucer in its present sense. The technical meaning, creating a raised design or pattern by pressing a die having the raised negative image of the design against the material to be patterned, has not changed during the ensuing 650 years. What has changed are the applications, materials used, and the scale of the design. Today's embossed patterns may have features measured in nanometers, bringing new capabilities to integrated optical systems. Nanoembossing, often referred to as nano-imprint lithography (NIL), can rapidly pattern relatively large area substrates with feature sizes below 100 nm. Creating patterns at this scale previously required far slower and more costly techniques such as electron-beam writing. NIL, used in combination with related technologies, can create an entire wafer of integrated optical structures with nanometer features in minutes. An array of replicated lenses formed on the surface of a 4-in. silicon wafer. Wafer-level NIL promises reductions in the size, cost and complexity of integrated optics, while improving performance. Integrating embossed components on a wafer offers major size reductions. Producing truly "monolithic" optical systems could reduce centimeter-sized systems to millimeter-sized die. Optical performance is improved by locating components physically closer together, reducing the optical path length and the attendant losses. For example, optical elements such as lenses can be fabricated directly onto optical sources, such as lasers.

Surprising to the non-expert, the small physical size of the nano-structures also creates optical elements that differ significantly in some characteristics from normal macro-optics. Optical elements with feature sizes smaller than the wavelength of the incident light, sometimes referred to as sub-wavelength elements, interact differently with the light than optical elements with larger feature sizes. More specifically, light interacts in different ways with periodic structures having spacing smaller than its wavelength. The altered interactions create a variety of devices with characteristics that differ from similar but larger structures. The sub-wavelength optical elements offer some new, desirable characteristics. Sub-wavelength grating structures, for example, may have a wider acceptance angle than conventional gratings. Wider acceptance angles relieve manufacturing tolerances and may permit automated assembly, further reducing costs. Sub-wavelength gratings also perform more consistently over a broader range of wavelengths than conventional gratings. NIL has demonstrated a growing variety of nano-optical components, including gratings, filters, anti-reflective coatings, refractive and diffractive components, fresnel lenses, waveguides, and lens arrays. With all of these potential products and proven benefits, why has nano-optics languished so long in the laboratories? Because formidable difficulties must be surmounted to replicate these tiny dimensions in volume manufacturing processes. To understand the manufacturing challenges that are now being overcome, look at the basic processing techniques that have been developed. NIL in its various forms has five proçessing steps:

(i) Design and fabricate a master stamp with features patterned by high-resolution techniques.

(ii) Impress the master stamp on a layer of suitably prepared material.

(iii) Maintain the master stamp pressure while the material first flows to duplicate the stamp, and then solidifies to preserve the duplicate pattern.

(iv) Separate the patterned replica from the master stamp, without damaging nanoscale features.

(v) Post-process the replica further for its desired optical function.

The starting point for NIL is the creation of a master stamp or mold. The ultimate resolution capability of the embossing depends solely on the master stamp. The design and fabrication of the master stamp is critical to minimum feature size, uniformity, fluid flow, and ease of separation. Designing a stamp to allow quick and complete fluid flow of the softened material is challenging. Once the design is satisfactory, it must be transferred to the master. Fine-line approaches, such as electron beam, are used to create nanoscale features on the master stamp. Electron beams have fine-line capability, but direct e-beam writing has a working area only a few millimeters square, and is relatively slow. Stamps used for hot embossing or for injection molding are commonly made from nickel or steel. Stamps for UV embossing must be of UV-transparent materials. While the completed master stamp could be directly used to produce many replicas, wear or deterioration through repeated use would require creating a new master stamp. In practice, to preserve its pristine quality, the completed master stamp generally is cloned, using the replication techniques described here. Cloned sub-masters are used to produce the nanoscale features in volume production, either simultaneously over a whole substrate, or repetitively, at chip size, onto a substrate. The substrates may be silicon, quartz, glass, or other materials. Replication layers are generally coatings of heat-curable or UV curable polymers, with carefully controlled coating uniformity and thickness. While polymers are common materials for embossing, they may not offer the dimensional precision or the long-term stability under challenging environmental conditions demanded of some system components. This need has been filled by the development of special materials, such as of sol-gel organic/inorganic composite materials. These composites can be UV cured or thermally cured like polymers, and after curing display optical and mechanical properties close to those of glass.

Embossing and curing the pattern on the substrate varies with different processes. Once the coating is cured, the stamp must be separated from the cured material, without damaging the replicated pattern. As feature size becomes smaller, clean separations are more difficult. A special coating may be applied to the stamp before molding, to facilitate separation. The embossed wafer may be post-processed by etching, deposition, or by repeating further lithographic processes. The cured patterned material can be the etch resist, for reactive ion etching of the underlying substrate. Alternatively, the patterned material may be used as a resist for metal deposition and lift-off. Post-processing might also include additional lithography, further depositions, and the addition of antireflective or protective coatings. Hot embossing of wafers uses the cloned master die as the wafer stamp. Wafers are first coated with a thermoplastic polymer, that is, a polymer that softens when heated. The wafer is positioned in a controlled-atmosphere stamping chamber. After heating the wafer, the stamp is brought down with pre-determined pressure and temperature to soften the polymer and simultaneously form the image on the whole wafer. After allowing time for polymer flow during heating, the stamp and wafer are cooled to harden the image. To maintain the pattern, pressure must continue to be applied until both stamp and substrate are cooled below the thermoplastic softening point.

The hot-embossed wafer may be used for its embossed features, or it may be post-processed as a resist for etching, material deposition, and lift-off patterning. For resist use, curing may be followed by a light etch to remove the remaining traces of materials from the clear areas. An alternative hot embossing approach is step-and-repeat stamping, using a smaller than wafer size master stamp. Step-and-stamp imitates the action of step-and-repeat optical steppers used in semiconductor fabrication. The smaller master stamp is simpler to design, produce, and control. It is also less costly than a full wafer stamp. The step-and-repeat approach mitigates the dimensional, planarity, flow, and separation anxieties inherent in wafer embossing, by literally reducing the problem. The tradeoffs include the added time required for multiple repetitions of the stamping process, the need to confine heating to the stamp area, and the cost and complexity of a precision step-and-repeat apparatus. Cold embossing uses UV light to polymerize and solidify materials after forming by the stamp. The starting materials are thin films of UV-curable materials. Typical materials may be monomers, or inorganic-organic hybrid polymers, which are polymerised by UV to polymers with desirable optical properties. The films may be deposited onto silicon, gallium arsenide, or similar wafers. The UV-transmissive patterned master template is pressed against the film, imprinting the pattern, which is UV cured before pressure is released. UV embossing allows accurately aligning and overlaying multiple levels of features.

UV embossing has a clear advantage over hot embossing in allowing multi-level applications, since the first level is polymerised after exposure, making it irrevocably formed and dimensionally stable, regardless of subsequent layer applications and exposures. UV embossing is a low pressure, room temperature process, with little shrinkage of the pattern. It also allows partial UV casting, by limiting the illumination to maintain clear areas on the wafer. Cold embossing may be done on one or both sides of the wafer. Double-sided embossing allows micromolding. The laser lens array was formed by cold embossing. As with hot embossing, UV embossing may be used with a single wafer size stamp, or with a smaller stamp in a "step and flash" procedure similar to the "step and stamp" described above. Injection molding of optical components offers throughput and cost advantages for high-volume production. Injection molding can produce 3-D objects and integrated multicomponent systems. To move beyond conventional injection molding and capture nanofeatures, electroformed metal inserts are nanopatterned and included as part of the master mold. Double-sided injection molding at wafer scale has been used to create optical structures on both sides of a supporting substrate. These structures may be the same on both sides of the substrate, or they may be different; for example, combining a diffractive lens on one side with a refractive lens on the other. The two sides may be processed separately, or simultaneously. In either case, precise alignment must be made between the features through the intervening substrate. Injection molding has been used to produce optical components in hundreds of thousands. However, the high tooling cost can only be justified for these large quantities. Injection molding draws upon a wider range of materials than direct embossing. Acrylics or polycarbonates may be used for all-plastic systems in low-cost, high-volume injection molding, producing inexpensive commercial products in the hundreds of thousands.

A summary comparison of the three embossing processes. Hot embossing, at wafer level or by step-and-stamp, lends itself to single side patterning on thermoplastic materials. Process temperatures are typically 90°C above the glassivation temperature.

Throughput is limited by the time required to heat, hold, and cool the die and substrate. Step and stamp avoids the high cost and tolerances of a wafer-size stamp, at a penalty in multiplying the time delays. Hot embossing has some well established and perhaps more profitable commercial applications. Music CDs are commonly replicated in volume by embossing. Roll-to-roll hot embossing has been used to add interesting color effects to wrapping paper, and to produce material with difficult to counterfeit optical patterns for security documents, such as identification cards. Cold (UV) embossing may be preferred for critically dimensioned materials. It can be a room temperature process, eliminating the heating and cooling times and thermal expansions/contractions of hot embossing. It allows single or repeated multiple patterns to be layered on one or both sides of a substrate with accurate alignments. Both sides of the substrate may be embossed with identical or different patterns. For applications with more critical planarity, dimensional tolerances, or environmental requirements, cold embossing of photosensitive organic-inorganic sol-gel materials gives results close to those of glass. Injection molding produces 3-D molding, allowing entire systems to be molded as a unit. It is not a room temperature process; processing time and temperature depend on the materials used. These may be inexpensive commercial plastics, or highly accurate specialty materials.

Patterning can be single- or double-sided, in 3-D, although not generally multilayered as in cold embossing. The high tooling cost makes injection molding practical for high-volume (e.g. 100,000 units and up) products. A mixture of techniques and equipment from the semiconductor and micropackaging industries has been adapted to nano-optics fabrication. Much of the development work reported above used one-of-a-kind laboratory equipment, built for that purpose, or modified equipment, originally developed for other purposes. One report referred to a modified semiconductor mask aligner. Another recent paper reported demonstrating hot embossing using a standard flip chip aligner-bonder.

One nano-optics manufacturer has developed customised robotic embossing systems specifically for volume production. Only recently have major equipment manufacturers announced production equipment intended to meet the challenges of nanometer optics. Dedicated aligner-steppers with sub-20-nm embossing resolution are now available. Product applications based upon NIL include wavelength monitors, beam splitters, pattern generators, polarisation filters, switches, and many others. Also commercially available now are replicated components that are high performance building blocks for fabricating optical assemblies. These offer manufacturing advantages in comprising a common family of similar products, allowing standardisation of manufacturing operations and equipment. A typical component size. Embossing has come a long way since Chaucer. After decades of laboratory development, nanoembossing promises to open new optical frontiers with truly monolithic production systems. The growing commercialisation of nano-optic systems will subject them to the scrutiny and judgment of the marketplace. Success in surmounting the manufacturing challenges will determine whether nanoembossed optical systems remain a technology niche, or become a broad new technology opportunity.

Optoelectronic Space Applications

Diffractive optical elements, optoelectronic transducers and photonic components, which play an important role in optical data communication, can be substantially improved by lateral nanostructures. With the development of lateral optoelectronic nanostructures the way to controllable diffractive optics is paved. For this, elements with specific interference structures are necessary, which act as specific and possibly controllable transmission or reflection filters. Nanostructured optoelectronic components (e.g. quantum well or quantum dot lasers, photonic crystals) offer large market potentials in the future, e.g. for optical data communication or in the range of consumer electronics (for example laser television).

Nanostructured optoelectronic components offer promising application in space in the fields of optical satellite telecommunications or sensor technology (infrared sensors, high resolution charge-coupled devices, etc). With optical wireless data links (OWL) for intrasatellite communication as well as optical intersatellite links, significantly smaller and lighter devices a higher bandwidth could be realised in comparison to conventional microwave communications. Optical intersatellite links were demonstrated in the frame of the ARTEMIS mission of the ESA. For the data transmission, extreme frequency-stable solid state lasers (Nd:YAG lasers) are used, which are pumped with diode lasers.

The German company, Tesat, is a leading manufacturer of laser terminals for optical intersatellite communication. Such laser terminals are also interesting for scientific applications; for example, as an injection seeder for a satellite-based Doppler-Lidar (ALADDIN), as a satellite-based measuring device for gravitation wave detectors (LISA, with SMART as demonstration mission), or as a frequency normal for a satellite-based FT-spectrometer (POISON).

Semiconductor quantum dots, which have been manufactured in high quality by means of self-organisation for about the last 5 years, offer a new degree of freedom in selecting the working wavelength of photonic elements. They allow manufacturers to cover almost completely the entire spectral region from the ultraviolet to the far infrared, with a small number of substrate materials. Further advantages of quantum dot lasers are a small energy consumption through low threshold current densities, a high modulation range for high-speed applications as well as an improved temperature stability. Beyond that, an improved radiation hardness has already been proven by quantum dot lasers compared with quantum well lasers. First commercial uses of quantum dot lasers are expected in 2003. Due to their radiation hardness and the low energy consumption, quantum dot lasers in principle are relevant for space applications, e.g. as pump lasers for solid state lasers, which are needed for different applications. In order to realise the potential of quantum dot lasers in space applications, appropriate measures have to be accomplished by the space industry for the specification, system integration and space qualification.

Photonic crystals are a further example of nano-optoelectronic components with application potential in optical data communication. Photonic crystals exhibit a periodic refractive index and possess an analogy to semiconductors in electronics, a "photonic band gap" for certain frequency in the visible and IR wavelength ranges. The lattice constant of photonic crystals lies in the range of half the wavelength of the light in the medium. For visible light this means that for the production of photonic crystals, a precision within the range of 10 nm is necessary. Two-dimensional structures today can be routinely manufactured with high precision.

At present, intensified efforts are made for the development of three-dimensional photonic crystals, e.g. with utilisation of lithography and self-organisation procedures, in which nanoscale colloids (e.g. from polymers or silicates) arrange spontaneously to a cubic lattice. These lattices are used as templates for lattices from more interesting materials such as metals and metal oxides. Three-dimensional photonic crystals would open up new possibilities in optical data communication (light could be guided and branched to arbitrary directions) and offer in principle the potential for the realisation of purely optical circuits (optical computing). Such photonic transistors are however at present still very far from realisation. In the long run, photonic crystals will find applications in optical satellite communications.

IR sensors offer a multiplicity of application potential in space, e.g. for the satellite-based earth observation and atmosphere research, for astronomy, as navigation aid for space systems or for optical data communication. Approaches for the miniaturisation and further improvement of infrared sensors are based among other things on the application of two- (quantum well), one- (quantum wire) or zero-dimensional (quantum dot) nanostructures. With the help of quantum well or quantum dot structures the detection characteristics of IR sensors can be adjusted selectively to the relevant spectral region (band gap engineering). Quantum well IR sensors, based on GaAs are developed, for example, by the Center for Space Microelectronics Technology of NASA for special space applications. This quantum well infrared photodetector (QWIP) consists of a GaAs-layer, which is embedded, sandwich-like, in two $Al_xGa_{1-x}As$ layers. The characteristics of the quantum wells can be adjusted by varying the thickness of the GaAs layer and the composition of the barrier layer. By means of molecular beam epitaxy, nm-thick layers can be produced on large areas with atomic precision. Ga-As QWIP can also be used for long-wave IR radiation > 6 μm.

The Technical University of Munich pursues a different approach in the frame of a BMBF (Bundesministerium für Bildung und Forschung—Federal Ministry of Education and Research) joint project on the self organisation of Si/Ge islands on silicon. The research activities are focused on controlling the characteristics of epitactical, defect free Si/Ge islands on silicon substrates, which can be produced with the self-organising, parallel Stranski-Krastanow procedure in the material system Si/SiGe. The objective here is to develop coupled systems with several layers of quantum dots in a homogeneous layer system, which exhibit new functions by charge transfer and electrostatic coupling and can be used as optical detectors particularly in the mid IR (MIR) range. With combined QD/QW structures a 50-fold increased photoresponse can be obtained compared with sole quantum dot structures. Expected advantages of QD MIR (mid-infrared) detectors to be mentioned are the extended spectral range, a high durability and reproducibility, radiation hardness and a low dark current. For the year 2003, the realisation of a prototype 2-colour IR detector in co-operation with DaimlerChrysler was done.

Optical Processing & Integrated Optical Circuits

A new class of devices is smashing the size, manufacturing, cost and reliability hang-ups that have hampered optical elements to date. The commercial introduction of wafer-based, nanofabrication techniques now makes it possible to create a new class of optical components called subwavelength optical elements (SOEs). With physical structures far smaller than the wavelength of light, the physics of the interaction of these fine-scale surface structures with light yields new arrangements of optical-processing functions.

These arrangements have greater density, more robust performance and greater levels of integration when compared with many existing technologies and could fundamentally alter approaches to optical system design. However, as with any new technology, the designer must clearly understand the general concepts underlying the technology as well as its potential impact (in this case, on optical component design), as well as implications for the evolution of the technology. Integrated optical component and module design has achieved a high degree of sophistication in the combination of discrete optical elements. The properties of optical materials and their combinations, clever optics design, and advances in alignment and assembly are applied to increase component and module density, reduce costs and increase reliability. But there are limits. The natural properties of many discrete optical elements are relatively fixed, limiting the flexibility of the optical component designer and reducing transferability from design to design. This is where SOEs come in. SOEs are the realisation of nanotechnology as applied to optical elements. As such, they offer excellent optical properties while being easy to integrate with other optical materials in a broad range of configurations. In addition, they self-integrate, allowing the combination of optical functions to reduce parts count and increase reliability, and to allow increased flexibility in optical component design. A broad range of optical effects can be achieved by the manipulation of nanostructures in the optical beam path. The interaction of light with subwavelength grating structures-gratings with one or more dimensions one or more orders of magnitude smaller than the wavelength of the incident light-produces a range of controllable effects. To explain those effects, a simple subwavelength grating structure. When light is normally incident on the structure, the transmitted portion of the light-that which passes through the grating structure-is affected. The effect on the passed light can be varied by adjusting the size, shape and spacing of the grating structure. In particular, for the one-dimensional grating structure, appropriate selection of dimensions will result in behaviour as a polarizer, waveplate or polarisation-dependent filter.

More sophisticated effects, such as polarisation-independent filters, can be achieved by two-dimensional gratings. The general class of these structures is referred to as SOEs. Because the diameter of the incident light beam is often much larger than the structures of the grating, the effect on the transmitted light is really the statistical sum of numerous local interactions between the light and the grating. For example, if a light beam measuring 300 microns in diameter is incident on a two-dimensional grating such as that shown, where the structure size is roughly 100 x 100 nanometers, more than a million nanostructures will be illuminated. Therefore, varying the grating dimensions spatially across the incident beam front allows additional control over optical processing effects. The physics of the SOE depends on rigorous application of the boundary conditions of Maxwell's equations to describe the interaction of light with the structures. At the wavelengths of light used in telecommunications-980 through 1,800 nm-the structures required to achieve those effects have some dimensions on the order of tens to a few hundred nanometers. At the lower end of the scale, single electron or quantum effects may also be observed. Whereas the principles of reflection, refraction, diffraction and interference describe the behaviour of traditional optical elements, with SOEs the equations describing conventional optical behaviour do not fully cover the resultant phenomena because quantum mechanical effects now come into play.

In many applications, subwavelength structures act as a nanoscale diffraction grating whose interaction with incident light can be modeled by rigorous application of diffraction-grating theory and the abovementioned boundary conditions of Maxwell's equations. Consider refraction, a key property exploited in optical components. Traditionally, different materials must be used to obtain different refractive indices. In SOEs, however, different refractive indices can be obtained with the same material by adjusting the physical structure. As an example, SOE structures can be used to create an "artificial" birefringence effect. If a is the grating period and t is the grating width. With that sort of flexibility, why are SOEs just now becoming available for telecommunications applications? The primary reason has been manufacturability. While these optical effects have been researched for at least 20 years, cost-effective manufacturing of the optical elements has not been available. Building subwavelength grating structures in a research environment has generally required high-energy techniques, such as electron-beam lithography, or extremely precise process control, for instance, to grow nanostructures via "self-alignment." Second, there must be a consistent process for creating a broad range of nanostructure patterns. Many techniques for creating nanostructure patterns are limited in the range of possible patterns. Nanoimprint lithography overcomes those limits and has four key steps: making a mold inscribed with the complement of the desired nanostructure, impressing this mold into a resist-coated wafer, separating the mold and selectively removing the resist with reactive ion etching to transfer the nanopattern to the target material. After that, postimprint processing may be used to add metal layers to enhance performance and to provide a protective overcoat for handling in a standard manufacturing environment. Testing and dicing follow. Because nanoimprint lithography forms the nanostructures of SOEs by a direct, physical process rather than by beams of energy, the wave diffraction, scattering and interference in the resist do not limit fabrication resolution. Silicon dioxide is commonly used to make a mold patterned with the desired nanostructures. Various techniques, including electron beam lithography, can be used to create the desired negative image of a nanostructure. Because the mold can be replicated and reused, complex multistep and multiprocess methods can be used to create the desired nanostructure. Since those initial tooling steps do not need to be repeated for each production wafer, they are amortised over the full production run of a particular SOE. Using different molds with different nanostructure patterns allows the same manufacturing process to create the full range of SOEs. Unpackaged for use in free-space applications, the final SOE is an optical chip with a subwavelength grating on one side of a substrate. The overall thickness of the element is dependent on the thickness of the substrate. The (x,y) dimensions depend on application requirements.

The practical consequences of using SOEs can be explored by looking at a specific realisation—an SOE polarisation beam splitter/combiner, or PBS/C. By appropriate selection of the dimensions of a one-dimensional subwavelength grating structure, one polarisation will be transmitted, while the orthogonal polarisation will be reflected, as illustrated. At that scale, the subwavelength-diffraction grating exhibits only zero-order diffraction, resulting in a number of useful element characteristics. These include consistent performance across a broad range of wavelengths, the grating structure performs the same for wavelengths from 980 nm to 1,800 nm, and wide variations in the angle of incidence of the light beam that are up to 20 degrees of deviation from normal. Typical performance parameters for an SOE PBS reveal insertion loss of less than 0.13 dB on both the reflected and the transmitted beams. Extinction ratios are greater than 40 dB and 20 dB for the transmitted and reflected beams, respectively. The flexibility of SOEs allows trade-offs around those parameters. The benefits of a PBS/C SOE can be realised in two areas. First, there are architectural benefits in optical component design. Because of their small form factor, they allow more compact component designs. Reduced insertion losses are also realised because SOEs can be abutted to other elements with minimum free-space separation.

There are differences in the way that the elements handle light, for instance, the polarisation beam splitter is a reflective device with 180 degrees of effective separation obtained in less than 1 micron. Because of those differences, the beam path can be laid out to simplify device design and possibly eliminate other optical elements. Further, the reflective nature of the SOE PBS lets it support high-energy applications associated with laser transmitters and fiber amplifiers. Second, there are advantages for manufacturing. The wide acceptance angle simplifies the alignment process, reducing manufacturing time and cost. It is also possible to use automated "pick-and-place" manufacturing technologies. The SOEs are robust: with proper selection of materials they can be subjected to temperature ranges of –200°C to 400°C.

Consequently, they can be subjected to a broad range of manufacturing processes. Finally, reduced size may both simplify packaging and reduce packaging costs. Applications of the SOE PBS/C have been explored for fiber amplifiers, circulators and isolators, interleavers, optical switches and variable optical attenuators. Here, their size and low power consumption are attractive relative to traditional technologies. Current devices are the initial applications of SOEs, but the possibilities are expanding in several directions. SOE-based building-block functionality will be introduced both as chips and as packaged devices. Demonstrated SOE functionality spans polarizers, polarisation beam splitters/combiners, filters, photodetectors and photonic bandgap devices. Dynamic control for switching, attenuation and tuning is also possible. Monolithic integration of SOEs can be achieved by stacking SOE layers to create aggregated optical effects. Nanoimprint lithography permits direct layering on of SOEs without resorting to lamination techniques. Combining SOEs with optically active layers allows optical control circuits to be built, resulting in complex optical components "on-a-chip." Multilayer SOE integration has already been demonstrated by combining filters with photodetector arrays to create dynamic optical feedback loops. Because SOEs are are self-compatible and fabricated via wafer-scale manufacturing with relatively minor differences from device to device, the means exist to implement them.

Nano-optics is a valuably novel class of technology that takes advantage of light's unique interaction with subwavelength, nanoscale patterned materials, and nanotechnology enabled fabrication methods to create a broadly applicable optical device and manufacturing platform. In optical circuits used in consumer electronics, communications, and security applications, nano-optic devices offer significant benefits by virtue of their novel behaviour, small size, ease of integration, and lower overall costs. Through these virtues, nano-optics delivers compelling alternatives to conventional bulk optical solutions, and offers functionality and form factors that conventional optics cannot match. The following is a basic review of nano-optics theory and applications. The unique optical properties of nanooptic devices derive from appropriately patterning material on the nanoscale with structures whose critical dimensions are sized to be several times less than the wavelength of light at which they operate, thereby creating optical materials with highly useful modified optical functions. Through a proper combination of materials and structures, nano-optic devices can perform any passive optical function including polarisation filtering, phase retardation, spectral filtering, and propagation management (e.g., lenses and beam splitters). Various functions can be designed for both free-space- and waveguide-based applications.

The former is far and away the most common in commercial applications. Nano-optic devices also can be designed to operate over any wavelength range. The basic methodology is applicable to UV, visible, and IR wavelengths with appropriate variations in structural dimensions and materials. Additionally, they can form the core of an optical system. Practical nano-optic applications require a complete optical system comprised of the nano-patterned material and adjacent materials including the optical substrate and thin-film coatings. The nano-optic structures are defined as a combination of material and physical attributes, as illustrated. Attributes can be divided in to three types:

(i) *Material:* nano-structure material (e.g., dielectrics, metals, and polymers), interstructure fill material, adjacent thin film materials, and substrate material (e.g., glass, plastics, dielectrics, and crystalline material).

(ii) *Structural:* pattern (e.g., linear, mesh, and circular), dimensions (e.g., period, thickness, and duty cycle), and spatial variations (e.g., "chirping", arrays, and multiple layers).

(iii) *System:* relationship of the nano-optic structures with the optical beam path and any actuator materials and structures (e.g., liquid crystal, solid state materials, electro-optic polymers, and MEMS).

Subwavelength structure size is a key to nano-optics' optical and physical advantages. Nano-optic elements employ microstructures one or more orders of magnitude smaller than the wavelengths of the incident light—with dimensions typically on the order of 10s to a few 100s of nanometers. At such dimensions, these structures may interact with light both according to the principles of classical optics and also in a realm of comparatively small-scale local interactions; rigorous application of the boundary conditions of Maxwell's equation describes the interactions of light with nano-optic structures. For very small structures, single electron or quantum effects may be observed. As a consequence, many nano-optic devices exhibit unique optical behaviours—such as high polarisation discrimination in a micron-thin space, achromatic phase retardance, and single layer combinations of optical functions like polarisation and spectral filtering—either by coincidence or intent. These nano-optics provide useful alternatives for optical circuit design; others can simply be designed as drop-in replacements for classical bulk optics.

In addition to unique behaviours, the devices also achieve their optical effects over relatively short distances—nano-optic devices are generally very thin, usually one micron or less in thickness. This has strong positive repercussions for integrated optics since nano-optic structures can often be applied, similar to thin films, as a "coating layer" during a complex manufacturing process. Single electron and quantum effects also enable optical performance to be readily customised. Modifying material properties by adding nano-structures results in a continuous spectrum of optical functionality—essentially this combination creates hybrid, "artificial" optical materials—that allows performance specifications and optical wavelengths to be incrementally and accurately varied to meet specific application requirements. Nano-optic structures can be realised through a variety of manufacturing methods. For commercial usefulness, an optical device is not optimally practical unless it can be manufactured in high volumes, in multiple application specific variations, and readily integrated with other relevant technologies. The first criterion is necessary to achieve economies of scale; the second, to ensure that the relevant market opportunity is broad enough to be interesting; and the third, to facilitate easy integration of nano-optic elements into more complex systems. The available fabrication approaches (self-assembly and nano-lithography) achieve these objectives to varying degrees.

Some materials can be caused to form regular, nanoscale structures under appropriate, controlled conditions—a.k.a. self-assembly. The difficulty with this approach is lack of flexibility in achievable structures and usable materials, which limits the functions that can be realised. Nanolithography, which is akin to standard semi-conductor manufacturing, generally uses some method to create an image in a polymer resist layer; this image is then used as an etching mask to transfer the nanoscale pattern into the target material. Nanolithography methods include interference lithography, e-beam (electron beam) lithography, and nanopattern replication. Overall, each of these methods is optimal for some set of nanostructure patterns, materials, and volume requirements. As a manufacturing method, nanolithography supports a range of integration modes for optical circuits:

(i) Monolithic "self-integration" can be achieved by layering one atop the other to create aggregate optical effects.

(ii) Optimisation of traditional multi-element, multi-technology optical circuit architectures is possible by taking advantage of the unique optical behaviour of nano-optics.

(iii) Spatial integration can be accomplished by organising varying optical functions into an array structure via nanopattern replication.

(iv) Hybrid integration is achieved by adding a nano-optic layer(s) to functional optical materials.

(v) Integration of nano-optics with optically active layers, such as liquid crystal or MEMS structures, can create electronically controllable optical devices.

Integrated optical devices and subassemblies based on nano-optic elements realise a range of benefits. Compared to conventional bulk optics, their size and optical properties make alignment in assembly more forgiving, less labor-intensive, and less costly. When physically combined with other nano-optics or other materials in manufacturing to create monolithic integrated optics, this eliminates the complication and cost of multidevice lamination, while improving reliability. Also, arrayed nano-optic devices can be applied in multi-beam or multi-path optical circuits, eliminating the need to individually align discrete optics. In general, the process by which nano-optic elements are fabricated is flexible and robust, allowing for cost-saving manufacturing process integration in creating hybrid and monolithic optics; again, the resultant optics are often smaller, more robust, more functional, and easier to assemble. Because nano-optic devices are manufactured utilising wafer-based processes proven in semi-conductor fabrication, this enables flexible sharing of manufacturing capacity and supports high-volume manufacturing throughput.

Nano-optics draws on a growing "design vocabulary" that is functionally equivalent to the design libraries used in semiconductor applications. Multiple design elements can be used with complex, multi-step, and multi-process methods to create particularly sophisticated nano-patterns and multi-layer devices. The broad range of different nano-structure patterns, materials, and modes of integration means that there are no inherent limits in the functionality and operational wavelength of nanooptic devices. These attributes are solely driven by the needs of the application and the optical circuit. To date, the commercialisation of nano-optics has demonstrated the portability of the manufacturing platform and its products across multiple markets including optical data storage, digital imaging, projection display, and telecommunications. As nano-optic technology continues to propagate, it promises to further expand both in applications and in benefits realised by users of nano-optic based devices.

Digital Imaging, Communications and Sensors: Physically, a nano-optic enabled variable optical attenuator (VOA) consists of a crossed pair of polarising nanostructures sandwiching a liquid crystal cell; applying an electrical field across the liquid crystal will rotate the liquid crystal molecules and control the fraction of light that is passed through the structure. This tunable optic addresses the frequent need to control and block the output power of polarised light sources, such as optical transceivers, in order to dynamically optimize their signals and to facilitate maintenance. Because the nano-optic polarizer is directly integrated into the liquid crystal cell construction, this allows the entire component to be less than 1 mm thick.

Writes arbitrary, complex patterns using a focused beam of electrons. The benefit is that almost any pattern can be created; a drawback for commercial production is that individual wafer processing times can be fairly long (tens to thousands of hours).

Relies on other methods, e.g., the two above—to create a master plate capturing the inverse of the desired pattern. A printing-like transfer method then patterns the polymer resist. The benefit is high-fidelity reproduction allowing extremely small feature size, complex patterns, effective material independence during repetitive production, and for different structures in the same production line. It is not applicable for limited runs or when only a fixed interference pattern is required.

A nano-optic polarisation beam splitter/combiner (PBS/C) provides 180 degrees of effective separation—meaning that one polarisation is transmitted and the other is reflected—via a submicron-thick nanoustructural layer. This device's beam separation and thin form allow adjacent components to be closer coupled compared with conventional optics, reducing existing optical path lengths by up to one-half.

In turn, shorter optical path lengths reduce beam divergence, improving coupling efficiency and reducing power loss. Many optical circuit designs also incorporate a wave plate adjacent to the PBS/C.

Readily creates interference patterns with useful dimensions using UV light sources. The benefit of this method is simplicity; the difficulty is in creating complex shapes and arrays.

Science Fiction

In 1997, Francis Collins, the spokesperson for the US Human Genome Project (HGP), claimed that, "the project's Ethical, Legal and Social Implications (ELSI) programme [was] unique among technology programmes in its mandate to consider and deal with these issues alongside the development of the technology" (cited in McCain 2003, p. 112). Recent assessments of its impact have been far from celebratory (*e.g.* Evans 2002, Huijer 2003, McCain 2003). Indeed, it is claimed that the ELSI programme insulated the HGP from criticism rather than facilitating negotiations between scientists and non-scientists. Against this background, when Mihail Roco, a key promoter of nanoscience and technology (NST) in the U.S. and director of the National Nanotechnology Initiative (NNI), claims that societal implications have been an integral component of the NNI from the start and argues that the National Science Foundation (NSF) "has made support for social, ethical and economic research studies a priority" (Roco 2003a, p. 185), it is reasonable to wonder to what extent this represents a genuine invitation to the agora or a façade that merely disguises science's traditional agoraphobia. As Nik Brown has recently argued, "The 'post-normal science' thesis [...] which sees science increasingly dependent on wider political and public aspirations should, it appears, be received with caution" (Brown 2003, p. 18). However, notwithstanding these reservations, it would be wrong to dismiss the opportunities created by the current social and political exigencies requiring technoscience to explore the ethical and social implications of its activities. Even if it is just a façade, it represents a surface that at the very least can be tagged with critical graffiti.

Having said this, it would be equally problematic to think that the deliberative space in which discussions of the social and ethical implications of nanotechnology are unfolding, or will unfold, is an empty one. At the moment, this space is being structured by a form of extrapolation that draws on narrative elements from the science fiction (SF) genre. In this paper, I argue that there are important limitations associated with trying to understand the ethical and social implications of NST within this discursive space, and that an understanding of these limitations must precede or should be taken into account in a more reflexive debate on NST. First, I consider the implications of arguing that SF is not an external but an internal aspect of NST discourse. Following this, I show that the central metaphor in NST discourse – nanotechnoscientists as master builders – provides a semantic link to SF narrative elements. This link allows NST authors to extrapolate by drawing on SF world-building techniques. I, then, provide a detailed analysis of this process by examining the key role that the SF literary device of the *novum* plays in two NST texts. A number of scholars have already drawn attention to the important function that the SF literary device of the *novum* plays in NST discourse, especially Milburn (2002) and Marshall (2004) and to a lesser extent Miksanek (2001) and Landon (2004). What my discussion adds to these is a more detailed textual analysis of the functioning of the *novum* and an exploration of the implications of these discursive strategies for debates on the ethical and social implications of NST. The first text is Drexler's *Engines of Creation*, the second an edited book on the convergence of nano, bio, information technology and cognitive science (NBIC), *Converging Technologies for Improving Human Performance*. It is edited by Mihail C. Roco and Williams Sims Bainbridge, both active promoters of NST.

Although many in the NST community might argue that Drexler's vision is both dated and outside the mainstream, the editors of and contributors to the second text are very much part of the NST mainstream. In bringing these two texts together, I show that there are more similarities than would be initially expected, not necessarily in terms of their substantive claims but in terms of the formal narrative structures through which their claims are engendered. This is followed by a discussion of the limitations associated with the framing of ethical and social implications by current NST discourse. In the conclusion, I consider some further implications of the way NST discourse mobilizes the future.

Science Fiction and Nanoscience and Technology

Technoscientists in the NST field frequently draw on SF in order to construct a binary opposition that is deployed to police the boundary (Gieryn 1999) between science and non-science. A well-known instance is the debate initiated by Gary Stix (1996), a staff writer for *Scientific American*, who wrote a highly critical piece on Eric Drexler's agenda for nanotechnology. Amongst other things, as Milburn notes, Stix compares "Drexler's writing to the scientific romances of Jules Verne and H.G. Wells, suggesting that 'real nanotechnology' is not to be found in these science fiction stories" (Milburn 2002, p. 265). However, as Fogelberg and Glimell argue in their analysis of the debate, one of the key issues at stake is the meaning of scientific practice (Fogelberg & Glimell 2003, pp. 10-12). Drexler's more speculative extrapolative approach, based on theoretical computational modeling, is seen to be at odds with the experimentally based work that is taken to be the hallmark of good science. Instead, I argue that narrative elements from the SF genre are not external to but contribute to the constitution of NST discourse itself. By drawing attention to the shifting and permeable border between science and SF in NST, it is not my intention to either put in question the scientific credentials of nanotechnoscientists by insinuating that they are not doing 'real science' or, more generally, to undermine the credibility of science due to its reliance on narrative techniques found in fiction. As Donna Haraway argues, "Not only is no language, including mathematics, ever free of troping; not only is facticity always saturated by metaphoricity; but also, any sustained account of the world is dense with storytelling. 'Reality' is not compromised by the pervasiveness of narrative; one gives up nothing except the illusion of epistemological transcendence, by attending closely to stories." [Haraway 1997, p. 64] Science is not possible despite narrative but precisely because of it. However, not all narratives are the same; they draw on different naratological devices. Discourses that extrapolate technoscientific developments into the future, through SF narrative elements, contain assumptions about, amongst other things, the nature of being, the dynamics of historical change, the aspirations of citizens, and the relationship between society, culture and technoscience.

With this in mind, I will now discuss more specifically how SF narrative elements are incorporated into NST. Definitions of NST are highly contested (Fogelberg & Glimell 2003, pp. 5-26). This is due to its status as an 'emergent science', that is to say a science whose truth claims remain to be settled by scientific or public consensus (Hamilton 2003, p. 268). What is more, given the heterogeneous and interdisciplinary nature of the NST field, it is not likely that definitional closure will be achieved soon. Definitional problems also arise because NST, as we shall see below, is radically future oriented; thus, it is also defined by how its potential is refracted towards competing futures (Wood *et al.* 2003, p. 3). However, a minimum definition would draw attention to the significance of its length scale:

> One nanometer (one billionth of a meter) is a magical point on the dimensional scale. Nanostructures are at the confluence of the smallest of human-made devices and the largest molecules of living things. Nanoscale science and engineering here refer to the fundamental understanding and resulting technological advances arising from the exploitation of new physical, chemical and biological properties of systems that are intermediate in size, between isolated atoms and molecules and bulk materials, where the transitional properties between the two limits can be controlled. [Roco cited in Ratner & Ratner 2003, p. 7] The 'newness' of the nanoscale refers to the difference between the macroscopic and nanoscopic properties of materials. To take Ratner and Ratner's (2003) example, although a metric ton, a kilogram, and gram of gold all have the same physical properties, the same is not true when one scales down to the nano length. Gold's color, melting point, and chemical properties are different at the nano length scale as a result of the nature of atomic interactions and the fact that these are not averaged out as they are in bulk material. In other words, "Nano gold doesn't act like bulk gold" (Ratner & Ratner 2003, p. 2). Thus as Roco and Bainbridge argue, "The nanoscale is not just another step toward miniaturisation, but a qualitatively new scale. The new behaviour is dominated by quantum mechanics, material confinement in small structures, large interfaces, and other unique properties" (2001, pp. 4-5). When these unique properties are combined with the prevalent and dominant metaphor that reigns in NST discourse – nanostructures as the *building blocks* of matter and the nanotechnoscientist as the *master builder* – we can begin to appreciate the radical transformative powers that NST not only denotes but also connotes. For instance, the Nobel laureate for physics, Horst Stormer, suggests that when we are empowered by nanotechnology to "play with the ultimate toy box of nature – atoms and molecules [...] the possibilities to create new things appear limitless" (cited in NSTC 1999, p. 1). Indeed it is not infrequent to encounter references to NST's radical transformative

> powers. For instance, claims such as "By anyone's measure, nanotechnology is the next big thing. In fact, according to government R&D planners, nanotechnology is nothing short of the next Industrial Revolution" (Schulz 2000, p. 41) are rather common.

In fact, these claims are foundational for analyses of the ethical, legal, and social implications that have been initiated by the NST community itself. One might even argue that the fact that the social implications have been central to the NNI represents not so much a belief in the legitimacy of submitting NST to social and ethical analysis as much as the conviction that NST is like no other technoscientific practice in its ability to impact and transform both the social and natural world. As Wood *et al.* argue in their review of the emerging field, "Nanotechnology is being heralded as a new technological revolution, one so profound that it will touch all aspects of human society" (2003, p. 1).

Yet, while it is certainly the case that there have been important developments that make possible the manipulation of matter with precision at the nanoscale, as commentators have also noted, NST is only here as a trace of a future yet to be produced (Fogelberg & Glimell 2003, Milburn 2002, Wood et al. 2003). Even NST's most energetic promoters have to admit that "nanotechnology is still in its infancy, because only rudimentary nanostructures can be created with some control" (Roco & Bainbridge 2001, p. 1). There is a rather significant gap between what can be achieved with NST today and what is imagined that will be achievable in the future; predictions of revolutionary transformations seem premature.

This gap, of course, is not specific to NST and can be found in other fields. It is typically sutured rhetorically through hype that not only mobilizes meaning but also social, political, and economic resources by promising breathtaking advances, miracle cures, and virtually unimaginable wealth. However, in the case of NST the hype is different. For instance, although biotechnology hype promises wealth, global food abundance, and intimations of immortality through genetic therapy and enhancement, NST's hype promises more! By drawing on the metaphor of the nanotechnoscientist as the master builder and NST as the toolbox that makes possible the manipulation of the fundamental stuff that makes up the world, NST claims nothing less than to be able to rebuild the world. This ultimate conceit, which feeds NST's molecular speculations, is elegantly captured by the title of the U.S. National Science and Technology Council brochure on nanotechnology: "Nanotechnology: Shaping the World Atom by Atom" (NSTC 1999).

It is instructive to compare the semantic suppleness associated with the metaphors used in the HGP with those deployed in NST. The HGP promised to produce a "plan", "blueprint", "encyclopaedia", or a "program" of life (Rothman 1998, p. 25). In all instances its vision of the future was limited by the frontier between the organic and the inorganic. However, NST's 'shaping-the-world-one-atom-at-a-time' metaphor makes it possible to transcend this boundary: Nobel laureate and nanotechnoscientist Richard Smalley claims, "Nanotechnology is the builder's final frontier" (cited in NSTC 1999, p. 1).

Recognizing the centrality of the world-building metaphor is important because its semantic connotations also make it possible for NST discourse to draw on narrative elements from the SF literary genre that is, in part, characterized by its ability to produce radically different future or parallel worlds. This creates the discursive conditions for what, following Landon, we can call SF thinking. SF thinking generates the rhetoric that bridges the gap between the givens of science and the goals of the imaginary marvelous, the emphasis always on 'explaining' the marvelous with rhetoric that makes it seem plausible, or at least not yet impossible. [cited in Gerlach & Hamilton 2000, p. 465] By incorporating SF thinking, NST discourse overcomes the gap between what is possible today and what might be possible in the future. This is achieved by using extrapolative narrative techniques that are well established in the genre. In other words, or in other worlds, it is able to solve the tension inherent in claiming that we are already living in a nano-era while also recognizing that the dawning of the nano-era depends on much that has yet to happen.

Mainstream and Periphery in Nanoscience and Technology

Although Drexler is credited with having coined the term 'nanotechnology', it is undeniable that his status in the field is problematic. As noted above, he is one of the targets of the science fiction/fact opposition deployed to locate certain NST activities outside the boundary of real science. It is frequently argued that his vision of nanotechnology remains peripheral and outside the mainstream. As evidence for this, one might point to the fact that his agenda has certainly not been explicitly endorsed by the NNI. Still, even though his book, *Engines of Creation*, may be frequently criticized, it has also introduced a generation of scientists and engineers to a nanotech 'futurescape'. Thus even a staunch critic such as Smalley, who believes that there are insurmountable objections to Drexler's proposed molecular assemblers, has conceded that Drexler "has had tremendous effect on the field through his books" (cited in Milburn 2002, p. 280). Moreover, the disagreement between Smalley and Drexler is not about the revolutionary or social transformative impact of NST (*i.e.* its capacity to rebuild the world): "Smalley acknowledges that nanotechnology, even in the more modest form of his own nanotubes of carbon, eventually 'may change the future of humankind' and that nanotechnology from chemistry on a nanometerscale 'may make even Drexler blush'." [Fogelberg & Glimell 2003, p. 19] In making these points, it is not my intention to shore up Drexler's scientific credibility or to undermine those of his opponents. Rather, it is to note that we should not allow the controversies over the viability, or not, of molecular assemblers to obscure the similarities that exist in terms of how SF narrative elements are used to negotiate the gap between current technoscientific capabilities and their future development.

In *Engines of Creation*, Drexler introduces us to a future where molecular manufacturing will be capable of making "virtually anything from common materials without labor, replacing smoking factories with systems as clean as forests. They will transform technology and the economy at their roots [...] They will indeed be engines of abundance." [Drexler 1990, p. 63]

Engines of health, or cell repair machines, will cure disease and prolong life; other engines will contribute to the launching of a new space programme. All of this, and more, will be possible on the journey towards a positive-sum society that will culminate in an "open future of wealth, room and diversity, [where] groups will be free to form almost any sort of society they wish, free to fail or set a shinning example for the world" (Drexler 1990, p. 237). The only problem is that these "engines of creation" or molecular assemblers have yet to be produced. Still, Drexler's writing narrates their coming as unavoidable.

The arrival of assemblers is to follow a path already initiated by current bio and molecular technology. Protein machines will combine the cutting and pasting abilities of enzymes with the programmability of ribosomes to produce new nanoscale non-protein materials that will in turn be used to create second-generation nanomachines or *universal assemblers*. When these assemblers are combined with the astronomical computing power of nanocomputers, the knowledge of molecular and atomic architecture collated by nano-reverse-engineering-machines or *disassemblers*, and the ability to self-replicate in order to achieve economies of nanoscale, the nano-era will finally be upon us (Drexler 1990, pp. 3-20).

If we leave aside the technical argument regarding the viability of molecular assemblers, there are a number of narrative devices that make nanomachines both credible and inevitable in Drexler's text. Following SF convention, his text constructs a "sublime chronotope" in which the action unfolds (*i.e.* the romance of how molecular assemblers will rebuild the world). SF critic Istvan Csicsery-Ronay, Jr. defines the sublime chronotope as a "literary 'space-time' where fictional things work according to their own particular laws of time and space. SF works generally depict one or more special chronotopes that are wonderfully strange and ultimately vast and powerful" (Csicsery-Ronay 1996, p. 386). In Drexler's text, the chronotope has two dimensions: synchronic (*i.e.* at one point in time) and diachronic (*i.e.* across time or historical). Synchronically, Drexler's nano-chronotope invites us to see a world that has been thoroughly overhauled and reconstituted through the tropes of the atomic, the molecular, and the machinic. Diachronically, he narrates an 'engine' of historical change that links the past with a de-familiarized present. This present promises the future, as the acorn promises the oak.

In synchronic mode, the chronotope is woven through the narration of a space that is both familiar and alien. It is our world but its landscapes, scale, structures, rules, and action are all atomic. If Marx claimed that "Value [...] does not have its description branded on its forehead; it rather transforms every product of labor into a social hieroglyphic" (cited in Graham 2002, p. 236), Drexler might argue that value is inscribed as an atomic hieroglyphic. Indeed, he begins his book by claiming that "throughout history, variations in the arrangement of atoms have distinguished the cheap from the cherished, the diseased from the healthy" (Drexler 1990, p. 3), and he goes on to write, "Our ability to arrange atoms lies at the foundation of technology. We have come far in our atom arranging, from chipping flint for arrowheads to machining aluminum for spaceships" (*ibid.*).

Framed by Drexler's nano-chronotope, human interaction with nature and the development of technology is nothing more than the attempt to manipulate atoms, initially clumsily but increasingly with more precision (*i.e.* bulk versus molecular technology). The chronotope that stages the plot in *Engines of Creation* not only invites us to reconsider our relationship vis-à-vis nature, it also demands that we develop a molecular conception of our bodily selves: "The ill, the old and the injured all suffer from misarranged patterns of atoms, whether misarranged by invading viruses, passing time or swerving cars. Devices able to rearrange atoms will be able to set them right" (*ibid.*, p. 99). The figure of 'the machine' is the second key discursive element in the chronotope.

Drexler argues that there is no more incontrovertible evidence of the viability of nanomachines or molecular assemblers than the existence of protein machines or ribosomes that assemble proteins in our cells (*ibid.*, p. 6). Thus, "molecular machines in the cell demonstrate that molecular machines work" (*ibid.*).

A little later in the text, he first redefines life as a "special structure" which is governed by the "machinery of life" (*ibid.*, p. 17) and adds, "The history of life is the history of an arms race based on molecular machinery" (*ibid.*, p. 26). By combining a world where reality is reduced to myriads of atomic configurations and the bonds that hold or fail to bind them together, where technology is crude or precise atomic manipulation, health a harmonious atomic arrangement and disease an atomic cacophony, with vital molecular machinery as the basis of life, Drexler's text re-ontologizes the world. In doing so, he creates a sublime chronotope that provides an ideal habitat for his nanomachines because they straddle the two significant dimensions of this new domain, *i.e.* the atomic and the *machinic*. Grab your atomic force microscopes. We have entered the age of assembler and the book of the world is written in the language of atomic bonds.

The synchronic dimension of the chronotope is traversed by a diachronic or historical vector. It narrates how we have arrived at the stage where the assembler revolution is already contained in our present, making it inevitable. Drexler claims that it is possible to isolate the principles of change whose explanatory domain span "molecules, cells, beasts, minds, and machines [and] should endure even in an age of biotechnology, nanomachines and artificial minds" (*ibid.*, p. 21). After identifying molecular replicators – *i.e.* RNA, viral genes, human genes, *etc.* – as the chronotope's principal historical actors (*ibid.*, p. 25), he argues that through the evolutionary mechanisms of mutation and selection there is a continuity between *The Rise of the Replicators* (RNA molecules) and the rise of all other things that populate the earth: "Mutation and selection of genes has, through long ages, filled the world with grass and trees, with insects, fish and people. More recently other things have appeared and multiplied – tools, houses, aircraft, and computers. And like the lifeless RNA molecules, this hardware has evolved." [*Ibid.*, p. 30] He further embeds the production of 'hardware' within evolutionary semantics by arguing that the principles of engineering can be understood in terms of mutation and selection: "In engineering, enlightened trial and error, not the planning of flawless intellects, has brought most advances" (*ibid.*, p. 31).

If synchronically we have seen how the atomic and machinic nano-chronotope provides an ideal space for molecular assemblers, diachronically the chronotope locates the engineer as the hero whose practice embodies the principles of change that govern the nano-chronotope. Moreover, by inserting the engineer in the context of evolutionary trans-historical forces, the molecular-assembler revolution becomes unstoppable. Thus, it is not surprising that he concludes his book by interrogating the present with questions that originate in the future: "If we succeed (and if you survive) then you may be honored with endless questions from pesky great-grandchildren: 'What was it like when you were a kid, back before the Breakthrough?' and 'What was it like growing old?' and 'What did you think when you heard the Breakthrough was coming?' and 'What did you do then?' By your answers you will tell once more the tale of how the future was won." [*Ibid.*, p. 239]

This is not only a call for 'nano-engineers of the world to unite' and take their place in a world historical event that has already been determined, it also provides the key to the functioning of the chronotope.

What Drexler's text achieves, unwittingly or not, is a narrative that re-ontologizes the past, present, and future. This is achieved by rebuilding the world synchronically and diachronically around the *Breakthrough*, the arrival of the molecular assembler.

The narrative process whereby a single element is used as the axis around which a future alternative world is generated is a key discursive element of SF. Though there is much debate of the status of SF as a genre, there is some consensus on the centrality of the device of the *novum*: "A *novum* is a deliberately introduced change made to the world as experienced by author and reader, but a change based on scientific or other logic; it is such a significant part of the SF that the *novum* frequently determines the subsequent narrative." [James 1994, p. 108, italics added]

The novum is a variation of the "What if..." question that is used as a world-building device by extrapolating the potential ramifications of the interruption to reality contained in the question (*e.g.* time travel, artificial intelligence, a parallel universe, molecular assembly). The assembled world derives its coherence not from the logic or validity of the *novum* itself but from the way all its dimensions have been processed by the machinery of the *novum*. This is precisely the discursive scaffolding that underpins Drexler's sublime chronotope and in turn provides the stage that projects nanotechnology into the future, or retracts the future into the present. In other words, this is how Drexler bridges the gap between what is possible now and what he envisions will be possible in the future. Without this SF discursive device, Drexler's vision of nanotechnology could not be assembled.

Converging Technologies for Improving Human Performance: Nanotechnology, Biotechnology, Information Technology and Cognitive Science is a published report that derives from a 2001 workshop sponsored by the NSF and the Department of Commerce (DOC). Since then a number of NBIC meetings have taken place. The report is of interest for a variety of reasons. First, it is the most recent and sustained effort to construct a broad vision of NST by making NBIC central to the achievement of a variety of technoscientific, social, economic, and political goals. Moreover, given the scope and transdisciplinary nature of the NNI, and the variety of agencies that it mobilized, a broad integrative vision is likely to remain a crucial element in future national NST initiatives. Second, the editors and contributors are drawn from NST's mainstream. For Mihail Roco, a key figure in the NNI, NBIC represents a continuation of the work already begun. Finally, both editors have consistently championed 'analyses of ethical and social implications'; consequently, it provides an ideal site to read the framing of these questions within NST discourse. NBIC supporters argue that the impetus for convergence is driven by "the integration and synergy of the four technologies (nano-bio-info-cogno) [that] originate from the nanoscale, where the building blocks of matter are established" (Roco & Bainbridge 2003, p. vii). The integration between bio and nano is possible because the unity of matter at the nanoscale means that the structure of both organic and inorganic materials is determined by the same fundamental principles. Consequently, it becomes possible for technology to "harness natural processes to engineer new materials, biological products, and machines from the nanoscale up to the scale of meters" (*ibid.*, p. 2). The integration and synergies between info nano and bio are diverse. On the one hand, the enhancement of computing power (*i.e.* speed and memory) is expected to derive from new nano-engineered materials as well as from novel architectures in the form of quantum and biological (DNA based) computing (Theis 2001; Ratner & Ratner 2003, pp. 130-39; Wood *et al.* 2003, pp. 19-24). On the other hand, developments in NST and biotechnology themselves depend on computer based modeling and visualisation made possible by the digitalisation of molecular processes.

The integration of the cognitive science component is tied to the development of the "Human Cognome Project" whose goal would be to map "the structure and function of the mind". It is argued that cognitive science would be able to explain "the mind and human behaviour" by understanding their "physico-chemical- biological processes at the nanoscale". This would be made possible by the convergence of bio, computer, and nanotechnology. In turn, the ability to enhance cognition and communication, as a result of the accrued knowledge, would make new scientific and technological discoveries possible. Ultimately, the multiple synergistic pathways in NBIC herald a new renaissance based on a comprehensive understanding of the structure and behaviour of matter from the nanoscale up to the most complex systems yet discovered, the human brain. Unification of science based on unity in nature and its holistic investigation will lead to technological convergence and a more efficient social structure for reaching human goals.

It is interesting to note that amongst the goals reported in the volume are items that would not be out of place even in Eric Drexler's nano-chronotope. If Drexler entices us with visions of abundance, a contributor to the NBIC volume defines poverty as a technological challenge and predicts that intelligent machines will "eradicate poverty and usher in a golden age for all humankind". Indeed, such will be the magnitude of the wealth produced that "new economic theories based on abundance may emerge to replace current theories based on scarcity".

Engines of Creation, as noted above, presents the possibility of harnessing the design principles and mechanisms of biological molecular machines to create nanomachines capable of producing inorganic materials. In similar fashion, Roco and Bainbridge report that "fundamental knowledge about molecular-level processes essential to the growth and metabolism of living cells may be applied, through analogy, to development of new organic materials". If the previous two scenarios are conceivable in a Drexlerian world, the next one, brain-to-brain communication, is nudging towards SF, even by Drexler's standards. The NBIC programme foresees the development of *The Communicator*, a device that will enhance individual attributes and remove barriers to group communication such as [...] user's physical disabilities, language differences, geographic distance and disparity in the knowledge possessed by group members [...] Improving group interactions via brain-to-brain and brain-machine-brain interfaces will also be explored. For Roco and Bainbridge, brain-to-brain communication provides a stepping stone towards a networked society capable of sustaining "a global intelligence", where "humanity would become like a single distributed and interconnected 'brain' based on new core pathways of society". There are more examples that could be cited, but the point is not to isolate individual objects or scenarios that appear to be plucked out of SF novels, but to understand how NST discourse can circulate such inflated future currency as current technoscientific tender. Not unlike Drexler's book, the NBIC text embeds its promissory notes in a sublime chronotope that re-ontologizes the world both synchronically and diachronically.

In synchronic mode, the NBIC-chronotope is constituted as a continuous and unified space-time, which stretches from the nanoscale to the scale of meters and beyond. The coherence of this space is underwritten by the unity of matter at the nanoscale and is thus regulated by a hierarchy of causality that operates from the bottom up. The fundamental properties of matter are determined by its constituent molecular dynamics. Phenomena such as memory, emotion, and thought are to be explained by reference to a hierarchy that privileges the nanoscale organisation of atoms and constructs a causal explanatory pathway that links nanostructures to the structure of DNA that in turn extends the link to the interaction of neurons in the brain. The great chain of being begins at the bottom but does not end with the body or the brain. In the NBIC-chronotope, the conceptualisation of the brain serves to fuse what would seem to be different phenomenological domains (*i.e.* the material and the social-cultural), allowing NBIC to expand its ontological prospecting claims. The brain is operationalized as a communicative and information processing system; social interaction and group behaviour are defined by the same operators. This opens the way to re-ontologizing the brain's neural network as the cognitive 'nanostructure' of social life through which the bottom-up causal hierarchy can be replicated in the social and cultural domains. Thus, the ability to enhance these functions (*i.e.* information processing and communication) by drawing on a cognitive science leveraged by bio, nano, and information technology makes it possible to conceive of the "improvement of collective behaviour and productivity". It is in this sense that convergence would lead to devices like the *Communicator* that would provide the basis for a "more efficient social structure for reaching human goals" and even for envisioning the "bond of humanity driven by an interconnected virtual brain of the Earth's communities searching for intellectual comprehension and conquest of nature".

Fuelling this vision of the ability to manage everything from the nanoscale to the interactions of humanity as a whole is an explicit and profound reductionism that arises from the conviction that "all disciplines share a common ability to work at the molecular and nano length scales using information technology and biology concepts" (Roco 2003b, p. 93). As a result, "partisans" who argue for the "independence of biology, psychology, and the social sciences [...] against 'reductionism', asserting that their fields had discovered autonomous truths that should not be reduced" are utterly self-defeating (Roco & Bainbridge 2003, p. 13). Underpinning this reductionism is an ontology in which distinct phenomenological domains lack domain-specific principles of organisation, thus "a networked society of billions of human beings" is to the human being what "a human being is to a single nerve cell" (*ibid.*, p. 22). Consequently, a "collective social system may be compared to a larger form of a biological organism" (*ibid.*). The unification of the natural and social sciences would make possible the development of an explanatory regime capable of encompassing "collective behaviour in physics, chemistry, biology, engineering, astronomy, and society" (Roco 2003b, p. 84). Indeed, in a chronotope where every internal ontological border is disassembled to its constituent molecular configurations, it becomes possible to conceive of a "predictive science of society and to apply corrective actions based on the convergence ideas of NBIC", and to re-ontologize culture as the product of the brain's physiology thus leading to the dual evolution of human culture and physiology.

Underpinning the NBIC-chronotope are two distinct but interlinked tropes: 'communication' and 'unity'. That the ontologisation of DNA as an informational code and the conceptualisation of the genome as a biological computer (Thacker 2004, p. 40) makes it possible to think of molecular intervention in terms of reprogramming, is already well established. By asserting the unity of matter at the nanoscale and erasing the distinction between the organic and inorganic, NST is able to extend the informational paradigm to all matter: "Programmable matter is a technical approach to the physical world in which the distinction between information and materiality is effaced. For nanotech, the entire apparatus of nanomachines [...] is itself built out of the same molecular and atomic elements that compose the physical world. Nanotech's ultimate engineering fantasy – that of the nanocomputer or a computer hardware apparatus that is assembled from atoms – is a direct example of its will to materialize information."

NBIC incorporates society and culture into this informational logic by conceptualizing the brain itself as the programmable matter that underpins social behaviour and interaction. Thus there is no longer an outside of the NBIC chronotope. The corollary to this unified world is the existence of a common molecular syntax: once deciphered, it will make just about anything possible. It is this universal machine language of matter that allows the conversion between bits, atoms, neurons, and genes and the seamless integration of people, technologies, societies, and humanity.

If synchronically it is the ontological unity of the social and natural world arising from a common molecular grammar that makes NBIC convergence inevitable, diachronically it is the ontologisation of the history of humanity as a transhistorical quest for improvements in human performance. This is presented diagrammatically with a table which begins with the development of the cell, body, brain, *etc.* includes universities, printing, the industrial revolution, *etc.* and inexorably moves to NBIC in order to predict "societal and business reorganisation" and even "evolution transcending human cell, body and brain". This historical trajectory opens up the space for a new type of historical actor, a new renaissance man or woman, the scientist engineer, capable of mastering the unified language of the world, in others words capable of punctuating the current equilibrium by completing a process that has already begun: NBIC convergence. The sense of inevitability, however, is the product of how the gap between the present and future has been overcome. Like Drexler's text, the NBIC text draws on the narrative device of the *novum* (*i.e.* NBIC convergence) that constructs a discursively coherent world that stretches from the past to the future. However, the price of coherence is that everything must be traced back to the interruption that the *novum* introduces. Thus, all of history converges towards the *novum* that in turn gives birth to the future. In the narrative developed in the NBIC text, every dimension of the world, both diachronic and synchronic, has been processed through the NBIC filter and colored by the trope of convergence and unity. Thus, the principles governing the structure of organic and inorganic matter converge, the technologies of different disciplines converge, the natural sciences converge, the natural and social sciences converge, individuals and technology converge, individuals converge into networks, societies converge, and humanity finally becomes unified. Environmental degradation, poverty, disease, cultural misunderstanding, war, *etc.* can all be solved through NBIC convergence. The NBIC world is akin to a hall of mirrors with NBIC convergence at the center: though stretched, contorted, and deformed, every reflection refers back to the principle of NBIC.

SF in NST: Bridging the Gaps

The reasons why the device of the *novum* fails to generate a propitious context for the consideration of the ethical and social implications of NST are the very same reasons that explain its success and centrality as a narrative device in the literary genre of SF. In the later, its function is the construction of a coherent and plausible world that is separated from our own world in time and/or space. This is achieved by making the *novum* the crucible on which all aspects of the extrapolated world are forged. It functions through a viral logic by replicating itself in all the principal phenomenological domains of the chronotope. The price of plausibility and coherence is unidimensionality – *i.e.* organising the structure of the world around one principle. This constructed world in turn provides the ontological stage in which the characters are embedded and the plot unfolds. However, when this same narrative technique is used in NST discourse to extrapolate from current technoscientific abilities to the future, a number of problematic effects are produced.

First, SF literature typically incorporates a historical account, or future history, that explains how the fictional world has come about. It normally contains the period before, during, and after the *novum*. If the narrated world is to be credible, the relationship between the three periods must be one of inevitability. This sense of historical necessity is also reproduced, as I have shown above, when the *novum* structures NST discourse. Thus, as a result of how their respective *nova* have generated the diachronic and synchronic dimensions of their sublime chronotopes, Drexler and Roco and Bainbridge's accounts create a sense of inevitability. However, if the inevitability of these processes are accepted, then there is logically and discursively a rather limited role for ethical reflection or analysis of social implications.

Second, to the extent that the *novum* used to extrapolate a future world is a technoscientific innovation, as is the case in NST discourse, then the extrapolation will take on a technological determinist logic. Technological determinism explains social, cultural, political, and economic change in terms of technoscientific development. However, this logic is a poor operationalisation of the dynamics between technoscience and society. The framing of technoscience as the explanatory cause of social phenomena fails to register the complex processes that embed technoscientific practice in specific social, cultural, political, and economic relations. Indeed, as the body of scholarship developing around the social studies of science reveals, technoscience is a social achievement dependent on, for instance, economic rationalities, contests for legitimacy and authority, micro-interactions in the laboratory, social organisation, and the development of social networks (Gieryn 1999, Latour & Woolgar 1986). Thus, technoscientific practice relies on the simultaneous production and/or mobilisation of social, economic, political, and cultural conditions through which it is invested with legitimacy and effectivity (Latour 1986, Turnbull 2000). The specific ways in which these social processes are negotiated will open up certain developmental pathways while closing off others. It is for this reason that Latour claims that technologies "far from primarily fulfilling a purpose [...] start by exploring heterogeneous universes that nothing, up to that point, could have foreseen and behind which trail new functions" (Latour 2002, p. 250). The extrapolative structure of the *novum* erases the contingencies inherent in technoscientific development by projecting it along a linear developmental path that will most certainly be frustrated. As Brown (2003, p. 4) argues, "In the short term we tend to completely overestimate the practical capabilities of technologies. In the longer-term we tend to get it wrong altogether, with technologies occasionally taking us completely by surprise". This becomes particularly problematic when these developmental paths are invested, as they are within a technological determinist logic made possible by the *novum*, with the ability to resolve all manner of social, cultural, and political problems. Potential non-technological solutions become marginalized and are not pursued.

However, the most fundamental shortcoming of deploying the *novum* as a device for framing discussions on the ethical and social implications of NST is that the *novum* bridges far too many gaps! It not only bridges the technical gap, but also the social and ethical gaps by generating a (fictional) future social world which contains beneficent social implications with only minor ethical complications. As we saw above, the technologies extrapolated from molecular assemblers and NBIC convergence promise a future of prosperity, peace, and physical well-being. Framed in this way, not to promote these technologies and their alleged beneficent social impacts becomes politically negligent if not utterly unethical. However, this momentum towards action obscures the fact that the credibility of the beneficent social implications and the lack of serious ethical conundrums are secured by the narrative structure of the *novum*, not through a critical analysis of social outcomes or serious ethical or normative discussion. Moreover, the *novum* also assigns the social sciences and humanities the function of analyzing and contributing to the management of the social processes necessary to arrive at the proposed future. In this way, they are divested of their potential critical role. For instance, social scientists are asked to analyze public opinion with a view to overcoming public resistance through the effective communication of nano-benefits and promises: *i.e.* by including the public in the political economy of desire and hope generated by the *novum*. They are also asked to aid nano-development by analyzing the mechanisms and procedures which will streamline processes of nano-innovation. In all these contexts, social scientists and humanities scholars are not invited to test the assumptions that underpin the social future generated by the *novum*. Thus, it becomes difficult to envision how a truly critical space is to develop. Moreover, the totalizing utopian vision produced by the *novum* invites similarly generated counter-visions. The latter deploy the structure of the *novum* much as do the former; they differ only in the malevolent logic with which the narrative is invested. Consequently, it is extremely likely that the production of dystopian NST futures may arise not so much from technophobia, fear-mongering, or inadequate knowledge but from the difficulty of criticizing the seemingly impenetrable utopian futures projected through the *novum*. In this context, the most effective critical maneuver is to insert a dystopian virus into a pro-NST programme and use its *novum* to assemble a dystopian future.

An understanding of this phenomenon is particularly important because it is the dystopian *novum* that has drawn the attention of popular culture and has, as argued by Marshall (2004), contributed as much to the development of NST as has its utopian counterpart. However, for many of the reasons listed above, in the context of the utopian *novum*, the dystopian register fails as a constructive critique of the social and ethical consequences at stake in NST.

In this study, I have argued that the relation between SF narrative elements and NST is not external but internal. This is due to NST's radical future orientation, which opens up a gap between what is technoscientifically possible today and its inflated promises for the future. I have argued that this gap is bridged by linking the dominant metaphor in NST discourse – the nanotechnoscientist as the *master builder* – to SF narrative techniques used to build future or parallel worlds. I have examined these techniques in detail in two NST texts: Drexler's *Engines of Creation* and Roco and Bainbridge's text on NBIC convergence. I have tried to show how narrative techniques are used in order to extrapolate credible and plausible futures through a synchronic and diachronic re-ontologisation of the world. *First,* both the sense of inevitability and technological determinism, associated with the *novum*, tend to erode the necessity of analyses of ethical and social implications. *Second,* I have drawn attention to the fact that the *novum* does more than merely bridge the technical gap; it also bridges the ethical gap by narrating a desirable fictional social world organised around the extrapolated technology. This contains the moral imperative to realise the extrapolated society while simultaneously cloaking the extent to which the plausibility of the extrapolated society is a function of the narrative device of the *novum*. In this context, the role for the humanities and the social sciences is to facilitate the development of the technology rather than to critically engage with it. *Finally,* I have suggested that the radical immunity to critique that is constructed through the *novum* creates an ideal medium for dystopian counter visions that in turn display many of the same shortcomings in their apocalyptic rendering of NST. I would argue that an understanding of these discursive tendencies must be borne in mind in the attempt to open up a space for a more open and critical analysis of the ethical and social implications of NST.

However, in highlighting how SF narrative elements in NST discourse fail to facilitate effective critical engagement, I am not arguing that SF as a literary genre is not a suitable vehicle for critical reflection on technoscientific developments. It is necessary to be clear about the fact that the existence of SF narrative elements in NST discourse does not make the latter a work of literary SF. What is more, SF as a literary genre is, in fact, better at opening up a space for critical reflection than is the NST discourse described and analyzed in this paper. In other words, ironically literary SF succeeds where NST discourse fails. This is because, as SF writer and critic Samuel R. Delaney argues, "Science fiction is not about the future; it uses the future as a narrative convention to present significant distortions of the present" (1984, p. 47). Similarly, Frederick Jameson (1982) argues that these distortions serve to defamiliarize the present and open up the exploration of alternative social, cultural, and political arrangements. The plausibility of the extrapolated 'future' in SF need only be sufficient to stage the exploration of scientific, political, social, and cultural questions in dramatic form. As a result, SF is less concerned with the 'objective' factors that give rise to a specific future, less concerned with forecasting or describing possible future societies, than [...] with presenting a specific future and discovering what it means to act in specific ways in terms of the belief that those ways of acting are necessary for accepting, rejecting or doubting the principles upon which a particular future social order rests. [Elkins 1979, p. 24]

Thus, the 'future' in SF is only a dramatic device for exploring the present. In contrast, NST discourse confuses the effect of the 'future' produced by the *novum* and its related narrative strategies for the future itself. It confuses the suspended disbelief that is evoked by a world organised around a single principle as a vehicle for a dramatic enactment with foresight. Whereas in SF the extrapolated future is a stepping-stone for critical reflection, in NST discourse the extrapolated future is the endpoint of the reflection. Given the range of techniques, tools, instruments, machines, algorithms, materials, hardware, processes, projects, disciplines, actors, economic interests, and governance agendas that are included under the rubric of NST, it is unlikely that the fields of NST will all develop in unison. More likely than not they will produce varied ethical, legal, and social implications that will have to be monitored and analyzed as they unfold in different social, cultural, political, and economic contexts. An ethical lag is only a problem if we lack the social and political institutions to restrain technoscience when we deem it necessary. A necessary corrective to the unrealistic task of trying to understand the ethical and social implications of NST as if they were one process is to ask ourselves: Does it make sense to group all our macro-technologies in the same way? Moreover, if it is true that the extrapolated future made possible by SF narrative speaks more to the present than the future, we might ask ourselves what are the ethical and social implications of an organisation of technoscientific activity that needs to claim such clairvoyance and promise so much to merely function?

Working Towards an Integrated Optical Circuits

The mid 1980s as a means to break the diffraction limit on spatial resolution attainable with optical measurement. Traditional NSOM utilizes a tapered, metal-coated optical fiber with a small aperture as either an excitation source or collection device. The spatial resolution attainable with this technique is roughly defined by the size of the aperture itself. Unfortunately, rough metallic coatings on glass fibers, coupled with the finite skin depth of metals at optical frequencies, place a limit on the spatial resolution attainable with fiber probes. Typical spatial resolution for the fiber approach to NSOM is ~50-100 nm, though careful probe and sample preparation have yielded ~30 nm resolution. Research groups have long been interested in mapping out the spatial distribution of the proteins in photosynthetic membranes. These proteins are approximately 10 nm in diameter, and electron microscopy indicates that they are closley packed in their native configuration. The density of this arrangement precludes optical mapping with either far-field measurement or traditional aperture NSOM. In the mid-1990's, a number of groups began working on what would prove to be the solution to the fundamental resolution limitations in aperture NSOM. The idea was to spatially map an optical signal from the surface of a sample by scattering that signal with a solid probe.

The choice of probe ranged from tungsten STM probes to metallised AFM tips to silicon probes. While optical image interpretation was complicated by topographic coupling, the spatial resolution in the apertureless technique was improved to ~10's of nm. In 1999 the Xie group began work on a type of apertureless NSOM that takes advantage of the phenomenon of field enhancement by noble metal probes. With the appropriate combination of excitation polarisation and probe geometry, a gold or silver probe is capable of strongly enhancing the magnitude of the incident illumination. This strong, localised field can be used as an excitation source for a spectroscopic transition of interest, such as fluorescence or Raman scattering. By locally exciting a spectroscopic transition in the sample and detecting the Stokes shifted light in the far field, the interpretation of apertureless NSOM images is greatly simplified. They utilize the nonlinear response of the sample by exciting two-photon fluorescence in the sample.

This tip enhanced nonlinear optical microscopy (TENOM) amplifies the contrast. The enhancements achievable with specifically fabricated gold probes are sufficient to yield near-field two photon fluorescence signal above the background generated in the diffraction-limited far-field region. Furthermore, TENOM has improved spatial resolution to the sub-20 nm regime. This high resolution and strong signal generation should be capable of, for example, mapping the spatial distribution of PSI and PSII in the thylakoid membrane. While TENOM has presented a solution to issues of high resolution optical imaging, it does have several intrinsic drawbacks.

These drawbacks include the quenching of non-energy-transferred fluorescence, the necessity to carefully tailor the probe geometry to yield high field enhancement, and the practical requirement of a nonlinear signal source. Currently exploration is on to get a variety of approaches to eliminate the few drawbacks of TENOM and to expand the capabilities of apertureless NSOM in general. The first concern has been to address the problem of fluorescence quenching by the metal probe. To this end, an ion- and electron-beam assited deposition (IBAD and EBAD, respectively) of dielectrics has been applied to the apertureless probes. The growth of glass with these techniques allows us to build a spacer layer between the probe and the sample, reducing quenching while still maintaining high field enhancement at the sample. The theoretical treatment of fluorescence quenching by probes of complex geometries is currently being explored through the use of finite difference time domain (FDTD) simulation and Poynting vector integration.

Due to the complex nature of light-metal interactions, electromagnetic simulation is vital in the TENOM and apertureless NSOM research. The FDTD technique is a flexible, powerful method for fully three dimensional simulation of the response of metal probes to incident radiation. The power of the FDTD technique lies largely in its conceptual simplicity. A geometry of interest is defined by a number of cells, called Yee cells. The complex permittivities of each cell are assigned. After a geometry has been defined, an incident optical field is simulated, and the Maxwell's equations are integrated in a leapfrog fashion.

The result is a method that can calculate the resultant fields for any arbitrary geometry illuminated with optical sources. FDTD has been applied to the study of NSOM in the past. Unfortunately, the large computational burden associated with this technique made three dimensional simulations quite difficult. Two dimensional calculations have generated insight into the aperture method of NSOM. In this research a commercial FDTD package (RemCom's XFDTD) was utilised to simulate apertureless probes in three dimensions, working to adapt this microwave-focused package to the complex permittivities intrinsic to the optical regime behaviour of real metals. This commercial software, coupled with the recent tremendous improvement in computational power, has provided a general method for simulating complex geometries, such as groups of spheres. In order to expand apertureless NSOM to linear signal sources, intensity enhancements on the order of 10,000 were required in these experiments. In order to develop a physical intuition for the nature of strong field enhancement, and to rationally design probes for high field enhancement, FDTD was used to model apertureless NSOM. The simulations indicated that probes with finite size on the scale of the excitation light were preferable to the "quasi-infinite" tips used in the past. Furthermore, by properly tuning the size of a right trigonal pyramid to the excitation light, field enhancements sufficient for exciting linear signal sources and yielding signal above the far field background were predicted.

The simulated high field enhancement has been experimentally verified. A focused ion beam instrument was used to make apertureless NSOM probes. This instrument (FEI Strata DB-235) allowed to mill materials with ~10 nm resolution, image with ~5 nm resolution, and deposit glass or platinum with ~20 nm resolution. This powerful tool for nano-optics and near field microscopy applications, including surface enhanced Raman scattering (SERS) and chemical sensors was utilised. Electromagnetic coherence theory for optical near fields refers to the electromagnetic field at the immediate vicinity of the sample where the evanescent or non-propagating waves are strong. The near field is a primary example of a highly non-paraxial field and, in general, it is composed of three orthogonal electric field components. The conventional tools of electromagnetic coherence theory are, however, valid mainly for beam-like fields. We have put forward the definitions for the degree of polarisation of general three-dimensional fields and for the electromagnetic degree of coherence. We have shown that in the limit of complete coherence, the electromagnetic coherence tensors, both in the space-time and space-frequency domains, behave analogously to the corresponding scalar functions.

Furthermore, we have rigorously formulated the electromagnetic theory of coherent modes, and constructed the mode expansion for a class of statistically homogeneous and isotropic electromagnetic fields. Currently, we are investigating the optical coherence in a near-field detection system, and tightly focused electromagnetic fields. We study the state of partial polarisation of tightly focused optical fields. In the vicinity of such a focus the electromagnetic field displays a complicated three dimensional structure. For example, the light is fully polarised on concentric rings around the optical axis even when the incident light is unpolarised in the 2D sense. The distribution of the degree of polarisation in the image plane of an imaging system of high numerical aperture. The rings of full polarisation are clearly resolved in the image. We have modeled the electromagnetic field in various nanostructures using a boundary-integral method in two dimensions. In the boundary-integral method the electromagnetic scattering problem is casted into a form of an integral equation involving integrations only over the boundaries of the system. We have applied the method to analyze changes in the degree of polarisation in light transmission through a tip aperture of a scanning near-field optical microscope. Furthermore, we have carried out spectral analysis of light transmission through sub-wavelength slits in metal films. We study experimentally the detection and spectroscopy of very small gold nanoparticles. Such particles are an interesting model system to study the transition from individual atoms to the optical properties of bulk matter. Nanoparticles also have important applications in chemical sensing and as labels in biology. In our first experiments we used supercontinuum light produced in a photonic crystal fibre to detect sub 10-nm gold particles. A confocal microscope image of 15 nm gold particles.

The cross section shows that the supercontinuum light can be focused to almost a diffraction limited spot. The spectrum of the light scattered by an individual particle. The scattering spectrum displays a clear resonance which can be used to distinguish gold nanoparticles from other scatterers in a biological medium. This work was performed in the nano-optics group of professor Vahid Sandoghdar at ETH Zürich. We are currently continuing this work at HUT with experiments on the absorption spectroscopy of individual metallic nanoparticles. Single walled carbon nanotubes (SWNTs) are tubular graphitic molecules that can be microns long and less than one nanometer wide. The combination of the unique properties of the carbon–carbon bond and their interesting molecular structure have given SWNTs unique physical properties and provided the potential for revolutionary applications. SWNT structure is characterised by two integers (n,m) that can be easily related to both its diameter and chirality. SWNTs have an unusual electronic configuration: each SWNT with a distinct (n,m), where n–m divisible by three is metallic, while the rest are semiconducting Further, SWNT electronic properties exhibit a strong structural dependence which helps drive potential applications.

Simple one electron models indicate that the density of states (DOS) of a nanotube is characterised by singularities, with each singularity corresponding to a quantum subband. Recent studies of individual SWNT electronic structure using scanning tunneling spectroscopy (STS) agree with these general predictions. Simple one electron models also imply that the dipole-allowed optical transitions occur between mirror image peaks in the DOS. Absorption spectra of SWNT bundles contain three broad absorption regions corresponding to the first and second allowed transitions for semiconductor SWNTs, and the first allowed transition for metallic SWNTs, respectively.

However, these spectra do not exhibit any features identified with a particular nanotube structure, since strong electronic coupling mixes the energy states from the different SWNTs in the bundle. This strong intertube coupling, combined with the large inhomogeneity in nanotube structures present in an ensemble, completely obscures any fine structure in the absorption spectra. Isolation of individual nanotubes is critical for decoupling the optical response of one nanotube from another, and was recently demonstrated by O'Connell et al. for nanotubes manufactured by the high-pressure CO chemical vapor deposition (HiPCo) process (H-SWNTs). Absorption spectra of an ensemble of isolated SWNTs (Mic-SWNTs) show many more highly resolved features than the corresponding corresponding absorption spectrum from nanotube bundles. These distinct features correspond to dipole-allowed optical transitions from SWNTs with different (n,m) values and/or from higher interband transitions for a nanotube with a given (n,m) value. The general isolation procedure can be applied using numerous surfactants, as well as nanotubes produced by various methods such as the laser oven (L-SWNTs), and arc discharge (A-SWNTs) processes. Bundled SWNTs do not fluoresce since any metallic tubes present in the bundle provide an efficient nonradiative pathway for photoexcited electrons.

However, for HiPCo SWNTs, Mic-SWNT samples show strong, dipole allowed fluorescence at the bandedge. The measured Stokes shift between the absorbing and emitting lines is quite small (_40 cm_1). SWNTs are poor emitters; the fluorescence quantum yield (QY) of SWNTs is on the order of QY _10_4. The dependence of the SWNT electronic structure on diameter and chirality [(n,m) values] is fundamentally important and is also of technical interest. Simple one electron models predict that the energy of the dipole-allowed optical transitions depends linearly on the nanotube diameter. However, the observation of structurally dependent photoluminescence from SWNTs represented a major breakthrough, as the dependence of electronic transition energies was shown to depend on (n,m) values explicitly and not simply on the SWNT diameter. In particular, the magnitude and sign of deviations from the predicted linear dependence were observed to depend on the value of (n-m) mod 3, suggesting that trigonal warping effects were playing a significant role. Empirical fitting parameters have been introduced modifying the simple one-electron picture to compensate for these effects, thus allowing for a precise prediction of the transition energies for different tube (n,m) structures. However, despite the observation of fluorescence spectral features attributable to specific nanotube (n,m) structures, it is not fully understood why the transition frequencies differ significantly from simple theoretical predictions, although many-particle interactions and/or trigonal warping effects likely play a role in this discrepancy. Notwithstanding these recent advances in SWNT separation and isolation, the optical spectra clearly suffer from severe inhomogeneous broadening due to the distribution of different nanotube structures in the ensemble.

Thus, early optical studies could only determine an averaged overall behaviour that obscured crucial electronic characteristics of the individual nanotube. For example, since spectra from SWNTs with different (n,m) values will overlap, it is difficult to make definite conclusions regarding important emission features such as the spectral line width or details of the band shape, or to observe evidence of coupling to vibrations. Therefore, new approaches are needed to further advance the understanding of SWNT optical properties and accelerate the de velopment of novel SWNT photonic applications, such as nanometer scale integrated electroluminescent devices. Spin-casting the Mic-SWNT suspension onto a cover slip allows for isolation of individual nanotubes by standard confocal microscopy. A typical fluorescence image from the spun Mic-SWNT sample is shown, with a spatial resolution of 300 nm. The image was obtained by collecting an optical spectrum at every pixel and then binning regions of the individual spectra where strong fluorescence signals are expected according to the ensemble spectrum. Since atomic force microscopy (AFM) measurements reveal predominantly short SWNTs (length 200–300 nm) lying on top of residual surfactant patches, distinct tubular features in the Mic-SWNT fluorescence images are typically unresolved.

Alternatively, by dispersing SWNTs directly in organic solvents such as dichloroethylene, followed by spin coating, occasionally individual nanotubes that fluoresce can be isolated, both the fluorescence and Raman images clearly show the characteristic one-dimensional features of the SWNT thus proving that the fluorescence in fact results from the single nanotube. Fluorescence images from the isolated Mic-SWNTs show distinct, bright spots at different positions for different emission wavelengths. Displays representative spectra detected from one of these bright spots. Each spectrum now exhibits a single emission band with a smooth line shape; all traces of contributions to the fluorescence spectra from the different SWNT (n,m) structures have been eliminated. The fluorescence maximum for all three spectra matches three transitions observed in the ensemble spectrum.

In so much as the ensemble fluorescence spectrum is a weighted superposition of spectra from individual SWNTs. Single nanotube spectroscopy clearly eliminates the overlapping spectral features associated with ensemble averaging and thus allows for the potential discovery of crucial SWNT optical and electronic properties. For example, SWNT fluorescence spectra are described by a single, featureless, symmetrical band corresponding to a single Lorentzian line shape function The fluorescence from individual semiconductor quantum dots and from small molecules exhibit an emission intermittency or on-off "blinking" behaviour for all excitation intensities, all temperatures, and on time-scales spanning many orders of magnitude.

In fact, fluorescence intermittency is commonly used as the conclusive indicator of single chromophore optical emission. Surprisingly, the fluorescence intensity from a single SWNT at 300 K recorded using moderate excitation intensities (<70 kWcm_2) is constant over many seconds with no indication of fluctuations on the 40-ms to 100-second time scale. At extremely high excitation powers (>100 kWcm_2), fluorescence intensity fluctuations can be observed, likely due to laser-induced sample heating. The use of single-photon sources is extremely desirable for applications in quantum optics, such as quantum cryptography, since the presence of even two photons in a signal pulse is an opportunity for an eavesdropper to steal or manipulate information. Thus, single molecules or single nanoparticles are attractive as single-photon "on demand" sources in quantum optics. However, their unavoidable fluorescence intermittency drastically hampers their potential utility. For example, in quantum cryptography it is difficult for the receiver to determine whether an "off" period is related to the information being transferred, or just the result of a random blinking event. Some stable sources of single photons do exist, such as nitrogen vacancies in diamond nanoparticles. However, these sources have a very broad linewidth typically over 100 nm, thus limiting the bandwidth of any potential devices. SWNTs have many advantages for possible use as single-photon sources for quantum optics applications. On the millisecond to minutes time scale, and at room temperature, SWNT fluorescence is stable showing no signs of intensity or spectral blinking. SWNTs also have an extremely narrow linewidth, and can be pumped electrically which allows for the exciting possibility of future nanometer scale integrated single-photon photonic devices. SWNTs also are resistant to photobleaching, as their fluorescence intensity does not change despite hours of intense photoillumination. Finally, unlike dye molecules or impurities in diamond, SWNTs have an easily size-tunable emission that is in the infrared at technologically relevant wavelengths (1.3–1.5 mm). Single molecule sources are also highly desirable for applications in biology and biotechnology. For example, the use of single fluorophores to track individual molecules inside a cell is a compelling method for studying intracellular processes.

However, photobleaching of organic dye labels severely restricts the single molecule processes that can be studied in cells. Detection of a single analyte as reported through changes in the optical emission of a single fluorophore is also a promising approach to achieving the ultimate in detection sensitivity. On the other hand the fluorescence intermittency of single molecules makes the fluorometric detection of single analytes extremely difficult, since the modulation of fluorescence upon binding of the analyte cannot be distinguished from a random blinking event. SWNTs hold great promise for possible use as single-molecule fluorophores. First, SWNT fluorescence does not photobleach at 300 K even after hours of continuous laser excitation. Thus, extremely long-lived biological processes on the singlemolecule level can be followed much more easily. Second, the fact that SWNTs do not exhibit fluorescence blinking allows them to be used as fluorometric reporters of a single biological agent. SWNT photonic devices show strong near-infrared (NIR) electroluminescence, allowing for easy "on-chip" integration of single SWNT sensors with silicon technology. Finally, SWNTs are optically active at wavelengths in the range of 800–1100 nm, which allows for NIR excitation and detection. This NIR optical activity has the tremendous advantage that background fluorescence from the sample can be dramatically reduced allowing for much improved sensitivity on the single fluorophore level. For some nanoparticles the processes that contribute to fluorescence blinking events slow down as the temperature is lowered. For example, in CdSe semiconductor nanocrystals a higher fraction of nanoparticles are "on" (or bright) at low temperatures versus at room temperature. Interestingly, at low temperatures (T=1.8 K) a significant fraction (about 2/3) of SWNTs show an intermittency in their fluorescence intensity, and also a spectral wandering of their fluorescence energy as shown. Spectral diffusion cannot be observed at room temperature, since the fluorescence linewidth is much larger than the change in energy during a spectral jump.

However, it is currently unclear whether the processes that cause blinking "turn on" at low temperatures, or whether SWNTs blink inherently on timescales much faster than 10 ms. With the time resolution of past experiments, fluctuations in the sub-millisecond regime were not observable. Combined fluorescence and Raman scattering measurements on the single molecule level open up new ways to characterize SWNTs. Single nanotube studies enable the connection between fluorescence energy, (n,m) structural parameters, and RBM frequency of a particular SWNT with a single measurement, and without any complicated fitting procedures necessitated by working with macroscopic samples. Single nanotube studies also reveal new insights into nanotube optical properties, such as a variation of the fluorescence energies of nanotubes having the same structure. Unlike for single molecules, for SWNTs the photoluminescence unexpectedly does not show any intensity or spectral fluctuations at room temperature.

The fact that SWNTs show no emission intensity blinking or bleaching demonstrates that SWNTs have the po tential to provide a stable, single molecule infrared photon source, with an extremely narrow linewidth, which would have a significant impact for future nanometer scale integrated photonic devices and which shows promise for applications in quantum optics and biological sensing.

Nanoparticles and Implications of Nanoscience

Nanoscience has taken scientists around the world by storm. It claims to revolutionize the world we live in with radical breakthroughs in areas such as materials and manufacturing, electronics, medicine and healthcare, environment and energy, chemical and pharmaceutical, biotechnology and agriculture, computation and information technology.

Nanoparticles are small clusters of atoms about 1 to 100 nanometers long. 'Nano' derives from the Greek word "*nanos*", which means dwarf or extremely small. It can be used as a prefix for any unit like a second or a liter to mean a billionth of that unit. A nanosecond is a billionth of a second. A nanoliter is a billionth of a liter. And therefore a nanometer is a billionth of a meter or 10^{-9} m.

- If a baseball is the size of Earth, a nanoparticle would be the size of an apple.
- We can also compare it with things in the natural world.
- The smallest synthetic nanotechnology made so far are the individual components of an Intel Pentium III microprocessor and they span about 200 nanometers.

Nanoparticles are larger than individual atoms and molecules but are smaller than bulk solid. Hence, they obey neither absolute quantum chemistry nor laws of classical physics and have properties that differ markedly from those expected. There are two major phenomenons that are responsible for these differences. First is the high dispersity of nanocrystalline systems. As the size of a crystal is reduced, the number of atoms at the surface of the crystal compared to the number of atoms in the crystal itself, increases. For example, a 4 nm diameter CdS nanoparticle has about 1500 atoms, of which about a third are on the surface. Properties, which are usually determined by the molecular structure of the bulk lattice, now become increasingly dominated by the defect structure of the surface. The second phenomenon occurs noticeably only in metals and semiconductors. It is called size quantisation and arises because the size of a nanoparticle is comparable to the de Broglie wavelength of its charge carriers (*i.e.* electrons and holes). Due to the spatial confinement of the charge carriers, the edge of the valance and conduction bands split into discrete, quantized, electronic levels. These electronic levels are similar to those in atoms and molecules. The spacing of the electronic levels and the bandgap increases with decreasing particle size. This is because the electron hole pairs are now much closer together and the Coulombic interaction between them can no longer be neglected giving an overall higher kinetic energy. This increase in bandgap can be observed experimentally by the blue-shift in the absorption spectrum or sometimes even visually by the colour of the samples. A larger bandgap means that more energy is required to excite an electron from the valance band to the conduction band and hence light of a higher frequency and lower wavelength would be absorbed. The increase in molar absorption coefficient shows that the oscillator strength also increases as particle size decreases. This is due to the strong overlapping of the wave functions of confined charge carriers. Experiments carried out in the past have found the absorption spectrum to be potential dependent, with decreasing wavelengths as applied potential decreases. This change is fully reversible before the threshold potential.

The first generation of nanofillers is emerging from R&D into commercial application in composites. Federal funding for nanotechnology research and development in USA has increased six-fold, from $116 million (USD) in 1997 to an estimated $961 million in 2004. In the U.S. alone, the FY 2005 budget request that President George W. Bush has sent to Congress calls for a total National Nanotechnology Initiative (NNI) budget of $982 million, which represents a 2 per cent increase. With all this research, what has come out of the R&D pipeline ready for commercial application? As it turns out, plenty. The following are new and, surprisingly, not-so-new applications of two types of nanotechnology: ceramic nanoparticles and carbon nanotubes used as highly functional fillers/additives in polymer products of interest to composites manufacturers. Introduction of these materials can result in the improvement of the polymer's mechanical properties in molded parts and coatings on a wide variety of fronts, including increased part strength, ductility and dimensional stability as well as resistance to damage by abrasion and exposure to caustic chemicals and/or extreme heat. Nanotechnology involves the control and manipulation of materials at the nanoscale—particle sizes from 1 to 100 nanometers (nm), where 1 nm equals 1 *billionth* of a meter— to create new materials and structures that have novel properties due to their small size. Conventional multifunctional materials, such as fiber-reinforced or filled polymers, require a trade-off between various mechanical and physical properties, cost and processability. However, combinations of properties derived at the nanoscale provide opportunities to circumvent these traditional trade-offs. The transition from microparticles to nanoparticles leads to a number of changes in physical properties. One of the major factors is the increase in the ratio of surface area to volume. As the surface area of a particle increases, the portion of its constituent atoms at or near the surface increases exponentially, creating more sites for bonding, catalysis or reaction with surrounding materials, resulting in improved properties such as increased strength or chemical or heat resistance. Additionally, the fact that nanoparticles have dimensions below the critical wavelength of light can render them transparent, which has implications for part aesthetics. One of the most common groups of the new generation of commercial nanotechnologies is ceramics. Ceramics can be categorized into two subgroups: silicates (or silicon oxide), generally in the form of nanoscale flakes of clay, and metal oxide ceramics such as titanium, zinc, aluminum and iron oxides.

Nanoclays—The most common type is montmorillonite (MMT), a layered aluminosilicate in the smectite family of clays. MMT is a component of bentonite clay, formed when volcanic ash settled into ancient brine seas, and now mined in Wyoming, western Europe, China, Australia, Iran and elsewhere worldwide. Unlike clay minerals such as talc and mica that have been used as fillers for years, MMT can be delaminated and dispersed into individual layers only one nanometer thick by about 70 nm to 150 nm across. The result is a radical increase in the surface area-to-volume ratio: With a surface area of 750 m2/g, 20g of MMT platelets could cover a football field! Southern Clay Products Inc. (Austin, Texas), a manufacturer of functional clay-based additives, began MMT product development in the early 1990s. The company's Cloisite additives, commercially available since 1999, consist of organically modified, nanometer-sized MMT platelets. Unseparated MMT particles are often referred to as tactoids. When tactoids are partially separated in a polymer, they are referred to as being intercalated. When thoroughly separated, individual platelets are said to be exfoliated. To form a nanocomposite, the clay must first be modified with an appropriate organic surface treatment to enable complete dispersion into and miscibility with the polymer matrix. Then the tactoids must be exfoliated into platelets either through the shear stresses exerted during melt compounding or via chemical processes during polymerisation of the matrix. As Karl Kamena, Southern Clay's commercial manager of Cloisite nanoclays, describes it, "The incorporation of single-digit percentages of high surface-area nanoclays into host polymer systems improves the performance of the specific polymer matrix without significant deterioration of properties." The numerous benefits that Cloisite can confer on the host polymer include increased mechanical properties, such as modulus, melt strength and scratch resistance, as well as increased heat resistance, dimensional stability and flame retardation.

The extent of enhancement is related directly to the degree of exfoliation. Currently, about 40 per cent of the MMT platelets can be fully exfoliated, with the remainder forming tactoids of 2 to 10 platelets. Southern Clay continues to improve exfoliation methods, anticipating that, eventually, remaining tactoids will contain as few as one to three platelets. In 1998, Southern Clay began working with General Motors (GM, Detroit, Mich.) and resin manufacturer Basell (Hoofddorp, The Netherlands) to develop a nanoclay-reinforced thermoplastic olefin (TPO) that would outperform the talc-filled TPO then in use. An 8-micron Cloisite clay, each particle of which contains one million platelets, was selected as the filler. Work carried out jointly by the companies allowed the new material to be formulated to different shrinkage rates and to match colour with conventionally filled molded-in-colour parts. Early on, the 8-micron particles had a tendency during the extrusion stage (when the clay was mixed into the TPO) to aggregate into 200-micron clumps rather than exfoliate, causing surface defects and painting problems in the injection molded parts. The team then proved that this issue could be resolved by making adjustments to the extrusion temperature, the extruder screw rotation speed, the material feed rate and other variables during the nanocomposite manufacturing process. GM first introduced a 3 per cent nanoclay-filled TPO on the step assists for the 2002 GMC *Safari* and Chevrolet *Astro* vans, which had been using a 20 per cent talc-filled TPO. The same material moved into a higher volume application in the body side trim moldings for the 2004 and 2005 Chevrolet *Impala*, molded at Blackhawk Automotive Plastics (Troy, Mich.). Will Rodgers, GM staff research scientist, says GM saw weight savings of 3 to 21 per cent on these redesigned parts. With increased confidence in the material, GM now is using 7 lb of molded-in-colour nanocomposite parts to make this year's *Hummer H2 SUT* cargo bed trim (the center bridge, the sail panels and the box rail protectors), all molded by SportRack Automotive (Sterling Heights, Mich.). These are the first parts that GM has designed as nanocomposites from conception, choosing the material primarily because of its low coefficient of thermal expansion (CTE) and improved appearance. For example, by varying the surface finish of the tool, three levels of gloss were molded into the box rail protectors to match adjoining parts, something that was never possible with talc-filled TPO. GM also reports better scratch/mar performance, increased cohesive strength between the paint and the part, greater ductility at low temperatures, lower flammability, a larger injection molding process window, sharper feature lines, improved knit line appearance and the elimination of tiger striping, (Tiger striping is caused when a filled material groups itself into lines of higher and lower fill weight or viscosity as it flows into a mold, giving the part a striped appearance.)

Although the MMT-filled TPO is about an order of magnitude more expensive than the talc-filled material (about \$3/lb vs. \$0.35/lb), the improvements in part performance and appearance apparently justify the cost.

Metal oxides—Scratch resistance is a highly sought quality in coatings, such as automotive finishes. However, use of silica or clay fillers in quantities sufficient to improve scratch resistance tends to make the coating hazy and reduce its viscosity. Alumina particles are harder than silica but also lead to haziness. This is particularly troublesome in clear coats. Nanophase Technologies Corp. (Romeoville, Ill.) and BYK-Chemie (Wesel, Germany) have teamed to address this problem with two new additives marketed under the tradenames Nanobyk-3600 and Nanobyk-3601. Both additives contain Nanophase's 37-nm diameter Nanodur aluminum oxide particles in concentrations of 30 to 50 per cent by weight, and are intended for use with aqueous (Nanobyk-3600) and nonaqueous UV-curable coatings (Nanobyk 3601). Nanophase Technologies produces its aluminum oxide nanoparticles via physical vapor synthesis: Arc energy is applied to a metallic precursor to generate a vapor at a high temperature. A reactant gas is introduced to the vapor and then the mixture is cooled at a controlled rate so that it condenses and forms nanoparticles, which then are coated with a thin polymeric shell that ensures their compatibility with various polymers and fluids. When the additives are blended into resins and coatings at very low 1.5 to 6 per cent concentrations, scratch resistance reportedly goes up dramatically. In Nanophase tests, particle loadings of 1 per cent resulted in five- to ten-fold increases in scratch resistance, depending on the resin formulation tested, with less than 0.5 percent haze. (At greater than 1 per cent haze, or about a 2 per cent loading of Nanodur alumina particles, a resin is no longer considered "clear.")

Exploration of carbon nanotubes is relatively new. Single wall nanotubes (SWNTs) are long, thin cylinders made of carbon atoms arranged in a graphitic lattice structure that may or may not have closed ends. Multiwall nanotubes (MWNTs) are concentric cylinders of carbon in a similar graphite structure, held together by weak intermolecular-level forces. SWNTs and MWNTs have length-to-diameter ratios that can range from 100:1 to 10,000:1. Their large surface-area-to-volume ratios mean that carbon nanotubes (CNTs) affect matrix properties far beyond what would be predicted from rule-of-mixtures calculations. Nanotubes are currently being commercialized in applications where electrical conductivity and increased mechanical performance are primary design drivers.

Electrical properties. SWNTs can possess electrical conductivity of 1 billion amps/cm2, Young's modulus of more than 1 TeraPascal and a maximum strain of 10 per cent. MWNTs have lesser properties but are easier to produce, though they also are less well understood, due to their greater complexity and variety. Orders of magnitude less costly than SWNTs, MWNTs are more attractive for commercial applications. While they have been used for the last 10 years in thermoplastics as a direct replacement for carbon black in electrical applications, recent developments are enabling their use in thermosets, and significant efforts are underway to exploit their mechanical properties as well. Carbon black and carbon nanotubes are both pure graphitic carbon and so have the same electrical properties. Carbon black particles, however, are spherical, and the rod-like CNTs have a higher aspect ratio, which means far fewer CNT particles are needed to form a conductive "bridge" through a nonconductive polymer. Typical carbon black loading in polymers is 7 to 18 percent (depending on the polymer and the level of conductivity needed), which will make the polymer brittle. CNT's achieve the same conductivity in 1 to 3 percent loadings. Moreover, the CNT's conductivity-versus-loading curve is more gradual and more easily tailored to a specific level of conductivity than carbon black's curve.

In 1984, Hyperion Catalysis (Cambridge, Mass.) was the first company to file a patent on a nanotube manufacturing method. Hyperion's MWNTs have 8 to 12 walls and are about 10 nm wide by 10 micrometers long, an aspect ratio of 1,000:1. They are made through a process which Hyperion's sales and marketing manager Pat Collins describes as "continuous, catalyzed vapor phase decomposition and recombination, using light hydrocarbon feedstock." When an atomized metallic catalyst comes in contact with a hydrocarbon gas in a reactor under appropriate time, pressure, and temperature conditions, nanotubes will grow on the catalyst. Each nanotube has a tiny remnant of catalyst inside it. The MWNTs, as formed, tend to clump together in aggregates, attracted to each other by intermolecular Van der Waals forces. The high shear forces which are necessary to disperse the CNTs into the resin are obtained by using a high-viscosity resin and a twin-screw extruder typically used to compound thermoplastics. The disentangled and dispersed nanotube/resin mixture is formulated in concentrations of up to 20 per cent. A wide variety of nanotube-filled thermoplastics can be purchased, either ready to use with low CNT loadings or as a master batch with higher nanotube concentrations, which then can be let down (diluted) to the appropriate usage level. The ready-to-use resins are offered in antistatic, electrostatic dissipative and conductive grades, with volume resistivities ranging from 1010 ohm/cm for antistatic grades to 1 ohm/cm for conductive grades (as compared to 1.68 x 10-6 ohm/cm for copper). Because the nanotubes are in such low concentrations and are so small, their addition has little effect on the base resin's properties, allowing the material to be processed on the injection molder's existing equipment under conditions similar to those used to process the base resin. Hyperion's continuous process runs 24 hours a day, seven days a week, producing tons of pure (no byproducts, such as amorphous carbon) nanotubes per week.

Automotive fuel systems have to be designed for electrostatic discharge. The turbulent flow of fast-moving fuel creates a static charge which must be dissipated. If charge is allowed to build up it can cause pinholes to develop in the fuel system and may damage electronic sensors. On rare occasions, sparks may even discharge during refueling. The fuel system, which includes the filler neck, inlet check valves, fuel module (pump/sender unit) inside the tank, the fuel lines and the fuel filter, must all, therefore, be conductive. (The Society of Automotive Engineers (SAE) will soon issue its newly revised *Recommended Practice, SAE J1645*, which will outline how all materials that touch fast-moving fuel must be grounded.) Fuel systems static dissipation is a rapidly growing segment of Hyperion's business, according to Collins. BASF (Ludwigshafen, Germany) has already commercialized a conductive polyoxymethylene (POM) with Hyperion nanotubes, specifically for automotive fuel systems. CNT-reinforced nylon 6, manufactured by Degussa (Parsippany, N.J.), is used for the tubing in the flexible portion of the fuel line in 70 percent of cars made in North America.

O-rings in the system also must be conductive or they will build unwanted capacitance. Precix (New Bedford, Mass.) recently introduced its F98 series of O-rings, made by incorporating Hyperion's multiwall nanotubes into Precix's own proprietary flouroelastomer resin system. Compared to the company's carbon black-filled O-rings, the F98 compound is said to have three to five times greater resistance to fuel vapor permeation, with significantly reduced fuel swell (the expansion of part dimensions due to absorption of the fuel) as well as a durometer of 79 versus 88, allowing for much lower insertion pressure when assembling parts. Hyperion also supplies various conductive thermoplastics to manufacturers such as GE Plastics (Pittsfield, Mass) for use as injection molding material for electrostatically painted automotive body parts, such as mirror housings, door handles and fenders.

According to a *CT* web search, the Renault *Clio* and *Mégane* currently sport fenders made from a nanotube-filled polyamide/polyphenylene ether (PA/PPE) blend. Since CNT loadings are typically 2 to 3 percent rather than the 12 to 15 per cent necessary with carbon black, the plastics do not become brittle, thus negating the need for impact modifiers which would produce the undesirable side effect of reducing the plastic's glass transition temperature (Tg). The CNT-reinforced parts need no primer and can be painted at the same temperature and on the same electrostatic line used with metal parts, reducing cost and improving colour matching. Because the CNT concentrations are low, the surface finish also is smoother than plastics with other types of conductive fillers. Collins adds that Hyperion now has branched out beyond thermoplastics into thermosets, with nanotube-filled versions of Resolution Performance Products' (Ennis, Texas) Epon 1009 and Epon 828 resins. Epon 1009 is a high molecular weight, solid pellet form of bisphenol-A, which can be compounded into liquid epoxies or vinyl esters. Hyperion sells the Epon 828 nanocomposite in paste form, which is made by grinding the Epon 1009 nanocomposite pellets into a powder and blending the powder into the 828 with a multishaft double planetary mixer in a liquid dispersion process. Collins also notes that Hyperion will introduce easier-to-use liquid master batches of other nanotube-filled epoxies in the near future, primarily for electrical applications, such as conductive epoxies, potting compounds and connectors. Although the nanotubes may add $5/lb to $6/lb to the final polymer mix, Collins says there are many applications where the cost is justified either through cost savings in the final product manufacturing process or because the necessary properties cannot be obtained any other way.

Mechanical properties. Zyvex Corp. (Richardson, Texas) approaches the market from a different angle, seeking to use specially modified carbon nanotubes to increase the mechanical performance in sporting goods, aerospace and defense applications. Zyvex does not make its own nanotubes, but instead uses SWNTs and MWNTs from other suppliers, and certifies them for quality and batch-to-batch consistency using a series of standard analytical tests. Zyvex uses a patented process to functionalize the surface of the nanomaterials with a non-covalently bonded "Kentera" molecule which non-covalently bonds the nanomaterial surface via pi-pi bonding and also adheres well to the host matrix. Pi-pi bonding, which involves a slightly stronger molecular force than the intermolecular forces that cause the nanotubes to clump together, enables the Kentera molecule to slip between the nanomaterials and disperse them. Zyvex offers its Nanosolve functionalized nanotubes and nanofibers as dispersions in solvents or water additives, or in solvent-free concentrates, which can be used in thermoplastics and a subset of thermosets comprising polyurethanes and epoxies. This includes concentrates in Resolution Performance Products' Epon 826, 828 and 862 at customized weight loadings specific to an application, whether it is thermal, electrical or mechanical. Marni Rutkofsky, Nanosolve chemist and technical sales representative, is overseeing third-party ASTM physical and mechanical testing on several materials. She says preliminary results at less than 5 per cent loading of MWNTs in thermoplastic polycarbonate show increases of 80, 50 and 150 per cent, respectively, in tensile strength, elongation and Young's modulus. Results at similar loading in polyurethane show a four-fold increase in thermal conductivity. Data on epoxy is expected later this year.

One beneficiary of Zyvex research is Easton Sports (Van Nuys, Calif.) a leading manufacturer of bicycling, hockey, baseball, softball and archery equipment. Three years ago, Easton embarked on an ambitious project to explore the advantages that nanotubes could bring to bicycle components bladder molded at its production facility in Dongguan City, Guangdong Province, China. Easton drum winds its own prepreg and planned to incorporate CNTs into the resin before introducing the resin to the fiber. The challenge was to find a nanotube chemistry that would remain well dispersed in the resin during the course of Easton's manufacturing process and yield improved properties at a reasonable cost. Zyvex and Easton tried several iterations of the resin filled with the CNT/Kentera chemistry before developing a formulation that met the challenge. "We were looking more for strength increase than weight decrease," says John Harrington, VP of Easton's Bicycle Div. "Our products are already some of the lightest on the market, and we want to be conservative with weight reductions when we are using a new material." However, Easton has shaved 10g to 20g from most of the CNT-filled prepreg components, while achieving an average 15 per cent increase in product flexural strength. Road, mountain, and triathlon handlebars, forks, seatposts and OEM seatstays and chainstays are already on the market. The CNT Technology logo distinguishes components which have incorporated nanotubes, and most of the rest of Easton's bicycle components will follow suit. Retail prices on the logoed components have gone up an average of only 10 per cent—an increase that Harrington says consumers are willing to pay for the increase in performance. While nanosized fillers aren't an overnight revolution, it is clear that there is significant activity on many fronts, with increasing use on all. Beyond the applications highlighted here, nanofillers have been incorporated into in golf clubs and golf balls, tennis rackets and a number of other products, with many more to come. The main barriers to nanomaterials use now seem to be development cost and the end-user's ability to obtain proper dispersion in and compatibility with resins. However, all of the nanosuppliers interviewed above are focused on making materials less expensive and more accessible by scaling up production, standardizing products, issuing test data and offering them in user-friendly forms.

You must be wondering by now after all the 'over-the-head' theory about how different nanoparticles are from bulk and atomic material, what this really mean to us average everyday individuals. A lot!! Due to the two unique phenomenon which occurs in nanoparticles, their properties (electrical, optical, chemical, mechanical, magnetic, etc.) can be selectively controlled by engineering the size, morphology, and composition of the particles. These new substances will have enhanced or entirely different properties from their parent materials. Nanoparticles can make metals stronger and harder, give ceramics enhanced ductility and formability, enable normally insulating materials to conduct heat or electricity, and make protective coatings transparent. Hence, industries can re-engineer many existing products and design novel new products/processes that function at unprecedented levels. Use the menu bar on the top to read about more about some examples of the possible applications.

Solar cells are photcvoltaic devices, which are devices that can convert light into electricity. They are much in demand as they offer an inexhaustible and environmentally benign energy source. Most solar cells are made of amorphous silicon. The problem with this is that the silicon must be of a very high purity and have a near perfect crystal structure. This makes it very expensive to produce. The efficiency of such a cell is also very small, typically converting only 13-18 per cent of sunlight to electricity. However, low efficiency wouldn't matter if huge arrays of cells could be produced cheaply. After all, nature's solar cells, chloroplasts in plants are less than 1 per cent efficient. Most solar powered devices rely on the same principle: a photon of sunlight boosts an electron in the material into a mobile state so that it can be used to generate electricity. The problem with this simple mechanism is that the electrons are negatively charged and will leave a positive charge. These opposite charges attract one another and therefore will tend to recombine, squandering the absorbed energy as heat or as re-emitted light. Silicon solar cells use an electric field to push the negatively charged electrons and positive charges apart. While chloroplasts adopt a more subtle approach of separate charges by making a distinction between the units that generate the electron and those that transport it away. Deciding to copy nature's trick, Michael Gratzel and Brian O'Regan at the Swiss Federal Institute of Technology began research and produced the Grätzel cell in 1991. The Grätzel cell uses intensely coloured organic dye molecules to capture light energy to inject an electron from the dye into a semiconductor such as titanium dioxide (TiO_2). This remarkably efficient charge separation reaction initiates current flow and the output of electrical energy by the cell.

The ideal material used in the cell must have a high surface area for light absorption and charge separation. Nanoparticles, having a comparable surface area to volume ratio, provides for just that. Titanium dioxide nanoparticles are used to make nanoporous thin film supported upon a glass substrate. The material obtained has optical transparency, excellent stability and good electrical conductivity. The benefit of these novel photoelectrical solar cells is that they can be fabricated from cheap, low purity materials by simple and low cost procedures. Contrary to expectation, some of the new devices also have strikingly high conversion efficiency. The size-tunable bandgaps of the semiconductor nanoparticles, due to size quantisation, also means more efficient solar cells can be produced for photovoltaics (electricity production) and water splitting (hydrogen production) processes.

Electrochromism is defined as a reversible colour change of a material caused by the application of an electrical current or potential. It has potential applications in making smart windows for the modulation of incoming light, as well as displays, a field in which competition with liquid crystal displays (LCD) is possible.

There used to be only two types of electrochromic devices. In the first type, the electrodes are made of conducting glass and covered with an organic or inorganic polymer. The two materials would usually display complementary electrochromism and hence produce the same colour change when one is oxidized while the other is reduced. This type of construction is used in electrochromic windows. It is bistable, which means that once the colour change has occurred, the state of the device remains even in absence of applied voltage. The limitation of this typ e of electrochromic devicesis the slowness of the colour change, due to thelow migration rate of the counterions in the bulk polymer. It is also difficult to obtain strong colour changes or bright colours. In the second type, two complementary electrochromic molecules are dissolved in a solvent. This type of system is very simple to build, reacts very fast and can produce dark or bright colours. The drawback of this type is that an electrical current is needed to maintain the coloured state because the two types of coloured molecules diffuse through the system and react with each other to restore the bleached states. It therefore cannot be used for large area devices or for battery-powered displays.

We therefore need a system which is both bistable and changes colour rapidly. This can be done by attaching a suitable molecule that is colourless in the oxidized state and coloured in the reduced state onto the surface of a colourless semiconductor on conducting glass. When a sufficiently negative potential is applied, electrons are injected from the conducting glass into the conduction band of the semiconductor and reduce the adsorbed molecules. The reverse process occurs when a positive potential is applied and the molecules get bleached. The advantage of this system is that combines immobility of the electrochromic material with the rapidity and colouration efficiency of molecular systems. However, a single molecular monolayer does not absorb a perceptible amount of light. This is where nanoparticles come in!

By using nanocrystalline semiconductor films, it was possible to amplify the light absorption of the molecular monolayer to obtain visible colour changes. The nanocrystalline layer are highly porous and usually the real surface of a 5 μm thick layer is about 500 times the projected area. Therefore, as light passes through the layer, it crosses several hundreds of monolayers of coloured molecules giving a strong absorption.

Some commercial products have already been produced! The most successful is the automatic dimming rear-view mirror sense ambient light and glare from behind, and reduce annoying glare from bright lights. While the company NTera has produced paper quality displays. These displays are bi-stable, not requiring much power keep to a pixel coloured, and reflective, not needing power for backlighting, and therefore can operate with very little power. Coupled with the fast switching, (*i.e.* changing of colour) due to amplification with nanoparticles, this technology could in most flat panel displays.

The invention of the transistor in 1948 by John Bardeen and William Shockley, triggered a new era in electronics. The trend since then has been to create smaller and smaller product using fewer chips of greater complexity and smaller 'feature' sizes. The development of integrated circuits and storage devices have continued to progress at an exponential rate. At present it takes two or three years for each successive halving of component size. However, the technologies currently in use have fundamental limits below, which the devices no longer function in a predictable manner. For instance, oxide layers used in Complimentary Metal Oxide Semiconductors (CMOS) devices are becoming so thin that leakage currents are conducted quantum mechanically by electron tunneling. Therefore, replacement technology to advance miniaturisation even further down, eventually to the dimensions of single atoms and molecules, is keenly sought. In 1985, Dmitri Averin and Konstantin Likharev, proposed the idea of a new three-terminal device called a single electron tunneling (SET) transistor. Two years later Theodore Fulton and Gerald Dolan at Bell Labs in the US fabricated such a device and demonstrated how it operates. Since then much research has gone into this area and better single electron devices has been fabricated. These devices usual have a molecule or atom as a functional unit and operate based on the quantum effect. Due to the small dimensions of the device (a couple of nanometers and hence is a nanoparticle), there are repulsive Coulombic interactions between the electrons and energy is required to add an extra electron. This energy cost can be tuned to zero by applying a voltage to the gate electrode. The electrostatic potential at this gate will be such that an extra electron can hop from the source onto the molecule. However, due to Coulomb repulsion, a second, extra, electron cannot hop on at the same time. The first electron must leave the molecule, moving into the drain before the next electron can enter. This one-by-one electron motion, is known as the single electron quantum tunneling. The creation an ultrafast, single electron transistor would lead to the development of "quantum" computers with supercomputer powers and the size of a thumbtack. However, there is still a long road ahead before atomic or molecular transistors can be assembled into viable, dense, fast logic-circuits. Currently these single electron devices are still no competition for silicon transistors. But they will serve both scientifically, for studying electron motion through nanoscale objects, and technologically, for developing chemical techniques with which to fabricate electronic devices on single molecules.

Advances in nanoscience and nanotechnology promise to have major implications for health, wealth, and peace in the upcoming decades. Knowledge in this field is growing worldwide, leading to fundamental scientific advances. In turn, this will lead to dramatic changes in the ways that materials, devices, and systems are understood and created. The National Nanotechnology Initiative (NNI) seeks to accelerate that progress and to facilitate its incorporation into beneficial technologies. Among the expected breakthroughs are orders-of-magnitude increases in computer efficiency, human organ restoration using engineered tissue, "designer" materials created from directed assembly of atoms and molecules, and the emergence of entirely new phenomena in chemistry and physics.

The study of the societal implications of nanotechnology must be an integral part of the NNI. An interagency effort within the U.S. Government, the NNI supports a broad programme of nanoscale research in materials science, physics, chemistry, and biology; it explicitly seeks to create new opportunities for interdisciplinary work. It is balanced across five broad activities: fundamental research; grand challenges; centers and networks of excellence; research infrastructure; and, the ethical, legal, and social implications, including educational and workforce programmes. This report outlines some potential areas for research into societal implications of nanotechnology. It has been prepared just as the NNI is commencing, when there is greater opportunity to affect the NNI investment strategy. Research on societal implications will boost the chances for NNI's success and help the nation take advantage of new technology sooner, better, and with greater confidence. Moreover, sober, technically competent research on the interactions between nanotechnology and society will help mute speculative hype and dispel some of the unfounded fears that sometimes accompany dramatic advances in scientific understanding.

Toward this end, the National Science and Technology Council (NSTC), Committee on Technology (CT), Subcommittee on Nanoscale Science, Engineering, and Technology (NSET) the Federal interagency group coordinating the NNI ¾ sponsored a workshop on "Societal Implications of Nanoscience and Nanotechnology." Held September 28-29, 2000 at the National Science Foundation, this workshop brought together nanotechnology researchers, social scientists, and policy makers representing academia, government, and the private sector. Their charge was to: (1) survey current studies on the societal implications of nanotechnology (educational, technological, economic, medical, environmental, ethical, legal, etc.); (2) identify investigative and assessment methods for future studies of societal implications; (3) propose a vision for accomplishing nanotechnology's promise while minimizing undesirable consequences.

This report sponsored by NSF incorporates fully the views, opinions and presentations contributed by workshop participants and other leading experts. The NSET report to the NSTC Committee on Technology presents a more concise perspective. The workshop participants offered recommendations to: (a) accelerate the beneficial use of nanotechnology while diminishing the risks, (b) improve research and education, and (c) guide the contributions of key organisations. These recommendations, summarized below, serve as a basis for both the NNI participants and the public to begin addressing societal issues of nanotechnology:

- Make support for social and economic research studies on nanotechnology a high priority.
- Include social science research on the societal implications in the nanotechnology research centers, and consider creation of a distributed research center for social and economic research. Build openness, disclosure, and public participation into the process of developing nanotechnology research and development programme direction.
- The National Nanotechnology Coordination Office (NNCO) should establish a mechanism to inform, educate, and involve the public regarding potential impacts of nanotechnology. The NNCO should receive feedback from the nanotechnology community, social scientists, the private sector, and the public with the goals of (a) continuously monitoring the potential societal opportunities and challenges; and (b) providing timely input to responsible organisations.
- Create the knowledge base and institutional infrastructure to evaluate nanotechnology's scientific, technological, and societal impacts and implications from short-term (3 to 5 year), medium-term (5 to 20 year), and long-term (over 20 year) perspectives. This must include interdisciplinary research that incorporates a systems approach (research-technology development-societal impacts), life cycle analysis, and real time monitoring and assessment.
- Educate and train a new generation of scientists and workers skilled in nanoscience and nanotechnology at all levels. Develop specific curricula and programmes designed to:
  a. introduce nanoscale concepts into mathematics, science, engineering, and technological education;
  b. include societal implications and ethical sensitivity in the training of nanotechnologists;

c. produce a sufficient number and variety of well-trained social and economic scientists prepared to work in the nanotechnology area;

d. develop effective means for giving nanotechnology students an interdisciplinary perspective while strengthening the disciplinary expertise they will need to make maximum professional contributions; and

e. establish fruitful partnerships between industry and educational institutions to provide nanotechnology students adequate experience with nanoscale fabrication, manipulation, and characterisation techniques.

- Encourage professional societies to develop forums and continuing education activities to inform, educate, and involve professionals in nanoscience and nanotechnology.

Over the next 10 to 20 years, nanotechnology will fundamentally transform science, technology, and society. However, to take full advantage of opportunities, the entire scientific and technology community must set broad goals; creatively envision the possibilities for meeting societal needs; and involve all participants, including the general public, in exploiting them.

A revolution is occurring in science and technology, based on the recently developed ability to measure, manipulate and organise matter on the nanoscale — 1 to 100 billionths of a meter. At the nanoscale, physics, chemistry, biology, materials science, and engineering converge toward the same principles and tools. As a result, progress in nanoscience will have very far-reaching impact. The nanoscale is not just another step toward miniaturisation, but a qualitatively new scale. The new behaviour is dominated by quantum mechanics, material confinement in small structures, large interfacial volume fraction, and other unique properties, phenomena and processes. Many current theories of matter at the microscale have critical lengths of nanometer dimensions. These theories will be inadequate to describe the new phenomena at the nanoscale. As knowledge in nanoscience increases worldwide, there will likely be fundamental scientific advances. In turn, this will lead to dramatic changes in the ways materials, devices, and systems are understood and created. Innovative nanoscale properties and functions will be achieved through the control of matter at its building blocks: atom-byatom, molecule-by-molecule, and nanostructure-by-nanostructure. Nanotechnology will include the integration of these nanoscale structures into larger material components, systems, and architectures. However, within these larger scale systems the control and construction will remain at the nanoscale. Today, nanotechnology is still in its infancy, because only rudimentary nanostructures can be created with some control. However, among the envisioned breakthroughs are orders-of-magnitude increases in computer efficiency, human organ restoration using engineered tissue, "designer" materials created from directed assembly of atoms and molecules, as well as emergence of entirely new phenomena in chemistry and physics. Nanotechnology has captured the imaginations of scientists, engineers and economists not only because of the explosion of discoveries at the nanoscale, but also because of the potential societal implications. A White House letter (from the Office of Science and Technology Policy and Office of Management and Budget) sent in the fall of 2000 to all Federal agencies has placed nanotechnology at the top of the list of emerging fields of research and development in the United States.

The National Nanotechnology Initiative was approved by Congress in November 2000, providing a total of $422 million spread over six departments and agencies. Nanotechnology's relevance is underlined by the importance of controlling matter at the nanoscale for healthcare, the environment, sustainability, and almost every industry. There is little doubt that the broader implications of this nanoscience and nanotechnology revolution for society at large will be profound.

National Nanotechnology Initiative

The National Nanotechnology Initiative (NNI, http://nano.gov) is a multi-agency effort within the U.S. Government that supports a broad programme of Federal nanoscale research in materials, physics, chemistry, and biology. It explicitly seeks to create opportunities for interdisciplinary work integrating these traditional disciplines. The NNI will accelerate the pace of fundamental research in nanoscale science and engineering, creating the knowledge needed to enable technological innovation, training the workforce needed to exploit that knowledge, and providing the manufacturing science base needed for future commercial production. Potential breakthroughs are possible in areas such as materials and manufacturing, medicine and healthcare, environment and energy, biotechnology and agriculture, electronics and information technology, and national security.

The effect of nanotechnology on the health, wealth, and standard of living for people in this century could be at least as significant as the combined influences of microelectronics, medical imaging, computer-aided engineering, and man-made polymers developed in the past century. The NNI is balanced across five broad activities: fundamental research; grand challenges; centers and networks of excellence; research infrastructure; and societal/workforce implications. Under this last activity, nanotechnology's effect on society – legal, ethical, social, economic, and workforce preparation – will be studied to help identify potential concerns and ways to address them. As the NNI is commencing, there is a *rare opportunity to integrate the societal studies and dialogues from the very beginning and to include societal studies as a core part of the NNI investment strategy.*

NSET Workshop on "Societal Implications of Nanoscience and Nanotechnology"

Research on societal implications will boost the NNI's success and help us to take advantage of the new technology sooner, better, and with greater confidence. Toward this end, the National Science and Technology Council (NSTC), Committee on Technology (CT), Subcommittee on Nanoscale Science, Engineering, and Technology (NSET).... the Federal interagency group that coordinates the NNI, sponsored a workshop on "Societal Implications of Nanoscience and Nanotechnology."

Held September 28-29, 2000 at the National Science Foundation, this workshop brought together nanotechnology researchers, social scientists, and policy makers representing academia, government, and the private sector. It had four principal objectives:

- Survey current studies on the societal implications of nanotechnology (educational, technological, economic, medical, environmental, ethical, legal, cultural, etc.).
- Identify investigative and assessment methods for future studies of societal implications.
- Propose a vision and alternative pathways toward that vision integrating short-term (3 to 5 year), medium-term (5 to 20 year), and long-term (more than 20 year) perspectives.
- Recommend areas for research investment and education improvement. This report addresses issues far broader than science and engineering, such as how nanotechnology will change society and the measures to be taken to prepare for these transformations. The conclusions and recommendations in this report will provide a basis for the NNI participants and the public to address future societal implications issues.

Nanotechnology Goals

Nanoscale science and engineering will lead to better understanding of nature; advances in fundamental research and education; and significant changes in industrial manufacturing, the economy, healthcare, and environmental management and sustainability. Examples of the promise of nanotechnology, with projected total worldwide market size of over $1 trillion annually in 10 to 15 years, include the following:

- *Manufacturing*: The nanometer scale is expected to become a highly efficient length scale for manufacturing once nanoscience provides the understanding and nanoengineering develops the tools. *Materials* with high performance, unique properties and functions will be produced that traditional chemistry could not create. Nanostructured materials and processes are estimated to increase their market impact to about $340 billion per year in the next 10 years (Hitachi Research Institute, personal communication, 2001).

- *Electronics*: Nanotechnology is projected to yield annual production of about $300 billion for the semiconductor industry and about the same amount more for global integrated circuits sales within 10 to 15 years (see R. Doering, page 74-75 of this report).
- *Improved Healthcare*: Nanotechnology will help prolong life, improve its quality, and extend human physical capabilities.
- *Pharmaceuticals*: About half of all production will be dependent on nanotechnology— affecting over $180 billion per year in 10 to 15 years (E. Cooper, Elan/Nanosystems, personal communication, 2000).
- *Chemical Plants*: Nanostructured catalysts have applications in the petroleum and chemical processing industries, with an estimated annual impact of $100 billion in 10 to 15 years (assuming a historical rate of increase of about 10% from $30 billion in 1999; "NNI: The Initiative and Its Implementation Plan," page 84).
- *Transportation*: Nanomaterials and nanoelectronics will yield lighter, faster, and safer vehicles and more durable, reliable, and cost-effective roads, bridges, runways, pipelines, and rail systems. Nanotechnology-enabled aerospace products alone are projected to have an annual market value of about $70 billion in ten years (Hitachi Research Institute, personal communication, 2001).
- *Sustainability*: Nanotechnology will improve agricultural yields for an increased population, provide more economical water filtration and desalination, and enable renewable energy sources such as highly efficient solar energy conversion; it will reduce the need for scarce material resources and diminish pollution for a cleaner environment. For example, in 10 to 15 years, projections indicate that nanotechnology-based lighting advances have the potential to reduce worldwide consumption of energy by more than 10%, reflecting a savings of $100 billion dollars per year and a corresponding reduction of 200 million tons of carbon emissions ("NNI: The Initiative and Its Implementation Plan," page 93).

Knowledge and Scientific Understanding of Nature

The study of nanoscale systems promises to lead to fundamentally new advances in science and engineering and in our understanding of biological, environmental, and planetary systems. It also will redirect our scientific approach toward more generic and interdisciplinary research. Nanoscience is at the unexplored frontiers of science and engineering, and it offers one of the most exciting opportunities for innovation in technology. Nanotechnology will provide the capacity to create affordable products with dramatically improved performance. This will come through a basic understanding of ways to control and manipulate matter at the nanometer scale and through the incorporation of nanostructures and nanoprocesses into technological innovations. It will be a center of intense international competition when it lives up to its promise as a generator of technology. Nanotechnology promises to be a dominant force in our society in the coming decades. Commercial inroads in the hard disk, coating, photographic, and pharmaceutical industries have already shown how new scientific breakthroughs at this scale can change production paradigms and revolutionize multibillion-dollar businesses. However, formidable challenges remain in fundamental understanding of systems on this scale before the potential of nanotechnology can be realised.

In modern time, nanotechnology is still in its infancy, and only rudimentary nanostructures can be created with some control. The science of atoms and simple molecules, on one end, and the science of matter from microstructures to larger scales, on the other, are generally established. The remaining size-related challenge is at the nanoscale — roughly between 1 and 100 molecular diameters — where the fundamental properties of materials are determined and can be engineered. A revolution has been occurring in science and technology, based on the recently developed ability to measure, manipulate and organise matter on this scale. Recently discovered organised structures of matter (such as carbo nanotubes, molecular motors, DNA-based assemblies, quantum dots, and molecular switches) and new phenomena (such as giant magnetoresistance, coulomb blockade, and those caused by size confinement) are scientific breakthroughs that merely hint at possible future developments. The nanoscale is not just another step toward miniaturisation, but a qualitatively new scale. The new behaviour is dominated by quantum mechanics, material confinement in small structures, large interfaces, and other unique properties, phenomena and processes. Many current theories of matter at the microscale have critical lengths of nanometer dimensions; these theories will be inadequate to describe the new phenomena at the nanoscale. Nanoscience will be an essential component in better understanding of nature in the next decades. Important issues include greater interdisciplinary research collaborations, specific education and training, and transition of ideas and people to industry.

Industrial Manufacturing, Materials and Products

Nanotechnology is fundamentally changing the way materials and devices will be produced in the future. The ability to synthesize nanoscale building blocks with precisely controlled size and composition and then to assemble them into larger structures with unique properties and functions will revolutionize materials and manufacturing. Researchers will be able to develop material structures not previously observed in nature, beyond what classical chemistry can offer. Some of the benefits that nanostructuring can bring include lighter, stronger, and programmable materials; reductions in life-cycle costs through lower failure rates; innovative devices based on new principles and architectures; and use of molecular/cluster manufacturing, which takes advantage of assembly at the nanoscale level for a given purpose.

The Semiconductor Industry Association (SIA) has developed a roadmap for continued improvements in miniaturisation, speed, and power reduction in information processing devices — sensors for signal acquisition, logic devices for processing, storage devices for memory, displays for visualisation, and transmission devices for communication. The SIA roadmap projects the future of nanoelectronics and computer technology to approximately 2010 and to 0.1 micron (100 nanometer) structures, just short of fully nanostructured devices. The roadmap ends short of true nanostructured devices because the principles, fabrication methods, and techniques for integrating devices into systems at the nanoscale are generally unknown. New approaches such as chemical and biomolecular computing, and quantum computing making use of nanoscale phenomena and nanostructures, are expected to emerge. The molecular building blocks of life — proteins, nucleic acids, lipids, carbohydrates, and their non-biological mimics — are examples of materials that possess unique properties determined by their size, folding, and patterns at the nanoscale. Biosynthesis and bioprocessing offer fundamentally new ways to manufacture chemicals and pharmaceutical products. Integration of biological building blocks into synthetic materials and devices will allow the combination of biological functions with other desirable materials properties. Imitation of biological systems provides a major area of research in several disciplines. For example, the active area of bio-mimetic chemistry is based on this approach.

Medicine and the Human Body

Living systems are governed by molecular behaviour at the nanometer scale, where chemistry, physics, biology, and computer simulation all now converge. Recent insights into the uses of nanofabricated devices and systems suggest that today's laborious process of genome sequencing and detecting the genes' expression can be made dramatically more efficient through use of nanofabricated surfaces and devices. Expanding our ability to characterize an individual's genetic makeup will revolutionize diagnostics and therapeutics. Beyond facilitating optimal drug usage, nanotechnology can provide new formulations and routes for drug delivery, enormously broadening the drugs' therapeutic potential. Increasing nanotechnological capabilities will also markedly benefit basic studies of cell biology and pathology. As a result of the development of new analytical tools capable of probing the world of the nanometer, it is becoming increasingly possible to characterize the chemical and mechanical properties of cells (including processes such as cell division and locomotion) and to measure properties of single molecules. These capabilities complement (and largely supplant) the ensemble average techniques presently used in the life sciences. Moreover, biocompatible, high-performance materials will result from the ability to control their nanostructure. Artificial inorganic and organic nanoscale materials can be introduced into cells to play roles in diagnostics (e.g., quantum dots in visualisation), but also potentially as active components. Finally, nanotechnologyenabled increases in computational power will permit the characterisation of macromolecular networks in realistic environments. Such simulations will be essential for developing biocompatible implants and for studying the drug discovery process. An open issue is how the healthcare system would change with such large changes in medical technology.

Agriculture, Water and Clean Environment

Nanotechnology will lead to dramatic changes in the use of natural resources, energy, and water, as outlined in the following paragraphs. Waste and pollution will be minimized. Moreover, new technologies will allow recovery and reuse of materials, energy, and water.

Agriculture

Nanotechnology will contribute directly to advancements in agriculture in a number of ways: (1) molecularly engineered biodegradable chemicals for nourishing plants and protecting against insects; (2) genetic improvement for animals and plants; (3) delivery of genes and drugs to animals; and (4) nano-array-based technologies for DNA testing, which, for example, will allow a scientist to know which genes are expressed in a plant when it is exposed to salt or drought stress. The application of nanotechnology in agriculture has only begun to be appreciated.

Water

Global population is increasing while fresh water supplies are decreasing. The United Nations predicts that by the year 2025 that 48 countries will be short of fresh water accounting for 32% of the world's population ("NNI: The Initiative and Its Implementation Plan", page 95). Water purification and desalinisation are some of the focus areas of preventative defense and environmental security since they can meet future water demands globally. Consumptive water use has been increasing twice as fast as the population and the resulting shortages have been worsened by contamination. Nanotechnology-based devices for water desalinisation have been designed to desalt sea water using at least 10 times less energy than state-of–the art reverse osmosis and at least 100 times less energy than distillation. The critical experiments underpinning these estimations are underway now. This energy-efficient process is possible by fabricating of very high surface area electrodes that are electrically conductive using aligned carbon nanotubes, and by other innovations in the system design.

Clean Environment

Nanoscience and engineering could significantly affect molecular understanding of nanoscale processes that take place in the environment; the generation and remediation of environmental problems through control of emissions; the development of new "green" technologies that minimize the production of undesirable by-products; and the remediation of existing waste sites and streams. Nanotechnology also will afford the removal of the smallest contaminants from water supplies (less than 200 nanometers) and air (under 20 nanometers) and the continuous measurement and mitigation of pollution in large areas. In order to hasten the integrated understanding of the environmental role of nanoscale phenomena, scientists and engineers studying the fundamental properties of nanostructures will need to work together with those attempting to understand complex processes in the environment. Model nanostructures can be studied, but in all cases the research must be justified by its connection to naturally occurring systems or to environmentally beneficial uses.

Environments for investigations are not limited and might include terrestrial locations such as acid mines, subsurface aquifers, or polar environments. *Energy:* Nanotechnology has the potential to significantly impact energy efficiency, storage, and production. Several new technologies that utilize the power of nanostructuring, but developed without benefit of the new nanoscale analytical capabilities, illustrate this potential:

- Increasing the efficiency of converting solar energy into useful forms.
- High efficiency fuel cells, including hydrogen storage in nanotubes.
- A long-term research programme in the chemical industry on the use of crystalline materials as catalyst supports has yielded catalysts with well-defined pore sizes in the range of 1 nanometer to reduce energy consumption and waste; their use is now the basis of an industry that exceeds $30 billion a year ("NNI: The Initiative and Its Implementation Plan,").
- Developed by the oil industry, the ordered mesoporous material MCM-41 (known also as "self-assembled monolayers on mesoporous supports," SAMMS), with pore sizes in the range of 10-100 nanometers, is now widely used for the removal of ultrafine contaminants (see work performed at Pacific Northwest National Laboratory in *Nanotechnology Research Directions,* Kluwer Academic Publishers, 2000, pp. 216-218).

- Several chemical manufacturing companies are developing a nanoparticle-reinforced polymeric material that can replace structural metallic components in automobiles; widespread use of those nanocomposites could lead to a reduction of 1.5 billion liters of gasoline consumption over the life of one year's production of vehicles, thereby reducing carbon dioxide emissions annually by more than 5 billion kilograms ("NNI: The Initiative and Its Implementation Plan,").
- Significant changes in lighting technologies are expected in the next ten years. Semiconductors used in the preparation of light emitting diodes (LEDs) for lighting can increasingly be sculpted on nanoscale dimensions. In the United States, roughly 20% of all electricity is consumed for lighting, including both incandescent and fluorescent lights. In 10 to 15 years, projections indicate that such nanotechnologybased lighting advances have the potential to reduce worldwide consumption of energy by more than 10%, reflecting a savings of $100 billion dollars per year and a corresponding reduction of 200 million tons of carbon emissions ("NNI: The Initiative and Its Implementation Plan," pages 92-93).
- The replacement of carbon black in tires by nanometer-scale particles of inorganic clays and polymers is a new technology that is leading to the production of environmentally friendly, wear-resistant tires.

National Security

Defense applications include (1) continued information dominance through advanced nanoelectronics, identified as an important capability for the military; (2) more sophisticated virtual reality systems based on nanostructured electronics that enable more affordable, effective training; (3) increased use of enhanced automation and robotics to offset reductions in military manpower, reduce risks to troops, and improve vehicle performance; (4) achievement of the higher performance (lighter weight, higher strength) needed in military platforms while providing diminished failure rates and lower life-cycle costs; (5) needed improvements in chemical/biological/nuclear sensing and in casualty care; (6) design improvements in systems used for nuclear non-proliferation monitoring and management; and (7) combined nanomechanical and micromechanical devices for control of nuclear defense systems. In many cases economic and military opportunities are considered to be complementary. Strong applications of nanotechnology in other areas would provide support for national security in the long term, and vice versa.

Space Exploration

The stringent fuel constraints for lifting payloads into earth orbit and beyond, and the desire to send spacecraft away from the sun for extended missions (where solar power would be greatly diminished) compel continued reduction in size, weight, and power consumption of payloads. Nanostructured materials and devices promise solutions to these challenges. Nanostructuring is also critical to the design and manufacture of lightweight, high-strength, thermally stable materials for aircraft, rockets, space stations, and planetary/solar exploratory platforms. The augmented utilisation of miniaturized, highly automated systems will also lead to dramatic improvements in manufacturing technology. Moreover, the low-gravity, high-vacuum space environment may aid the development of nanostructures and nanoscale systems that cannot be created on Earth.

Moving into the Market

Since economists have not yet really begun research on nanotechnology, their insights are somewhat tentative and based on experience with earlier technologies. A common paradigm is that new applications will be initially more costly than existing technologies, but will achieve better performance. However, completely new technologies may be cheaper, such as chemical manufacturing to mass produce nanoelectronic circuits as opposed to current methods using lithography in microelectronics. Overall, nanotechnology will offer substantial advantages, being smaller, faster, stronger, safer, and more reliable. At the same time, it will require investments in new production facilities and in a host of ancillary industries supplying the raw materials, components, and manufacturing machines. Because it will take time to achieve economies of scale and to develop the most efficient fabrication methods, costs are likely to be relatively high in the beginning.

For this reason, nanotechnology-based goods and services will probably be introduced earlier in those markets where performance characteristics are especially important and price is a secondary consideration. Examples are medical applications and space exploration. The experience gained will reduce technical and production uncertainties and prepare these technologies for deployment into the market place. Similarly, in the private sector, technology transfer is likely to occur from performance-oriented areas (such as medicine) to price-oriented ones (such as agriculture). As a given technology matures, its cost may decline, leading to greater penetration of the market even where performance is not decisive.

The displacement of an old technology by a new one tends to be both slow and incomplete. Displacement of older methods will accelerate to the extent that nanotechnology extends its technical range and perhaps lowers its relative price. However, nanotechnology also is likely to stimulate innovations in older technologies that make them better able to compete — an ironic but potentially beneficial secondorder effect. The diffusion and impact of nanotechnology will be partly a function of the development of complementary technologies and of a network of users. Whole new industries may have to be developed — along with the trained scientists and technicians to staff them. There may be many obstacles along this road that ordinary market processes cannot easily overcome.

An important role for the government will be to invest in the long-term, high-risk, high-gain research needed to create these new industries and to ensure that they are consistent with broader societal objectives. Federal support of the nanotechnology initiative is necessary to enable the United States to take advantage of this strategic technology and remain competitive in the global marketplace well into the future. Focused research programmes on nanotechnology have been initiated in other industrialized countries. Currently, the United States has a lead on synthesis, chemicals, and biological aspects; it lags in research on nanodevices, production of nano-instruments, ultra-precision engineering, ceramics, and other structural materials. Japan has an advantage in nanodevices and consolidated nanostructures; Europe is strong in dispersions, coatings, and new instrumentation.

8

Towards Explaining the Differences Among Nanoscale Scanning, Nanosensors, Nanowires and Nanoribbons: An Overview

Meaning of Nanosensors, Gas Sensors and Nanoscale Scanning

Medical nanorobots need to acquire information from their environment to properly execute their assigned tasks. Such acquisition is achieved using onboard nanoscale sensors, or nanosensors, of various types. Nanosensors allow for medical nanodevices to monitor environmental states at three different operational levels:

(i) Internal nanorobot states.

(ii) Local and global somatic states (inside the human body).

(iii) Extrasomatic states (sensory data originating outside the human body).

The general physical limits to sensory perception are reviewed by Block and Bialek. The specific nanosensor technologies required include sensors to detect chemical substances, displacement and motion, force and mass, and acoustic, thermal, and electromagnetic stimuli. The area of chemical microsensors is well developed, and there is increasing research interest in nanoscale chemical sensors, "chemosensors" and "biosensors". The most common nanomedical application of chemical sensors will be to measure the concentration of specific molecules and biopolymers in aqueous solvent—whether in blood serum, interstitial fluid, or cytosol. This may be accomplished using ordered arrays of receptors or by using sorting rotors to directly count molecular populations in known sample volumes. Spatial and temporal concentration gradient sensing is also essential in navigation, communications, and in mediating rapid response to environmental stimuli. Chemotactic sensors may be used to sample the chemical composition of surfaces.

One simple broadband concentration sensor, shown schematically consists of a graduated series of receptors having uniformly high specificity but engineered with progressively greater affinities (successively smaller equilibrium dissociation constants K_d (molecules/nm^3); for the target molecule, which is present in the test sample at concentration c_{ligand}. Exposure of the receptor array to the test sample for a time t_{EQ} necessary to reach diffusion-driven equilibrium gives a probability of receptor occupancy $P_{occ} \sim 0.91$ in a receptor with $K_d \sim 0.1c_{ligand}$, $P_{occ} \sim 0.50$ in a receptor with $K_d \sim c_{ligand}$, and $P_{occ} \sim 0.09$ in a receptor with $K_d \sim 10c_{ligand}$. During each measurement cycle, all steric probes are simultaneously extended into their associated receptor volumes at a time t_{EQ} after presentation of sample to the array. Probes which reach full extension register an empty receptor; those which cannot fully extend register an occupied receptor.

After registration, the probes are retracted and ejection rods are thrust into all receptors (typically requiring ~1 nanonewton (nN) of force for occupied receptors) to empty them while the test chamber is flushed clear in preparation for the next cycle to begin. Ideally, the probe rod pushing into the binding site should push the bound molecule further into the binding site to avoid complications that may arise due to competing kinetic barriers, and ejection rods should push molecules far enough to ensure that the next resample is independent. Consider a series of N steric probe units as proposed by Drexler. Each probe measures 8 nm x (2.5 nm)^2 ~ 50 nm^3 and has mass ~ 2 x 10^{-22} kg.

The ratio of the dissociation constants of adjacent units is $k = K_{di} / K_{di+1} > 1$, with $K_{d1} \sim 10^{-4}$ molecules/nm^3 and $K_{dN} \sim 10^{-13}$ molecules/nm^3 typically in the human body, and so $N = 1 + \{\log_{10} (K_{d1}/K_{dN}) / \log_{10} (k)\}$. Minimum measurement error occurs when c_{ligand} exactly matches the K_d of a probe unit, that is, when $c_{ligand} = K_{di}$ (i.e., $P_{occupiedi}$ = 0.5). Maximum measurement error occurs when c_{ligand} lies exactly midway between two probes such that $K_{di} > c_{ligand} > K_{di+1}$, or, more specifically, at the geometric midpoint $c_{ligand} = (K_{di} K_{di+1})^{1/2} = K_{di} k^{-1/2}$. For common small molecules, $t_{EQ} \sim 10^{-6}$ sec; for common large molecules, $t_{EQ} \sim 0.2 \times 10^{-6}$ sec, but a sample chamber ~(25 nm)^3 in size is diffusion-limited to a minimum $t_{EQ} \sim 10^{-5}$ sec for $c_{ligand} >\sim 10^{-4}$ nm^{-3}. For the rarest small molecules in the human body, $t_{EQ} \sim 800$ sec, ~200 sec for rare large molecules, and a larger chamber may also be required. Preconcentration of the test sample by dehydration using sorting rotors to rapidly remove water from the sample chamber reduces t_{EQ} to 10-40 sec for very rare molecules.

Steric probes using a driving force of ~100 pN, over 2 nm of motion per cycle require 200 zJ/measurement. If 1000 measurement cycles of a 10-probe sensor suffice to determine concentration to ~2 significant figures, this information costs ~0.002 pJ to acquire, a power consumption of from ~10 pW for large common molecules to ~10^{-9} pW (~0.6 kT/sec) for small rare molecules while the sensor is working. Power dissipation may run up to ~10 times higher, depending upon the number of occupied receptors that must be cleared via ejection rods during each cycle. If the concentration of the target ligand varies only slightly within known limits, similar results may be obtained using a narrowband sensor that employs a repeating array of a single receptor type tuned to the maximum expected concentration c_{max} of the target ligand. If the sensor consists of N_r receptors, choosing K_d >~ c_{max} $N_r^{-1}/(1 - N_r^{-1})$ ensures that all receptors are occupied at $c = c_{max}$, thus avoiding array saturation at concentrations below c_{max}. At minimum sensitivity, $K_d \sim c_{min} (1 - N_r^{-1})/ N_r^{-1}$, and so $(1 + (Dc/c)) = (c_{max}/c_{min})^{(Nr-1)}$ for the entire array.

Thus a sensor with N_r = 100 receptor units achieves Dc/c = 0.10 (10 per cent) with a $(46\ nm)^3$ sensor; N_r = 1400 receptor units achieves Dc/c = 0.01 (1 per cent) with a $(112\ nm)^3$ sensor. A dedicated micron-scale nanodevice employing a $(1.44\ micron)^3$ sensor array having 3 million receptor cells could at best distinguish 0.001 per cent concentration differentials in ~ 3 sec for common molecules like serum glucose. The ion channel switch (ICS) biosensor (aka. the "Australian sensor") can detect small molecule concentrations as low as $\sim 10^{-14}\ nm^{-3}$ in a ~600 sec measurement time. Concentration may be determined by varying exposure time to the sample. One energy-efficient designuses an input sorting rotor running at varying speeds (according to target concentration) synchronised with a counting rotor (linked by rods and ratchets to a data storage register device) to assay the number of molecules of the desired type that are present in a known volume of fluid. The counting rotor uses a steric probe, moved up and down in synchrony with the arrival of a binding site under the test probe, to count the number of molecules transferred by the sorting rotor. The fluid sample is drawn from the environment into a 4000 nm^3 reservoir ($L_{reservoir}$ ~ 16 nm) with the equivalent of 10^4 refills/sec on a flow-through basis, using paddlewheel pumps or reciprocating pistons; not shown) operated slowly enough to avoid damaging molecules in the sample volume. This sensor, which measures 45 nm × 45 nm × 10 nm comprising ~500,000 atoms ($\sim 10^{-20}$ kg), should count 'N ~ $10^4 - 10^5$ molecules/sec of small common molecules present at typical human serum concentrations of $10^{-3} - 10^{-2}\ nm^{-3}$, such as glucose, with the rotor spinning at ~1 per cent of normal speed. For rare molecules with c ~ $10^{-11}\ nm^{-3}$, the count rate for a single sensing rotor falls to 'N ~ 10^{-4} molecules/sec. Sample preconcentration may allow further improvement by a factor of up to 10-100. Energy requirements are comparable to those for sorting rotors except for the need to retrieve ~all target molecules from a known sample volume, which slows the sorting process, requiring ~10 zJ/count for common molecules and ~100 zJ/count for the rarest molecules in the human body. Large-molecule concentrations may be measured using a shuttle pump and a counting device. It may also be useful to design a sensor that is capable of detecting as many different chemical species as possible, in contrast to detecting the precise concentration of a single species, for use in chemomessaging, chemonavigation, cellular diagnosis, and other in vivo assay work. Counting rotors may be used for this purpose. Each 12-receptor sorting rotor/counting rotor pair has a minimum 400 nm^3 volume. Allowing an additional ~800 nm^3 per pair to account for power, control, mechanical attachments and housings, which includes ~300 nm^3 per pair for sample chamber volume and sample access, requires ~100 nm^3 per receptor, roughly the same as for the steric probe units described earlier. A dedicated 1 $micron^3$ chemical assay nanorobot, which includes ~0.25 $micron^3$ sample volume, with ~10^7 pairs (receptors) could thus continuously scan for as many as ~10^7 different chemical species, counting 'N ~ 1 large molecule/sec at concentrations of c_{ligand} ~ 10^{-10} molecules/nm^3.

For a purely sensory spherical nanodevice, a uniform distribution of a relatively small number of chemoreceptors confers optimal sensitivity. However, in applications where sensors (which attract and then release) are in competition with pumps (which strongly attract and then remove) for the same target molecules at the nanodevice surface, greater sensing efficiency may be achieved by clustering the sensors, as in a "nose" organ. Proteins may be modified as they age within biological cells. For instance, under steady state conditions, ~10 per cent of protein molecules may exhibit carbonyl (oxidation) modifications, glycosylation (non-enzymatic), and dysfunctional proteins may be ubiquitinised and phosphorylated under enzyme control. Heat shock also modifies proteins. Chemosensor receptors must be carefully designed to accommodate changes in target ligand structures or sites that are likely to be subject to such alterations. Medical nanodevices can detect both extremely steep and extremely shallow spatial concentration gradients. For example, a receptor array sensor consisting of 10 steric probe units has a total volume of ~1000 nm^3 including its prorated share of drive systems, housings, etc. In just a few measurement cycles, adjacent sensors can detect a maximum concentration change from 10^{-2} nm^{-3} to 10^{-11} nm^{-3} across a mean nanosensor separation of 10 nm on a nanodevice surface, a steep chemical spatial gradient of ~10^{10} per cent per nanometer. J. Soreff points out that maintaining a steep concentration gradient of a freely diffusing species implies a power dissipation of ~kT per molecule per time to diffuse a concentration-halving distance. For a common, small, tethered molecule with c_{ligand} ~ 3 × 10^{-3} molecules/nm^3 that diffuses ~1 nm in t_{diff} ~ 1 nanosec, implied power density D_{mol} ~ kT c_{ligand}/t_{diff} ~ 10^{13} watts/m^3, a power density that might possibly be encountered in the vicinity of chemoelectric transducers.

Similarly, counting rotor sensors for small common serum molecules like glucose deployed at opposite ends of a 1-micron nanodevice can detect a concentration differential of ~10^{-4} over a ~1 sec measurement period, a shallow chemical spatial gradient of only ~10^{-5} per cent per nanometer. Even lower gradients may be detected by communication between physically separated cooperating nanodevices. Koshland and Macnab have shown that bacteria possess a mechanism for sensing small temporal concentration gradients. A bacterium detects a spatial gradient of attractant not by comparing the concentration at its head and tail, but rather by traveling through space and comparing its observations over time. *E. coli* can measure a ~0.01 per cent gradient during the time it travels one body length (~2 microns) at top speed

(20-40 microns/sec), a minimum temporal gradient of 0.1 per cent per second. This appears very efficient, since the counting rotor sensor proposed above is not much better, achieving ~0.01 per cent per second. As for peak gradients, the 10-probe receptor array sensor can detect a maximum concentration change (e.g., from an already-known rare concentration) in ~10^{-4} sec, a peak temporal gradient of ~10^{15} per cent per second. Besides concentration sensing, chemical sensors may be used to determine the chemical characteristics of surfaces. A binding pad coated with an array of reversible, possibly reconfigurable, artificial receptors may be pressed against test surfaces such as cell membranes, or the perimeters of very large molecules, in order to detect the presence of specific chemical ligands. If pad receptor specificity is high for the target ligand, then the degree of pad "stickiness" to the test surface is a good measure of the population of target ligands on that surface. Probing 2-D surfaces with receptors may be constrained by orientation effects, since binding may require a specific relative orientation of ligand and receptor.

Compliant pad receptors can reduce this problem but may impair lateral resolution—there is a three-way tradeoff among angular tolerance, lateral resolution, and multioriented multicopy receptors. A physical contact sensor prevents confusion of surface-bound and free-solvated ligands of the same species. The selective and reversible stickiness of such pads can also be exploited for locomotion. Molecular adhesion experiments in chemical force microscopy and interfacial force microscopy using AFM probes with functionalised tips typically detect maximum adhesive noncovalent force differentials of 40-160 piconewton (pN)/ligand for competing ligands at a peak-force distance of ~0.2 nm. Forces >10 pN are readily measured by nanosensors, so binding pad sensors should give an exact count of the number of target ligands present per unit area of test surface if (receptor areal number density) > (ligand areal number density) > (1 ligand/pad area), and if contact time is adequate. Receptors for many different ligands can be mounted on moving rotors or belts, or employed as interchangeable tool tips, thus allowing a single sensor pad to reconfigure its specificity or sensitivity under external control. Ligand mimics may also be used to allow a medical nanodevice to detect a particular receptor in its environment. Each such "receptor sensor" is a protruding physical structure that mimics all or part of the electrosteric structure of the ligand to which the target receptor has maximum specificity.

For example, probes consisting of artificial structures that simulate antigen peptides (8-15 amino acids long) bound in an MHC-like structure could be used to sense the presence of specific T-cell receptors. Sensing pads displaying ligand mimic structures are brought into contact with the test surface. Any target receptors present will bind to a mimic structure, thus may be counted as the sensor sweeps slowly across the surface. Immediately after recognition has occurred, the mimic structure may be reversibly fragmented, destroying receptor/mimic affinity and allowing the receptor to release the mimic. Mimics are mounted on the sensor pad with an appropriately elastic compliance to prevent structure detachment during a measurement scan. Receptor sensing is most useful in detecting receptors for large molecules.

In the case of receptors for small molecules, there may sometimes be insufficient space in the target molecular volume to match the surface electrosteric structure, to accommodate ligand fragmentation control rods or cables, or to build internal boundaries sufficient to permit ligand fragmentation without creating undesirable reactive species. Some of these problems with small-molecule receptors may be partially overcome by employing an annular repulsive ring through which a mimic ligand may be forcibly retracted without fragmentation, but receptor orientation effects will present additional constraints. The ability to precisely measure position and displacement, velocity and acceleration, rotation and orientation are essential requirements for nanoscale locomotion, physical manipulation, and navigation. Diamondoid parts may contain features, or be positioned, in picometer (1 pm = 10^{-12} m) steps, much smaller than the unavoidable RMS thermal vibrations. The nanomanipulator robot arm may be moved in picometer increments, and the RMS thermal longitudinal displacements at the tip of a 20-nm long, 10-nm wide diamondoid lever is ~1 pm. Displacement sensitivity of 10 pm is routinely achieved in STMs. It is claimed this can be improved down to the ~1 pm level; STM resolution of 2 pm has been demonstrated experimentally. A C_{60} molecule used as the active element in an electromechanical amplifier transmits ~100 times more electrical current when physically compressed by 100 pm, suggesting a minimum deformation detection limit of a few picometers. Assuming a very stiff k_s = 600 N/m, a minimal SNR = 1 gives a minimum detectable displacement of 6 pm, or 10 pm at a more reasonable SNR = 2 (20 dB).

The conservative conclusion is that Dx_{min} ~ 10 pm displacements should be reliably detectable by medical nanosensors in vivo, in a measurement time t_{meas} ~ 10^{-9} sec. This compares favourably with the stereocilia of the inner ear, which can only detect 100 pm displacements in 10^{-5} sec. Typical thermal velocities of nanoscale diamondoid components at 310 K, are ~60 m/sec for 1 nm^3 objects, ~2 m/sec for $(10\ nm)^3$ objects, 0.06 m/sec for $(100\ nm)^3$ objects, and 0.02 m/sec for 1 $micron^3$ objects. These thermal motions are in degrees of freedom whose range of motion is largely confined, by design, to sub-nanometer distances within nanoscale measuring devices. However, objects of nanomedical interest may have velocities ranging from ~10^{-9} m/sec to ~10^3 m/sec (e.g., the speed of sound in water), and may include biological elements in the environment or internal machine components. Thus it is useful to explore the fundamental limitations involved in measuring the full range of this physical variable. There are at least two simple ways to determine a velocity. First, velocity may be measured by transferring the kinetic energy of a moving object or fluid volume to a sensor element. The movement of the sensor element is then clocked as it traverses a known distance. *Second,* and more efficiently, the passage of the moving object may sequentially trip two latches located a known distance apart, again revealing the moving object's velocity; alternatively, detecting the presence or absence of a protruding feature on a passing body using probe rods can determine velocity without any requirement for large kinetic energy transfers. The need for physical gating movements by sensor components and computational register-shift operations using diamondoid rods moving at ~GHz clocking speeds suggests a conservative minimum stable sensor cycle time of ~10^{-9} sec. The two principal restrictions on the maximum detectable velocity v_{max} are the maximum linear dimension of the sensor element, or L_{sensor}, and the minimum event time that may be accurately discriminated, or Dt_{min}. For a sensor element with thermal noise energy DE ~ 10 kT = 43 zJ at 310 K, the quantum mechanical limit is Dt_{min} >~$\hbar$/(2 DE) ~ 10^{-15} sec. Dt_{min} is also limited by the thermal variation in length of the individual latches. If the minimum detectable latch displacement is Dx ~ 10 pm, and the maximum speed of latch displacement is approximately the speed of sound in the latch material (e.g., v_{sound} ~ 17,300 m/sec for diamond), then Dt ~ Dx/v_{sound} ~ 10^{-15} sec, near the quantum mechanical limit. Dt is further restricted by the minimum detectable phase variance between latches located at either end of the sensor.

An acoustically-transmitted clock pulse emitted from a centrally-located source that is Dx closer to one of the two latches can discriminate a phase variance of >~Dx/L_{sensor} ~ 10^{-5} taking L_{sensor} ~ 1 micron, so the minimum discriminable time is again Dt_{min} ~ 10^{-5} L_{sensor}/v_{sound} ~ 10^{-15} sec; hence v_{max} ~ L_{sensor}/Dt_{min} ~ 10^{9} m/sec ~ c (= 3 × 10^{8} m/sec, the speed of light). However, unlike photons, physical objects moving in an aqueous medium are normally limited to the speed of sound in that medium, e.g., v_{sound} ~ 1500 m/sec in water at 310 K If v_{max} is taken as ~10^{3} m/sec, then for a sensor of size L_{sensor} ~ 1 micron the required time discrimination is only Dt_{min} ~ L_{sensor}/v_{max} ~ 10^{-9} sec. The minimum detectable velocity v_{min} is limited by the minimum measurable linear displacement Dx ~ 10 pm and the limits of our patience in making the measurement. For a maximum measurement time of t_{meas} ~ 1 sec, v_{min} ~ Dx/t_{meas} ~ 0.01 nm/sec in vacuo. In a nanomedically-relevant aqueous medium, the systematic motion of a free-floating (untethered) object of radius R is masked by added Brownian noise, giving the limiting $v_{brownian}$ = DX/t >~ 260 nm/sec for R = 10 micron and $v_{brownian}$ >~ 2.6 micron/sec for R = 100 nm, where t = t_{meas} = 1 sec. The thermal motions of tethered objects moving along surfaces (e.g., vesicles transported along microtubules by kinesin motors; are more restricted, hence Brownian velocity masking is reduced. Modest accuracy may be achieved using acoustic reflection in a Doppler velocimeter. Employing acoustic waves of frequency n, the minimum measurable velocity change Dv = v_{sound}/(n t_{meas}) = 150 microns/sec in an aqueous medium at 310 K with v_{sound} = 1500 m/sec, n ~ 10 MHz, and t_{meas} = 1 sec, corresponding to the shifted frequency n (1

+ (Dv/v_{sound})) having enough time to make one extra cycle during t_{meas}. Even at such long wavelengths (~150 microns), significant backscattering of ultrasound from ~7.82 micron red cells is observed experimentally.

Accelerations inside nanoscale devices are often measured in the trillions of g's (g = 9.81 m/sec^2). For example, an L = 1 nm diamondoid block (m = 3.5×10^{-24} kg) with thermal energy ~kT experiences random thermal accelerations of a_{therm} ~ kT/g m L = 10^{11}g's; a 10-nm logic rod with vibrational energy ~kT experiences ~10^{12} g oscillations; a 200-nm wide hoop spinning axially with 1000 m/sec rim speed has rotational acceleration of a ~ 10^{12} g's. As one member of a class of devices, a simple nanoscale omni-directional accelerometer is conceptualised as a reaction mass suspended by three pairs of reaction springs (one pair per orthogonal directional axis) with each spring connected to the walls of an evacuated box. Acceleration in any direction extends some springs and compresses others, which is measured by force or displacement sensors attached to each spring. The signal/noise ratio for this sensor is Sensor mass exchanges reciprocally with acceleration sensitivity, so the lower limit even for a dedicated micron-scale platinum-core box-spring accelerometer is a_{min} ~1000 g's at SNR ~ 2 (undamped resonant acceleration ~ 10^6 g's for L = 2 microns). The maximum detectable acceleration requires m ~ 10^{-23} kg (~1.4 nm diamond cube) and k_s ~ 1000 N/m, giving a_{max} ~5×10^{13} g's at SNR = 1.

The center of mass of a nanorobot, the sensor may avoid confusing centrifugal and translation forces. Otherwise, information from multiple sensors must be combined to obtain an unambiguous measurement. Multiple measurements can improve the SNR for constant accelerations of long duration. In 1998, ~100 nm vibrating-beam silicon accelerometers were under development by Analog Devices and Daimler-Benz. Tunneling accelerometers had been fabricated with ~10^6 g resolution at a 350 Hz bandwidth or ~10^{-8} g/Hz$^{1/2}$. Silicon-based capacitive accelerometers (5-1000 g full scale) manufactured by Silicon Designs (Issaquah, WA) were widely used in automotive airbag safety systems; in 1998, Motorola shipped its ten-millionth accelerometer, and the global market for automotive silicon accelerometers was expected to reach $463 million by 2002. Formed elements in the human bloodstream (e.g., red cells, white cells, platelets) typically are buffeted by much smaller accelerations. For example at 310 K, instantaneous random thermal accelerations of $a_{thermal}$ ~ 500 g's are experienced by a 0.2-micron virus particle, ~0.05 g's by a 2-micron platelet, ~10^{-4} g's by a 10-micron neutrophil—and ~0.4 g's by a 1 micron^3 spherical nanorobot. Force-based sensors cannot easily detect such low accelerations, since the smaller the sensor, the larger the a_{min} for a given SNR.

An alternative approach uses displacement sensors to determine an object's velocity twice in rapid sequence, allowing acceleration to be computed from the difference. Consider a constantly accelerating object that triggers three clocking latches at positions x_1, x_2, and x_3, in sequence. Time and position of the object are measured as (x_1, t_1), (x_2, t_2), and (x_3, t_3). Assume that a slowly accelerating external object passes three clocking latches that are fixed to the exterior of a nanorobot of radius R. Measurements of the object's velocity incur an unavoidable error due to nanorobot thermal motion. The object is observed to pass between latch pairs during a measurement time ~t_{meas}. During this same interval, the nanorobot translates a distance DX = (kT t_{meas}/3 phR)$^{1/2}$, incurring a maximum velocity measurement error of ~DX/t_{meas}, and thus from maximum acceleration measurement error of ~2 DX/t_{meas}^2 if (t_3 – t_2) ~ (t_2 - t_1) ~ t_{meas}.

The mitral and aortic valves in the heart (DP/Dx ~ 10^5 N/m^3, a_{fluid} ~ 10 g's, Dx ~ a few cm), as well as valves in the veins and lymphatics, operate as bistable analog decelerometers employing this principle. Micron-scale medical nanorobots traveling across blood vessel surfaces could use pressure sensors capable of 10^{-6} atm resolution to infer local blood flow accelerations to an accuracy of ± 0.01 g from sequential measurements taken 1 mm apart. Consider a cylindrical gyroscope spinning at an angular velocity w (radians/sec), with radius r and thickness h, whose center of gravity lies a distance L_p from a frictionless pivot point along the rotational axis. The pivot point defines the center of a spherical sensor grid of radius R ($r^2 + 4h^2$)$^{1/2}$ having distinguishable grid elements located a distance Dx ~ 1 nm apart. Under uniform acceleration a, the gyroscope precesses around its spin axis at an angular velocity w_p = 2 a L_p/w r^2 = Dx/R t_{meas}, where t_{meas} is measurement duration. In choosing an w with which to measure a_{min}, there is an upper and a lower limit. The lower bound is the minimum spin angular velocity below which the pivoted gyroscope cannot spin stably about the vertical axis and begins to wobble, thus ruining the measurement. If w_{max}/w_{min} > 1, then nonwobbling spin velocities are available below the bursting speed of the gyroscope cylinder.

Since this ratio scales with R, there is a minimum sensor size R_{min} below which no useful spin velocities are available. For t_{meas} = 10^{-3} sec, R_{min} >~ 69 microns (optimum r = 48 microns, h = 25 microns); for t_{meas} = 1 sec, R_{min} >~ 6.9 microns (optimum r = 4.8 microns, h = 2.5 microns). Thus it appears that pivoted gyroscopic accelerometers likely will be employed only in devices >> 10 microns in size. Microscale gyroscopes implemented in silicon have been discussed by Greiff and others. Aside from pure acceleration, the rate of accelerative onset—sometimes called jolt, surge, or jerk—can produce additional damaging effects on macroscale mechanical and biological structures. For example, the effects of accelerative onset ('a) on the human body were studied by Col. John Paul Stapp in a series of aggressive rocket sled experiments conducted during the 1950s. For human subjects, cardiovascular shock was entirely absent for 'a >~ 500-600 g/sec but began to appear at 'a >~ 1100-1400 g/sec, at 1-40 g accelerations. Human tolerance to linear decelerative force is determined both by rate of onset and peak acceleration; the endurance limit for well-restrained healthy young males has been given as <~500 g/sec and <~50 g for >~0.2 sec durations. By 1998, the effects of accelerative onset on nanomechanical structures had not been extensively studied, but similar limits might apply to nanomachines. For example, a diamond rod of length L_{rod} = 1 micron, cross-sectional area S_{rod} = 0.01 micron^2, mass $m_{rod} = 3.5 \times 10^{-17}$ kg, and working strength $W_{rod} \sim 5 \times 10^{10}$ N/m^2 fails when subjected to a static acceleration a_{fail} >~ W_{rod} S_{rod}/m_{rod} ~ 10^{12} g. But accelerative shocks can produce dynamic amplification in the response spectrum of elastic systems. Dynamic stress rises to a peak of roughly twice the corresponding static value when 'a_{peak} >~ 4 n_{res} a_{max} for a trapezoidal pulse shape, where n_{res} is the frequency of natural vibration of the system, because the stresses from the free oscillations of the system, induced by the changes in the acceleration of the system, can add to the static stresses from the acceleration, with damage done by the total stress. Taking k_s ~ 1 N/m for the rod described above, then n_{res} ~ 30 MHz and 'a_{peak} ~ 10^{20} g/sec at a_{max} ~ 10^{12} g. For comparison, the teragravity nanocentrifuge has mass m ~ 10^{-17} kg and is spun up to $a_{max} = 5 \times 10^{11}$ g's in >~10^{-6} sec, giving 'a <~ 5×10^{17} g/sec << 'a_{peak}. There are circumstances in which it is necessary for a medical nanodevice to measure, establish or maintain a preferred orientation, as for example nanorobots with a clearly defined "top" and "bottom" or nanodevices performing lengthy intracellular operations requiring precision stationkeeping. Measurement of rotation rates may also be required. Pivoted gyroscopes of the type are impractical for orientational sensing in all but the largest medical nanodevices. Setting w_{max} equal to w_{min}, taking a_{min} = g = 9.81 m/sec^2 (gravitational acceleration), and optimising for minimum sensor size gives a critical nonwobbling nonbursting spin velocity w_{crit} ~ 6.8×10^7 radians/sec at a minimum device radius of ~35 microns. If a pivoted gyroscope is smaller than this minimum radius, then spinning it slower than w_{crit} causes the gyro to tumble helplessly, while spinning it faster than w_{crit} tears the gyro apart.

However, a triaxially gimballed nanogyroscope may be regarded as having its pivot point near its center of mass, so L_p becomes very small and w_{min} becomes very slow, making nanogyroscopes feasible in medical nanorobots. Taking h = 1 micron, r = 0.5 micron, r = 3510 kg/m^3 and a_{min} = g, then if the gimbals are aligned such that L_p = 1 nm between the pivot point and the effective center of gravity along the rotation axis, then w_{min} ~ 400 rad/sec. If gimbal tolerances are improved to L_p = 0.1 nm, then w_{min} ~ 125 rad/sec. How stable is a nanogyroscopic orientation standard? There are two principal considerations. First, Brownian thermal rotation in all three angular degrees of freedom (q, j and y) gives rise to a small nutation around the nanogyroscopic precession axis of magnitude Dq_{nutate} ~ p_j/p_y, where p_j and p_y are angular momenta of the gyro. The minimum value for p_j is approximated by (1/2) kT ~ (1/2) p_j^2/I_1, where I_1 = (1/4) m r^2 + (1/12) m h^2, the moment of inertia around j, and m = p r^2 h r. Second, the gimbal bearings have small frictional losses, exerting small torques on the gyro and causing a small precession away from the original orientation.

The angular velocity of precession $w_{precess} \sim T_{gimbal}/(I_3\ w)$ (rad/sec) where T_{gimbal} is the frictional torque caused by an imperfect gimbal bearing. Assume that the gimbal is driven by external forces to oscillate at some frequency n_{gimbal}, and that a gimbal bearing dissipates power at the rate of P_{gimbal} (watts), so that each oscillatory motion of the gimbal dissipates $E_{gimbal} = P_{gimbal}/n_{gimbal}$ (joules). If the angular amplitude of the gimbal motion is a_{gimbal} (radians) and the gimbal radius is r_{gimbal}, then the gimbal bearing travels $X_{gimbal} = a_{gimbal}\ r_{gimbal}$ (meters) per gimbal oscillation and the force applied in dissipating an energy E_{gimbal} is $F_{gimbal} = E_{gimbal}/X_{gimbal}$. Thus, the torque applied by each gimbal oscillation event is $T_{gimbal} = F_{gimbal}\ r_{gimbal} = P_{gimbal}/(n_{gimbal}\ a_{gimbal})$; $P_{gimbal} = k_p\ v_{gimbal}^2$ (see Drexler), where $v_{gimbal} = a_{gimbal}\ n_{gimbal}\ r_{gimbal}$ is the sliding speed of the gimbal bearing surfaces during movement and k_p is a constant that depends solely on the geometry of the bearing. The change in orientation angle caused by each gimbal oscillation event is $Dq_{osc} = w_{precess}/n_{gimbal}$. The number of gimbal oscillations N = t n_{gimbal}, where t is elapsed time since the gyro was last calibrated. If gimbal oscillations are independent and randomly distributed, then the total change in orientation angle is very roughly. From Drexler, $k_p = 2.7 \times 10^{-14}$ watt-sec²/m² as a conservative value for a small (~2 nm) stiff bearing. Random hydrodynamic flows induced by thermal fluctuations inside biological cells have a characteristic duration $t_{fluct} \sim 10$ millisec, which suggests $a_{gimbal} \sim 0.06$ radians (~3°), for an R ~ 1 micron in cyto nanorobot. Taking r = 0.5 micron, h = 1 micron, r = 3510 kg/m³ for diamond (giving m ~ 3×10^{-15} kg), w = 1×10^9 rad/sec (just below the $w_{max} \sim 4.8 \times 10^9$ rad/sec bursting speed), and assuming $n_{gimbal} \sim t_{fluct}^{1} = 100$ Hz (to which the results are not terribly sensitive), then $Dq_{precess} \sim 10^{-8}\ t^{1/2}$ (radians). Thus a $Dq_{precess} \sim 1$ microradian orientation shift takes ~1 hour; a $Dq_{precess} \sim 4$ microradian shift takes ~1 day; and a $Dq_{precess} \sim 70$ microradian (~10 arcsec) shift takes ~1 year.

Thus a gimballed nanogyroscope may be sufficiently stable to serve as a readily storable, easily transportable, and highly accurate onboard orientation standard. A nanopendulum may also be employed for orientational sensing. As one example of a large class of related devices, consider a simple rigid spherical nanopendulum of radius r and mass m, with a large hemispherical bob free to swing through the entire solid angle about a central pivot point, with a gimballed housing to allow active avoidance of the pivot support beam. The nanopendulum bob moves along a spherical surface located far enough from a concentric spherical sensor grid of radius R = r + z_{sep} to minimize van der Waals interactions, with bob and sensor surfaces electrically neutral or uniformly mutually repulsive. Much like the pin cushion receptor model, the sensor grid consists of a spherical array of diamondoid sensor rods Dx ~ 1 nm wide that are rapidly and simultaneously extended into the evacuated sensor cavity to determine bob position during a measurement cycle of duration t_{meas}. Radial sensor rod velocity should exceed typical tangential bob velocity by at least 1-2 orders of magnitude. For the example given below, a sensor stud on a 500-nm bob that is turning at v_{turn} ~ $v_{thermal}$ = 3.6 mm/sec passes a ~1 nm sensor rod in ~280 nanosec; a sensor rod displacing ~1 nm in the radial direction in ~1 nanosec travels at v_{rod} ~ 1 m/sec. This nanopendulum may serve both as an orientation sensor and as a rotation rate sensor, using principles similar to vestibular (semicircular canal) mechanics. If a simple nanopendulum spherical sensor housing with bob initially at rest in a uniform vertical gravity field g is rotated through the smallest detectable angle q_{min} (radians) around the center pivot, then ignoring friction, the bob's position over the sensor grid moves. For Dx = 1 nm and r = 500 nm, Dq_{min} ~ 2 milliradians (~0.1 deg). For a practical sliding interface (e.g., between bob and sensor surface) of contact area S and turning velocity v_{turn}, Drexler estimates frictional dissipation, primarily driven by shear-reflection drag and band-stiffness scattering, as P_{drag} ~ k_1 $v_{turn}{}^2$ S, where k_1 ~ 400 kg/m^2-sec, hence the energy dissipated between the bob and the nearest sensor rod (the one interacting with the bob) is E_{drag} ~ k_1 v_{turn} S Dx. Bob gravitational energy E_{bob} = m g r sin Dq_{min} ~ m g r Dq_{min} for small displacements. The bob swings freely at the minimum displacement if E_{bob}/E_{drag} >1. If m = 10^{-15} kg, r = 500 nm, and S = 1 nm^2, then v_{turn} < 2.5 m/sec or w < 49 megaradians/sec for the bob to swing freely (vs. $v_{thermal}$ = 3.6 mm/sec and w_{res} ~ 9900 rad/sec for this pendulum). The bob-up/bob-down energy differential is ~2 m g r = 9.8 zJ ~2.3 kT for m = 10^{-15} kg and r = 500 nm. To distinguish opposite orientations (Dq ~ p) requires N_{meas} ~ (kT/2 m g r)^2 ~ 1 measurement for this sensor. To distinguish Dq_{min} ~ 2 milliradians requires averaging N_{meas} ~ (kT/E_{bob})^2 ~ 190,000 independent measurements (of ~0.002 kT/measurement) or t_{meas} ~ 190 microsec using a sensor rod clock cycle of ~1 nanosec.

Thus, after subtracting out the effects of translation displacements as determined by linear accelerometers, a nanodevice can measure angular displacements in its orientation to within ± Dq_{min} in a measurement time t_{meas} ~10^{-4} sec. With sensor rods moving at ~1 m/sec, energy dissipation for each rod sliding in its housing is ~0.01 zJ/cycle. If only ~100 test rods need be triggered to continuously track the moving stud, then a sensor rod clock cycle of ~1 nanosec implies a continuous nanopendulum power dissipation of ~1 pW; additional drag power loss between stud and contacting rod is only P_{drag} ~0.0004 pW. Many other configurations of this basic device are readily conceived and may be more volume-efficient, including nested spheres or ovoids, two-dimensional rocking disk sections for in-plane rotational measurements (three devices required for monitoring all angular degrees of freedom in a rigid rotator), and multiaxial circumferential tracks or tubes containing rolling balls or sliding plugs. J. Logajan suggests that a pair of linear three-dimensional translation accelerometers should be able to unambiguously extract the 3 translation and 3 rotation components of all possible acceleration vectors. Rotation about any axis will result in distinctly different mass displacements while accelerated translation along any axis will result in identical displacements. The smallest angular velocity detectable using a simple nanopendulum is given approximately by w_{min} ~ Dx/r t_{meas} = 2 milliradian/sec (~0.1 deg/sec) for Dx = 1 nm, r = 500 nm, and t_{meas} = 1 sec. The largest detectable angular velocity w_{max} ~ 2 p/t_{meas} if t_{meas} is the minimum time required to count one complete rotation of the sensor; the smallest t_{meas} ~ 2 p/w_{max} ~100 nanosec for w_{max} ~ 49 megaradians/secor ~5 per cent of the bursting speed for an r = 1 micron spherical diamondoid nanodevice. What rotational displacements will normally be encountered by an in vivo medical nanorobot? A 1-micron device experiences instantaneous Brownian rotations of $w_B = (kT/4\ p\ h\ R^3\ t)^{1/2}$ radians/sec, where t = $(MW_{kg}/32\ p\ kT\ N_A\ c_{H2O}\ R^4)^{1/2}$, the mean time between molecular collisions with the device (mostly by water molecules), where c_{H2O} ~ 3.3×10^{28} molecules/m^3. For R = 0.5 microns and T = 310 K, t ~ 6×10^{-20} sec/collision and w_B ~ 7×10^9 rad/sec, with each displacement of ~ t w_B = 4×10^{-10} radian, none of which are detectable. However, the net of all Brownian rotational displacements is one rotation every 16 sec, ~0.4 radian/sec.

The largest measurable angular displacements in vivo are caused by tumbling due to differential shear forces in the bloodstream. Normal vessel wall shear rates in physiological bloodflow range from 100-1400 radians/sec in the larger arteries to 500-4000 radians/sec in the smaller arteries and capillaries. The measured rolling velocity of white cells during vascular "tank tread" locomotion on venule walls is ~40 micron/sec, or ~8 radians/sec, which may be encountered by nanodevices employing cytocarriage. The human vestibular mechanism has a frequency response range of 0.048-260 radians/sec. Thus, the maximum range of physiologically relevant rotation rates is 10^{-2} to 10^4 radians/sec. Since the natural harmonic frequency of a simple nanopendulum is $w_{res} = (g/r)^{1/2}$ = 4400 radians/sec for r = 500 nm, the possibility of forced oscillations and resonance need be faced only when measuring the highest likely physiological rotation rates, in which case the addition of a damping force F_{damp} ~ m v_{turn} w_{res} during resonance events should eliminate the problem. F_{damp} for critical damping peaks at ~mg ~ 0.01 pN for the m = 10^{-15} kg, r = 500 nm case (v_{turn} ~ 2 mm/sec). Piconewton forces may be applied and must be detected by medical nanodevices. Pendulum- and cantilever-based sensors may be used to probe nanoscale forces, to monitor the ambient gravitational field; to measure molecular-scale masses to single-proton accuracy, and to discriminate among molecules having different isotopic composition. Extremely small forces have been measured using macroscale instrumentation, particularly the atomic force microscope (AFM). In 1998, experimental forces of 1 nN in the contact regime and 1 pN in the noncontact regime were routinely measured. In 1994, D. Rugar achieved a force sensitivity of 5×10^{-16} N using an AFM with a 90-nm thick cantilever at room temperature, and in 1998 Rugar reported a force resolution of 7×10^{-18} N using a magnetic resonance force microscope (MRFM) with a 230-micron long, 60-nm thick, silicon cantilever. The theoretical limit on the force sensitivity of the AFM has been estimated as 10^{-18} – 10^{-19} N for temperatures near absolute zero using fiber optic interferometry or other techniques. However, many of these methods are impractical or impossible to employ in medical nanodevices. What is the minimum force F_{min} likely to be detectable by nanorobot sensors? Several lines of reasoning produce consistent answers. For example, force may be measured by the displacement Dx of a beam against the known resistive pressure provided by a spring of minimum stiffness k_s; that is, $F_{min} \sim k_s$ Dx. However, the energy stored in the spring must exceed thermal noise energy, with SNR ~ ln ($k_s Dx^2$ /2 kT, giving F_{min} = 2 kT e^{SNR}/Dx. For T = 310 K and SNR = 2, F_{min} = 100 pN at Dx = 0.6 nm, 10 pN at 6 nm, and 1 pN at 60 nm. Indeed, the natural molecular motors (kinesin, dynein, and actin-myosin) have isometric force outputs on the order of 1-14 pN per protein molecule, which are routinely observed using microscale cantilever spring sensors measuring total forces of 10-100 pN. Breaking antigen-antibody or receptor-ligand interactions generally requires rupture forces in the 50-300 pN range.

A force of ~160 pN separates biotin that is seated in its avidin receptor; ~85 pN pulls iminobiotin out of the avidin receptor. A force of ~300 pN ruptures a hydrogen bond between two isolated water molecules, but >3000 pN are likely to be needed to rupture a single C-C covalent bond. The rupture strength of single covalent bonds has been directly measured experimentally by AFM as, for instance, 2000 ± 300 pN for Si-C (528 zJ/molecule dissociation enthalpy, 0.185 nm bond length) and 1400 ± 300 pN for Au-S. Force sensing may also require using measurement devices that involve sliding interfaces of two mutually inert periodic surfaces. This motion may be characterised by the potential energy of interaction, ~0.1-1 zJ as determined by Monte Carlo simulations of typical diamondoid material at 0.2-0.5 nm separations. Assuming a minimum 0.5 zJ energy barrier, for small sliding contacts the typical force amplitude $F_{min} \sim 1.7 \times 10^{10} DV_{barrier} \sim 8.5$ pN. A force of 10 piconewtons across a small-molecule receptor $(0.3 \text{ nm})^3$ of surface area 0.09 nm^2 represents an energy density of ~10^8 J/m^3 × 0.027 nm^3 = 2.7 zJ ~ 0.6 kT. And the nanorobotic manipulator arm is capable of ~picometer step sizes, representing a minimum applicable force of F_{min} ~ 10 pN assuming 0.04 m/N compliance for the entire mechanism. Thus as a conservative estimate, the smallest force that is reliably detectable (and applicable) by nanodevices is probably ~10 pN. Research in space medicine has discovered that micron-scale elements of the immune system have extraordinary sensitivity to gravity. One study found that bone marrow-derived (B6MP102) macrophage single cells respond after 8 seconds to exposure to a 10^{-2} g hypogravity environment as indicated by increased cell spreading, a well-known marker of cell activation. Individual immune system cells placed in microgravity exhibit enhanced growth but depressed differentiation, decreased activation by concanavalin A, a 20 per cent reduction in glucose metabolism, increased resistance to antibiotics, increased number of pseudopodia on monocytes, and changes in cytoplasmic streaming velocity and frequency. As of 1998, the exact mechanism by which the leukocyte "gravity sensor" operates remained unknown. It has been speculated that differences in densities of several organelles, such as the nucleolus, the ribosomes, and the centrioles, could generate detectable pressures on the structure of the cytoskeleton under gravitational loading mediated through mechanosensitive stretch-activated ion channels or the extracellular matrix.

While indirect gravitational effects that modify the cellular environment (such as reduced sedimentation or decreased thermal convection due to hypogravity) could also be responsible, a direct interaction of the gravitational force with cellular structures appears to be the most likely sensor mechanism. For instance, if a change in gravitational loading equivalent to the force generated by a natural molecular motor (F ~ 1 pN) is detectable by the cytoskeleton, and this force is applied to the entire mass of a (10 micron)^3 cell (m ~ 10^{-12} kg), then the minimum detectable gravitational acceleration is g ~ F/m ~ 1 m/sec^2 = 0.1 g, which should be adequate to detect the onset of hypogravity. A nanomechanical gravity sensor of similar size to a macrophage cell may be at least 100,000 times more sensitive. For example, the energy of a tethered mass swinging freely in a uniform gravity field is approximately E_g ~ m g Dh, where Dh is the maximum vertical amplitude of the motion. To be detectable, oscillator energy must exceed thermal noise energy, so SNR ~ ln (m g Dh $N_{meas}{}^{1/2}$/kT), giving g_{min} = kT e^{SNR}/m Dh $N_{meas}{}^{1/2}$ ~ kT e^{SNR}/r L^4 $N_{meas}{}^{1/2}$ for a bob mass of density r and size L^3, N_{meas} independent measurements, and assuming Dh ~ L. If T = 310 K, SNR = 2, N_{meas} = 1, and r = 21,450 kg/m^3 (Pt), then g_{min} = 0.1 g for L ~ 1 micron, 0.01 g for L ~ 2 microns, 10^{-5} g for L ~ 11 microns, and 10^{-6} g for L = 20 microns, the size of the average human tissue cell. Regarding nanogravimeter design, for a pendulum of length L in a gravity field g ~ 9.81 m/sec^2 (1 g), the resonant period of the motion T_{res} = 2 p (L/g)$^{1/2}$ = 2 × 10^{-3} sec for L = 1 micron, 9 × 10^{-3} sec for L = 20 microns. Beam and rod are coiled in two dimensions (although three-dimensional coiling, more difficult to depict, provides maximum compactness). L_{beam} and l_{rod} are chosen to minimize total coil area A_{coil} = 2 R_{beam} L_{beam} + 2 r_{rod} l_{rod}. Using aluminum for the rod and beam (E = 6.9 × 10^{10} N/m^2, G = 2.6 × 10^{10} N/m^2, r = 2700 kg/m^3), L_{beam} = 1200 nm, l_{rod} = 1920 nm, r_{rod} = 0.5 nm, V_{sample} = 1 nm^3, and Dt_{min} = 10^{-9} sec, then DMW ~ 1 proton for L_{spring} = 1 nm >> Dx_{min} = 0.01 nm, the minimum measurable displacement. (For L_{spring} > 1 nm, physiological ~1 g accelerative loads on a ~1 micron nanorobot in which the sensor is embedded produce undetectable accelerative beam displacements < Dx_{min}).

Moving the suspension rod closer to the sample holder reduces device sensitivity. For L_{spring} = L_{beam}, minimum DMW ~ 1200 protons, permitting moderately large molecules to be weighed in minimum time (~10^{-6} sec). Using a very thin sensor configuration (h_{sensor} ~20 nm) and coils occupying ~50 per cent of sensor cavity space, sensor dimensions are (79 nm)^2 × 20 nm giving a massometer volume ~10^5 nm^3 or ~0.01 per cent of the total volume of a 1 micron^3 nanorobot. Sensor mass is ~ 3 × 10^{-19} kg, including diamondoid housing and prorated share of support mechanisms. If one measurement cycle including sample loading/unloading operations can be completed every ~10 microsec, then the device processes its own mass of glucose molecules in ~10 sec. Note that biological molecules weighed in air or in vacuo may lose water of hydration, whereas weighing in water presents other complications that merit further investigation. J. Soreff points out that for a molecule on a spring, the precision of the measurement is linked to the sharpness of the resonance, which is optimised to the degree the vibrational frequency of the spring/molecular mass system is kept well below the vibrational resonance frequencies of the molecule itself.

The following discussion reports the results of a simple lumped-element analysis of the effects of internal vibration on the massometer which confirms the previously described performance estimates for the massometer. Suppose that it is desired to weigh some mass M to a precision DM by attaching it to a weak measuring spring (spring constant k_{weak}) and measuring the shift in resonant frequency. The shift in frequency due to coupling to an internal mode of the sample must be less than (1/2) (DM/M) (1/2 p) $(k_{weak}/M)^{1/2}$. If the sample has some lowest internal mode frequency n_{int}, then approximating the sample as two M/2 masses separated by x_{strong} connected by a strong spring (k_{strong}) gives $n_{int} = (1/p)(k_{strong}/M)^{1/2}$. Setting $x_{weak} = e^{st}$ and $x_{strong} = R\, e^{st}$ where t = time, s = time constant, and $R = -(1 + (2\, k_{weak}/M\, s^2))$, and since the angular frequencies of the two springs are $w_{strong} = 2\, p\, n_{int} = 2\, (k_{strong}/M)^{1/2}$ and $w_{weak} = (k_{weak}/M)^{1/2}$, then the lower frequency solution for $w_{weak} << w_{strong}$ may be approximated by $s = i\, w_{weak}\, (1 - w_{weak}^2/2\, w_{strong}^2)$ and so the (undesired) fractional change in frequency due to the flexibility of the object to be weighed. For a molecule sample size of L_{sample}, $n_{int} \sim v_{sound}/L_{sample} \sim 500$ GHz for $v_{sound} \sim 1500$ m/sec and $L_{sample} \sim 3$ nm; $M = r\, L_{sample}^3 = 2.7 \times 10^{-23}$ kg for r = 1000 kg/m^3. To detect the mass difference of a single proton in a sample molecule, sensor $Q = M/m_p = 16{,}000$; the sensor may operate at a factor of $Q^{1/2}$ below n_{int}, or $n_{spring} \sim n_{int}/Q^{1/2} \sim 4$ GHz. A period of 0.25 nanosec is slightly too short to allow a phase shift of $Dt_{min} \sim 1$ nanosec, since the period is only one-quarter of this. However, a phase shift of p radians should accumulate in $t_{meas} = Q/2\, n_{spring} \sim 2$ microsec and should be detectable by looking at magnitudes of forces. Interestingly, an SPM microprobe of circular cross-section and exponential profile with a 0.1 nm^2 tip has been proposed that could also reach ~proton mass detection sensitivity at room temperature and 1 atm pressure with an eigenmode frequency of 10 GHz; an electrically-driven carbon nanotube-based resonant-beam balance demonstrated ~10^{-17} kg (~10^{10} proton or ~0.01 micron^3 mass) sensitivity in 1999. Existing surface acoustic wave (SAW) devices are chemical sensors that can measure areal mass densities as small as 80 picogram/cm^2, or 0.01 N_2 molecule per nm^2 of surface, though current devices are limited to ~micron^2 areas.

A proposed quartz microresonator microbalance could detect a 10^{-6} molecular monolayer, the equivalent of 5×10^{-9} molecule per nm^2. Detection of ~400 million protons using an MRFM vibrating rod ~60 nm thick was demonstrated by Rugar in 1998. It will often be useful in nanomedicine to distinguish molecules containing different isotopes of the same chemical elements, for example to count the number of organic molecules in a sample that contain radioactive C^{14} atoms in place of stable C^{12} atoms or to make direct assays of heavy radiometal-contaminated tissues. Nanodevices may also sort isotopically heteronuclear molecules into homonuclear fractions, for example to separate deuterochemicals from hydrochemicals or to acquire isotopically pure materials with which to construct:

(i) Small, high-speed rotors with precisely balanced mass distributions.

(ii) Components with extremely fine size differences.

(iii) Nanostructures with maximum thermal conductivity, or mixed isotopes to reduce thermal conductivity.

(iv) Long-lived isotopically pure structures that generate no internal radioactivity, thus promoting longer device lifetimes.

The most reliable sensor for discriminating among isotopes within target molecules is the single-proton massometer. Isotopes of any chemical element differ in mass by at least ~1 neutron (~protonic) increments. Thus a massometer with single-proton resolution can resolve all isotopes starting from chemically pure samples. Isobars (atoms having the same mass number) and isotones (atoms having the same number of neutrons) may be finally resolved chemically; nuclear isomers differing in mass by ~0.001 amu cannot be resolved.

Teragravity centrifugation and time-of-flight particle-beam deflection techniques may also prove useful. Isotopes vary in many subtle ways due to their mass differences, although it is not yet clear how a potential isotope nanosensor might exploit some or all of these distinctions:

(i) *Line Shifts*—the electronic energy states of atoms or molecules depend on the reduced nuclear mass, causing a spectral line shift of ~1.0005 between hydrogen and deuterium, less among heavier isotopes, detectable spectroscopically.

(ii) Hyperfine Structure—isotopes differ in electrical quadrupole moment, nuclear spin and magnetic moment, giving rise to hyperfine structure in the optical spectra, especially in the heavier elements. For example, the magnetic moments of hydrogen and deuterium (2.79268 and 0.857387 nuclear magnetons, respectively) are distinguishable via NMR. The NMR frequency in a 10 kilogauss magnetic field is 42.5759 MHz for H, 6.53566 MHz for D. Of course, not all isotopes are NMR sensitive.

(iii) *Vibrations and Rotations*—molecular spectra are especially sensitive to isotopic changes, since the quanta of vibrational and rotational energies are directly dependent upon the masses of the atoms involved, and there are also some subtle effects from differences in nuclear spin. Separations are commonly achieved by laser excitation.

(iv) *Cross-Sections*—isotopes have different thermal neutron capture cross sections; for instance, the cross-section of C^{12} is 3.7×10^3 barns (1 barn = 10^{-24} cm^2) compared to 9×10^{-4} barns for C^{13} and $<10^{-6}$ barns for C^{14}. (Electron scattering might be better, since the electron is less likely to cause the nucleus to fission and it is generally difficult to use neutrons to probe samples on the molecular scale.) Similarly, the neutron cross-section for H_2O is 0.6 barns compared to 9.2×10^{-4} barns for D_2O.

(v) *Other Properties*—isotopes differ slightly in various bulk properties, many of which could be measured in submicron-scale aliquots by nanosensors. The ability of both large mammals and micron-size photosynthetic plant cells to preferentially concentrate C^{12} or C^{13} in organic matter, and the predicted ability of carbon nanotubes to selectively extract tritium from low-concentration H_2-T_2 mixtures, constitute proofs of principle that at least some of the bulk properties.

This section describes the ability of medical nanodevices to detect and measure fluid pressure variations while functioning inside the human body. Generation of acoustic signals for power transmission, communication, and navigation are considered elsewhere. A change in the ambient pressure in a fluid environment (DP) may be determined by measuring the change in volume (DV) of a fixed quantity of gas at constant temperature, relative to a known sensor volume at a known reference pressure. (Thermal sensors elsewhere in the nanorobot provide readings by which corrections can be made for temperature variations.) If the sensor volume consists of a cylinder of fixed cross-sectional area, the change in volume is converted into a linear displacement of a piston located at the interface between gas and fluid environment.

At SNR = 2, a $(22\ nm)^3 = 10^4\ nm^3$ sensor ($\sim 5 \times 10^5$ atoms or $\sim 10^{-20}$ kg for a cube with 1 nm thick walls) can detect a minimum pressure change $DP_{min} \sim 0.03$ atm; a $(680\ nm)^3 = 0.3\ micron^3$ sensor ($\sim 3 \times 10^9$ atoms, $\sim 6 \times 10^{-17}$ kg) detects $DP_{min} \sim 10^{-6}$ atm, probably near the practical limit for mobile in vivo medical nanodevices. Note that DP_{min} scales approximately as the inverse cube of mean sensor dimension, and immunity to thermal noise scales exponentially with sensor volume. Current silicon micromachined pressure sensors are $\sim 1\ mm^3$ with 10^{-3} atm sensitivity, a factor of $\sim 10^{12}$ poorer sensitivity than the theoretical minimum for their size. (A field-effect chemical sensor with an active sensing volume of $0.1\ mm^3$ that measures pressure indirectly has detected a 10^{-11} atm step in hydrogen gas with $t_{meas} \sim 3$ sec, only a factor of 1000 above the theoretical minimum for a sensor of that size.) The human ear can detect pressure waves as small as 2×10^{-10} atm.

Two 0.3-micron^3 pressure sensors located at either end of a 1 micron nanorobot could detect pressure changes as small as 10^{-6} atm per micron, a ~1 atm/m spatial gradient. The maximum detectable gradient would involve adjacent measurements of minimum (10^{-6} atm) and maximum (~10^4 atm) absolute pressure, a theoretical peak spatial gradient of ~10^{16} atm/m, though a practical device could not survive exposure to such extreme shear force. For an L = 22 nm sensor driven to full throw by a minimum detectable DP_{min} = 0.03 atm pressure spike (again taking h/L = 0.1 and r ~ 1000 kg/m^3), n_{driven} = 34 MHz; n_{driven} = 45 GHz when the sensor is driven to full throw by the maximum possible pressure spike P_{max} = 39,000 atm consistent with water remaining in the liquid state. The piston may be driven at these frequencies, or slower, using triangular-wave pulse trains of pressure spikes of these magnitudes. For an L = 680 nm sensor driven to full throw by a minimum detectable DP_{min} = 10^{-6} atm pressure spike, n_{driven} = 6 KHz; n_{driven} = 1.5 GHz when the sensor is driven to full throw by the maximum pressure spike P_{max} = 39,000 atm. What is the maximum tolerable pressure spike in aqueous media? The requirement of subsonic piston motion in water gives the highest maximum: $P_{max} < (1/2)\, r\, v_{sound}^2$ (N/m^2), if r = 3510 kg/m^3 for a solid diamondoid piston and the speed of sound v_s ~ 1450 m/sec in water for bulk modulus B_{water} = 2.2×10^9 N/m^2, r_{water} = 1000 kg/m^3, giving P_{max} < 39,000 atm. However, rapid dissipative pressure spikes P_{max} > 26,000 atm add sufficient energy to boil 310 K water, and isothermal static compression of pure water at 310 K causes crystallisation into Ice VI near ~11,500 atm. Additional restrictions on (and features of) high-frequency pressure transducers are described later. It will often be necessary to determine the degree to which a fluid storage tank is full or empty. In most circumstances, an ullage sensor may consist of a simple static pressure sensor embedded in the tank wall, with readout calibrated according to the van der Waals equation for gases and liquids.

A fractal network of chemical sensors permeating the internal volume of the tank could provide continuous readouts of local concentration which could be integrated to estimate total storage in the tank. The force required to rotate a paddle inside the tank would increase as a tank was filled, due to the increase in net viscosity of the contents. Transmission time of thermal or acoustic pulses might also give some measure of fluid concentration and physical phase. C. Phoenix suggests storing liquids in a piston, allowing direct and precise molecular volume readout, in designs where this is convenient; in theory, a piston that slides the minimum measurable distance of ~10 pm within the 2-nm-diameter throat of a water-filled cylinder could detect the removal of a single water molecule from the column. The ability of medical nanodevices to measure both absolute temperature and changes in temperature is crucial for monitoring in vivo physiological thermoregulatory mechanisms and intracellular energy transactions. Precision thermal sensing is also important within nanoscale devices to provide corrective input for pressure, chemical, and displacement sensors, and to improve the stability of onboard clocks. General limits of thermal sensing are considered, followed by several proposals for specific implementations. A sensor system consisting of N atoms has (3N-6) internal coordinates and thus ~(3N-6) oscillators, each with an average energy of ~kT assuming no modes are high enough to be frozen out. However, the standard deviation of the energy of the oscillators around the mean energy is ~kT, and these energy fluctuations are uncorrelated (ignoring coupling between the oscillators), so (3N-6) of them will have an average fluctuation of kT $(3N-6)^{1/2}$. (By comparison, the energy in one 10-micron infrared photon is ~5 kT.) Thermal equilibration time $t_{EQ} \sim L^2 C_V/K_t$ for a sensor of size L, of heat capacity C_V (1.8×10^6 joules/m^3-K for diamond) and of thermal conductivity K_t (2000 watts/m-K for diamond).

Thus, $t_{EQ} \sim 10^{-13}$ sec for a sensor of size L = 10 nm, 10^{-11} sec for L = 100 nm, and 10^{-9} sec for L = 1 micron, so thermal sensors up to 1 micron in size should easily achieve thermal equilibration within a typical measurement cycle time $Dt_{min} \sim 10^{-9}$ sec. Sensor measurement time $t_{meas} = N_{meas}\, Dt_{min}$. A (57 nm)^3 sensor can detect $DT_{min}/T = 10^{-4}$ (~31 millikelvins at 310 K) in a single measurement ($N_{meas} = 1$), with measurement time $t_{meas} \sim 1$ nanosec. A sensitivity of $DT_{min}/T = 10^{-6}$ (~310 microkelvins at 310 K) may be achieved using either a ~1 micron^3 sensor with a single measurement ($N_{meas} = 1$, $t_{meas} = 1$ nanosec) or a (124 nm)^3 sensor with $N_{meas} = 1000$ independent measurement cycles and $t_{meas} = 1$ microsec. Finally, a 1 micron^3 sensor can detect $DT_{min}/T = 3 \times 10^{-9}$ (~1 microkelvin at 310 K) with $N_{meas} = 100{,}000$ independent sensor cycles, giving a measurement time $t_{meas} \sim 100$ microsec. These figures are confirmed by experimental estimates of detector noise temperature, such as $DT_{min}/T = (k/L^3\, C_V)^{1/2} \sim 10^{-6}$ for a silicon nitride detector with $C_V = 5.2 \times 10^6$ joules/m^3-K and L ~1 micron. It has been calculated that sensitivity of ~1 microkelvin is theoretically possible using quartz electronic microresonators as precision thermometers. Thermocouple probes with 100 nm tips have already shown ~100 microkelvin sensitivity, tunneling thermometers capable of measuring thermoelectric potential localised to atomic-scale dimensions have been proposed, and experiments leading toward “yoctocalorimetry” were being pursued in 1998. For comparison, heat sensors in human skin have $DT_{min}/T \sim 3 \times 10^{-4}$ (~90 millikelvins), and the infrared sensor pit of the rattlesnake is sensitive to an energy intensity of ~0.8 pJ/micron^2 in a measurement time t_{meas} ~35 millisec and achieves $DT_{min}/T \sim 3 \times 10^{-6}$; mosquitos register $DT_{min}/T \sim 6 \times 10^{-6}$ at a distance of ~1 cm. Two (200-400 nm)^3 thermal sensors located at either end of a 1 micron nanorobot can detect temperature changes as small as 10^6 K per micron, a ~1 kelvin/m spatial gradient. The maximum detectable gradient in nanomedical systems would involve adjacent measurements differing by ~100°C, a peak spatial gradient of ~10^8 kelvins/m. The detection event must be completed before the two sensors can thermally equilibrate, e.g., $t_{meas} < t_{EQ} = L^2 C_V/K_t$. If the material between the sensors is mostly diamondoid, then $t_{EQ} \sim 10^{-9}$ sec; if mostly water, then $t_{EQ} \sim 10^{-5}$ sec, for L = 1 micron. On the timescale of nanomachine operations, typically 10^{-9}-10^{-6} sec, the human body provides an essentially isothermal operating regime. However, medical nanorobots circulating in the bloodstream may encounter DT of ~3 K during one circulation time ~60 sec under normal circumstances, a temporal thermal gradient of ~0.1 kelvin/sec, or >10 kelvin/sec at 1 mm tissue depth for a fingertip placed on a 600 K stove. Thermal nanosensors can take independent measurements differing by as much as ~100°C in ~10^{-9} sec, giving a maximum detectable thermal gradient of 10^{11} kelvins/sec, far in excess of physiological rates. Detection of the minimum ~1 microkelvin change between two measurements taken 1 sec apart probably represents the minimum practical temperature gradient, ~1 microkelvin/sec.

For comparison, the record-holder in the insect world is the eyeless cave beetle *Speophyes lucidulus*, whose antennae can detect thermal change rates as small as ~3000 microkelvins/sec. The extraordinary thermal conductivity of diamond ensures that endogenous nanorobot waste heat is rapidly conducted to the surrounding aqueous medium. For example, a 1 micron^3 diamondoid nanorobot generating 10 picowatts of onboard power at its core experiences a temperature differential of ~10^{-8} K between core and perimeter (using thermal conductivity K_t = 2000 watts/m-K for diamond), well below minimum detectable limits. The most important internal electrical sources are the electrochemical gradients caused by gated channel and transporter molecular pump operations at the intracellular level; muscular, membrane, digestive, and neural activity at the intercellular and organ level; and piezoelectric fields generated by movement of collagenous tissues (e.g., tendon and elastin), bones, and many other biological materials.

Typically these sources generate potentials of 10-100 millivolts over distances of 0.01-10 microns, producing local fields in the range of 10^3-10^7 volts/m. For example, the 50 millivolt transmembrane potential necessary to open a sodium channel corresponds to an electric field of 5×10^6 volts/m. Interestingly, implanted electrical devices such as pacemakers may produce up to 600 volts/m one millimeter from the casing. External sources may also contribute to detectable nanomedical electric fields within the human body, though usually to a lesser degree. Placing a hand flat between two plates charged to $E_p = 3 \times 10^6$ volts/m (near the breakdown voltage for air) induces an instantaneous field of $E \sim E_p/k_{water}$ = 42,000 volts/m transversely through the limb. (A DC field induces ion current flows which accumulate surface charge and cancel out the internal field after ~one RC time.) Electric fields directly beneath high-tension power lines are 2,000-11,000 volts/m, or 50-1000 volts/m at a lateral displacement of 25 meters along the ground. Household appliances produce 200-300 volt/m fields; 20 amperes flowing through standard Romex copper wiring in the walls of a house produces 100-600 volts/m in the middle of a room, up to 24,000 volts/m at the wall ($E = B/m_0 e_0 c$; c = the speed of light, B); static sparks from carpet walking are ~10,000 volts/m, and ~10^5 volts/m makes hair stand on end. Quiescent atmospheric charge is ~100 volts/m, rising to 5000 volts/m during severe thunderstorms oscillating from positive to negative over a ~10,000 volt/m range. Lightning bolts (~20 coul 0.002-sec discharge) produce 60,000 volt/m spikes at a distance of 10 meters from the strike, 600 volts/m at a range of 1 kilometer.

Electron-hole avalanches in semiconductors are initiated by fields >~5 × 10^7 volts/m. Penal electrocutions use ~1500 volts/m. Direct force sensing of these electric fields is ineffective; the minimum detectable field $E_{min} = F_{min}/q = 6 \times 10^7$ volts/m using F_{min} = 10 pN and a single-charge sensor. Particle deflection seems more efficient. Consider a collimated stream of singly-charged particles, each of mass m and initial velocity V_0 entering an evacuated test chamber of length L and experiencing a uniform electric field E, causing a lateral displacement $Dx = q E L^2 / 2 m V_0^2$ measured by the sensor. For Dx_{min} = 10 pm, T = 310 K and SNR = 2, E_{min} = 10,000 volts/m for a sensor of size L = 28 nm, or 100 volts/m for L = 280 nm, allowing measurement of all nanomedically relevant fields that are homogeneous across the entire device. Using the larger 280-nm sensor, time of flight is ~10^{-12} sec for electrons or ~10^{-9} sec for U^{238} ions, permitting t_{meas} ~ 10^{-9} sec and dissipating at least ~30 pW waste heat (due to kinetic impact) in continuous operation. Time-varying electric fields can be measured up to t_{meas}^{-1} ~ GHz frequencies, and electric current may be determined using standard techniques because current density is scale-invariant. Care must be taken to eliminate measurement error due to exogenous magnetic fields. Piezoelectric or other electromechanical transducers can convert electricity into mechanical signals appropriate for all-mechanical computation and control systems. Another class of electric field detector of similar size and sensitivity is the electrosensitive gel sensor. For example, the negatively charged heparan sulfate proteoglycan network found in the secretory granule matrix expands 50 per cent in volume within milliseconds of exposure to a -2 volt/micron field, and the response appears linear with voltage.

Thus the length of a 400-nm long gel-filled sensor of fixed cross-sectional area should expand a minimally-detectable 10 pm (= Dx_{min}) in a 100 volt/m field. Similar sensitivities to small periodic fields, analogous to shark electroreceptors (maximum sensitivity ~10^{-7} volts/m,), are found in ion-channel embedded artificial membranes that selectively amplify electric signals in noisy environments via stochastic resonance. Using n_e = 1 MHz and t_{meas} = 1 sec gives E_{minAC} = 0.2 volt/m. Cylindrical sensors are 10-100 times more sensitive, and sensitivity may possibly be further enhanced by coupling the periodic electric potential to a Michaelis-Menten type enzyme or a similar nanomechanism embedded in the sensor shell to achieve tuning. Fields as small as ~10^{-4} volts/m may be detectable by ~20-micron mammalian living cells; such detection may trigger biochemical observables which may be monitored or eavesdropped by resident nanomedical devices. Submicron-scale electrometers capable of ~0.1

electrons/$Hz^{1/2}$ (charge per unit bandwidth) have been demonstrated experimentally, and also single-electron transistors (SETs) to 10^{-4} electron/$Hz^{1/2}$, with the potential to reach 10^{-6} electron/$Hz^{1/2}$. Galvanotropic bacteria have been reported in the literature. Radiofrequency receivers for power transmission and communication in vivo are described in Sections 6.4.2 and 7.2.3. Antenna radiation patterns may also be useful in diagnostics and macrosensing applications.

Hence B ~ 0.04 microtesla for r = 1 micron, adjacent to an axonal conductor. For comparison, magnetic fields directly beneath high-tension power lines are 10-40 microtesla, or 2-8 microtesla at a lateral displacement of 25 meters along the ground. Household appliances produce 1-1000 microtesla at a distance of 3 cm and 0.01-2 microtesla at a 1-meter range (electric shavers produce 0.4-60 microtesla at a range of 15 cm). 20-amp Romex wiring in the walls of a house can produce 2 microtesla in mid-room, up to 80 microtesla at the wall, though in 1998 the average strength in 98 per cent of U.S. homes from artificial sources was 0.05-0.09 microtesla, typical exposures throughout the day were 0.1-1 microtesla, and the federal workplace standard was a maximum of 100 microtesla. Magnetotactic marsh bacteria, birds and insects can detect the geomagnetic field vector (identifying Earth's magnetic poles), which is 30-70 microtesla, depending on geographic position. Behavioural thresholds have been measured as 0.005-0.02 microtesla for pigeons and 0.001-0.01 microtesla for honeybees, and no more than 1000-2000 microteslas for human subjects.

Some biochemical changes have been observed in human cells exposed to ~10-100 microtesla 60 Hz fields, although the data are controversial; it appears the fields pump iron atoms through cell membranes, potentially causing peroxidation or other damage to occur. In 1998, transcranial magnetic stimulation was used to expedite recovery of motor skills for stroke patients, and was being explored as a treatment for depression and a means of altering patient mood. Lightning bolts (i ~ 10,000 amp/discharge) produce 200 microtesla spikes at a distance of 10 meters from the strike, 2 microtesla at a range of 1 kilometer. Computer simulations of a fully heterogeneous human body divided into 37,000 $(1.31\ cm)^3$ voxels produced a maximum induced current of 8.84 amp/m^2 (average ~1 amp/m^2) when the body was exposed to a uniform 60 Hz 1-tesla magnetic field. Patients undergoing Echo-Planar MRI scanning report "tickling" and "pain" when exposed to field changes exceeding 60 tesla/sec; ventricular fibrillation may result from myocardial current densities of 4 amp/m^2, which is induced at 250 tesla/sec. The simplest magnetosensor is a freely pivoting permanent magnet that experiences a restoring force when the magnet's axis is not aligned with the external field. This "compass effect" is also employed by magnetotactic bacteria, which have ~0.1 micron ferromagnets in magnetosomes in their cells that allow them to detect and track the geomagnetic field vector.

Permanent magnets of cross-sectional area A_{magnet} and length l_{magnet} made of Alnico or Alcomax (r ~ 8000 kg/m^3) can provide B_{magnet} ~ 1.4 tesla. (In 1998, the record flux density was 13.5 tesla for dipole magnets, ~27 tesla for resistive magnets.) If the minimum detectable force is F_{min} = 10 pN, B_{min} = 50 microtesla for A_{magnet} = 0.4 micron^2, about right for magnetotactic bacteria, and B_{min} = 2 microtesla for A_{magnet} = 2 micron^2. Another class of magnetosensor detects the change in dimensions when a magnetic body is magnetised, which is called magnetostriction. For instance, iron-cobalt ferrite expands by DL/L ~ 7.5 × 10^{-4} at saturation magnetostriction in a ~1 tesla field, permitting detection of ~10 millitesla field changes using a 1 micron sensor and Dx_{min} ~ 10 pm. Small magnetostrictive volume changes make possible a sensor of even greater sensitivity. A field of ~0.2 tesla causes a fractional volume increase of ~10^{-6} in iron and up to 40 × 10^{-6} in some iron-nickel (~30 per cent Ni) alloys. A (370 nm)^3 magnetostriction sensor monitored by stretch sensors with displacement sensitivity Dx_{min} = 0.1 nm (~1 atomic radius) can in theory detect B_{min} ~ 0.1 microtesla. Magnetostriction is nonlinear with applied field, and there are significant hysteresis effects which may confuse data interpretation.

However, multisensor arrays combining positive- and negative-magnetostriction materials should enhance signal extraction. Since the change in volume is small, temperature compensation may be required. Hall effect probe microscopes have demonstrated 10 microtesla sensitivity with 350 nm spatial resolution, and alternative nanoscale magnetic detectors such as direct-current SQUID (superconducting quantum interference device) magnetometers, stray field devices, and magnetic spin sensors are being developed. In 1999, micro SQUID could measure magnetic moments as small as ~10^4 Bohr magnetons or ~10^{-19} J/tesla, roughly 7 × 10^6 protons or a 14-nm iron cube. Magnetoencephalographs using SQUID magnetometers to measure external electromagnetic fields produced by neural traffic in the brain can register 10^{-6}–10^{-3} microtesla signals. Other biomagnetic measurement techniques have helped to investigate electrophysiological disturbances in heart muscle and in the brain. By 1998, microcoils as small as 100 microns in diameter had been constructed using microcontact printing of 25 micron wide silver wires, producing magnetic flux densities >0.4 tesla with a 10 milliamp current, switchable in <10^{-3} sec. In biosystems and nanosystems alike, the detection of an optical photon is a high-energy event. A single photon of red light at l = 700 nm carries an energy of 280 zJ, a blue photon at 400 nm conveys 500 zJ, and an ultraviolet (UV) photon at 200 nm transmits 1000 zJ to its receptor. These energies, ranging from 65-234 kT, exceed thermal noise by a wide margin, hence are all easily detectable reliably by suitable nanosensors. Note that optical photon detection does not require a nanosensor the size of one wavelength or larger—for example, 1-nm chlorophyll molecules have no difficulty absorbing 660-nm photons.

Biological systems provide numerous examples of single-photon transducers, including photopigment molecules such as chlorophyll and rhodopsin which mediate electron transfers in <~1 picosecond by undergoing conformational changes, triggering a series of biochemical events. An estimated 100-500 cell membrane channels are affected by the detection of a single optical photon. The first documented case of a physiological role for chlorophyll in animals has been found in the fish *Malacosteus niger*, which uses a chlorophyll derivative to see far-infrared light. Mammalian cells also act as indirect UV sensors as for instance when energy absorption induces perturbations of membrane structure or a conformational change in membrane proteins, activating the JNK cascade below 300 pJ/micron^2, and centrosomes may serve as cellular directional IR sensors in single fibroblast cells. Integration times for photon sensors are briefly treated.

Artificial "bioelectronic" cyanine-quinone chromophore molecules ~1 nm in size can detect optical photons of wavelengths l = 580-630 nm, and can trigger and reset in 1-3 picosec (governed by the tunneling times associated with reformation of the zwitterionic states of the inputs), in theory permitting ~100 GHz modulations to be received. Exciton-generating poly(p-phenylene vinylene) polymer switches have the potential to achieve ~THz switching speed; a birefringent Kerr medium also switches in a few picoseconds. Chalcogenide glass has viscosity $>10^{13}$ kg/m-sec in darkness, but ~5×10^{11} kg/m-sec under illumination. Proposals for specific implementations of photoergic transducers and optical cables are elsewhere. High energy photons including X-rays (l ~ 2 nm, ~10^5 zJ or 0.0001 picojoule) and gamma rays (l ~ 0.002 nm, ~0.1 picojoule), and ejecta from the decay of radioactive nuclei such as a-particles (helium nuclei, ~1 picojoule) and b-particles (free electrons, 0.01-0.1 picojoule) may be extremely destructive to nanomechanical systems and are difficult to reliably measure quantitatively. For example, a charged particle with 0.01 pJ of kinetic energy (K.E.) traversing a decelerative electrostatic field of 10^9 volts/m requires K.E./qE ~ 60 microns to slow to a halt. However, particle-induced visual sensations have been observed by dark-adapted astronauts in space, and Ruderfer has suggested that even solar neutrinos may be detectable by the human brain (though most physicists would find this implausible). The goal of cellular bioscanning is the noninvasive and non-destructive in vivo examination of interior biological structures. One of the most common nanomedical sensor tasks is the scanning of cellular and subcellular structures. Such tasks may include localisation and examination of cytoplasmic and nuclear membranes, as well as the identification and diagnostic measurement of cellular contents including organelles and other natural molecular devices, cytoskeletal structures, biochemical composition and the kinetics of the cytoplasm. The precision and speed of medical nanodevices is so great that they can provide a surfeit of detailed diagnostic information well beyond that which is normally needed in classical medicine for a complete analysis of somatic status. Except in the most subtle cases a malfunction in any one component of the cellular machinery normally provokes a cascade of pathological observables in many other subsystems.

Detection of any one of these cascade observables, if sufficiently unique and well-defined, may provide adequate diagnostic information (see) to plan a proper reparative procedure. DNA is one of the few cellular components that is regularly inspected and repaired. Most homeostatic systems adopt a more simple philosophy of periodic replacement of components regardless of functionality. Nanomedicine allows the philosophy of inspection and repair to be extended to all cellular components. The following discussion briefly describes a few of the sensory techniques that may be useful in achieving this objective.

Using Tactile Topographic Scanning

Tactile topographic scanning provides the most direct means for examining cellular structures in vivo. For example, ex vivo live cell scanning in air by commercially available atomic force microscopy using a 20-40 nm radius tip allows nondestructive feature resolutions of ~50 nm across the top 10 nm of the cell. Investigators have used the scanning tip to punch a hole in the cell, then pull back and scan the breach, observing the membrane heal itself via self-assembly in real time. Up to ~KHz scanning frequencies are possible, with up to 1024 data points per scan line. The nanomechanical manipulator arm achieves comparable tip velocities and positional accuracies. Assuming a scan rate of ~10^6 pixels/sec, a micron-scale nanodevice, once securely anchored to the surface of a (~20 micron)^3 human cell, could employ a tactile scanning probe to image the 0.1 per cent of plasma membrane lying within its (1.4 micron)^2 vicinity in ~2 sec to ~1 nm^2 resolution (~1 mm/sec tip velocity), or ~50 sec to ~0.2 nm (e.g., atomic) resolution (~0.2 mm/sec tip velocity). Inside the cell, and again post-anchoring, an entire 6 micron^2 mitochondrial surface could be imaged to atomic resolution in ~100 sec; the surface of a 100 nm length of 25 nm diameter microtubule could be atomically resolved in 0.2 sec—though of course in a living cell these structures may be changing dynamically during the scanning process.

The continuous power dissipation of a (0.1 micron)^2 scan head moving at 1 mm/sec through water is ~0.002 pW (~kT/pixel at ~10^6 pixels/sec), though of course the energy cost of sensing, recording, and processing each pixel must be at least ~10 kT/pixel (ignoring the possibility of pre-computational image compression), so the total scanner power draw could be as high as 0.02-0.1 pW in continuous operation. Special scanning tips and techniques should allow topographic, roughness, elastic, adhesive, chemical, electrostatic (charge density), conductance, capacitance, magnetic, or thermal surface properties to be measured. For large (~0.1-1 micron) cellular components, identification and preliminary diagnosis of improper structure may be possible using measured surface characteristics which may be matched to entries in an extensive onboard library, perhaps combined with dynamic monitoring of anomalies in cross-membrane molecular traffic using chemical nanosensors.

Most membranes are self-sealing, so it should also be possible to gently insert a telescoping member into the target organelle, which member then slowly reticulates and extrudes smaller probes with sensory tips in a fixed pattern and step size, allowing the acquisition of detailed internal structural and compositional information. Small (~1-10 nm) protein-based components of the cell such as enzymes, MHC carriers, and ribosomes are self-assembling or require only modest assistance (e.g., molecular chaperones) to self-assemble. Reversible mechanical denaturation of these smaller proteins, using diamondoid probe structure to displace water molecules (thus reducing hydrophobic forces) and by using specialised handling tools akin to functionalised AFM tips and molecular clamps, may be followed by precise nondestructive amino acid sequencing to allow identification and diagnostic compositional analysis. Afterward, the protein molecule may be refolded back into its original minimum-energy conformation, possibly with the assistance of chaperone-like structures in some cases. Acoustic microscopy is another noninvasive scanning technique that can provide nanomedically useful spatial resolutions. Frequencies are in the GHz range, far exceeding the relaxation time of the protoplasm.

A cryogenic acoustic microscope operated at 8 GHz has demonstrated 20 nm lateral resolution in liquid helium. By 1998, the best resolution achieved with water as the coupling fluid has been 240 nm at a frequency of 4.4 GHz. Operated in water at 310 K, a nanomechanical 1.5 GHz acoustic microscope would achieve a far-field minimum lateral resolution l/2 ~ 500 nm, $\sim 10^5$ voxels per human cell, sufficient to locate and count all major organelles and a few intermediate-scale structures. Scanning acoustic microscopes (SAM) operated at 2 GHz in reflection mode (which allows detection of interference effects) achieve 30-50 nm resolution in the direction of the acoustical axis, and the Subtraction SAM approach reveals topographical deviations of 7.5 nm at 1 GHz. It has been proposed that picosecond ultrasonics could be used to obtain an image of the cytoskeleton with detail comparable to that of conventional X-ray images of a human skeleton. Power requirements are a significant constraint on acoustic reflection microscopy (echolocation). In the simplest case, consider an acoustic emitter of radius r_E which converts an input power of P_{in} into acoustic power of intensity $I_E < I_{max}$ = 1000 watts/m^2 with efficiency e per cent, conservatively taken here to be 0.50 (50 per cent). The emitter produces a train of omnidirectional pressure pulses of amplitude $A_p = (2 \, r \, v_{sound} \, I_E)^{1/2}$ (N/m^2) and frequency n which travels a distance X_{path} to a target of radius r_T.

As the signal reaches the target, its amplitude has been reduced by (r_E/X_{path}) due to the $1/r^2$ dependence of intensity in spherical waves and by $e^{-atiss \, n \, Xpath}$ by attenuation with $a_{tiss} = 8.3 \times 10^{-6}$ sec/m for soft tissue. Upon reaching the target, a fraction $f_{reflect}$

of the signal is reflected as a point source echo; if the target surface has an acoustic impedance similar to liver tissue and the medium is similar to water, $f_{reflect} \sim 0.05$ (5 per cent). The echo then travels a distance X_{path} back to a receiver (which may or may not be located near the emitter), the amplitude again losing (r_T/X_{path}) by geometry and $e^{-atiss\ n\ Xpath}$ by attenuation. To scan the entire interior of a (20 micron)^3 cell, whether from within or without, requires $X_{path} >\sim 20$ microns. A frequency of n = 1.5 GHz allows $v_{sound}/n \sim 1$ micron spatial resolution. The fastest possible pulse repetition time is $X_{path}/v_{sound} \sim 13$ nanosec. Assuming $r_E = r_T = 0.5$ micron, $P_{in} \sim 7$ pW or ~5 watts/m^2, well within the safe intensity range for acoustic radiation. At maximum safe transmitter intensity I_{max}, the longest scannable path length at 1.5 GHz is $X_{path} \sim 46$ microns. Power constraints are somewhat less severe in the case of acoustic transmission microscopy (acoustic tomography), a technique which requires a minimum of two physically separated components (e.g., a transmitter and a receiver). Consider an acoustic emitter radiating a series of omnidirectional pressure pulses of frequency n which travel across a tissue cell of width X_{path}, to be detected by a receiver on the other side after a signal transit time $t_{cell} \sim X_{path}/v_{cell}$, where v_{cell} is the speed of sound in the cytosol. In the simplest case, assume that the cytosol is clear except for one target of interest along the path (e.g., an organelle) in which the speed of sound is v_{target} # v_{cell}.

The minimum detectable difference in travel times between a signal that passes through the target and one that does not is Dt (~10^{-9} sec for diamondoid systems), so the minimum resolvable target size is $r_{min} \sim v_{target}\ v_{cell}\ Dt/2\ abs(v_{target} - v_{cell})$ when $r_{min} >\sim v_{cell}/n$. For soft tissue targets in water at 1.5 GHz, $r_{min} \sim 30$ microns; for tooth enamel in water, $r_{min} \sim 1$ micron. However, the use of sampling gates will permit the detection of phase shifts that are much less than the characteristic time constant of the system in incoming waves, hence the minimum thickness of objects in theory detectable by acoustic tomography can be far smaller than the values for r_{min} estimated above.

In either mode of operation, data gathering is accelerated by positioning more than one receiver around the target. Sensitivity is boosted by increasing emitter size, taking multiple measurements, or by illuminating the target with some large external source located elsewhere in the tissue or even outside of the organ. However, cells and body tissues are mesoscopic "junkyards"—highly heterogeneous media which may produce large numbers of nontarget scattering events, thus increasing the difficulty of extracting signal from noise. Additional complications arise due to:

(i) Scattering on rough surfaces.

(ii) Rapid pressure variations with range in the Fresnel zone or near field of the transmitted signal.

(iii) Cytoplasmic viscosity inhomogeneities due to the asymmetric arrangement of cytoskeletal structure, granules, vacuoles, and the endomembrane system.

(iv) Variation in cytoplasmic Young's modulus due to time-varying tensions in semirandomly-distributed cytosolic fibrillar elements (e.g., a relaxed or contracted state) which alter elasticity and thus the local speed of sound.

Electrostatic scanning is largely ineffective because of Debye-Huckel shielding. Magnetic stray field probes allow resolution of 10 nm physical features, but only for materials with substantial magnetic domains—which most biological substances lack. But nuclear magnetic resonance (NMR) imaging may allow cellular tomography by creating 3-D proton (hydrogen atom) density maps. Atomic density maps of other biologically important elements with nonzero nuclear magnetic moments (including D, Li^{6}, B^{10}, B^{11}, C^{13}, N^{14}, N^{15}, O^{17}, F^{19}, Na^{23}, Mg^{25}, P^{31}, Cl^{35}, K^{39}, Fe^{57}, Cu^{63} and Cu^{65}) may also be compiled. For example, sodium imaging is already used clinically to assess brain damage in patients with strokes, epilepsy, and tumors. In a hypothetical NMR cytotomographic nanoinstrument, a large permanent magnet is positioned near the surface of the cell or organelle to be examined. This creates a large static background magnetic field that polarizes the protons. The spatial gradient of this field establishes a unique resonant frequency, called the Larmor frequency, within each isomagnetic surface throughout the test volume. A second time-varying magnetic driver field (e.g., vibrating permanent magnet or weak rf field) is then scanned through the full range of resonant frequencies, exciting into resonance the protons (nuclear spins) in each isomagnetic surface, in turn.

Depending on the sensor implementation chosen, each resonance detected may cause absorption of driver field energy, an increase in measured impedance, or even a return echo of magnetic energy (if the driver field is operated in pulse echo mode) as the excited protons relax to equilibrium in ~1 sec. The large polarising magnet is then rotated to a new orientation, moving the isomagnetic surfaces to new positions within the test volume, and the scan is repeated. A 3D map of the spatial proton distribution may be computed after several scan cycles. Only those regions within a linewidth of resonance DB will generate an appreciable signal, and the expected spatial resolution (say, along the z-axis) Dz = DB/(dB/dz), where dB/dz is the spatial flux gradient of B. For a polarising field B = 1.4 tesla placed adjacent to a (20 micron)^3 cell, the cross-cell gradient dB/dz ~ 7 × 10^4 tesla/m; placed next to a (1 micron)^3 organelle, dB/dz ~ 1.4 × 10^6 tesla/m. Assuming an effective NMR linewidth ~ 0.002 tesla, minimum spatial resolutions are 29 nm and 1.4 nm, respectively. Unfortunately, the smallest reliably detectable energy in the sensor element is ~ kT e^{SNR}. The energy required to flip a single proton is E_{flip} = 4 p n_L L_{proton}, where L_{proton} = 5.28 × 10^{35} joule-sec (the quantised spin angular momentum of the proton) and the Larmor resonance frequency n_L = g_{proton} B; g_{proton} is the gyromagnetic ratio (4.26 × 10^7 Hz/tesla for protons) and B is the polarising magnetic flux. However, the population difference between spin-up and spin-down nuclei in NMR is very small. For small energy differences, the Boltzmann distribution only allows a fraction $E_{flip}/2$ kT to be flipped before the upper and lower state populations are made equal and the absorption disappears. When scanning a cell or organelle, the vast majority of the protons present are in water molecules. Water at 310 K has a proton density $n_{water} = 6.7 \times 10^{28}$ protons/m^3, compared to $n_{fat} \sim 6.4 \times 10^{28}$ protons/m^3 (palmitic acid). Distinguishing fat from water thus requires a minimum sample volume of $N_{min}/(n_{water} - n_{fat}) \sim 55$ micron^3. Proteins range from 4.4×10^{28} protons/m^3 (aspartic acid minus one water) to 7.6×10^{28} protons/m^3 (leucine minus one water), average $n_{protein} \sim 5.6 \times 10^{28}$ protons/m^3; $n_{carbohydrate} \sim 5.8 \times 10^{28}$ protons/m^3 (glucose minus one water), allowing NMR cellular tomography to resolve minimum feature sizes of 2-4 microns (560 micron^3).

This resolution may include intracellular structures such as the endoplasmic reticulum (rough ER ~1100 micron^3, smooth ER ~400 micron^3), the Golgi complex (~500 micron^3), the nucleus (~270 micron^3), possibly the nucleolus (~50 micron^3) and even large localised ferritin granule concentrations. In 1998, the minimum sample size for microcoil (~470 micron diameter) NMR scanning was ~10,000 micron^3 or ~10^{16} protons for a 1-minute measurement cycle. Individual T-cell vacuoles ~1 micron in diameter labeled with dextran-coated iron oxide particles have been imaged by MRI.

Understanding Electromagnetic Waves

Electromagnetic waves of optical wavelength l that interact with an object are diffracted into two components, called "far-field" and "near-field." The propagation of electromagnetic radiation over distances z > l acts as a spatial filter of finite bandwidth, resulting in the familiar diffraction-limited resolution ~l/2. Classical optics is concerned with this far-field regime with low spatial frequencies < 2/l, and conventional optical imaging will be difficult at the cellular level in vivo. (Short-wave X-rays will damage living biological cells.) Information about the high spatial frequency components of the diffracted waves is lost in the far-field regime, so information about sub-wavelength features of the object cannot be retrieved in classical microscopy. However, for propagation over distances z << l, far higher spatial frequencies can be detected because their amplitudes are then of the same order as the sample (z = 0). This second diffraction component is the "near-field" evanescent waves with high spatial frequencies >2/l. Evanescent waves are confined to subwavelength distances from the object. Thus a localised optical probe, such as a subwavelength aperture in an opaque screen, can be scanned raster fashion in this regime to generate an image with a resolution on the order of the probe size. The original Near-field Scanning Optical Microscope (NSOM) surpassed the classical diffraction limit by operating an optical probe at close proximity to the object. The

NSOM probe uses an aluminum-coated "light funnel" scanned over the sample. Visible light emanates from the narrow end (~20 nm in diameter) of the light funnel and either reflects off the sample or travels through the sample into a detector, producing a visible light image of the surface with ~12 nm resolution at l = 514.5 nm provided the distance between light source and sample is very short, about 5 nm, with signal intensities up to 10^{11} photons/sec (~50 nanowatts). This represents a resolution of ~l/40. The near-field acoustic equivalent is found in the medical stethoscope, which exhibits a resolution of ~l/100. Applications include dynamical studies at video scanning rates, low-noise high-resolution spectroscopy, and differential absorption measurements. Optical imaging of individual dye molecules has already been demonstrated, with the ability to determine the orientation and depth of each target molecule located within ~30 nm of the scanned surface. Molecules with nanometer-scale packing densities have been resolved to ~0.4-nm diameters using STMs to create photon emission maps (an electric current generates photons in the sample), and laser interferometric NSOMs have produced clear optical images of dispersed oil drops on mica to ~1 nm resolution. Live specimen 250 nm optical sectioning for three-dimensional dynamic imaging has also been demonstrated.

Submicron laser emitters have been available since the late 1980s. It should be possible to use NSOM-like nanoprobes to optically scan the surfaces of cells or organelles to <~1 nm resolution, mapping their topography and spectroscopic characteristics to depths of tens of nanometers without penetrating the surface. However, a proper membrane-sealing invasive light funnel ~20 nm in diameter might not seriously disrupt some cellular or organelle membranes and thus could be inserted into the interiors of these bodies or through the cytoskeletal interstices to permit deeper volumetric scanning. The thermal conductivity of water at 310 K is 0.623 watts/m-K and the energy per 500 nm photon is 400 zJ, so for 1 micron^3 of watery tissue the maximum scan rate is ~35 micron^3/sec-K, if e^{SNR} photons are used to image each 1 nm^3 voxel with SNR = 2. Thus a 1-sec volumetric optical scan of an aqueous ~1-micron^3 sample volume to 1 nm^3 resolution requires a ~1 nanowatt scanner running at ~GHz bit rates, raising sample volume temperature by ~0.03 K. NSOM permits the determination of five of the six degrees of freedom for each molecule, lacking only the optically inactive rotation around the dipole axis. In principle, it should be possible to produce <~1 nm resolution near-field optical scans of in situ protein molecules, since with atomically precise fabrication and single lines of atoms as conductors a minimum light guide (metal-dielectric-metal) is 3 atoms wide. Given a molecular laser and adequate collimation, photons passing through folded proteins will scatter according to the molecular structure. Detection of sufficient photons comprising these scattering patterns should allow the noninvasive determination of protein structure; polarised photons provide information on chirality. Absorption and fluorescent signals will be visible from phenyl rings, tryptophan, and bound cofactors such as ATP and adenine (which is fluorescent).

Positions of monoclonal antibodies on virus surfaces are now identifiable experimentally using NSOM; it should be possible to map binding sites on virus and cell surfaces using fluorescently labelled antibodies. Single molecule detection has been proposed as a tool for rapid base-sequencing of DNA. Other optically-based cellular imaging techniques must be distinguished. Optical Coherence Tomography (OCT) uses Michelson interferometry to achieve ~10 micron spatial resolutions over tissue depths of 2-3 mm in nontransparent tissue in the near infrared. However, OCT requires numerous physical components not easily implemented on a micron-size detector (e.g., beam splitter, lens-grating pair, galvanometer mirror, optical prism), femtosecond pulse shaping, and illumination power levels of $\sim 10^7$ watts/m^2 >> 100 watts/m^2 "safe" continuous limit in tissue. Bioluminescence techniques in which the light source is placed inside the tissue (e.g., a transgenic mouse with a luciferase gene in every cell of its body) has a spatial resolution limited to ~10 per cent of the depth, or ~10 microns resolution at a ~100 micron depth.

Coherent anti-Stokes Raman scattering (CARS) already permits organelle imaging in living cells and can in theory create a point-by-point chemical map of a cell using two intersecting lasers. Three-dimensional observation of microscopic biological nonliving structures by means of X-ray holography requires a high degree of spatial coherence and good contrast between target and surroundings. Good contrast may be achieved in the wavelength range between the K absorption edges of carbon (l = 4.37 nm) and oxygen (l =2.33 nm), the "water window" where carbon-containing biological objects absorb radiation efficiently but water is relatively transparent. A 50-micron diameter emitter has been tested that uses near-IR 5-femtosecond laser pulses impinging upon a helium gas target to create a well-collimated (<1 milliradian) beam of coherent soft X-rays at a 1 KHz repetition rate producing a brightness of 5×10^8 photons/mm^2-milliradian^2-sec, a peak X-ray intensity of $>10^{10}$ watts/m^2 on the propagation axis behind the He target. Many cellular parameters need not be measured directly in order to be detected by a medical nanodevice. Cell volume sensing is a case in point. An intracellular nanodevice can indirectly monitor changes in the volume of the cell in which it resides by one of two methods. First, measurements of the mechanical deformation of the cellular membrane, stretch-activated channels, or cytoskeletal strains and structural changes are quite sensitive to alterations in total cell volume. Second, concentration or dilution of the cytoplasmic environment through cell shrinkage or swelling leads to the activation of various volume-regulation responses which may be detected by the nanorobot. Changes in the concentration of soluble cytosolic proteins may nonspecifically affect enzyme activity via "macromolecular crowding".

Minor changes in cell volume can cause severalfold changes in ion transport. Cellular signalling entities that have been linked to the transduction and amplification of the primary volume signal include Ca^{++} transients, phosphoinositide turnover, eicosanoid metabolism, kinase/phosphatase systems such as JNK and p38, cAMP, and G-proteins. These signal amplification pathways may be monitored using chemical concentration sensors aboard the medical nanodevice, allowing the nanorobot to eavesdrop on the natural sensory channel traffic of the cell. In many ways the neuron is the most nanomedically important class of cell in the human body. Nanomedical applications regarding neurons and the brain are addressed in.

The following is a brief summary of noninvasive (i.e., no axonal membrane penetration) nanotechnological methods for monitoring the electrical traffic of individual neurons. Noninvasive measurement of axonal traffic within nerve bundles will require multiple sensors and greater sensitivity to compensate for shielding by the perineurium, a tight resistive sheath enclosing the bundle ~20 microns thick with resistivity ~4000 ohm-cm. The "typical" ~20 micron human neuron discharges 5-100 times per second, moving from -60 mV potential to +30 mV potential in ~10^{-3} sec. Thus the variation in electric field at the axonal surface is ± 4500 volts/m. Since electric field sensors can detect 100 volt/m fields up to GHz frequencies, an electric sensor attached by circumaxonal cuff or pressed against the axonal surface (possibly at the node of Ranvier) should readily detect each action potential discharge. (The diameter of human nerve axons is 0.1-20 microns.)

By 1998 silicon-to-neuron extracellular junctions already permitted direct stimulation of individual nerve cells in vitro without killing the cells, and extracellular electrodes were commonly used to detect neuronal electrical activity noninvasively, both in vivo and in vitro. Artificial electric fields may also be employed to trigger or moderate neural signals. Cell membrane capacitance is typically ~0.01 picofarads/micron^2, and varies with the state of the health of the cell. The magnetic flux density caused by a single action potential discharge is ~0.1 microtesla at the axonal surface, which may be detected by a "compass oscillator" type magnetosensor with a ~KHz maximum sampling rate. It might be possible for artificial magnetic fields to directly influence neural transmissions. Even a static 65 millitesla field has been shown to reduce frog skin Na^+ transport by 10-30 per cent. Each neuronal discharge develops an electrical energy of ~20 picojoule (~10^{10} kT), far smaller than the magnetic energy stored in a B = 1.4 tesla field of a permanent micromagnet traversing an L^3 = (20 micron)^3 volume which is $B^2 L^3/2 m_0$ ~ 6000 pJ. If properly manipulated, such a field may be sufficient to enhance, modulate, or extinguish a passing neural signal. While neurons come in many shapes and sizes our "exemplar" ~14,000-micron^3 neuron discharging ~90 mV into an input impedance of ~500 Kohms produces ~0.2 microampere current per pulse and generates a continuous (average) 100-300 pW of waste heat as measured experimentally. The discharge rate of 5-100 Hz can produce brief surges up to ~2000 pW during a high-frequency train, but the duty cycle of such trains is far less than 100 per cent, reducing time-averaged dissipation to the observed 100-300 pW range. Single impulses are measured experimentally to produce 2-7 microkelvin temperature spikes in cold or room-temperature mammalian non-myelinated nerve fibers and ~23 microkelvins in non-myelinated garfish olfactory nerve fibers at an energy density ranging from 270-1670 joules/m^3-impulse from 0-20°C. In non-myelinated fibers the initial heat occurs in two temperature-dependent phases: a burst of positive heat, followed by rapid heat reabsorption (called the negative heat). The positive heat derives from the dissipation of free energy stored in the membrane capacity, and from the decrease in entropy of the membrane dielectric with depolarisation.

An L ~ 20-micron neuron in good thermal contact with an aqueous heat sink at 310 K has thermal conductance L K_t ~ 10^{-5} watts/K, so trains of 5-100 Hz impulses lasting 1 second should raise cellular temperature by 10-30 microkelvins; up to ~200 microkelvin thermal spikes from such trains have been observed experimentally. These events are easily detectable by nanoscale thermal sensors capable of ~1 microkelvin sensitivity up to ~1 KHz. The ~microkelvin heat signature of individual impulses or very short pulse trains can probably be temporally resolved because the minimum pulse repetition time is ~10 millisec (at 100 Hz) which is much longer than the thermal time constant for an L ~ 20-micron neuron (thermal conductivity K_t = 0.528 watts/m-K and heat capacity $C_V = 3.86 \times 10^6$ joules/m^3-K for brain tissue; which is t_{EQ} ~ $L^2 C_V/K_t$ ~ 3 millisec. The synaptic cleft between the axonal presynaptic terminal and the dendritic postsynaptic membrane is 10-20 nm in most synapses, although in the vertebrate myoneural junction it may be as large as 100 nm. Contact area per bouton is ~1 micron^2, giving a total gap volume of ~10^7-10^8 nm^3. The density of acetylcholine receptors is highest in muscles along the crests and upper thirds of the junctional folds (~10,000/micron^2), and is lowest in the extrasynaptic regions (~5/micron^2).

Each action potential discharge triggers the release of ~10^4-10^5 molecules of acetylcholine into the gap volume of an active neuromuscular junction (diffusion time ~1 microsec), raising c_{ligand} from near zero to ~3×10^{-4} molecules/nm^3 (~0.0005 M) in ~1 millisec, followed by near-complete hydrolysation by acetylcholinesterase during the 1-2 millisec refractory period. Into the gap volume may easily be inserted a ~10^5 nm^3 neurotransmitter concentration sensor able to measure ~100 acetylcholine molecules in ~1 millisec, thus detecting pulses at the fastest discharge rate. A similar device could be used to precisely regulate neuro-transmitter concentration at the junction, and hence the neural signal itself, under nanodevice control. Simple electrochemical and mechanochemical artificial synapses have been demonstrated. Many other noninvasive neurosensing techniques are readily conceivable. For instance, counting rotors or sodium magnetic resonance imaging could detect changes in the ionic composition (e.g., Na^+, K^+) of the periaxonal fluid before, during, and after discharge. Starting in 1968, H. Frohlich, observing that millivolt electrical potentials maintained across cell membranes ~10 nm thick give rise to huge fields ~10^7 volts/

m possibly producing an electret state, theorised that membrane molecules must be highly electrically polarised and thus could interact to produce coherent surface acoustic vibrational modes in the 10-100 GHz (microwave) frequency range; the longest wavelength is about twice the membrane thickness. Interestingly, this frequency span is very close to the maximum trigger/reset frequency for bioelectronic molecules).

• Note that (~100 mV) (1.6 × 10^{-19} coul) ~ 4 kT, so a membrane molecule with a single charge on either end should be reliably reoriented by a depolarisation wave, coupling pressure waves and electrostatic field fluctuations. However, the direct detection of 10-100 GHz millimeter radiation by non-nanotechnological means is experimentally difficult and controversial because the tests must be performed in vivo in close proximity to an actively metabolising cell in water—and water strongly absorbs microwaves over macroscale ranges (e.g., ~99 per cent absorption in 3 mm at 100 GHz). Nevertheless, active cells have shown enhanced Raman anti-Stokes scattering, an effect ascribed to the converse of the Frohlich oscillations. In one study, the normalised growth rate of yeast cultures was enhanced or inhibited when irradiated by CW microwave fields of ~30 watts/m^2 of various frequencies; growth rate data spanning 62 separate runs revealed a repeatable frequency-dependent spectral fine structure with six distinct peaks of width ~10 MHz near 42 GHz.

Investigations of related phenomena are ongoing and voluminous; the interested reader should peruse *Bioelectromagnetics*, the archival journal of this field. 100 GHz waves attenuate only ~1 per cent after passing through ~3 microns of soft tissue. A single electron injected into an integral membrane protein could act as an oscillating dipole, making a 300 volt/m signal 10 nm from the protein antenna with an energy transfer of ~0.004 kT per cycle; ~1000 oscillating electrons could produce a measurable field. A 20-micron diameter cell modeled as a nonuniform spherical dipole layer with transmembrane dipoles located 10 nm apart and embedded in a dissipative medium could produce 10^2-10^5 volt/m microwave fields 1-10 microns from the cell surface.

Thus, a variety of rf and microwave electromagnetic emanations may in theory be detectable both within and nearby living cells which could prove diagnostic of numerous internal states. Such states may include cytoskeletal dynamics, metabolic rates, plasmon-type excitations due to the collective motion of ions freed in chemical reactions, positional, rotational or conformational changes in biological macromolecules and membranes, internal movements of organelles and nerve traffic conduction, cellular pinocytosis, cellular reproduction events, cell membrane identity (e.g., distinguishing erythrocyte, Gram-positive and Gram-negative cell coat conductivities at 10 KHz), and cell-cell interactions. Macrosensing is the detection of global somatic states (inside the human body) and extrasomatic states (sensory data originating outside the human body). While the treatment here is necessarily incomplete, the discussion nevertheless gives a good feel for the kinds of environmental variables that internally-situated nanodevices could sense. Not all capabilities outlined here need be available on every nanorobot, since injection of a cocktail of numerous distinct but mutually cooperative machine species allows designers to take full advantage of the benefits of functional specialisation.

In many cases, a given environmental variable can be measured by several different classes of sensor device. However, since these devices are microscopic it is in theory possible to operationalize almost all of the macrosensing capabilities described below in one patient using just a billion devices (~10 mm^{-3} whole-body deployment density), a total volume of ~1 mm^3 of nanorobots or ~0.1 per cent of the typical ~1 cm^3 therapeutic dose. Can sounds generated by a single cell be detected, and thus be useful for diagnosis? Probably not, given that low frequency acoustic radiators are notoriously inefficient. For example, mitochondrion organelles of the giant amoeba *Reticulomyxa* are shuttled back and forth by 1-4 cytoplasmic dynein motors while riding on the outside of a bundle of 1-6 microtubules. Each dynein motor generates 2.6 pN of force and drives the mitochondria at up to ~10 micron/sec, developing 0.3-1.0 × 10^{-16} watts of mechanical power within each 320-nm diameter organelle.

Taking each organelle as a cylindrical acoustic radiator with a mechanical input power of P_{in} ~ 10^{-16} watts at n ~ 1000 Hz, the output acoustic pressure at the organelle surface is only ~10^{-9} atm an acoustic power intensity I ~ 10^{-15} watts/m^2 which is not detectable by micron-sized nanorobots. Nevertheless, cells and intracellular elements are capable of vibrating in a dynamic manner with complex harmonics that can be altered by growth factors and by the process of carcinogenesis, so the possibility cannot be completely ruled out. Blood pressure ranges from 0.1-0.2 atm in the arteries to as low as 0.005 atm in the veins. The systolic/diastolic differential ranges from 0.05-0.07 atm in the aorta and 0.01-0.02 atm in the pulmonary artery, falling to 0.001-0.003 atm in the microvessels, or 0.003-0.005 atm if the precapillary sphincter is dilated. In venous vessels, pulse fluctuations are 0.002-0.010 atm in the superior vena cava, 0.004-0.006 atm in the subclavian vein, ~0.004 atm in venules generally, and ~0.0005 atm in the brachial vein. There is also a ~0.05 Hz random fluctuation in the microvessels with amplitude on the order of 0.004-0.007 atm. Both blood pressure and pulse rate can be reliably monitored by a medical nanodevice virtually anywhere in the vascular system using a $(68\ nm)^3$ pressure sensor with ~0.001 atm sensitivity. Pulse propagation through body tissue is somewhat muted due to absorption in compressible fatty membranes, but most cells lie within 1-3 cell-widths of a capillary so the cardiac acoustic signal should still be measurable using more sensitive detectors. The time-averaged interstitial pressure in subcutaneous tissue is 0.001-0.004 atm.

Arterial pulse waves (vascular oscillations) carry subtle messages about the health of internal organs and the arterial tree. The idea of using pulse waves for diagnosis has a long history dating back 2000 years in China. For example, in the *Book on Pulse Waves* by Wang Shu-He (201-285 AD), waves detected by manual probing are classified using such subjective and qualitative descriptors as floating, deep, hidden, rapid, slow, moderate, feeble, replete, full, thready, faint, weak, soft, slippery, hesitant, hollow, firm, long, short, swift, running, intermittent, uneven, taut, string-tight, gigantic, or tremulous. Abnormal waves were empirically related to disease states. Wave data gathered by nanodevices could make possible a theoretically sound, quantitative system of noninvasive observation, classification, and diagnosis as a supplement to other nanomedical tools. The variation of mechanical pressure over a complete respiratory cycle is ~0.003 atm in the pleura, ~0.002 atm at the alveoli, detectable by nanomedical pressure sensors positioned in the vicinity of the respiratory organs. Holding a deep breath further stretches the pulmonary elastic tissue, up to 0.02 atm.

However, turbulent flows at Reynolds number $N_R > 2300$ in the trachea, main bronchus and lobar bronchus produce a whooshing noise that may be the loudest noncardiac sound in the human torso during conventional auscultation. Many other mechanical body noises should be globally audible to properly instrumented medical nanodevices. If normal chewing motions (of hard foods) release 1-10 milliwatts in a ~100 cm^3 oral volume with a ~1 sec jawstroke, power density is ~10 watts/m^3 or ~10^{-4} atm of tooth-crunching noise. A stomach growl registering 45 dB (vs. 30 dB whisper, 60 dB normal conversation) at 2 meters has a source power of 160 milliwatts; released from a 10 cm^3 gastric sphincter volume gives a ~2×10^{-6} atm acoustic wave, detectable throughout the body. Walking and running releases 20-100 joules/footfall for a 70 kg man; assuming the energy is absorbed within a ~1 cm thickness or within ~1 second by the sole of the foot, an upward-moving planar compression wave of 0.4-2.0 atm, easily detectable by acoustically instrumented nanodevices body-wide. (Shoe insoles dissipate energy and alter the shock wave pulse shape.) Hand-clapping generates 0.02-0.2 atm pulses, also easily detectable. Lesser noises including ~30 millisec hiccups at (4-60)/min, intestinal and ureteral peristalsis, sloshing of liquid stomach contents, heart murmurs, a tap on the shoulder by a friend, nasal sniffling and swallowing, clicks from picking or drumming fingernails, crepitations, manustuprations and ejaculations, the rustling noise of clothing against the skin, flapping eyelids, anal towelling, bruits (including murmurs and thrills) due to vascular lesions, dermal impact of water while showering, copulatory noises, urethral flow turbulence during urination, transmitted vibrations from musical instruments, creaking joints, and squeaking muscles can be detected locally if not globally. Implantation of significant interconnected in vivo diamondoid structures may produce increased sensitivity to internal noises, due to the extremely low acoustic absorption coefficient of diamond. Average source power for conversational speech in air is ~10 microwatts at the vocal cords (60 dB), up to ~1000 microwatts for shouting (90 dB) and as little as 0.1 microwatts (30 dB) for whispering. Vocal cord surface area ~1 cm^2, giving an acoustic intensity I ~ 0.001-10 watts/m^2. (Using the decibel notation, dB = 10 $\log_{10}$ (I/I_0), where $I_0 \sim 5 \times 10^{-13}$ watts/m^2 in air, $I_0 \sim 1 \times 10^{-16}$ watts/m^2 in water.) For water at 310 K, r = 993.4 kg/m^3 and v_{sound} = 1500 m/sec, therefore A_p = 0.0005-0.05 atm for speech, detectable by nanodevices throughout the body due to minimal attenuation at audible frequencies.

Other easily detectable vocalisations include whistling, humming, coughing, sneezing, rales, wheezing, expectorating, eructations, flatus, vomiting, hawking and noseblowing. In the case of spherical waves diverging from an omnidirectional transmitter of radius r at a distance X from the transmitter, average intensity declines inversely as the square of the distance and so. Of course, an ex vivo acoustic nanosensor may receive sound that has passed through no interface, hence may detect pressure waves ~3 orders of magnitude lower in amplitude. Assuming $A_{transmit}$ = $A_{incident}$, P_R ~ 600 microwatts, so ex vivo nanorobots with a 0.3 micron^3 sensor could hear people shouting at x_R = 2 meters. To hear talking (~10 microwatt source) requires a 2.4 micron^3 ex vivo pressure sensor (limit ~10^{-7} atm). Optimally positioned and calibrated nanomedical pressure sensors could directly measure changes in the ambient barometric pressure to within ± 10^{-6} atm. Normal atmospheric variation due to weather ranges from 0.94-1.05 atm; such slow moving changes are readily monitored. Very near the Earth's surface, the air pressure P at altitude h above sea level is approximated by $P = e^{-k_p h}$ (atm), where $k_p = 1.16 \times 10^{-4}$ m^{-1} at 20°C; at sea level, a 10^{-6} atm change in pressure reflects a change in altitude of only ~1 cm. However, the opening or closing of a door inside a (~5 m)^3 room that displaces >125 cm^3 of air also causes a minimally detectable >10^{-6} atm pressure pulse. Other sources of environmental pressure variation such as infrasonic (~0.2 Hz) microbaroms from offshore ocean storms, wind entering through open windows, forced-air currents from central heating or A/C systems, or even the movements of nearby people and pets may be detectable and thus, may further confuse the measurement, reducing absolute accuracy unless suitable corrections are made. Using a navigational transponder network, a population of ~10^{11} nanodevices each spaced an average ~100 microns apart throughout the body tissues can determine relative location to a positional accuracy of ~3 microns and an angular accuracy of ~2 milliradian, with the data updatable once every millisecond. Thus, dispersed, the network can continuously monitor and record the relative positions of all limbs with worst-case (cumulative error) ~0.8 mm accuracy over a 2-meter span. These devices can prepare high-resolution dynamic maps of body position, velocity, acceleration and rotation, allowing precise real-time digital kinesthesia during sports or artistic activities such as gymnastics, pole vaulting, or ballet; during transportative activities such as driving cars around hairpin turns, roller-coaster rides, military aircraft maneuvers and space launches; during precision tool-using such as needle-threading, antique watch repair, or while using the fingers as measurement calipers; during self-defense activities requiring complex motions such as karate or judo; and during emergency situations such as automobile crashes and tumbling motions during falls from great heights. All such sensory data is readily outmessaged to the patient or user in real time.

Nanorobots could assist in path integration or the reconstruction of experienced limb trajectories. Given the ability to determine rotation rates and to measure applied forces and torques, the network should also be able to forecast the anticipated future positions of limbs. By comparing these projections to actual results, the network can then infer the viscosity of the medium in which the activity is taking place (e.g., air or water), whether the environment is stable or is translating or rotating in some direction, and whether the human user is physically supported or in free-fall. Internal nanodevices can directly measure if a body is sitting, standing, laying prone, inverted, falling, floating or diving, and then communicate that information directly to the human user. Detection of patient activity states—e.g., the patient is sitting, standing, or walking—can be used to control nanorobot behaviours, activate or deactivate outmessaging displays, and so forth. A simple macroscale wearable tactile "compass belt" that would convey desired geographical directions has already been proposed. Since nanopendular sensors can reliably determine the direction of the local gravity field vector to within ~2 milliradians in ~10^{-4} sec, the navigational network can poll its members and arrive at an accurate consensus on which direction is up. Nanodevices affixed to relatively stable hard body parts will exhibit more consistent orientational readings. Measurement of the gravity vector allows the vertical orientation of the rest of the body to be precisely fixed in space, especially useful for gymnasts, trapeze artists, underwater divers in murky lakes, and vestibular-impaired individuals, to whom this information may be outmessaged directly. The gravitational vector may also be indirectly measured, albeit more slowly and less accurately, by monitoring the body's natural reactions to changes in the axis of gravitational loading. For instance, the variation in head-to-toe hydrostatic pressure for a 1.7 m tall, 70 kg human standing in a 1-g gravity field is ~0.17 atm; the ability to measure a systematic cross-body differential of 10^{-6} atm allows the detection of a ~10^{-5} g change along the lengthwise aspect, or ~10^{-4} g change along the transverse aspect. Thus a continuously updated whole-body barostatic map allows the human body to serve as a three-dimensional gravity/orientation sensor. A network of nanodevices can inventory the volume and density of each of the body's ~10^8 $(1 \text{ mm})^3$ voxels using a combination of acoustic ranging, somatic mapping, and flowmetry. Each voxel is identified as fat, muscle tissue, bone mass, interstitial fluid, and so forth.

These measurements allow the body's weight to be precisely computed as volume times density, rather than the usual method of determining weight using gravitational force data. The existence of a natural physiological "ponderostat," a crudely analogous humoral body-mass detector, has been proposed; indeed, the insulin-leptin system has been found to serve a related purpose, and is easily eavesdropped by medical nanorobots. To achieve such phenomenal positional accuracies, the nanodevice must be able to computationally resolve several complicating factors. First, localised mass concentrations representing nonuniformities in crustal density produce residuals of up to ± 0.0006 m/sec^2, which may be removed from the data using a standard map of known terrestrial isostatic variations and anomalies. Indeed, matching observations to such a map could provide useful longitudinal information as well. Another complication is the variation in gravity due to tidal forces amounting to $\sim 3 \times 10^{-7}$ g's, twice daily, which lies at the limits of detectability for a 20-micron gravimeter. Other minor geodesic and terrain-related corrections, too complicated for discussion here, may also need to be applied in certain circumstances. Note that the presence of nearby heavy objects does not influence measurement accuracy: a 100-ton building 10 meters away adds a lateral acceleration of only 7×10^{-9} g's to a human body. One final complication is that patient movements create kinematic accelerations that must be distinguished from the gravitational accelerations. Gravity readings can be corrected by taking derivatives of the signals from kinesthetic monitoring to give gravity in the reference frame of the patient's room, and many individual measurements may be averaged to improve accuracy since the gravity vector normally changes only very slowly. In vivo studies of the electric properties of blood vessels shows that the walls of veins and arteries present a specific electric resistance ~200 times greater than the conductive medium of blood (plasma)—roughly 200 ohm-m vs. 0.7 ohm-m. Thus blood vessels are properly regarded as relatively insulated conducting cables which can electrically connect an injured tissue with surrounding noninjured tissue. Capillaries form an electric junction with the interstitial fluid, so the electric gradient can be canceled by ionic transport. Cell membranes are insulating dielectrics with resistance and capacitance, penetrated by ionic channels or gates for ionic transport.

Additionally, in 1941 Szent-Gyorgyi suggested the possibility of semi-conduction in proteins, a theory which has since been elaborated by many investigators. B. Nordenstrom has proposed that the above schema represents a system for selective electrogenous mass transport of material between blood and tissue which he calls Vascular-Interstitial Closed (electric) Circuits (VICCs)—in effect, an additional electro-circulatory system operating in parallel with the well-known diffusive, osmotic, and hydrodynamic mechanisms of regular bloodstream materials transport. As long as ions can leak through open pores and migrate through ion channels, long distance transport cannot take place and the VICC system remains primarily a local (e.g., within tissues), not global, electrical circuit. Connections also exist with conductive media including cerebrospinal fluid, bile, and urine.

Nanorobots capable of monitoring the status of local VICCs may rapidly and efficiently acquire a wealth of systemic information without the need for direct inspection of the affected tissues. For example, when a working muscle produces catabolic products such as lactic acid, an electrochemical potential gradient develops between the muscle and surrounding tissue in a known manner. Injured tissues become polarised in relation to surrounding tissue, as initial catabolic degradation products acidify the tissue. Malignant neoplasms, benign neoplasms, and internally necrotic granulomas all may polarize electrically in relation to surrounding tissue; a vascular thrombus also contains ionised material. Large deflections in the diffusion potential of blood occur if blood is deoxygenated or is infected with Gram-negative bacteria. Blood changes its electric potential from +500 mV to +1000 mV during spontaneous coagulation, and the spontaneous electric potential at the site of a crush injury of the iliac crest in rats oscillated several times between +190 mV and -60 mV over a four day experimental trial. Nordenstrom proposes that in vivo electrophoresis via the VICC system may also play a role in mediating leukotaxis. When tissue is artificially polarised by electrodes, accumulation of granulocytes with margination and development of pseudopods is the result. In one experiment, exposure of a mesenterial membrane to a 1 microampere 1 volt field for 30 minutes stimulated diapedetic bleeding near the anode.

At higher power, the arterio-capillaries contracted, narrowed, and emptied of blood cells, while the veno-capillaries and venules widened and filled with granulocytes. Monitoring such actions of local VICCs experiencing natural voltage fluctuations could alert medical nanorobots to the presence of local injuries, tissue changes, or pathologies that otherwise might go unnoticed for a considerable time. Detection of electric fields to ~100 volts/m, the normal vertical atmospheric gradient, in theory might allow determination of aboveground altitude to ~1 m accuracy if the time between recalibrations is very brief. However, a human body (which is a good conductor), standing on level ground, acquires a slight negative surface charge and becomes "part of the ground" electrically. This distorts upward the equipotentials that usually run parallel to the ground, thoroughly corrupting an altitude measurement. Nearby lightning hits and storms are detectable, which is useful. However, medical nanodevices with appropriate magnetometers can measure variations in magnetic field to ~0.1 microtesla using a (660 nm)^3 permanent magnet sensor, in a measurement time t_{meas} ~ 0.6 millisec. Geomagnetic field maps of the Earth's surface are highly nonisotropic in both latitudinal and longitudinal directions; knowledge of the field plus the absolute direction of true north (e.g., using a nanogyroscope) gives longitude information from the separation of the planetary magnetic and spin poles. This implies that properly equipped in vivo nanodevices can establish their latitude and longitude on Earth's surface ~1000 times every second by this means. In particular, the horizontal component of the geomagnetic field vector ranges from 0-41 microtesla from magnetic pole to magnetic equator; the independent vertical field component ranges from 0-70 microtesla.

Thus a 0.1-microtesla sensor should resolve each component to an accuracy of ~11 arcmin or ~20 km in both latitude or longitude. Terrestrial surface magnetic anomalies (e.g., iron deposits) produce local variations of 0.03-30 microtesla and are thus detectable. Additional complications to these measurements include artificial magnetic field sources, the 0.01-0.1 microtesla solar daily variation, occasional magnetic storms producing erratic geomagnetic fluctuations of 0.01-5 microtesla, sudden-commencement ionospheric electrojets up to 0.3 microtesla, and various long-term secular variations in the geomagnetic field. Radio, television, and direct broadcast satellite signals probably are not detectable by individual nanodevices, but such detection may be indirectly achieved in vivo using nanorobot-accessible dedicated macroscale antennas implanted in the body. The piezoelectric effect is "the production of electrical polarisation in a material by the application of mechanical stress". Many materials in the human body are piezoelectric, including tendon and elastin, dentin and bone. This polarisation, or surface charge, varies as the physical stress imposed on the material changes over time. Thus, measurement of the piezoelectric effect in bone or tendon can provide information on the physical loads that are being carried by these materials. For example, a shear stress applied along the long axis of a bone alters the polarisation voltage that appears on a surface at right angles to the axis. In another experiment, piezoelectric surface charges on a loaded human femur were measured to range from -131 to +207 picocoulombs/cm^2 (-8 to +13 charges/micron^2) from one end of the bone to the other end, depending upon position along the shaft. Real-time monitoring of this data would permit specific inferences as to the amount of load the bone was carrying and from what direction, including bending, shearing, and twisting forces, from which whole-body activity states could subsequently be inferred. Knowledge of these surface charge variations might also be exploited in osteographic or functional navigation. In vivo medical nanodevices can gather little useful optical information from the external environment. Here's why. In biological soft tissues, scattering dominates absorption except in the pigmented layers of the epidermis and stratum corneum. Thus the propagation of light in tissues may be regarded as occurring in two steps. Over the optical band from 400-700 nm, incident intensity I_0 = 100-400 watts/m^2 when standing in direct sunlight; artificial lighting in homes and offices is typically 0.1-10 watts/m^2; moonlight provides only $I_0 \sim 10^{-4}$ watts/m^2; and the absolute threshold for human vision is ~10^{-8} watts/m^2. For $s_d \sim 900$ m^{-1}, the intensity ratio of transmitted/incident visible light falls to $I_d/I_0 = 0.1$ at $z = 2.5$ mm depth (~eyelids), ~10^{-4} at $z = 1$ cm depth, and ~10^{-11} at $z = 2.8$ cm—a depth at which the tissue would be completely dark to the human eye even under direct sunlight illumination at the outermost skin surface. Let us assume that a patient is standing in $I_0 \sim 400$ watt/m^2 direct unfiltered sunlight, and that $t_{meas} = 1$ sec, SNR = 2, $N_{sensor} = 25,000$ elements giving a nanorobot eyespot of $A_e N_{sensor} = 0.025$ micron^2, and that $s_d = 900$ m^{-1}.

The maximum tissue depth for optical photon detection is $z_{max} \sim 17$ mm ($I_d \sim 10^{-4}$ watt/m^2) which includes a volume of tissue comprising up to ~30 per cent of total body volume. Changes in illumination at the minimum indoor artificial level of ~0.1 watts/m^2 are visible to $z_{max} \sim 7$ mm depth. This eyespot, if exposed to air on the outermost surface of the skin, is just sensitive enough to detect full moonlight. Our general conclusion is that variations in normal indoor lighting may be directly measurable by nanorobots stationed within the outermost ~1 cm of body tissues. Imaging, as opposed to the simple illumination detection system described here, is a much more difficult design challenge The ability to detect individual neural cell electrical discharges noninvasively in many different ways, coupled with the abilities (A) to recognize and identify specific desired target nerve cells and (B) to pool data gathered independently by spatially separated nanodevices in real time, offers the possibility of indirect neural macrosensing of complex environmental stimuli by eavesdropping on the body's own regular sensory signal traffic. For example, facultatively mobile nanodevices may swim into the spiral artery of the ear and down through its bifurcations to reach the cochlear canal, then position themselves as neural monitors in the vicinity of the spiral nerve fibers and the nerves entering the epithelium of the organ of Corti (cochlear or auditory nerves) within the spiral ganglion. These monitors can detect, record, or rebroadcast to other nanodevices in the communications network all auditory neural traffic perceived by the human ear. Advanced speech recognition systems may permit recovery of spoken words and identification of individual speakers (including vocalisations by the user), recognition of background noises, reception and validation of cues or commands spoken directly to in vivo nanosystems by authorised medical personnel, and so forth. Since properly configured monitors can also modulate or stimulate nerve impulses, these devices may add audible signals to the audio traffic, thus may be employed as hearing aids (using feedback loops), real-time language translation mechanisms, continuous vocalisation/audition recorders, voice-stress analyzers, or nanodevice-user communications links. Nanomonitors positioned at the afferent nerve endings emanating from hair cells located in the otolithic membrane of the utricle and saccule and in the cristae ampullaris of the semicircular canals allow medical nanodevices to directly record, amplify, attenuate, or modulate the body's own sensations of gravity, rotation, and acceleration, although kinesthetic sensory management might also be required for complete control. Motor neurons likewise can be monitored to keep track of limb motions and positions, or specific muscle activities, and even to exert control. Neuron-resident nanorobots may detect auditory effects induced in the brain by pulsed microwaves from external sources.

Feline cochlear neurons sensitive to audible frequencies of >300 Hz respond to single microwave pulses at a threshold-specific absorption rate of 6-11 watts/kg-pulse, while in humans the threshold for effect, which depends on energy per pulse, may be as low as ~0.02 joule/m^2-pulse for people with low hearing threshold. Olfactory and gustatory sensory neural traffic similarly may be eavesdropped by nanosensory instruments. Nerve taps in the medulla oblongata or at the phrenic nerve that drives the diaphragmatic muscles allow direct monitoring of respiratory activity. Pain signals may be recorded or modified as required, as can mechanical and temperature nerve impulses from other receptors located in the skin. Even psychological variables such as emotionality, vigilance, and mental workload may be directly monitored by measuring ANS activity in sympathetic efferent fibers outside of the brain. The most complex and difficult challenge in neural macrosensing will be optic nerve taps.

The retina is thoroughly vascularised, permitting ready access to both photoreceptor (rod, cone, bipolar and ganglion) and integrator (horizontal, amacrine, and centrifugal bipolar) neurons. However, the optic nerve bundle itself has ~10^6 tightly bunched individual nerve fibers, a 10-100 MHz signal bandwidth, and significant natural data compression techniques which all must be untangled in real time. Developing algorithms capable of interpreting raw optical nerve traffic, say, to recognize a specific human face or a specific scene in the vision field, would prove a significant research challenge. (Direct monitoring of photo-receptors or retinal membrane potentials and other techniques may simplify untangling of the signal compression.)

Rapid visual field identification using artificial neural nets is also a subject of much current research interest. Eyeball rotations (e.g., via resident intradevice nanogyroscopes, eyelid position, pupil aperture, and lens accommodation under ciliary muscle control must be monitored to supplement the vision field analysis. We have only scratched the surface of the potential for nanorobot macrosensing. Medical nanodevices can quantitatively monitor variables not normally accessible to human consciousness, such as:

(i) Hormone and neurotransmitter levels.

(ii) Gastroelectric oscillations and skin conductivity.

(iii) Pupil dilations.

(iv) Drug and alcohol (and breakdown product) concentrations in the blood.

(v) Internal organ damage or malfunction.

(vi) Digital real-time performance data on vital organs or limbs (e.g., continuous kidney volume throughput, pancreatic insulin output, cholesterol metabolism in the liver, or lactic acid production in specific muscles during exercise).

(vii) Continuous mapping of thermoregulatory isotherms from which the temperature and heat capacity of the external medium (e.g., swimming pool water, cool night air) and the presence of sunlight or shade on the skin (via dermal thermal differentials) may be inferred.

The application of nanotechnology to sensors should allow improvements in functionality. In particular, new biosensor technology combined with micro and nanofabrication technology can deliver a huge range of applications. They should also lead to much decreased size, enabling the integration of 'nanosensors' into many other devices. Various chemical substances cyclically increase and decrease in serum concentration with time of day, permitting crude temporal macrosensing. Other substances significantly vary in serum concentration depending upon whether the patient is supine or standing, thus permitting limited biochemical postural macrosensing. If nanorobots can leave and re-enter the body, then macrosensing may include direct sampling of the external environment upon demand and the possibilities for remote data acquisition become virtually limitless. Sensors pervade many aspects of modern living. They are built into many consumer electronic devices, cars, medical devices, security and safety devices, and systems for monitoring pollution and environmental conditions. Many applications demand miniaturisation to reduce power consumption for integration into portable devices. Affordable mass production is also a prerequisite for sensors for consumer products, and for disposable devices such as sensors for medical diagnostics and pollution monitoring. Sensors support applications across the economy—industrial processes, and those in construction, extractive industries, agriculture, health care and so on—and can be incorporated into new or existing products. Sensors can model various parameters: physical parameters such as temperature, displacement, acceleration, flow and so on; and chemical and biochemical parameters, such as concentrations of gases, ions or molecules, and molecular interactions) We can also expect to see actuators that control movement on the nanoscale. Sensor/actuator combinations will deliver 'smart' and precise functions in products and processes. For example, nanofabrication and inspection tools require sensors and actuator systems that can position objects with nanometre accuracy. In this way, sensors and actuators constitute another enabling technology.

Defining Nanowires and Nanoribbons and Their Uses

Practical challenges with carbon nanotubes involving cost of synthesis, control of chirality and diameter, separation from bundles, and attachment of functional groups have prompted researchers to explore other types of one-dimensional nanostructures. Two of the most promising candidates in this regard are nanowires and nanoribbons. As the name suggests, nanoribbons are solid objects (unlike nanotubes, which are hollow) with a near-uniform rectangular cross-section. So far, nanoribbons have primarily been synthesised from the oxides of metals and semiconductors. In particular, SnO_2 and ZnO nanoribbons have been materials systems of great current interest because of potential applications as catalysts, in optoelectronic devices, and as chemical sensors for pollutant gas species and biomolecules. Although they grow to tens of microns long, the nanoribbons are remarkably single-crystalline and essentially free of dislocations. Thus they provide an ideal model for the systematic study of electrical, thermal, optical, and transport processes in one-dimensional semiconducting nanostructures, and their response to various external process conditions.

Towards Investigating the NO_2 Sensing Mechanism of SnO_2

To be able to fully commercialize their potential, it is important to better understand the sensing mechanism of such systems. With this goal in mind, researchers at the Brookhaven National Laboratory, Lawrence Berkeley National Laboratory, and Accelrys used MS Modeling's DMol^3 to investigate the nitrogen dioxide sensing mechanism of SnO_2 nanoribbons. Reporting in *Nano Letters*, the researchers examined the NO_2-sensing mechanism of SnO_2 nanoribbons with exposed (1 0'1) and (0 1 0) surfaces. The density functional theory (DFT) calculations revealed that:

- The most stable adsorbed species involved an unexpected NO_3 group doubly bonded to Sn centers
- An orders-of-magnitude drop in electrical conductance can be explained by significant electron transfer to the adatoms
- Computed binding energies were consistent with adsorbate stability up to 700 K, with X-ray absorption spectroscopy indicating predominantly NO_3 species on the nanoribbon surface.

Functions of Tin Dioxide Nanoribbons as Chemical Sensors

Recent experiments with SnO_2 nanoribbons indicate that these are highly effective in detecting even very small amounts of harmful gases like NO_2. Upon adsorption of these gases, the electrical conductance of the sample decreases by more than an order of magnitude. More interestingly, it is possible to get rid of the adsorbates by shining UV light, and the electrical conductance is completely restored to its original value. Such single-crystalline sensing elements have several advantages over conventional thin-film oxide sensors: low operating temperatures, no ill-defined coarse grain boundaries, and high active surface-to-volume ratio.

Towards Investigating the O_2 and CO Sensing Mechanism of Nanoribbons

The ability of the nanoribbons to sense O_2 and CO was also investigated. In the case of O_2, the response of the nanoribbon was highly sensitive to the concentration of O-vacancies on the surface. Thus, in the absence of any surface vacancies, the calculation predicted negligible charge transfer. However, when surface vacancies were present, an O_2 molecule can adsorb as a peroxide bridge, and withdraw a significant amount of electronic charge from the nanoribbon surface, thereby decreasing its electrical conductance. In the case of CO adsorption, there was a net electron transfer from the CO to the nanoribbon surface. Thus the calculation predicted an increase in nanoribbon electrical conductance upon CO-adsorption, in agreement with experimental results.

Addressing the Issues Related to Redesigning Platinum at the Nanoscale

Researchers from the Department of Energy's Sandia National Laboratories and the University of New Mexico (UNM) have developed a new way of mimicking photosynthetic proteins to manipulate platinum at the nanoscale. The method has the potential of changing the metal's properties and benefiting emerging technologies. While we are in the early stages of research, we see the possibility of manipulating the nanoscale structure of platinum so that we can have control over the size, porosity, composition, surface species, solubility, stability and other functional properties of these metal nanostructures," says John Shelnutt, the Sandia scientist leading the research effort. "Such control means that the redesigned platinum could be used in many new applications, including catalysis, sensors and optoelectronic and magnetic devices." He adds that while research groups have reported a few platinum nanostructures — including nanoparticles, nanowires, nanosheets and others — the addition of new types of nanostructures is "highly desirable and potentially technologically important." Working with Shelnutt in the research are Frank van Swol from Sandia, UNM graduate student Yujiang Song and Eulalia Pereira from the University of Porto in Portugal. The new method of manipulating platinum was detailed in a paper in the *Journal of American Chemical Society* published in December. The idea for the technique is similar to photosynthesis, in which plants use the energy from sunlight to produce sugar. But instead of manufacturing sugar, the new method changes a platinum ion to the neutral metal atoms. The photosynthetic protein mimicks this repeatedly, allowing metal to be deposited as desired at the nanoscale. The method involves putting porphyrins — the active part of photosynthetic proteins — along with the platinum salt in an aqueous solution of ascorbic acid at room temperature.

The porphyrins are placed in specific locations in the solution where it is intended that metal should be deposited. For example, the porphyrins may be confined to micelles, or liposomes. Micelles are spherical assemblies of detergent molecules in which the heads are exposed to the water and the tails stick together in the interior. Liposomes are similar structures, but they are larger and have water on the inside and outside separated by a closed membrane — sort of like a cell. The membrane is composed of two layers of detergent molecules, with the heads on the inner and outer surface facing the water and the tails forming the interior of the membrane. When light is shined on the porphyrins located in these detergent structures, the porphyrins excite, becoming catalysts for platinum reduction and deposition.

As this occurs, the metal grows onto the surfaces of the surfactant structures as a thin sheet or in other ways. In the case of micelles, the platinum grows into balls that look like the common toy "Koosh" ball. The ball size can be controlled by the amount of porphyrins and platinum in the solution, the amount of light illuminating the solution, and the amount of time the light is on. For the metals platinum and palladium that form these nanostructures, it is enough for the porphyrin molecule to grow only a small metal "seed" particle composed of about 500 atoms. When it reaches this size, the seed starts to catalyze its own rapid growth (by oxidation of ascorbic acid), budding off arms in all directions and creating the Koosh-ball-like nanostructures. The porphyrin provides a convenient method of making these seeds at the location and time desired, leading to a uniform and selectable nanostructure size.

The platinum nanostructures take on a different form when they are prepared under different conditions. When the porphyrin is in a micelle, the platinum nanostructures produced look like Koosh balls. When the porphyrin is in the bilayer membrane of a liposome, the platinum grows in 2-nanometer thick sheet or platinum lace on the outer surface of the membrane, giving circular sheets — sort of like two-dimensional Koosh balls. Under solution conditions for which the liposomes aggregate, growth can occur along the interfaces of the liposomes to give platinum foamlike materials and foam nanoballs. The type of nanostructure is mainly determined by the type of surfactant assembly upon which the platinum grows and the extent of growth from the individual seed nanoparticles. Since the porphyrin remains attached to the platinum nanostructure and active in the presence of light, it can also perform other functions besides growing itself. For example, when illuminated with light, the platinum nanostructure evolves hydrogen from water.

REFERENCES

Baker, R.T.K. Synthesis, properties and applications of graphite nanofibers. In *R&D status and trends*, ed. Siegel et al.

Bate, R., Frazier, G., Frensley, W., Reed., M., "An Overview of Nanoelectronics," *Texas Instruments Technical Journal*, July-August 1989, pp. 13-20.

Beck, J.S., J.C. Vartuli, W.J. Roth, M.E. Leonowicz, C.T. Kresge, K.D. Schmitt, C.T.-W. Chu, D.H. Olsen, E.W. Shepard, S.B. McCullen, J.B. Higgins, and J.L. Schlenker. 1992. *J. Am. Chem. Soc.* 114:10834.

Bensaude-Vincent, B.: 2001, 'The Construction of a Discipline: Materials Science in the United States', *Historical Studies in the Physical and Biological Sciences*, 31, 223-248.

Braun, T.; Schubert, A. & Zsindely, S.: 1997, 'Nanoscience and nanotechnology on the balance', *Scientometrics*, 38, 321-325.

Brinker, C.J. 1996. *Curr. Opin. Solid State Mater. Sci.* 1:798.

Brown, N.: 2004, 'Needle on the Real: Technoscience and Poetry at the Limits of Fabrication', in: N.K. Hayles (ed.), *Nanoculture: Implications of the New Technoscience*, Bristol, UK: Intellect Books, pp. 173-190.

Bueno, O.: 2004, 'The Drexler-Smalley Debated on Nanotechnology: Incommensurability at Work?', *Hyle: International Journal for Philosophy of Chemistry*, 10(2), 83-98.

Bueno, O.: 2004, 'Von Neumann, Self-Reproduction and the Constitution of Nanophenomena', in: D. Baird, A. Nordmann & J. Schummer (eds.), *Discovering the Nanoscale*, Amsterdam: IOS Press, pp. 101-115.

Crandall, B.C. and Lewis, J., *Nanotechnology: Research and Perspectives*, MIT Press, Cambridge, 1992.

Crow, M. & Sarewitz, D.: 2001, 'Nanotechnology and Societal Transformation', in: M.C. Roco and W.S. Bainbridge (eds.), *Societal Implications of Nanoscience and Nanotechnology*, Dordrecht: Kluwer, pp. 45-54.

Einsiedel, E.F.; Goldenberg, L.: 2004, 'Dwarfing the Social? Nanotechnology Lessons from the Biotechnology Front', *Bulletin of Science, Technology & Society*, 24, 28-33.

European Comission: *European Workshop on Social and Economic Research on Nanotechnologies and Nanosciences, Brussels, 14-15 April 2004* [www.stage-research.net/STAGE/PAGES/Nano.html].

European Commission (Community Health and Consumer Protection): 2004, *Nanotechnologies: A Preliminary Risk Analysis on theBasis of a Workshop, Brussels, 1-2 March 2004* [www.europa.eu.int/comm/health/ ph_risk/documents/ev_20040301_en.pdf].

European Commission: 2004, *Converging Technologies: Shaping the Future of European Societies (Report of the High Level Expert Group "Foresighting the New Technology Wave")*, Brussels: European Commission Research [http://europa.eu.int/comm/research/conferences/2004/ntw/index_en.html].

F. Buot, "Mesoscopic Physics and Nanoelectronics: Nanoscience and Nanotechnology," *Physics Reports*, pp.73-174, 1993.

F. Capasso and S. Datta, "Quantum Electron Devices," *Physics Today*, pp.74-82, February 1990.

Feynman, R., "There's Plenty of Room at the Bottom: An invitation to Enter a New Field of Physics," Talk at the Annual Meeting of the American Physical Society, 29 December 1959. Reprinted in Appendix B of Crandall and Lewis. See citation above.

Fielder, F.A.; Reynolds, G.H.: 1994, 'Legal Problems of Nanotechnology: An Overview', *Southern California Interdisciplinary Law Journal*, 3, 593-629.

Fleischer, Th.; Decker, M. & Fiedeler, U. (eds.): 2004, *Große Aufmerksamkeit für kleine Welten ' Nanotechnologie und ihre Folgen*, special issue of *Technikfolgenabschätzung ' Theorie und Praxis*, 13 (2), 5-85 [www.itas.fzk.de/tatup/042/inhalt.htm].

Fogelberg, H. & Glimell, H.: 2003, 'Molecular Matters: In Search of the Real Stuff', in: H. Fogelberg

& H. Glimell, *Bringing Visibility to the Invisible: Towards A Social Understanding of Nanotechnology*, Göteborg: Göteborg University, pp. 5-32 [www.sts.gu.se/publications/STS_report_6.pdf].

Fogelberg, H. & Glimell, H.: 2003, *Bringing Visibility to the Invisible: Towards A Social Understanding of Nanotechnology*, Göteborg: Göteborg University [www.sts.gu.se/publications/STS_report_6.pdf].

Fogelberg, H.: 2003, 'The Grand Politics of Technoscience: Contextualizing Nanotechnology', in: H. Fogelberg & H. Glimell, *Bringing Visibility to the Invisible: Towards A Social Understanding of Nanotechnology*, Göteborg: Göteborg University, pp. 33-54 [www.sts.gu.se/publications/STS_report_6.pdf].

Fogelberg, H.: 2003, 'The Material Culture of Nanotechnology', in: H. Fogelberg & H. Glimell, *Bringing Visibility to the Invisible: Towards A Social Understanding of Nanotechnology*, Göteborg: Göteborg University, pp. 99-114.

Frazier, G., "An Ideology For Nanoelectronics," in *Concurrent Computations: Algorithms, Architecture, and Technology*, Plenum Press, New York, 1988.

Goldhaber-Gordon, D., Montemerlo, M.S., Love, J.C., Opiteck, G.J., and Ellenbogen, J.C., "Overview of Nanoelectronic Devices," submitted to the *Proceedings of the IEEE*, February 1997. For more information, please send e-mail to nanotech@mitre.org.

Gorman, M.E.; Groves, J.F. & Shrager, J.: 2004, 'Societal Dimensions of Nanotechnology as a Trading Zone: Results from a Pilot Project', in: D. Baird, A. Nordmann & J. Schummer (eds.), *Discovering the Nanoscale*, Amsterdam: IOS Press, pp. 63-73.

Grinbaum, A. & Dupuy, J.-P.: 2004, 'Living with Uncertainty: Toward the Ongoing Normative Assessment of Nanotechnology', *Techne: Research in Philosophy and Technology*, 8(3) (forthcoming).

Grinbaum, A.: 2004, 'La condition de l'homme moderne et les nanotechnologies', in: G. Nivat (ed.), *Les limites de l´humain. 39èmes Rencontres Internationales de Genève*, L´Age d´Homme, Genève, p. 141.

Grunwald, A.: 2004, 'Ethische Aspekte der Nanotechnologie. Eine Felderkundung', *Technikfolgenabschätzung ' Theorie und Praxis*, 13 (2), 71-78.

Gupta, V.K. & Pangannaya, N.B.: 2000, 'Carbon nanotubes: bibliometric analysis of patents', *World Patent Information*, 22, 185-189.

Hansson, S.O.: 2004, 'Great Uncertainty about Small Things', *Techne: Research in Philosophy and Technology*, 8(3) (forthcoming).

Haruta, M. 1997. *Catalyst surveys of Japan* 1:61 and references therein.

Hayles, N.K. (ed.): 2004, *Nanoculture: Implications of the New Technoscience*, Bristol, UK: Intellect Books.

Hayles, N.K.: 2004, 'Connecting the Quantum Dots: Nanotechscience and Culture', in: N.K. Hayles (ed.), *Nanoculture: Implications of the New Technoscience*, Bristol, UK: Intellect Books, pp. 11-26.

Hemerén, G.: 2004, 'Nano Ethics Primer', in: European Commission (Community Health and Consumer Protection): *Nanotechnologies: A Preliminary Risk Analysis on theBasis of a Workshop, Brussels, 1-2 March 2004* (www.europa.eu.int/comm/health/ ph_risk/documents/ev_20040301_en.pdf), pp. 95-102.

Hennig, J.: 2004, 'Changes in the Design of Scanning Tunneling Microscopic Images from 1980 to 1990', *Techne: Research in Philosophy and Technology*, 8(3) (forthcoming).

Hessenbruch, A.: 2004, 'Nanotechnology and the Negotiation of Novelty', in: D. Baird, A. Nordmann & J. Schummer (eds.), *Discovering the Nanoscale*, Amsterdam: IOS Press, pp. 135-144.

Hullmann, A. & Meyer, M.: 2003, 'Publications and Patents in Nanotechnology. An overview of previous studies and the state of the art', *Scientometrics*, 58 (3) 507-527.

Huo, Q., D.I. Margolese, U. Ciesla, P. Feng, T.E. Gier, P. Sieger, R. Leon, P.M. Petroff, F. Schuth, and G.D. Stucky. 1994. *Nature* 368:317.

Jena, P., S.N. Khanna, and B.K. Rao. 1996. In *Science and technology of atomically engineered materials*, ed. P. Jena. River Edge, NJ: World Scientific.

Johansson, M.: 2003, 'Plenty of room at the bottom: Towards an anthropology of nanoscience', *Anthropology Today*, 19 (No. 6), 3-6.

Johnson, A.: 2004, 'The End of Pure Science: Science Policy from Bayh-Dole to the NNI', in: D. Baird, A. Nordmann & J. Schummer (eds.), *Discovering the Nanoscale*, Amsterdam: IOS Press, pp. 217-230.

Kresge, C.T., M.E. Leonowicz, W.J. Roth, J.C. Vartuli, and J.S. Beck. 1992. *Nature* 359:710.

Kupperman, A., S. Nadimi, S. Oliver, G. Ozin, J. Garcés, and M. Olken. 1993. *Nature* 365:239.

Kuusi, O.; Meyer, M.: 2002, 'Technological generalizations and leitbilder ' the anticipation of technological opportunities', *Technological Forecasting & Social Change*, 69, 625-639.

Landon, B.: 2004, 'Less is More: Much Less is Much More: The Insistent Allure of Nanotechnology Narratives in Science Fiction', in: N.K. Hayles (ed.), *Nanoculture: Implications of the New Technoscience*, Bristol, UK: Intellect Books, pp. 131-146.

Laszlo, P.: 2004, 'Is There Life After Partington?', *Hyle: International Journal for Philosophy of Chemistry*, 10(2), 169-178.

Lenhard, J.: 2004, 'Nanoscience and the Janus-Faced Character of Simulations', in: D. Baird, A. Nordmann & J. Schummer (eds.), *Discovering the Nanoscale*, Amsterdam: IOS Press, pp. 93-100.

Lent, C.S., Tougaw, P.D., Porod, W., Bernstein, G.H., "Quantum Cellular Automata," *Nanotechnology*, Vol. 4, p. 49, 1993.

Lewak, S.E.: 2004, 'What's the Buzz? Tell Me What's A-Happening: Wonder, Nanotechnology, and Alice's Adventures in Wonderland', in: N.K. Hayles (ed.), *Nanoculture: Implications of the New Technoscience*, Bristol, UK: Intellect Books, pp. 201-.

Lide, D.R., ed. 1993-1994. *CRC Handbook of Chemistry and Physics*, 74th ed.

Lin-Easton, P.C.: 2001, 'It's Time for Environmentalists to Think Small ' Real Small: A Call for the Involvement of Environmental Lawyers in Developing Precautionary Policies Molecular Nanotechnology', *Georgetown International Law Review*, 14, 106-134.

López, J.: 2004, 'Bridging the Gaps: Science Fiction in Nanotechnology', *Hyle: International Journal for Philosophy of Chemistry*, 10(2), 129-152.

Lösch, A.: 2004, 'Nanomedicine and Space: Discursive Orders of Mediating Innovations', in: D. Baird, A. Nordmann & J. Schummer (eds.), *Discovering the Nanoscale*, Amsterdam: IOS Press, pp. 193-202.

Paschen, H.; Coenen, C.; Fleischer, T.; Grünwald, R.; Oertel, D. & Revermann, C.: 2004,

Nanotechnologie: Forschung, Entwicklung, Anwendung, Berlin: Springer (*Nanotechnologie, TAB-Arbeitsbericht 92*, Berlin: Büro für Technikfolgen-Abschätzung beim Deutschen Bundestag).

Roberts, J.A.: 2004, 'Deciding the Future of Nanotechnologies: Legal Perspectives on Issues of Democracy and Technology', in: D. Baird, A. Nordmann & J. Schummer (eds.), *Discovering the Nanoscale*, Amsterdam: IOS Press, pp. 247-255.

Robinson, C.: 2004, 'Images in NanoScience/Technology', in: D. Baird, A. Nordmann & J. Schummer (eds.), *Discovering the Nanoscale*, Amsterdam: IOS Press, pp. 165-169.

Robison, W.L.: 2004, 'Nano-Ethics', in: D. Baird, A. Nordmann & J. Schummer (eds.), *Discovering the Nanoscale*, Amsterdam: IOS Press, pp. 285-300.

Roco, M.C. & Bainbridge, W.S. (eds.): 2001, *Societal implications of nanoscience and nanotechnology*, (Proceedings of a workshop organized by the National Science Foundation, September 28-29, 2000), Kluwer: Dordrecht [available online at http://itri.loyola.edu/nano/societalimpact/nanosi.pdf]

Roco, M.C. & Tomellini, R. (eds.): 2002, *Nanotechnology: Revolutionary Opportunities and Societal Implications, Workshop, Lecce (Italy), 31 January - 1 February 2002*, Luxemburg: Office for Official Publications of the European Communities, [200 pp.]

Roco, M.C.; Bainbridge, W.S. (eds.): 2002, *Converging Technologies for Improving Human Performance: Nanotechnology, Biotechnology, Information Technology and the Cognitive Science*, Arlington, VA: National Science Foundation.

Roher, H. 1993. *Jpn. J. Appl. Phys.* 32:1335.

Rohlfing, E.A., D.M. Cox, and A. Kaldor. 1984. *J. Chem. Phys.* 81:3846.

Rosen, A. 1998. A periodic table in three dimensions: A sightseeing tour in the nanometer world. In *Advances in quantum chemistry*. In press.

Ruthven, D.M., S. Farooq, K.S. Knaebel. 1994. *Pressure swing adsorption*. New York: VCH Publishers.

Sarewitz, D.; Woodhouse, E.: 2003, 'Small is Powerful', in: A. Lightman, D. Sarewitz & Chr. Desser, (eds.), *Living with the Genie: Essays on Technology and the Quest for Human Mastery*, Washington, DC: Island Press, pp. 63-83.

Schiemann, G.: 2004, 'Dissolution of the Nature-Technology Dichotomy? Perspectives on Nanotechnology from an Everyday Understanding of Nature', in: D. Baird, A. Nordmann & J. Schummer (eds.), *Discovering the Nanoscale*, Amsterdam: IOS Press, pp. 209-213.

Schmidt, J.C.: 2004, 'Unbounded Technologies: Working Through Technological Reductionism of Nanotechnology', in: D. Baird, A. Nordmann & J. Schummer (eds.), *Discovering the Nanoscale*, Amsterdam: IOS Press, pp. 35-50.

Schuler, E.: 2004, 'Perception of Risks and Nanotechnology', in: D. Baird, A. Nordmann & J. Schummer (eds.), *Discovering the Nanoscale*, Amsterdam: IOS Press, pp. 279-284.

Schummer, J.: 2003, 'The Notion of Nature in Chemistry', *Studies in History and Philosophy of Science*, 34 (2003), 705-736.

Schummer, J.: 2004, 'Naturverhältnisse in der modernen Wirkstoff-Forschung', in: K. Kornwachs (ed.), *Technik ' System ' Verantwortung*, Münster: LIT, pp. 629-638.

Schummer, J.: 2004, 'Why do Chemists Perform Experiments?', in: D. Sobczynska, P. Zeidler, E.

Zielonacka-Lis (eds.), *Chemistry in the Philosophical Melting Pot*, Peter Lang (Frankfurt/M.), 2004, pp. 395-410.

Schummer, J.; Baird, D. (eds.): 2004-5, *Nanotech Challenges*, joint special issue of *Hyle: International Journal for Philosophy of Chemistry & Techne: Research in Philosophy and Technology* (forthcoming).

Schwarz, A.E.: 2004, 'Shrinking the 'Ecological Footprint' with NanoTechnoScience?', in: D. Baird, A. Nordmann & J. Schummer (eds.), *Discovering the Nanoscale*, Amsterdam: IOS Press, pp. 203-208.

Appendix

RELATED INSTITUTIONS, UNIVERSITIES & LIBRARIES

- NANO.DOT, http://nanodot.org/ —A Foresight Institute website—News & Discussion
- Nanotechnology White Paper Library, http://nanobusiness.org/whitepaper.html—The NanoBusiness Association's "Nanotechnology White Paper Library" is a collection of links to more than 500 nanotechnology white papers, presentations and reports. Most are free; some may require registration with their host website.
- CRN, http://www.crnano.org/—Center for Responsible Nanotechnology
- The Institute of Nanotechnology, http://www.nano.org.uk/—The Institute of Nanotechnology has been created to foster, develop and promote all aspects of science and technology in those domains where dimensions and tolerances in the range of 0.1 nm to 100 nm play a critical role
- UCSC Engineering, http://www.cse.ucsc.edu/research/vision.html—University of California, Santa Cruz / Baskin School of Engineering
- Nanotechnology Initiative Research Group, http://www.me.berkeley.edu/nti/home.html—Department of Mechanical Engineering University of Berkeley
- Stanford Nanotechnology, http://scpd.stanford.edu/SCPD/courses/contentView/nanotechnology/—Stanford University is at the forefront in multidisciplinary research and innovation in nanoscience and nanotechnology. The Stanford Center for Professional Development, in its commitment to extend Stanford research and teaching to students in industry, has developed this portfolio of courses and programmes in this exciting new field.
- NSL NanoStructures Laboratory, http://nanoweb.mit.edu/ —The Nanostructures Laboratory (NSL) at MIT develops techniques for fabricating surface structures with feature sizes in the range from nanometers to micrometers, and uses these structures in a variety of research projects.

The NSL is closely coupled to the Space Nanotechnology Laboratory (SNL) with which it shares facilities and a variety of joint programmes.

- MIT Stanford UC Berkeley Nanotechnology Forum, http://www.quantuminsight.com/MSB/organization.html—The MIT Stanford UC Berkeley Nanotechnology Forum is an all-volunteer organization. It is organized under the MIT Club of Northern California and has affiliations with MIT, Stanford University, and the University of California at Berkeley. The Forum is run by the Chairman and is guided by the Steering Committee.
- The Foresight Institute—Non-profit institute focused on nanotechnology, the coming ability to build materials and products with atomic precision, and systems to aid knowledge exchange and critical discussion, thus improving public and private policy decisions.
- Nanotechnology Now—Comprehensive and informative nanotechnology portal. Provides introduction to nanotechnology, general information, images, interviews, news, events, research, books, glossary and links.
- Applied Computational Research Society—Nonprofit organization fostering interdisciplinary approaches to modeling and simulation in the advancement of computational science and high technology. Focused on nanoscience and MEMS applications.
- The ASME Nanotechnology Institute—American Society of Mechanical Engineers: nanotechnology resources, events, programmes, activities, online courses, webcasts, news and links.
- CMP Cientifica—News, networks, conferences and nanotechnology resources in Europe
- Euspen—European Society for Precision Engineering and Nanotechnology—contains society news, conference information, discussion forums and expertise search. Euspen is a not-for-profit company based at Cranfield University in the UK, with members around the world.
- Exploring the Nanoworld—Movies of nano-structured materials: ferrofluids, memory metals, amorphous metals, LEDs, re-entrant foams, self assembly, DNA, Magnetic Resonance Imaging, atomix imaging, Lego models, Giant Magneto Resistance, silicon, carbon nanotubes, polymers, and semiconductor devices. National Science Foundation supported Materials Research Science and Engineering Center on Nanostructured Materials and Interfaces at the University of Wisconsin-Madison.
- The gRobots Project—An open source project to build and run a globally distributed supercomputer for the simulation of evolvable nanotechnology.
- How Stuff Works: How Nanotechnology Will Work—Animated narrative

shows how Nanotechnology has the potential to totally change manufacturing, health care and many other areas.

- IBM Almaden STM Molecular Art—Some of the famous images of atoms and molecules made with IBM's scanning tunneling microscope.
- Institute for Atomic-Scale Engineering—Personal page of Forrest Bishop: published and unpublished concept articles.
- Institute of Nanotechnology (UK)—The Institute of Nanotechnology has been created to foster, develop and promote all aspects of science and technology in those domains where dimensions and tolerances in the range of 0.1 nm to 100nm play a critical role.
- Iran Nanotechnology Policy Studies Committee—A committee for studying nanotechnology trends in Iran and the world.
- Minatec—Centre for innovation in micro and nanotechnology. Teaching, research, and industrial development.
- Minatec—Centre for Innovation in Micro and Nanotechnology in Grenoble. Teaching, research, and industrial development.
- MITRE Corp. Nanosystems Group—Reviews current nanotechnology research and people behind it, downloadable copies of several widely-used articles. Nanoelectronics focus.
- Molecular Manufacturing Shortcut Group—Promotes developing nanotechnology as a way to facilitate space exploration and settlement.
- Moletronics and Nanotechnology references—Links to sites with information mainly on molecular electronics.
- Nano Science & Technology Institute—Events, publications and news in nanotechnology, microtechnology and biotechnology.
- Nano Space—Nano technology as it applies now and in the future to Space Exploration and Studies.
- NanoApex—Nanotechnology portal.
- Nanobase—Loyola College, Maryland. National Science Foundation sponsored site presenting up-to-date sources of information on nanotechnology in the areas of major research centers, funding agencies, major reports and books.
- NanoComputer Dream Team—Multidisciplinary project to make Earth's first nanometer supercomputer via Internet.
- Nanodot—News and Discussion of Coming Technologies: Weblog devoted to nanotechnology-related news stories, with attached discussion forum. (based on Slashdot software)

- Nanoforum.org: European Nanotechnology Gateway—Offers background information, news and developments in industry and research and a selection of publications.
- nanogloss—Online dictionary of nanotechnology.
- Nanoguys—Nanotechnology Career Consultants. Recruiting and placement of staff in nanotechnolgy related positions such as Organic, Inorganic, Biological, Project Management, Research Scientist, Engineer, and Lab Technician.
- Nanomagazine—Interviews with people in the Nanotechnology field.
- NanoManipulator: 3rdTech—Commercial version of the NanoManipulator system originally from University of North Carolina, Chapel Hill.
- NanoManipulator: University of North Carolina—The nanoManipulator provides an improved, natural interface to SPMs (STMs and ATMs) by coupling the microscope to a virtual-reality interface giving a virtual telepresence on the surface, scaled by a factor of about a million to one.
- Nanomedicine—Devoted to the medical applications and implications of molecular nanotechnology. Nanomedicine may be defined as the monitoring, repair, construction and control of human biological systems at the molecular level, using engineered nanodevices and nanostructures. Includes the Nanomedicine Art Gallery. (From the Foresight Institute.)
- Nanoscale Science and Engineering—National Science Foundation. Activities, Information & Education, Research & Development, Connections, and NSF & NNI (National Nanotechnology Initiative) Reports.
- NanoScout—Portal with forums, and links sorted by topic and popularity.
- NanoSpace—The Center for NanoSpace Technologies is a Texas-based nonprofit scientific research and education foundation chartered to conceive, establish, and conduct cutting-edge technology, research, and development in the areas of aerospace, education, energy, life sciences, and shipping & transportation.
- NanotechNews—News for nanotechnologists and investors: regular updates, many links to other nanotechnology websites, archives, search function, newswire.
- Nanotechnology—Institute of Physics quarterly journal
- Nanotechnology—Institute of Physics monthly journal for aspects of nanoscale science and technology
- The NanoTechnology Group—A consortium of nano companies, universities and organizations developing a nano science curriculum for K-12.

- Nanotechnology in Manufacturing—What Next? The Coming Revolution In Manufacturing: Autodesk Technology Forum Presentation by John Walker, May 10th, 1990. Somewhat dated but well worth reading.
- Nanotechnology Journals Database—References, and links to, papers about nanotechnology, from selected scientific journals (eg PRB, PRL, JACS, JAP, JPCB).
- Nanotechweb.org—A portal for nanoscience and nanotechnology online resources for academics, industrialists and investors, as well as for laymen interested in nanotechnnology.
- NanoTechWire.com—Provides information for nanotechnology and research.
- Nanotek—Nanotechnology: articles online, news, pictures, cvs
- nanoTEN—Nanotechnology consulting firm. Services include nanotechnology matchmaking, assessment of trends, feasibility studies, materials and products analysis, and presentation assistance.
- nanoTITAN—Nanotechnology Software, Data & Services: nanoTITAN offers Java libraries and applications, a nanodevice database, the nanoML specification, custom simulation and visualization to the nanotechnology and general scientific communities.
- Nanotube Patent Search—Results from a search of nanotube or nanotube related patents at the United States Patent & Trademark Office
- The Nanotube Site—Michigan State University's Library of Links for the Nanotube Research Community
- NanoWave—An MIT startup, Nanowave offers SPPE (Scanning Probe Position Encoder) technology for picometer-order (subnanometer) resolution over ranges greater than several inches.
- National Science Foundation—Partnership in Nanotechnology (US). Recent technical and policy reports, funding opportunities, profiles of projects currently supported, and links to other nanoscience/technology sites.
- Phantoms—European Union sponsored nano electronics network. Contains comprehensive information about European nanotechnology, and links to groups and events worldwide.
- Polymer Simulation—Computer simulation of a polymer sticking to a surface.
- Scanning Tunneling Microscope (STM)—Describes the invention of the topografiner, a precursor instrument, between 1965 and 1971, and also tells of the STM's development.
- Scientific American—Nanotechnology articles, some free and some archived.

- Self Replicating Systems—NASA and Self-Replicating Systems: Implications for Nanotechnology.
- Small Times—Daily articles covering MEMS, nanotechnology, and microsystems, with a business angle.
- Stony Brook Buckyballs—A virtual tour of fullerenes in Laszlo Mihaly's laboratory at the Physics Department at SUNY, Stony Brook.
- STT: Netherlands Study Center for Technology Trends—Somewhere in this confusing site is good material in English, though it is often labeled in Dutch, and Dutch material is often labeled in English. STT hopes to bring their project to the attention of a wide audience, thereby contributing to the development of nanotechnology.
- Technology Review: Nanotechnology and Materials—Articles, discussion and services for nanotech built on MIT's award-winning monthly magazine.
- Thermodynamics Research Laboratory—Classical, Quantum, Nanoscale & Statistical Thermodynamics & Mechanics data and property calculation sites.
- Virtual Journal of Nanoscale Science & Technology—A weekly multijournal compilation of the latest research on nanoscale systems. Published by the American Institute of Physics and the American Physical Society.
- VISION Online—Virtual Institute supported by the EC, committed to promotion and training in the precision engineering and nanotechnologies. Contains discussion forums, training programmes and patent library.
- Xerox/Zyvex Nanotechnology—Former Xerox Palo Alto Research Center Nanotechnology page: brief introduction to core concepts of molecular nanotechnology (MNT), and links for further reading.
- Yahoo Groups: Nanotechnology—Active nanotechnology discussion group.
- Yashnanotech—Provides marketing and consultancy services for products developed by nanotechnology. Promotes nanotech awareness through training, educational publications, seminars, workshops and conferences.
- Albany NanoTech "... a fully-integrated research, development, prototyping, pilot manufacturing and education resource managing a strategic portfolio of state-of-the-art laboratories, supercomputer and shared-user facilities and an array of research centers located at the University at Albany—SUNY."
- Arizona State University (ASU) Graduate Research Training Programme in Biomolecular Nanotechnology
- Arizona State University (ASU) Center for Applied NanoBioscience

- Arizona State University (ASU) Biodesign Institute
- Arizona State University (ASU) Lindsay Lab—Research in Biophysics, Molecular Electronics and Condensed Matter Physics
- Arizona State University (ASU) Center for BioOptical Nanotechnology
- Arizona State University (ASU) Nanostructures Research Group—Center for Solid State Electronics Research
- Asian Institue of Technology Nanotechnology Studies. Bangkok, Thailand
- Australian National University Research School of Physical Sciences and Engineering
- Australian National University Nanotechnology and Mesoscale Physics
- Ball State University Center for Computational Nanoscience
- Baylor College of Medicine Molecular Biology programmes
- Beckman Institute Biological Sensors Group (BSG)
- Berkeley Nanotechnology Club (BNC)
- Brigham Young University Cleanroom A class 10 cleanroom located on the BYU campus, used for research in solid state, optoelectronics, photonics, nanotechnology, microfabrication, MEMS, micro-fluidics, etc.
- Boston University College of Engineering Departmental programme in Nanotechnology. "... is concerned with the mechanical and electronic properties of nanometer scale systems as well as macroscopic systems that require nanometer tolerances in fabrication, control, and measurement."
- Boston University Laboratory for Nanometer Scale Engineering
- Boston University Mohanty Group. "The central theme of our research is the exploration of quantum mechanical effects in engineered nanoscale structures and devices with a goal to study fundamental physical phenomena."
- Brown University Nano & Micromechanics Laboratory, part of the Mechanics of Solids and Structures Group in the Division of Engineering
- California Institutes for Science and Innovation: California NanoSystems Institute (CNSI) University of California at Los Angeles and University of California at Santa Barbara California Institute for Telecommunications and Information Technology (Cal-(IT)2) University of California at San Diego and UC Irvine Center for Information Technology Research in the Interest of Society (CITRIS) University of California at Berkeley, University of California at Davis, University of California at Merced, University of California at Santa Cruz Institute for Bioengineering, Biotechnology and

Quantitative Biomedical Research (QB3) University of California at Berkely, UC San Francisco, UC Santa Cruz

- California Institute of Technology (Caltech) Materials and Process Simulation Center
- Caltech Roukes Group—Explorations with Three-Dimensional Nanostructures in Physics, Engineering and Biology
- Caltech Kavli Nanoscience Institute (KNI)
- Caltech Solid State Device Physics (SSDP) Research Group. Department of Applied Physics.
- Carnegie Mellon University Nanomaterials for Magnetic, Electronic and Optical Applications
- Carnegie Mellon University Matyjaszewski Group—Research—Organic/ Inorganic Nanocomposites
- Center for Applied Research and Technology at CMU The Central Michigan University Research Corporation (CMURC) is a not-for-profit organization established to facilitate innovative research and development opportunities between the university and high technology companies.
- Chalmers University of Technology, Sweden Nanoscale Science
- Chalmers University of Technology, Sweden Microtechnology Centre
- Chiang Mai University Nanomaterials Research Unit
- City College of New York New York Center for Biomedical Engineering
- City College of New York Molecular Beam Epitaxy (MBE) Program
- City University of New York IGERT: Nanostructural Materials & Devices (Integrative Graduate Education and Research Traineeship Program)
- City University of Hong Kong Center Of Super-Diamond and Advanced Films (COSDAF)
- Clarkson University Center for Advanced Materials Processing.
- Clemson University COMSET—Center for Optical Materials Science and Engineering Technologies.
- Clemson University Center for Advanced Engineering Fibers and Films (CAEFF).
- Clemson University Rao Group.
- Columbia University Center for Electronic Transport in Molecular Nanostructures NSF Funded "Nanoscale Science and Engineering Centers" Real Video (2:50) Professor James Yardley speaking about Nanotechnology

- Columbia University Materials Research Science and Engineering Center (MRSEC)
- Columbia University Environmental Molecular Science Institute
- Columbia University NANOWEB—Nanotechnology Resources at Columbia University
- Columbia University Adams Group
- Columbia University Brus Group
- Columbia University Herman Group—Laser Diagnostics & Solid State Physics
- Columbia University Nuckolls Laboratory
- Cornell University Nanofabrication Facility. Part of the National Nanotechnology Infrastructure Network (NNIN)
- Cornell University Craighead Research Group. "... research focuses on creating nanoscale devices..."
- Cornell University Alliance for Nanomedical Technologies
- Cornell University Center for Nanoscale Systems NSF Funded "Nanoscale Science and Engineering Centers"
- Cornell University Nanobiotechnology Center
- Cornell University Molecular BioEngineering Lab
- Cornell University Ober Group Research
- Cornell University Center for Materials Research
- Cornell University Laboratory of Atomic and Solid State Physics (LASSP)
- Cornell University McEuen Group—LASSP
- Cornell University Applied Physics—Nanoscience and Nanotechnology
- Cornell University Applied Physics—Condensed Matter, Solid State, and Materials Research
- Cornell University Applied Physics—Optical Physics, Quantum Electronics, and Photonics
- Cranfield University Nanotechnology Group, Advanced Materials Department
- Curtin University of Technology Nanochemistry Research Institute
- Dalhousie University Zwanziger Lab
- Darmstadt University of Technology Institute of Materials Science—Thin Films Division

- Dartmouth Molecular Materials Group
- Delft University of Technology Kavli Institute of Nanoscience Delft. The Kavli Institute of Nanoscience at Delft University of Technology (The Netherlands) consists of six research groups and a nanofacility cleanroom, all aimed at innovative research at the current frontier of science on a nano-scale.
- Delft University of Technology and Leiden University MSc NanoScience programme. Leiden University and the Delft University of Technology (The Netherlands) offer a unique MSc programme on NanoScience, aimed at students who are eager to transcend the traditional borders between scientific disciplines.
- Delft University of Technology Quantum Transport—one of the research groups in the department of Applied Physics
- Delft University of Technology Molecular Biophysics Group
- Drexel Nano Materials Group (Gogotsi's Group)
- Duke University Eom Research Group, Thin Films Laboratory
- Duke University Nanoscience Group
- Earnest — Alumni Magazine for CES at Clemson Nanotechnology research inspires industry collaboration
- Ecole Polytechnique Federale de Lausanne Nanomechanics and Tribology
- Eidgenössiche Technische Hochschule (ETH) NANO-PHYSICS group of the laboratory for solid state physics at ETH Zürich
- Emory University Nie Group. Focus on Drug Delivery, Cancer Therapeutics, Molecular Imaging, Molecular Profiling, and In-vitro Diagnostics.
- Florida State University Molecular Expressions Website
- Free University Berlin Physics Department — Atomic Manipulation Site
- George Washington University Wagner Lab. Synthesis, study and application of solid-state inorganic materials with technologically significant magnetic, electrical, optical, electrochemical or catalytic properties.
- Georgia Institute of Technology Microelectronics Research Center (MiRC). Part of the National Nanotechnology Infrastructure Network (NNIN)
- Georgia Institute of Technology Applied Sensors Laboratory
- Georgia Institute of Technology Zhong Lin (ZL) Wang's Nanoscience and Nanotechnology Group
- Georgia Institute of Technology Nanotechnology and Microfluidics
- Georgia Institute of Technology Nanostructure Research Laboratory

- Georgia Institute of Technology Nanostructure Optoelectronics Group
- Georgia Institute of Technology Molecular Design Institute
- Georgia Tech Research Institute GTRI Nanotechnology Laboratory
- Harvard University Nanoscale Science and Engineering Center—Science of Nanoscale Systems and their Device Applications The Nanoscale Science and Engineering Center (NSEC) is a collaboration among Harvard University, the Massachusetts Institute of Technology, the University of California at Santa Barbara and the Museum of Science in Boston with participation by Delft University of Technology (Netherlands), the University of Tokyo (Japan), and Brookhaven National Laboratory, Oak Ridge National Laboratory and Sandia National Laboratory. NSF Funded "Nanoscale Science and Engineering Centers"
- Harvard University Center for Imaging and Mesoscale Systems (CIMS). Part of the National Nanotechnology Infrastructure Network (NNIN)
- Harvard University Materials Science Group
- Harvard University Leiber Group—a Physical Chemistry research group in the Department of Chemistry and Chemical Biology of Harvard University
- Harvard University Hongkun Park—physics and chemistry of nanostructured materials
- Harvard University Westervelt group. Department of Physics and Division of Engineering and Applied Sciences
- Harvard University George M. Whitesides Research Group
- Harvard University Nanopore Group
- Harvard University Center for Molecular Imaging Research
- Harvard University Narayanamurti Group
- Haverford College "...to bring together faculty from biology, chemistry, physics, and math in the pursuit of an interdisciplinary effort to develop new protein-based biomaterials for the field of nanometer-scale electronic and mechanical device fabrication."
- Hebrew University of Jerusalem Center for Nanoscience and Nanotechnology
- Helsinki University of Technology Optics and Molecular Materials
- Helsinki University of Technology Materials Physics Laboratory
- Heriot-Watt University, Edinburgh, UK The Nano-Optics Group
- Hong Kong University of Science and Technology Institute of NanoMaterials and NanoTechnology (INMT)

- Howard University Materials Science Research Center. Part of the National Nanotechnology Infrastructure Network (NNIN)
- Illinois Institute of Technology Center on Nanotechnology and Society (CONAS) "... to catalyze informed, inter-disciplinary research and education on the implications of nanoscale science and technology for ethical, legal, policy, business, and wider social issues, and with a special focus on the human condition."
- Instituto de Carboquímica Group of Carbon Nanostructures and Nanotechnology
- Institutionen för Fysik och Mätteknik (IFM) Department of Physics and Measurement Technology, Biology and Chemistry—Materials Science Division
- Indiana University Advanced Research and Technology Institute
- Institut Für Neue Materialien, University of the Saarland in Saarbrücken (INM) (Germany) Institute for New Materials
- Institute for Solid State and Materials Research, Dresden Germany Materials Science Division
- Interuniversity Microelectronics Center (IMEC) IMEC today is the largest independent microelectronics research and development center in Europe.
- Iowa State University Condensed Matter Physics — Ames Lab
- Iowa State University Condensed Matter Physics — Canfield Research Group
- Iowa State University Surface Engineering and Molecular Assemblies (SEMA) Laboratory
- Institute for Physical High Technology Jena Molecular Nanotechnology Group
- Jawaharlal Nehru Centre for Advanced Scientific Research Carbon Lab. Prof C. N. R. Rao (FRS) Director.
- Joanneum Research—Institute of Nanostructured Materials and Photonics "Our 15 research units make Joanneum Research one of the largest non-university research institutions in Austria." Affliated with international institutions such as the University of California at Berkeley and the Max Planck Institute for Polymer Research in Germany.
- Johns Hopkins University Materials Research Science and Engineering Center (MRSEC). One of 26 MRSECs funded by the National Science Foundation, is composed of scientists at JHU, Brown University, and the National Institute of Standards and Technology (NIST). Research in the

Center focuses on nanostructures made from novel materials that exhibit enhanced magneto-electronic properties.

- Johns Hopkins University Dept of Mechanical Engineering
- Kansas State University Dept of Physics, Interdisciplinary Research
- Katholieke Universiteit Leuven Erasmus Mundus Master of Nanoscience and Nanotechnology
- Kaunas University of Technology, LITHUANIA Research Center for Microsystems and Nanotechnology
- Kobe University, Japan Molecular Electronics and Photonics
- Korea Advanced Institute of Science and Technology (KAIST) Superlattice Nanomaterials Laboratory
- Korea University Spintronic Materials Lab., Division of Materials Science and Engineering
- Kyushu University, Japan Division of Nanoelectronics, Dept. of Electronic Device Engineering
- Lehigh University Center for Advanced Materials and Nanotechnology.
- Lehigh University Department of Materials Science and Engineering.
- London Centre for Nanotechnology (LCN) "... designed to act as a focus for current interdisciplinary nanoscale materials and device research." A joint enterprise between University College London and Imperial College.
- Louisiana Tech University Nanosystems Engineering
- Loyola College—Nanobase International Technology Research Institute.
- Ludwig-Maximilians-Universität (LMU), Germany Center for NanoScience (CeNS)
- Ludwig-Maximilians-Universität (LMU), Germany Nanophysics Group
- Lund University The Nanometer Consortium
- Mahidol University Nanotechnology Center. "Building Capability on Molecular Engineering and Devices."
- Marshall University Norton Imaging Group.
- Massachusetts Institute of Technology (MIT)—BioInstrumentation Laboratory Technologies that are currently being developed in the BI Lab include a servo-controlled laser micro-surgical robot, artificial muscle technology based upon conducting polymer biomimetic materials, the Living Chip, a cell-based high throughput screening system for mass-scale discovery of new drugs, and the Nanowalker, a nano-stepping autonomous robotic instrumentation platform.

- FAQ for MIT Institute for Soldier Nanotechnology (ISN)
- MIT Microsystems Technology Laboratories
- MIT Laboratory for Experimental & Computational Micromechanics (LEXCOM)
- MIT Nanostructures Laboratory
- MIT Surfaces and Structures
- MIT NanoMechanical Technology Laboratory
- MIT Space Nanotechnology Laboratory
- Moungi Bawendi's Group MIT "...quantum mechanical behaviors demonstrated by semiconductor and metallic materials..."
- MIT Center for Materials Science and Engineering
- MIT Department of Materials Science and Engineering
- Max Planck Institute for Biophysical Chemistry Department of NanoBiophotonics
- Max Planck Institute for Solid State Research Synthetic Nanostructures Team
- Metroplex Research Consortium for Electronic Devices and Materials (MRCEDM) combines the research and development capabilities of The University of Texas at Arlington, Southern Methodist University, Texas Christian University and The University of North Texas to assist companies that innovate and manufacture advanced technical components and systems.
- Macdiarmid Institute for Advanced Materials and Nanotechnology A New Zealand research organisation concerned with high quality research and research education in materials science and nanotechnology.
- Michigan State University (MSU) Carbon Nanotubes
- MSU—David Tománek "The Nanotube Site"
- MSU—NSF Center on Low-Cost, High-Speed Polymer Composites Processing
- MSU—Composite Materials and Structures Center
- MSU—Advanced Materials Engineering Experiment Station (AMEES) is managed and directed as part of the Composite Materials and Structures Center
- Michigan Technological University Nanotechnology at Michigan Technological University
- McGill — NanoScience & Scanning Probe Microscopy... to apply the tools, techniques and materials developed in the past few years to the emerging new field of Nanoelectronics.

- McGill — NANOSCIENCE AND NANOTECHNOLOGY Course description
- Nagoya University Department of Molecular Design and Engineering
- Nanobiotechnology Center (NBTC) The Nanobiotechnology Center was established in January 2000 as a Science & Technology Center, with core funding from the National Science Foundation. Nanobiotechnology is an emerging area of scientific and technological opportunity that integrates nano/microfabrication and biosystems to the benefit of both. The Nanobiotechnology Center is characterized by its highly interdisciplinary nature and features a close collaboration between life scientists, physical scientists, and engineers. It has a fully integrated education and outreach effort in which all NBTC faculty participate. The Center brings together experts in their fields from Cornell University, the Wadsworth Center (New York State Health Department in Albany), Princeton University, Oregon Health & Science University, Clark Atlanta University, and Howard University. It also involves the active collaboration of K-12 educators, the Sciencenter Museum in Ithaca, NY, and representatives from industry and the government.
- Nanotechnology Institute (NTI) A collaboration led by Ben Franklin Technology Partners of Southeastern Pennsylvania, Drexel University and The University of Pennsylvania. "... to focus on the transfer of discoveries and intellectual knowledge in the area of nanotechnology from universities to industry partners and on the rapid application and commercialization of this technology to stimulate economic growth."
- Nanyang Technological University Precision Engineering & Nanotechnology Centre
- Nanyang Technological University Micro Electro Mechanical Systems Research
- Nanyang Technological University Advanced Material Research Centre
- Nanyang Technological University Microelectronics Centre
- Nara Institute of Science and Technology Graduate School of Material Science
- National Institute of Advanced Industrial Science and Technology (AIST) Nanomaterials Theory Group
- National Microelectronics Research Centre (NMRC) Ireland Nanotechnology Research
- National Taiwan University Center for Nano Science and Technology (NTU-CNST). "... integrated research projects of this programme include nanomaterials, nano-devices, NEMS, and Bio-nano."

- National University of Singapore
- New Jersey Institute Of Technology Device & Material Characterization Laboratory
- New Mexico State University Nano Physics
- New York University Center for Advanced Materials and Nanotechnology
- North Carolina Center for Nanoscale Materials (NCCNM) "...research activities in the center are directed towards understanding the fundamental science of nanoscale materials and utilizing their unique properties for commerical applications."
- North Carolina State University Triangle National Lithography Center. Part of the National Nanotechnology Infrastructure Network (NNIN) (a collaboration with the University of North Carolina at Chapel Hill)
- North Carolina State University Analytical Instrumentation Facility
- North Carolina State University Nanoindentation—studying mechanical properties on the atomic scale.
- North Carolina State University Nanoscale Tribology Laboratory.
- North Dakota State University Center for Nanoscale Science and Engineering
- Northeastern University NSF Center for High-rate Nanomanufacturing (CHN)
- Northwestern University Nanoscale Science and Engineering Center (NSEC) for Integrated Nanopatterning and Detection Technologies NSF Funded "Nanoscale Science and Engineering Centers"
- Northwestern University Mirkin Group
- Northwestern University — Institute for Nanotechnology Established as an umbrella organization for the multimillion dollar nanotechnology research efforts at Northwestern University.
- Northwestern University Center for Transportation Nanotechnology.
- Northwestern University MEMS / Nanotechnology.
- Northwestern University Stupp Laboratory
- Northwestern University Nanomechanics
- Northwestern University Mechanics of MEMS and Nanosystems
- Northwestern University Ruoff Group—Nanoscience and Technology Lab
- Northwestern University Center for Nanofabrication and Molecular Self-Assembly
- Northwestern University Professor Horacio Dante Espinosa—Micro and Nanomechanics Lab

- Northwestern University Center for Quantum Devices
- Northumbria University Advanced Materials Research Institute (AMRI). The Institute's technical focus is in the fields of advanced materials, surface engineering, corrosion and wear.
- Norwegian Institute of Technology (NTH) NMNL (Norwegian Micro and Nano Laboratories), the research infrastructure in Norway hosted by SINTEF (The Foundation for Scientific and Industrial Research at the Norwegian Institute of Technology), the Norwegian University of Science and Technology (NTNU), and the University of Oslo (UiO).
- Norwegian Institute of Technology (NTH) NANOCAPS (at SINTEF).
- Norwegian Institute of Technology (NTH) NANOMAT—Materials Science and Nanotechnology at the Norwegian Microtechnology Center (NMC-MRL) (at SINTEF).
- Nottingham University Nanoscience Group Nanometre scale structures and nanostructured materials play an increasingly pervasive role in a wide range of scientific disciplines ranging from solid state physics through to molecular biology.
- Nottingham Trent University Polymer Engineering Centre
- Oak Ridge National Laboratory (ORNL) Polymer Science Group Summary of ongoing research aimed at simulating nanomachines and aspects of fundamental chemical physics at the molecular scale.
- Ohio University Research Institute to Advance Studies in Nanotechnology
- Ohio University Nanoscale & Quantum Phenomena Institute
- Ohio University Nanospintronics and Nanomagnetics
- Ohio Supercomputer Center Nanotechnology at OSC
- Oklahoma State University NanoNet Probe Microscopy Laboratory
- Oklahoma State University Department of Physics
- Oregon State University College of Engineering
- Oregon State University Microtechnology-Based Energy, Chemical and Biological Systems (MECS)
- Osaka University Protonic NanoMachine Project.
- Oxford University Department of Materials
- Oxford University Begbroke Science Park Institute of Nanotechnology
- Paul Scherrer Institut Laboratory for Micro- and Nanotechnology
- Penn State Atomic-Scale Measurements and Control
- Penn State Center for Molecular Nanofabrication and Devices

- Penn State Nanofabrication Facility. Part of the National Nanotechnology Infrastructure Network (NNIN)
- Penn State Materials Research Institute
- Penn State Allara Research Group
- Penn State Department of Materials Science and Engineering
- PNNL/UW Joint Institute for Nanoscience
- Portland State University Center for Nanoscience and Nanotechnology
- Princeton University Nanostructure Laboratory.
- Princeton University Ceramic Materials Laboratory.
- Purdue Nanotechnology Initiative "...to demonstrate novel techniques for the design and fabrication of nanoelectronic devices by the chemical manipulation of nanometer (10-9 meters) sized clusters and molecular wires."
- Nanotechnology At Purdue (PDF) A visual guide to the various programmes and collaborations at Purdue. "This brochure highlights many of those projects across the categories of nanomaterials, nanodevices, nano/bio interfaces, nanomanufacturing, computational nanotechnology, and nanometrology/characterization."
- Purdue University Discovery Park—Birck Nanotechnology Center & Ultra-Performance Nanotechnology Center (ULTEC).
- Purdue University David Janes Group.
- Purdue University Laboratory for Chemical Nanotechnology (PLCN).
- Purdue University nanoHub The nanoHub is a new initiative designed to promote the application of computational science to nanotechnology. Using PUNCH, a network-computing software infrastructure, researchers and educators can access and operate simulation tools through standard Web browsers. Computational scientists can share tools with colleagues, experimentalists can operate simulations tools without downloading and installing them, and users can access supercomputing resources transparently. The focus is on state-of-the-art, research-grade tools designed to complement commercial simulation packages.
- Purdue University Center for Nanoscale Devices.
- Purdue University Nanoscale Physics.
- Purdue University Cluster-Based Materials Group.
- Queen Mary University of London Molecular and Materials Physics Group.
- Research Institute for Technical Physics and Materials Science Budapest, Hungary.

- Rensselaer Polytechnic Institute Nanotechnology Center.
- Rensselaer Polytechnic Institute Center for Directed Assembly of Nanostructures NSF Funded "Nanoscale Science and Engineering Centers"
- Rensselaer Polytechnic Institute Prof. Ajayan's Carbon Nanomaterials Technology Group.
- Rice University Center for Nanoscale Science and Technology
- Rice University Colvin Group
- Rice University Halas Nanophotonics Group
- Rice University Tour Group
- Rice University Laboratory for Nanophotonics
- Rice University Center for Biological and Environmental Nanotechnology NSF Funded "Nanoscale Science and Engineering Centers"
- Rice University Environmental & Energy Systems Insitute of Rice University (EESI)
- RMIT Royal Melbourne Institute of Technology—Nanotechnology Program
- Rochester Institute of Technology NanoPower Research Labs
- Rochester Institute of Technology Nano- and Micro-Electromechanical Systems, Nano- and Microsystems, Nanoarchitectronics
- Rutgers University Bio Nano Robotics
- Rutgers University Division of Computer & Information Sciences
- Rutgers University Laboratory for Surface Modification (LSM)
- Seoul National University Center for Science in Nanometer Scale
- Sheffield Hallam University Electronics Research Group, Microsystems & Machine Vision Laboratory (MMVL)
- South Dakota School of Mines and Technology Center for Accelerated Applications at the Nanoscale (CAAN).
- Stanford University Nanofabrication Facility—serves academic, industrial, and governmental researchers across the US in areas ranging from of optics, MEMS, biology, and chemistry, as well as traditional electronics device fabrication and process characterization. Part of the National Nanotechnology Infrastructure Network (NNIN)
- Nanoscience and Nanotechnology Online Seminars
- Explore the universe of nanotubes and bucky balls. Expand your vision of a changed world, from the quality of our goods to the quality of our lives. Presenting the latest nanoscience and nanotechnology concepts, the Stanford Engineering and Science Institute will explore the promise of a wide range

of exciting new products and applications capable of transforming and redefining many industries. Learn from Stanford faculty and industry experts the potentially broad impacts of nanotechnology for your business. Learn More

- Stanford University Department of Materials Science and Engineering
- Stanford University Center for Molecular and Genetic Medicine
- Stanford University Geballe Laboratory for Advanced Materials (LAM)
- Stanford University Chemical Engineering
- Stanford University Mechanical Engineering
- Stanford University S.C.Zhang Group
- http://buckminster.physics.sunysb.edu/ Department of Physics and Astronomy.... study the fundamental properties of Fullerenes (Carbon-60) and doped fullerenes
- Stockholm University Sol-Gel Group at the Department of Inorganic Chemistry
- http://www.matscieng.sunysb.edu/tsl/nano/ Thermal Spray Lab
- SUNY—University at Buffalo The Center for Advanced Photonic and Electronic Materials (CAPEM)—Laboratory for Spintronics Research in Semiconductors (LSRS)
- SUNY—University at Buffalo Electronic Packaging Laboratory
- SUNY — Molecular Structure Laboratory Department of Chemistry at SUNY Stony Brook.
- Swiss Federal Institute of Technology Zurich (ETH Zurich) FIRST! Center for Micro- and Nanoscience (Frontiers in Research: Space and Time)
- Tata Institute of Fundamental Research (TIFR) Department of Condensed Matter Physics and Materials Science
- Technical University of Denmark NANO@DTU—organizes the broad spectrum of nanotechnology research and education at DTU.
- Technical University of Denmark Nanotechnology Research Group.
- Technical University of Denmark Center for Atomic-scale Materials Physics.
- Technion—Israel Institute of Technology Microelectronics Research Center
- Technische Universitat Berlin Project Nanostructures
- Tel Aviv University TAU Research Institute for Nanoscience and Nanotechnology
- Texas Christian University Atom Optics Lab

- Texas Engineering Experiment Station (TEES) a member of Texas A&M University System
- Texas A&M University Texas Institute of Intelligent Bio-Nano Materials and Structures for Aerospace Vehicles (TiiMS)
- Texas State—Nanomaterials Applications Center—NAC
- Texas Nanotechnology Initiative A Consortium of industry, universities, government, and venture capitalists whose goal is to position Texas as the Nanotechnology State
- Towson University Nanotechnology Studies using Scanning Probe Microscopy
- Towson University Nanotechnology Laboratory
- Trinity College Dublin Nanomag Research Group
- Trinity College Dublin Centre for Research on Adaptive Nanostructures and Nanodevices (CRANN)
- TRI/Princeton "... TRI has extended its expertise to provide advanced research and education in polymers, fibers, films, human hair, and porous materials."
- Tufts University The Walt Group... nanosensors and nanostructures.
- University at Albany—SUNY Albany NanoTech "is a fully-integrated research, development, prototyping, pilot manufacturing and education resource managing a strategic portfolio of state-of-the-art laboratories, supercomputer and shared-user facilities and an array of research centers located at the University at Albany—SUNY." Formerly "The Center for Advanced Thin Film Technology".
- University at Albany College of Nanoscale Science and Engineering. "... cross-disciplinary curriculum integrates the fundamental science principles of physics, chemistry, computer science, and biology with the cross cutting fields of nanosciences, nanoengineering, nanoeconomics, and nanobiology."
- University at Albany—SUNY Advanced Computer Modeling for Nanosystems and Processes.
- University at Albany—SUNY Center for Advanced Thin Film Technology—Microsystems Integration Laboratory
- Université Catholique de Louvain UCL/CRMN—Centre de Recherche en Dispostifs et Matériaux Electroniques Micro- et Nanoscopiques (CeRMiN). Research Center in Micro and Nanoscopic Materials and Electronic Devices
- University College Cork Nanotechnology Group—Dr. Gareth Redmond Director. "The Group undertakes research aimed at development and innovative exploitation of nanoscale materials and devices within emerging

Information and Communication Technology (ICT) application areas."

- University of Aalborg Institute of Physics and Nanotechnology
- University of Aarhus and Aalborg University iNANO—interdisciplinary Nanoscience Center
- University of Akron Center for Molecular Design and Recognition
- University of Alberta Faculty—National Institute for Nanotechnology "... an integrated, multi-disciplinary institution involving researchers in physics, chemistry, engineering, biology, informatics, pharmacy and medicine. Established in 2001, it is operated as a partnership between the National Research Council and the University of Alberta, and is jointly funded by the Government of Canada, the Government of Alberta and the university."
- University of Alberta Buriak Group
- University of Alberta Centre for Nanoscale Physics
- University of Alberta Microsystems Technology Research Institute—NanoFab. "... an open access micro and nano fabrication facility for use by the University of Alberta research community and other groups including Canadian universities and companies involved in microsystem research and related prototyping development."
- University of Arizona Nanomechanics and Mesoscopic Physics
- University of Arizona AFM Lab
- University of Arkansas Semiconductor Fabrication and NanoScale Characterization Facility
- University of Basel National Center of Competence in Research Nanoscale Science
- University of Basel Nano-Electronics (Mesoscopic Physics)
- University of Bath, UK Materials Research Centre
- University of Bath, UK Centre for Electron Optical Studies
- University of Berlin — Project Nanostructures Focuses on the fabrication, characterization and theoretical modelling of quantum wires and quantum dots.
- University of Birmingham I[2] NanoTech Centre
- University of Birmingham Nanoscale Chemistry
- University of Birmingham Nanoscale Physics Research Laboratory
- University of Birmingham μ-Nano Consortium
- University of California at Berkeley • Davis • Merced • Santa Cruz The Center for Information Technology Research in the Interest of Society

(CITRIS). Includes as members/participants California High-Tech Industry & The State of California.

- University of California at Berkeley Stacy Group "... involved in the development of new synthetic routes that lead to the discovery of materials for emerging technologies."
- University of California at Berkeley Berkeley Sensor & Actuator Center—MEMS Reference Database
- University of California at Berkeley Microstructured Materials Group
- University of California at Berkeley Professor Connie Chang-Hasnain Optoelectronics Research Group
- University of California at Berkeley McEuen Group has moved to McEuen Group—LASSP Cornell University—Laboratory of Atomic and Solid State Physics—Department of Physics
- University of California at Berkeley Alivisatos Group
- University of California at Berkeley Zettl Research Group—Condensed Matter Physics—Department of Physics. An experimental solid state physics group.
- University of California at Berkeley, UC San Francisco, UC Santa Cruz Institute for Bioengineering, Biotechnology and Quantitative Biomedical Research (QB3). QB3 is one of four institutes established through the California Institutes for Science and Innovation
- University of California at Davis, NEAT Nanophases in the Environment, Agriculture, and Technology is a multidisciplinary research and education programme which links the fundamental physics, chemistry, and engineering of small particles and nanomaterials to several challenging areas of investigation.
- University of California, Irvine Integrated Nanosystems Research Facility (INRF)
- University of California at Los Angeles Heath Group—Metal & Semiconductor nanocrystal research.
- University of California at Los Angeles Nanoelectronics Research Facility
- University of California at Los Angeles Optoelectronics Group, headed by Prof. Eli Yablonovitch.
- University of California at Los Angeles Stoddart Research Group
- University of California at Los Angeles Wudl Group
- University of California, Riverside Center for Nanoscale Science and Engineering

- University of California, Riverside Nanotechnology Educational Outreach
- University of California, Riverside Biomedical Science & Nanotechnology Laboratory
- University of California at Los Angeles and University of California at Santa Barbara California NanoSystems Institute (CNSI). UCLA & UCSB have joined to build the California Nanosystems Institute, which will facilitate a multidisciplinary approach to develop the information, biomedical, and manufacturing technologies that will dominate science and the economy in the 21st century. CNSI is one of four institutes established through the California Institutes for Science and Innovation
- University of California at Santa Barbara Nanofabrication Facility. Part of the National Nanotechnology Infrastructure Network (NNIN)
- University of California at Santa Barbara Nanoscale Physics—Cleland Group
- University of California at Santa Barbara Awschalom Group
- University of California at Santa Barbara Institute for Quantum Engineering, Science and Technology (iQUEST) (formerly known as the Quantum Institute)
- University of California at Santa Barbara Department of Molecular, Cellular and Developmental Biology (MCDB)
- University of California Santa Cruz Quantum Electronic Group
- UCSC Extension General Interest Page (Current nanotechnology course offerings)
- University of California at San Diego Physics Department—Thin Film Lab—Nanoscience Group. Professor Ivan K. Schuller Director.
- University of California at San Diego Materials and Devices Layer at UCSD. Professor Ivan K. Schuller Director.
- University of California at San Diego MURI Integrated Nanostructured Supersensors. Professor Ivan K. Schuller Director.
- University of California at San Diego Silva Research Group
- University of California at San Diego and UC Irvine California Institute for Telecommunications and Information Technology (Cal-(IT)2). Cal-$(IT)^2$ is one of four institutes established through the California Institutes for Science and Innovation initiative proposed in the year 2000 by Governor Gray Davis. Cal-$(IT)^2$, a partnership between UC San Diego and UC Irvine, seeks to ensure that California maintains its leadership in the telecommunications and information technology marketplace.

- University of Cambridge CUED Nanoscale Science Laboratory
- University of Cambridge Centre of Molecular Materials for Photonics and Electronics
- University of Cambridge the Cavendish Laboratory is the Department of Physics
- University of Cambridge The Nanoscience Centre
- University of Cambridge IRC in Nanotechnology—The Interdisciplinary Research Collaboration (IRC) in Nanotechnology is funded by EPSRC, BBSRC, MRC and the MoD as a collaboration between the University of Cambridge, University College London and the University of Bristol.
- University of Canterbury—NEST Nanostructure Engineering Science and Technology Group
- University of Canterbury Cluster Assembled Nanostructures—Dr. Simon Brown, Research Interests
- Université Catholique de Louvain Research Center on Micro and Nanoscopic Electronic Devices and Materials (CERMIN)
- University of Chicago Materials Research Science and Engineering Center
- University of Cincinnati Nanoelectronics Laboratory
- University of Colorado Team Weimer
- University of Connecticut Nanostructured Materials Program
- University of Copenhagen, Denmark Center for Interdisciplinary Studies of Molecular Interactions.
- University of Copenhagen, Denmark Nano-Science Center
- University of Delaware Center for Molecular and Engineering Thermodynamics (CMET)
- University of Delaware Center for Nanomachined Surfaces
- University of Dublin Trinity College Centre for Research on Adaptive Nanostructures and Nanodevices.
- University of Durham Department of Physics.
- University of Durham Carbon Nanostructures.
- University of Durham Durham Nanoscale Magnetics Group.
- University of Durham Nanotechnology University Innovation Centre (Nanotechnology UIC).
- University of Glasgow Nanoelectronics Research Centre. Provides a focus for diverse research activities within the University of Glasgow linked by a common interest in Nanoelectronics and Nanofabrication.

- University of Glasgow Spectroscopy of Nanostructures
- University of Glasgow Kelvin Nanotechnology
- University of Greenwich NANO-SCIENCE SIMULATION GROUP
- University of Hamburg Nanostructure Physics Group
- University of Hannover, Germany Institute for Solid State Physics—Department of Nanostructures
- University of Houston Texas Center for Superconductivity (and Advanced Materials) at the University of Houston (TCSUH)
- University of Houston Sharma Group—Mechanical Engineering.
- University of Illinois at Chicago Materials Characterization Laboratory
- University of Illinois at Chicago Microfabrication Applications Laboratory (MAL)
- University of Illinois at Chicago Nanoscale structures of asphaltene steric-colloid and asphaltene micelles & vesicles.
- University of Illinois at Urbana-Champaign Center for Nanoscale Science & Technology (CNST)
- University of Illinois at Urbana-Champaign Beckman Institute for Advanced Science and Technology
- University of Illinois at Urbana-Champaign Beckman Institute for Advanced Science and Technology—Molecular and Electronic Nanostructures
- University of Illinois at Urbana-Champaign Micro and Nanotechnology Laboratory
- University of Illinois at Urbana-Champaign Frederick Seitz Materials Research Laboratory (MRL).
- University of Illinois at Urbana-Champaign MASS Group. "...research in all aspects of the emerging technology of Microelectromechanical Systems."
- University of Jyväskylä Nanoscience Center
- University of Kentucky Materials Research Science and Engineering Center (MRSEC) Advanced Carbon Materials Center
- University of Leeds Centre for Self-Organising Molecular Systems
- University of Leeds Centre for Nano-Device Modelling
- University of Leeds Masters Training Package (MTP) in Nanoscale Science and Technology, run jointly by the universities of Leeds and Sheffield.
- University of Leipzig (Universität Leipzig) Semiconductor Physics Group. Quantum dot physics and device applications, and hybrid organic-anorganic nanostructure research.

- University of Liverpool Centre for Nanoscale Science "... material aspects of nanotechnology and nanoparticle research."
- University of Liverpool Chemistry with Nanotechnology—MChem.
- University of London Nano4CE "... a specialist unit operated by the University with the remit of working with industry to develop product forms that utilise nanomaterials."
- University of Louisville Nanoscale Imaging Facility
- University of Louisville Chemical Engineering Department
- University of Louisville Electro-Optics Research Institute
- University of Louisville MicroTechnology Center
- University of Manchester Manchester Centre for Mesoscience & Nanotechnology
- University of Manchester Nanotechnology Research
- University of Maryland Maryland Center for Integrated Nano Science and Engineering (M-CINSE)
- University of Maryland, College Park — Robert H. Smith School of Business Asian Technology Information Programme.
- University of Maryland Materials Research Science and Engineering Center (MRSEC)
- University of Maryland Fuhrer Research Group / Nanoelectronics Research Group
- University of Maryland Drew Research Group / Nano-optics
- University of Maryland LPS Nanostructures Team
- University of Maryland The Williams Lab / Surface Physics
- University of Maryland Co-laboratory for Nanoparticle Based Manufacturing & Metrology—NM^2
- University of Massachusetts Amherst MassNanoTech
- University of Massachusetts Amherst Dinsmore Research Group.
- University of Massachusetts Amherst Materials Research Science and Engineering Center (MRSEC).
- University of Massachusetts Amherst Polymer Science & Engineering Department
- University of Massachusetts Lowell Nano Manufacturing.
- University of Michigan Solid State Electronics Laboratory. Part of the National Nanotechnology Infrastructure Network (NNIN)

- University of Michigan Center for Biologic Nanotechnology
- University of Michigan WIMS Engineering Research Center (ERC) an NSF-University-Industry Collaboration
- University of Michigan Glotzer Group—Laboratory for Computational Nanoscience and Soft Matter Simulation
- University of Michigan Materials Science & Engineering
- University of Michigan Kotov Group—Nano Research Laboratory
- University of Minnesota Minnesota Nano Technology Cluster. Part of the National Nanotechnology Infrastructure Network (NNIN)
- University of Minnesota Center for NanoEnergetics Research (CNER). CNER is an Army funded center created in the spring of 2001 and exists at four university sites, with the University of Minnesota as the lead institution.
- University of Minnesota Nanotechnology Coordinating Office
- University of Minnesota Particle Technology Laboratory (PTL)
- University of Minnesota Molecular Nanosciences Alliance for Interdisciplinary Studies and Activities (MONALISA)
- University of Minnesota Norris Research Group
- University of Montreal Pavillon J.-Armand Bombardier: Nanosciences and Nanotechnologies
- University of Muenster Interface Physics Group
- University of Nebraska—Lincoln Center for Materials Research and Analysis (CMRA)
- University of Nebraska—Lincoln Nanomachining/Nanofabrication at the Center for Electro-Optics.
- University of Nebraska—Lincoln Quantum Device Laboratory
- University of Nebraska—Lincoln Condensed Matter/Materials Physics and Nanotechnology.
- University of Nebraska Medical Center Nanomedicine Group
- University of Nevada, Reno Adams Group
- University of Newcastle upon Tyne Centre for Nanoscale Science. A Multidisciplinary research centre linking scientists, engineers, medical researchers, and clinicians developing micro- and nanodevices and technologies for applications in biology, biotechnology and medicine.
- University of Newcastle upon Tyne Institute for Nanoscale Science and Technology (INSAT)

- University of Newcastle upon Tyne INEX—a microsystems and nanotechnology R&D, commercialisation and manufacturing organisation.
- University of New Mexico Integrative Nanoscience and Microsystems
- University of New Hampshire The Nano Group
- University of New Hampshire Nanotechnology and Society Research Group
- University of New Hampshire Nanostructured Polymers Research Center
- University of New Hampshire Nanostructured Polymers Research Center—Polymer Research Group (PRG).
- University of New Hampshire Nanostructured Polymers Research Center—Polymer Nanoparticle Laboratory (PNL).
- University of New Hampshire Nanostructured Polymers Research Center—Advanced Polymer Laboratory (APL).
- University of New Mexico Nanomaterials and Nanocharacterization. Part of the National Nanotechnology Infrastructure Network (NNIN)
- University of New Mexico Brinker Group.
- University of North Carolina at Chapel Hill North Carolina Center for Nanoscale Materials (NCCNM)
- University of North Carolina at Chapel Hill Nanoscale Science Research Group (NSRG)
- University of North Carolina at Chapel Hill The nanoManipulator
- University of Notre Dame Center for Nano Science and Technology
- University of Notre Dame Nanofabrication Facility
- University of Notre Dame Nano Science and Technology Center
- University of Nottingham Nottingham Nanotechnology Centre (includes 6 specializations: Quantum Nanostructures, Synthesis of Nanoscale Vessels and Storage Devices, Single Molecule Nanoscience, Biofunctional Surfaces and Materials, Nanofabrication and Characterisation, Biological Application of Nanoparticles)
- University of Oregon Jim Hutchison Lab
- University of Oregon Materials Science Institute
- University of Oregon CAMCOR—Center for Advanced Materials Characterization in Oregon. "... a full-service, comprehensive materials characterization center available to research institutions and private industry."
- University of Oregon Oregon Center for Optics

- University of Oxford Carbon and Nanotechnology Group
- University of Oxford Carbon Nanotube Group
- University of Oxford Nanotechnology Group Pages
- University of Oxford Centre for Quantum Computation
- University of Pennsylvania LRSM / Laboratory for Research on the Structure of Matter
- University of Pennsylvania School of Engineering and Applied Science
- University of Pennsylvania Nano/Bio Interface Center
- University of Pittsburgh Swanson Center for Micro and Nano Systems
- University of Queensland (UQ) Nano Optics Group.
- University of Queensland Centre for Microscopy and Microanalysis.
- University of Queensland Soft Condensed Matter Group.
- University of Queensland NanoMaterials Centre.
- University of Queensland Nanotechnology & Biomaterials Centre.
- University of Queensland ARC Centre for Functional Nanomaterials (In partnership with the University of New South Wales, the Australian National University and the University of Western Sydney).
- University of Rochester Institute of Optics—Nano-Optics Group
- University of Sheffield Masters Training Package (MTP) in Nanoscale Science and Technology, run jointly by the universities of Leeds and Sheffield.
- University of Sheffield Nanorobotics Technologies for Multidimensional Imaging and Manipulation of Nanoobjects
- University of Southern California Laboratory for Molecular Robotics
- University of Southern California The Koel Group—Surface Science: Understanding Reactions at Nanoparticles and Catalysts
- University of Southern California Nanotechnology & Micro Electro-Mechanical Systems Research Area
- University of South Carolina NanoCenter
- University of South Carolina USC NIRT: Philosophical and Social Dimensions of Nanoscale Research
- University of Southhampton Nanoscale Systems Integration Group (NSI).
- University of Sussex Fullerene Research Centre.
- University of Sussex Nanoscience and Nanotechnology @ Sussex.

- University of Sydney Nanostructural Analysis Network Organisation—Major National Research Facility (NANO—MNRF). Australian facility for nanometric analysis of the structure and chemistry of materials in both physical and biological systems. An unincorporated joint venture between university, industry, state and federal governments.
- University of Technology Sydney Institute for Nanoscale Technology
- University of Technology Sydney Institute for Nanoscale Technology Nanohouse Initiative.
- University of Texas at Arlington NanoFAB Research and Teaching Facility
- University of Texas at Austin Microelectronics Research Center (MRC). Part of the National Nanotechnology Infrastructure Network (NNIN)
- University of Texas at Austin Center for Nano- and Molecular Science and Technology
- University of Texas at Austin Science, Technology and Society (STS) "Explore social impacts of rapid scientific and technological change."
- University of Texas at Dallas NanoTech Institute.
- University of Texas at Dallas Center for Quantum Electronics.
- University of Texas at Dallas MEMS Research Group
- University of Tokyo Quantum Microstructure Devices—Sakaki Laboratory
- University of Toronto Energenius Centre for Advanced Nanotechnology (ECAN)
- University of Toronto Electronic Materials Group
- University of Toronto Geoff A. Ozin Group
- University of Toronto Nanoclub aims to promote nanotechnology at the University of Toronto. We are entirely student-run, but receive support from the Department of Materials Science Engineering.
- University of Twente—Netherlands Strategic Research Orientation Nanolink.
- University of Ulster Nanotechnology Research Institute
- University of Vienna Institute of Materials Physics
- University of Virginia MRSEC—Center for Nanoscopic Materials Design
- University of WalloniaNanoWal—Wallonia Network for Nanotechnologies
- University of Warwick — The Centre for Nanotechnology and Microengineering "...a national centre for excellence in nanotechnology and ultraprecision engineering for both the academic community and manufacturing industries."

- University of Washington Center for Nanotechnology
- University of Washington Center for Nanotechnology—Nanotech Student Association. "NSA fosters interdisciplinary research in nanotechnology, provides a forum for exchange of ideas between students from different disciplines, and promotes nanotechnology at the university level."
- University of Washington NanoTech User Facility (NTUF). Part of the National Nanotechnology Infrastructure Network (NNIN)
- University of Washington Center for Applied Microtechnology (CAM)
- University of Washington Department of Materials Science & Engineering
- University of Washington Molecular Bioengineering & Nanotechnology
- University of Waterloo Nanotechnology Engineering Program
- University of Wisconsin, Madison Nanoscale Science and Engineering Center (NSEC)
- University of Wisconsin, Madison Center for NanoTechnology
- University of Wisconsin, Madison Nanowires
- University of Wisconsin, Madison Applied Superconductivity Center
- University of Wisconsin, Madison Center for Plasma-Aided Manufacturing
- University of Wisconsin, Madison Materials Research Science and Engineering Center on Nanostructured Materials and Interfaces (MRSEC)
- University of Wisconsin, Madison Learning how to Fabricate Nanowires
- University of Wisconsin, Madison Nealey Research Group
- University of Wollongong, Australia Fundamental Properties of Semiconductors Research Centre.
- University of Wollongong, Australia ARC Centre for Nanostructured Electromaterials.
- Uppsala University Division of Materials Science at the Ångström Laboratory.
- University of York York-JEOL Centre: a collaboration between the University of York, Yorkshire Forward and JEOL UK Ltd.
- University of York Nanoscale Quantum Simulations for Nanostructures and Advanced Materials—Nanoquanta
- Vanderbilt University Vanderbilt Institute of Nanoscale Science and Engineering (VINSE).
- Virginia Tech Physics Department—Research Areas—Centers & Institutes.
- Virginia Tech Fiber & Electro-Optics Research Center.

- Virginia Tech Center for Modeling and Simulation in Material Science.
- Wake Forest The Center for Nanotechnology.
- Walter Schottky Institute—Technische Universität München Experimental Semiconductor Physics 1 (E24)—Prof. Gerhard Abstreiter.
- WASEDA University Establishment of Molecular Nano-Engineering by Utilizing Nanostructure Arrays and its Development into Micro-Systems.
- Washington University in St. Louis Buhro Group.
- Wayne State University Brock Group—Inorganic, Solid State, and Materials Chemistry.
- Wayne State University Jaewu Choi's Group: nano Devices & Systems Laboratory.
- Weizmann Institute of Science Prof. Ehud Shapiro—Laboratory for Biological Nanocomputers & The BIOSPI Project.
- Weizmann Institute of Science Department of Condensed Matter Physics.
- Weizmann Institute of Science Department of Materials and Interfaces.
- Western Michigan University Nanotechnology Computation and Research Center (NRCC) "NRCC's unique niche area is "nanobioenvironmental chemistry" which will build on Western's existing research strengths in nanotechnology approaches to environmental chemistry, biotechnology, and interrelated phenomena and problems."
- Widener University Fullerenes At Widener.
- Albany Institute of Nanotechnology and Applied Sciences—Research facility affiliated with the University at Albany engaged in a variety of nanoscience fields, including MEMS. Pictures and facility descriptions, news, events, staff profiles, bibliography.
- Alberta University—Department of Physics, Centre for Nanoscale Physics: ultrafast microscopy, ultrafast spectroscopy, nanocrystals and ion implantation, modelling of dynamics at solid surfaces, superconductivity, phase transitions and biophysics.
- Analytical Instrumentation Facility—North carolina state university, providing studies in instruments electron & scanned probe microscopy.
- Australian National University—Nanotube group: producing boron nitride and carbon nanotubes using a new relatively low temperature milling process.
- Birmingham University—Nanoscale Physics Research Laboratory: surface modification, cluster physics, sensors, scanning probe microscopy, nano-optics, and thin films.

- California University—Department of Physics: Condensed Matter Physics: Zettl Research Group: research into the electronic, magnetic, and structural properties of novel materials, including nanostructures such as telescoping nanotube bearings. (Berkeley)
- Cambridge University—Department of Engineering: Nanoscale Science Laboratory: nanotubes, nanomagnetics, nanowires, spin dependent transport, nanotribology, sensors, scanning probe microscopy, electron beam lithography, and an introduction to nanotechnology.
- Center for Nano and Molecular Science and Technology at UT Austin—Center for Nano- and Molecular Science and Technology (CNM)is a multidisciplinary, collaborative research center in the University of Texas at Austin. The mission of the CNM is to foster education, science, and engineering in nanoscience and nanotechnology at the University of Texas.
- Cranfield University—Nanotechnology Group, Advanced Materials Department: Research, facilities and courses, including the MSc in Microsystems and Nanotechnology, the first nanotechnology degree course in Europe.
- Georgia Tech—Center for Nanoscience and Nanotechnology. Nanostructure synthesis, sensors, MEMS, photonic crystals, microscopy, nano-lithography.
- Illinois University—The Beckman Institute for Advanced Science and Technology: molecular and electronic nanostructures; computational electronics; scanning tunneling microscopy, including lithography and fabrication; semiconductor nanostructures and photonics; synthesizing and characterizing new materials.
- Imperial College—Undertakes programme of one-year interdisciplinary nanomaterials research masters course. Provides funds as well.
- Integrated Microfabrication Lab at BYU—Semiconductor process laboratory, used for research in solid state electronics, optoelectronics, and microfabrication..
- Laboratory for Thin Films Physics—Studies size effects in phase transitions, structure and surface phenomena in condensed films of various materials. Includes research interests, people, publications, photo gallery and links.
- Newcastle upon Tyne University—Centre for Nanoscale Science & Technology: multidisciplinary research centre linking scientists, engineers, medical researchers, and clinicians developing micro and nanodevices and technologies for applications in biology, biotechnology and medicine.
- Northwestern University—Transportation Center: nanotechnology as applied to the transportation sector.

- Northwestern University—Nanoscience and Technology Lab: synthesis and properties of carbon and boron nitride nanotubes and nanofibers, fullerenes, and how materials synthesis at the sub-micron scale can impact thc biological sciences.
- Nottingham University—Nanoscience Group: "bottom up" approaches to nanotechnology, scanning probe microscopes, properties of fullerenes and derivatives.
- Purdue University—Nanoscale Physics: research into nanotubes, nanoclusters, molecular wires, break junctions, scanning probe microscopy (STM and AFM), and field-ion microscopy.
- Rensselaer Polytechnic Institute—Advanced Materials & Nanotechnology Research
- Texas University—Malcolm Brown's Nanopage: ongoing research in the structure and synthesis of nanostructures and high resolution transmission electron microscopy and specialized light microscopy.
- Washington University—Center for study and research in nanotechnology, offering cross-discipline graduate courses in the field.

Bibliography

Baird, D.; Nordmann, A. & Schummer, J. (eds.): 2004, *Discovering the Nanoscale*, Amsterdam: IOS Press.

Bensaude-Vincent, B.: 2001, 'The Construction of a Discipline: Materials Science in the United States', *Historical Studies in the Physical and Biological Sciences*, 31, 223-248.

Berube, D.M.: 2004, 'The Rhetoric of Nanotechnology', in: D. Baird, A. Nordmann & J. Schummer (eds.), *Discovering the Nanoscale*, Amsterdam: IOS Press, pp. 173-192.

Brown, N.: 2004, 'Needle on the Real: Technoscience and Poetry at the Limits of Fabrication', in: N.K. Hayles (ed.), *Nanoculture: Implications of the New Technoscience*, Bristol, UK: Intellect Books, pp. 173-190.

Carroll, J.S.: 2001, 'Social Science Research Methods for Assessing Societal Implications of Nanotechnology', in: M.C. Roco and W.S. Bainbridge (eds.), *Societal Implications of Nanoscience and Nanotechnology*, Dordrecht: Kluwer, pp. 188-192.

Chianelli, R.R. 1998. Synthesis, fundamental properties and applications of nanocrystals, sheets, and fullerenes based on layered transition metal chalcogenides. In *R&D status and trends,* ed. Siegel et al.

Crandall, B.C. and Lewis, J.: 1992, *Nanotechnology: Research and Perspectives*, MIT Press, Cambridge.

Dupuy, J.P.: 2004, 'Pour une évaluation normative du programme nanotechnologique', *Annales des Mines*, (February 2004), 27-32.

Fielder, F.A.; Reynolds, G.H.: 1994, 'Legal Problems of Nanotechnology: An Overview', *Southern California Interdisciplinary Law Journal*, 3, 593-629.

Fogelberg, H. & Glimell, H.: 2003, 'Molecular Matters: In Search of the Real Stuff', in: H. Fogelberg & H. Glimell, *Bringing Visibility to the Invisible: Towards A Social Understanding of Nanotechnology*, Göteborg: Göteborg University, pp. 5-32 [www.sts.gu.se/publications/STS_report_6.pdf].

Fogelberg, H. & Glimell, H.: 2003, *Bringing Visibility to the Invisible: Towards A Social Understanding of Nanotechnology*, Göteborg: Göteborg University.

Fogelberg, H.: 2003, 'The Grand Politics of Technoscience: Contextualizing Nanotechnology', in: H. Fogelberg & H. Glimell, *Bringing Visibility to the Invisible: Towards A Social Understanding of Nanotechnology*, Göteborg: Göteborg University, pp. 33-54.

Fogelberg, H.: 2003, 'The Material Culture of Nanotechnology', in: H. Fogelberg & H. Glimell, *Bringing Visibility to the Invisible: Towards A Social Understanding of Nanotechnology*, Göteborg: Göteborg University, pp. 99-114.

Frazier, G., "An Ideology For Nanoelectronics," in *Concurrent Computations: Algorithms, Architecture, and Technology*, Plenum Press, New York, 1988.

Glimell, H.: 2001, 'Challenging Limits ' Excerpts from an Emerging Ethnography of Nano Physicists', in: H. Glimell & O. Johlin (eds.), *The Social Production of Technology: On the Everyday Life with Things*, Göteburg, SE: BAS Publisher, chapter 7, pp. 111-131 (reprinted in: H. Fogelberg & H. Glimell, *Bringing Visibility to the Invisible: Towards A Social Understanding of Nanotechnology*, Göteborg: Göteborg University, pp. 115-139.

Glimell, H.: 2001, 'Dynamics of the Emerging Field of Nanoscience', in: M.C. Roco and W.S. Bainbridge (eds.), *Societal Implications of Nanoscience and Nanotechnology*, Dordrecht: Kluwer, pp. 156-160.

Glimell, H.: 2003, 'A Nano Narrative: The Mircopolitics of a New Generic Enabling Technology', in: H. Fogelberg & H. Glimell, *Bringing Visibility to the Invisible: Towards A Social Understanding of Nanotechnology*, Göteborg: Göteborg University, pp. 55-77 [www.sts.gu.se/publications/STS_report_6.pdf].

Glimell, H.: 2004, 'Grand Visions and Lilliput Politics: Staging the Exploration of the 'Endless Frontier", in: D. Baird, A. Nordmann & J. Schummer (eds.), *Discovering the Nanoscale*, Amsterdam: IOS Press, pp. 231-246.

Gorman, M.E.; Groves, J.F. & Shrager, J.: 2004, 'Societal Dimensions of Nanotechnology as a Trading Zone: Results from a Pilot Project', in: D. Baird, A. Nordmann & J. Schummer (eds.), *Discovering the Nanoscale*, Amsterdam: IOS Press, pp. 63-73.

Grinbaum, A. & Dupuy, J.-P.: 2004, 'Living with Uncertainty: Toward the Ongoing Normative Assessment of Nanotechnology', *Techne: Research in Philosophy and Technology*, 8(3) (forthcoming).

Grinbaum, A.: 2004, 'La condition de l'homme moderne et les nanotechnologies', in: G. Nivat (ed.), *Les limites de l'humain. 39èmes Rencontres Internationales de Genève*, L'Age d'Homme, Genève, p. 141.

Keiper, A.: 2003, 'The Nanotechnology Revolution', *The New Atlantis*, 2, 17-34.

Khushf, G.: 2004, 'A Hierarchical Architecture for Nano-scale Science and Technology: Taking Stock of the Claims About Science Made By Advocates of NBIC Convergence', in: D. Baird, A. Nordmann & J. Schummer (eds.), *Discovering the Nanoscale*, Amsterdam: IOS Press, pp. 21-33.

Khushf, G.: 2004, 'Systems Theory and the Ethics of Human Enhancement: A Framework for NBIC Convergence', *Annals of the New York Academy of Sciences*, 1013, 124-149.

Kuusi, O.; Meyer, M.: 2002, 'Technological generalizations and leitbilder ' the anticipation of technological opportunities', *Technological Forecasting & Social Change*, 69, 625-639.

Landon, B.: 2004, 'Less is More: Much Less is Much More: The Insistent Allure of Nanotechnology Narratives in Science Fiction', in: N.K. Hayles (ed.), *Nanoculture: Implications of the New Technoscience*, Bristol, UK: Intellect Books, pp. 131-146.

Laszlo, P.: 2004, 'Is There Life After Partington?', *Hyle: International Journal for Philosophy of Chemistry*, 10(2), 169-178.

Lenhard, J.: 2004, 'Nanoscience and the Janus-Faced Character of Simulations', in: D. Baird, A. Nordmann & J. Schummer (eds.), *Discovering the Nanoscale*, Amsterdam: IOS Press, pp. 93-100.

Lewak, S.E.: 2004, 'What's the Buzz? Tell Me What's A-Happening: Wonder, Nanotechnology, and Alice's Adventures in Wonderland', in: N.K. Hayles (ed.), *Nanoculture: Implications of the New Technoscience*, Bristol, UK: Intellect Books, pp. 201-.

Lin-Easton, P.C.: 2001, 'It's Time for Environmentalists to Think Small ' Real Small: A Call for the Involvement of Environmental Lawyers in Developing Precautionary Policies Molecular Nanotechnology', *Georgetown International Law Review*, 14, 106-134.

López, J.: 2004, 'Bridging the Gaps: Science Fiction in Nanotechnology', *Hyle: International Journal for Philosophy of Chemistry*, 10(2), 129-152.

Lösch, A.: 2004, 'Nanomedicine and Space: Discursive Orders of Mediating Innovations', in: D. Baird, A. Nordmann & J. Schummer (eds.), *Discovering the Nanoscale*, Amsterdam: IOS Press, pp. 193-202.

Mahmood U., C.H. Tung, A. Bogdanov Jr., R. Weissleder. 1999. "Near-Infrared Optical Imaging of Protease Activity for Tumor Detection." *Radiology*, Dec; 213 (3): pp. 866-870.

Marty, A.: 2003, *Testimony before the Committee on Science on HR 766, Nanotechnology Research and Development Act*, US Government Printing Office, Washington, DC.

Mehta, M.D.: 2002, 'Nanoscience and Nanotechnology: Assessing the Nature of Innovation in These Fields', *Bulletin of Science, Technology & Society*, 22(4), 269-273.

Mehta, M.D.: 2004, 'From Biotechnology to Nanotechnology: What Can We Learn from Earlier Technologies?, *Bulletin of Science, Technology & Society*, 24, 34-39.

Merz, J.L. 1986. "The Optoelectronics Joint Research Laboratory: Light Shed on Cooperative

Research in Japan." *Scientific Bulletin* (ONR Far East Office). 11 (4), 1-30 (Oct.-Dec.).

Mody, C.C.M.: 2000, "A New Way of Flying': Différance, Rhetoric, and the Autogiro in Interwar Aviation', *Social Studies of Science*, 30, 513-543.

Mody, C.C.M.: 2001, 'A Little Dirt Never Hurt Anyone: Knowledge-Making and Contamination in Materials Science', *Social Studies of Science*, 31, 7-36.

Mody, C.C.M.: 2004, 'How Probe Microscopists Became Nanotechnologists', in: D. Baird, A. Nordmann & J. Schummer (eds.), *Discovering the Nanoscale*, Amsterdam: IOS Press, pp. 119-133.

Mody, C.C.M.: 2004, 'Instruments in Training: The Growth of American Probe Microscopy in the 1980s', in: D. Kaiser (ed.), *Pedagogy and the Practice of Science: Producing Physical Scientists, 1800-2000*, Cambridge, MA: MIT Press (forthcoming).

Mody, C.C.M.: 2004, 'Small, but Determined: Technological Determinism in Nanoscience', *Hyle: International Journal for Philosophy of Chemistry*, 10(2), 99-128.

Moor, J.H. & Weckert, J.: 2004, 'Nanoethics: Assessing the Nanoscale From an Ethical Point of View', in: D. Baird, A. Nordmann & J. Schummer (eds.), *Discovering the Nanoscale*, Amsterdam: IOS Press, pp. 301-310.

Munn Sanchez, E.: 2004, 'The Expert's Role in Nanoscience and Technology', in: D. Baird, A. Nordmann & J. Schummer (eds.), *Discovering the Nanoscale*, Amsterdam: IOS Press, pp. 257-266.

Naughton, G.: 1999. The Advanced Tissue Sciences story. *Sci. Am.* 280(4): p. 84-5.

Nordmann, A.: 2003, 'Shaping the World Atom by Atom: Eine nanowissenschaftliche WeltBildanalyse', in: A. Grunwald (ed.), *Technikgestaltung zwischen Wunsch und Wirklichkeit*, Berlin: Springer, pp. 191-199.

Nowotny, H.; Scott, P.; Gibbons, M.: 2001, *Re-thinking Science*, Polity, Cambridge.

Paschen, H.; Coenen, C.; Fleischer, T.; Grünwald, R.; Oertel, D. & Revermann, C.: 2004, *Nanotechnologie: Forschung, Entwicklung, Anwendung*, Berlin: Springer (*Nanotechnologie, TAB-Arbeitsbericht 92*, Berlin: Büro für Technikfolgen-Abschätzung beim Deutschen Bundestag).

Phoenix, C.; Drexler, E.: 2004, 'Safe Exponential Manufacturing', *Nanotechnology*, 15, 869-72.

Pitt, J.C.: 2004, 'The Epistemology of the Very Small', in: D. Baird, A. Nordmann & J. Schummer (eds.), *Discovering the Nanoscale*, Amsterdam: IOS Press, pp. 157-163.

Pressman, J.: 2004, 'Nano Narrative: A Parable from Electronic Literature', in: N.K. Hayles (ed.), *Nanoculture: Implications of the New Technoscience*, Bristol, UK: Intellect Books, pp. 191-200.

Pritchard, M.S.: 1996, "Conflicts of Interest: Conceptual and Normative Issues," *Academic Medicine*, 71:12, December, pp. 1305-1313.

R. T. Bate: 1990, "Nanoelectronics," *Nanotechnology*, Vol. 1, pp. 1-7.

Ramanan, V.R.: 1998. Nanocrystalline soft magnetic alloys for applications in electrical and electronic devices. In *R&D Status and Trends*, ed. Siegel et al.

Ramos, L.; Lubensky, T.C.; Dan, N.; Nelson, P.; Weitz, D.A.: 1999. "Surfactant-mediated two-dimensional crystallization of colloidal crystals", *Science*, 286, 2325-2328.

Ratner, M.; Ratner, D.: 2003, *Nanotechnology*, Prentice Hall, New Jersey.

Ratner, M.; Ratner, D.: 2003, *Nanotechnology: A Gentle Introduction to the Next Big Idea,* Prentice Hall, Upper Saddle River, NJ.

Ravetz, J.: 1999, 'What is Post-normal Science?', *Futures*, 31, 647-653.

Regis, E.: 1995, *Nano: The Emerging Science of Nanotechnology*, Little, Brown, Boston.

Rietman, E.A.: 2001, 'Drexler hypothesis of a universal assembler is supported not by theoretical arguments alone but by existence proof in the form of biological life', in: *Molecular Engineering of Nanosystems*, Springer, New York & Berlin.

Roberts, J.A.: 2004, 'Deciding the Future of Nanotechnologies: Legal Perspectives on Issues of Democracy and Technology', in: D. Baird, A. Nordmann & J. Schummer (eds.), *Discovering the Nanoscale*, Amsterdam: IOS Press, pp. 247-255.

Robinson, C.: 2004, 'Images in NanoScience/Technology', in: D. Baird, A. Nordmann & J. Schummer (eds.), *Discovering the Nanoscale*, Amsterdam: IOS Press, pp. 165-169.

Roco, M.C. & Bainbridge, W.S. (eds.): 2001, *Societal implications of nanoscience and nanotechnology*, (Proceedings of a workshop organized by the National Science Foundation, September 28-29, 2000), Kluwer: Dordrecht [available online at http://itri.loyola.edu/nano/societalimpact/nanosi.pdf]

Roco, M.C. & Bainbridge, W.S., (eds.): 2003, *Converging Technologies for Improving Human Performance: Nanotechnology, Biotechnology, Information Technology and Cognitive Science*, Kluwer, Dordrecht.

Roco, M.C. & Tomellini, R. (eds.): 2002, *Nanotechnology: Revolutionary Opportunities and Societal Implications, Workshop, Lecce (Italy), 31 January - 1 February 2002*, Luxemburg: Office for Official Publications of the European Communities, [200 pp.]

Roco, M.C.: 2003, 'Broader Societal Issues of Nanotechnology', *Journal of Nanoparticle Research*, 5, 181-189.

Roher, H.: 1993, *Jpn. J. Appl. Phys.* 32:1335. Rohlfing, E.A., D.M. Cox, and A. Kaldor. 1984. *J. Chem. Phys.* 81:3846.

Rohlfing, E.A., D.M. Cox, and A. Kaldor. 1984. *J. Chem. Phys.* 81:3846.

Rohrer, H.: 1995, 'The Nanometer Age – Challenge and Chance', *Microelectronic Engineering*, 27, 3-15.

Roland Pease, "'Living' robots powered by muscle," BBC News, January 17, 2005.

Romanov, A.E., V.I. Vladimirov. 1992. In *Dislocations in solids*, ed. F.R.N. Nabarro, Vol. 9. Amsterdam: North-Holland. Salishekev, G.A., O.R. Valiakhmetov, V.A. Valitov, and S.K. Mukhtarov. 1994. *Materials Science Forum*. 170-172:121.

Rosen, A. 1998, A periodic table in three dimensions: A sightseeing tour in the nanometer world. In *Advances in quantum chemistry*. In press.

Rosen, P.: 1993, 'The Social Construction of Mountain Bikes: Technology and Postmodernity in the Cycle Industry', *Social Studies of Science*, 23, 479-513.

Rosenberg, Nathan: 1994, *Exploring the Black Box: Technology, Economics, and History*. New York: Cambridge University Press.

Rothman, B.K.: 1998, *Genetic Maps and Human Imagination*, W.W. Norton, New York & London,

Schmidt, J.C.: 2004, 'Unbounded Technologies: Working Through Technological Reductionism of Nanotechnology', in: D. Baird, A. Nordmann & J. Schummer (eds.), *Discovering the Nanoscale*, Amsterdam: IOS Press, pp. 35-50.

Schuler, E.: 2004, 'Perception of Risks and Nanotechnology', in: D. Baird, A. Nordmann & J. Schummer (eds.), *Discovering the Nanoscale*, Amsterdam: IOS Press, pp. 279-284.

Star, S.L.; Griesemer, J.R.: 1989, 'Institutional Ecology, 'Translations' and Boundary Objects: Amateurs and Professionals in Berkeley's Museum of Vertebrate Zoology, 1907-1939', *Social Studies of Science*, 19, 387-420.

Steelman, John R.: 1947, *Science and Public Policy* (Washington, DC: US Government Printing Office). Stephenson, Neal. 1995. *The Diamond Age, or, Young Lady's Illustrated Primer* (New York: Bantam Books).

Tour, J.M.: 2003, *Molecular Electronics: Commercial Insights, Chemistry, Devices, Architecture and Programming*, World Scientific, Singapore.

Uyeda, R.: 1991, *Prog. in Mater. Sci.* 35:1. Van de Zande, B.M.I., M.R. Bohmer, L.G.J. Fokkink, and C. Shonenberger. 1997. *J. Phys. Chem.* 101:852.

Verbeek, A. & Callaert, J.: 2002, *Detailed analysis of the science-technology interaction in the field of Nanotechnology*, Luxemburg: Office for Official Publications of the European Communities (Linking Science to Technology. Bibliographic References in Patents, Vol. 9) [79 pp.].

Vermaas, P.E.: 2004, 'Nanoscale Technology: A Two-Sided Challenge for Interpretations of Quantum Mechanics', in: D. Baird, A. Nordmann & J. Schummer (eds.), *Discovering the Nanoscale*, Amsterdam: IOS Press, pp. 77-91.

Vincenti, W.G.: 1990, *What Engineers Know and How They Know It: Analytical Studies from Aeronautical History*, Johns Hopkins University Press, Baltimore.

Weckert, J.: 2001, 'The Control of Scientific Research: The Case of Nanotechnology', *Australian Journal of Professional and Applied Ethics*, 3, 29-44.

Weil, V. 1996. "Biotechnology: Societal Impact and Quandaries," in *Biotechnology and Ethics: A Blueprint for the Future, Report of NSF Workshop*, Northwestern University, Center for Biotechnology.

Weil, V. 2000. "Como Pueden Convivir Los Valores Universitarios Y Las Normes Empresariales?" in *Dos Ejes en La Vinculacion de Las Universidades a la Produccion*. ed. by Rosalba Casas and Giovanna

Weil, V.: 2001, 'Ethical Issues in Nanotechnology', in: M.C. Roco and W.S. Bainbridge (eds.), *Societal Implications of Nanoscience and Nanotechnology*, Dordrecht: Kluwer, pp. 193-198.

Index

D

E

F

G

M

N

O

P

Q

R

S

T

U

V

W

Y